# Immersive Sound Volume II



*Immersive Sound Volume II: The Design and Practice of Binaural and Multi-Channel Experiences* is a collection of interviews and essays providing a comprehensive guide to the history, theory, and creative applications of 3D and spatial audio across music, production, theater, installations, and film.

Offering both foundational knowledge and practical insights into the design and implementation of immersive sound experiences, expert contributors present each topic in depth, with chapters structured for both sequential learning and standalone reference. By combining historical context, cognitive perspectives, and real-world applications, this book connects theory with practice and helps readers expand their technical, academic, and creative potential in the rapidly evolving field of spatial audio.

This is a valuable resource for all spatial audio content creators, including audio engineers, sound designers, researchers, educators, composers, performers, and students seeking a comprehensive understanding of immersive sound and its applications across different disciplines. This book serves as a companion to *Immersive Sound: The Art and Science of Binaural and Multi-Channel Audio*.

**Agnieszka Roginska** conducts research in the simulation and applications of immersive and 3D audio including the capture, analysis, and synthesis of auditory environments. Applications of her work include AR/VR/XR, gaming, and mission critical. She is the author of numerous publications on the topics of acoustics and psychoacoustics of immersive audio and is Professor of Music Technology at New York University. Agnieszka is a Fellow of the Audio Engineering Society (AES) and a Past President of the AES.

**Paul Geluso** is a New York-based sound engineer, educator, and musician. He has worked professionally in many areas of sound and music production, being credited as engineer, producer, composer, and musician on hundreds of commercial releases including Grammy and Latin Grammy nominated recordings. He is Professor of Music Technology and founder of the Tonmeister program at New York University. He has served as Chief Sound Engineer at Harvestworks Digital Media Arts for over three decades.

Audio Engineering Society

## Audio Engineering Society Presents

www.aes.org



**Modern Recording Techniques, 10th Edition**
A Practical Guide to Modern Music Production
*David Miles Huber and Emiliano Caballero*

**Recording Orchestra and Other Classical Music Ensembles, 2nd Edition**
*Richard King*

**Sound Reproduction, 4th Edition**
The Acoustics and Psychoacoustics of Loudspeakers, Rooms and Headphones
*Floyd E. Toole, Sean Olive and Todd Welti*

**Classical Recording, 2nd Edition**
A Practical Guide in the Decca Tradition
*Caroline Haigh, John Dunkerley and Mark Rogers*

**Handbook for Sound Engineers, 6th Edition**
*Edited by Glen Ballou and Doug Jones*

***Audio Effects***
Theory, Implementation and Application
*Joshua D. Reiss and Andrew McPherson*

**Virtual Reality Audio**
A Guide to Immersive Sound in Virtual Experiences
*Anıl Çamcı and Jason Corey*

**Immersive Sound Volume II**
The Design and Practice of Binaural and Multi-Channel Experiences
*Edited by Agnieszka Roginska and Paul Geluso*

For more information about this series, please visit: www.routledge.com/Audio-Engineering-Society-Presents/book-series/AES

# Immersive Sound Volume II

## The Design and Practice of Binaural and Multi-Channel Experiences

Edited by Agnieszka Roginska and Paul Geluso

Routledge
Taylor & Francis Group
LONDON AND NEW YORK

Designed cover image: Getty Images

First published 2026
by Routledge
4 Park Square, Milton Park, Abingdon, Oxon OX14 4RN

and by Routledge
605 Third Avenue, New York, NY 10158

*Routledge is an imprint of the Taylor & Francis Group, an informa business*



*British Library Cataloguing-in-Publication Data*
A catalogue record for this book is available from the British Library

ISBN: 978-0-367-25947-1 (hbk)
ISBN: 978-0-367-25946-4 (pbk)
ISBN: 978-0-429-29752-6 (ebk)

DOI: 10.4324/9780429297526

Typeset in Sabon
by Apex CoVantage, LLC

# Contents

# Contributors

**Tom Ammerman**

Grammy-nominated music producer Tom Ammermann began his journey as a musician and music producer in the 1980s. At the turn of the 21st century, Tom produced unique surround audio productions for music and film projects as well as pioneering the very first surround mixes for headphones (binaural audio), which quickly established his binaural brand Headphone Surround 3D and reputation as an immersive audio producer. Currently, Tom is working on developing new technologies in spatial and next-generation audio while offering workshops and presentations at high-profile international audio conferences.

**Jim Anderson**

For nearly 50 years, Jim Anderson has set the standard for acoustic audio engineering and production, capturing pristine, high-definition stereo, surround sound and immersive recordings that have garnered 13 Grammy and Latin Grammy Awards, two Peabody Awards, and a pair of Emmy nominations, among others. Born and raised in western Pennsylvania, Anderson attended Duquesne University. Post graduation, he worked at NPR in Washington, DC, where he engineered broadcast sessions for everything from news and documentaries to performances by jazz icons, classical orchestras, folk singers, and contemporary composers. In 1980, Anderson moved to New York to launch his career as a freelance engineer and producer, and over the next several decades, he would go on to record studio albums, live concerts, and film scores across a diverse array of genres. In 2003, he became a professor at the prestigious Clive Davis Institute of Recorded Music in the Tisch School of the Arts at NYU and is now the Department's Professor Emeritus. In 2008, he was elected President of the Audio Engineering Society and is now the President of the Audio Engineering Society Educational Foundation.

**Vincent Gabriel Antonini**

Vincent Gabriel Antonini (aka Gabriel) is a musician/composer, broadcast engineer/editor, producer, author, FOH mixer, and studio consultant. Involved in design aspects for various projects such as NBC, PBS, Sports venues, and mixing many artists for studio and live. Gabriel has been nominated for 11 Emmys for sound design. Gabriel is currently a global support/sound designer for DPA microphones. Passionate about sound and the microphones that capture it, Gabriel strives to bring that appreciation to everyone he meets.

**Jane Ira Bloom**
Award winning soprano saxophonist/composer Jane Ira Bloom has been developing her unique voice on the soprano saxophone for over 45 years. She's a pioneer in the use of live electronics and movement in jazz, is the first musician ever commissioned by the NASA Art Program, and has an asteroid named in her honor by the International Astronomical Union (asteroid: 6083janeirabloom).

**Braxton Boren**
Braxton Boren is Associate Professor of Audio Technology at American University. His work uses acoustic simulation techniques to understand historical music and oratory. His work has been featured in the *New York Times*, BBC Radio, National Geographic, and Science. Professor Boren's research has been funded by the National Science Foundation, the National Endowment for the Humanities, and Emergent Ventures.

**David v.R. Bowles**
David v.R Bowles has produced and engineered recordings that have been GRAMMY- and JUNO-nominated. His releases in Dolby Atmos can be found on Avie Records, Out-Here Music (Delos), and Parma Recordings. He is a guest lecturer at NYU Steinhardt's Music Technology Tonmeister Seminar and Royal Danish Academy of Music. He was AES Western Regional Vice President 2023–2026.

**Thibaut Carpentier**
Thibaut Carpentier is a research engineer in the STMS Lab (Sciences and Technologies of Music and Sound) at IRCAM, Paris. Within the Acoustics & Cognition Team, his work focuses on sound spatialization, artificial reverberation, room acoustics, and computer tools for 3D composition and mixing. He leads the development of the Spat project as well as the Panoramix 3D mixing and post-production workstation.

**Rhiannon Catalyst**
Rhiannon Catalyst creates immersive theater and sound experiences where art, science, and technology converge. Beginning as an artist and music director in award-winning grassroots multimedia spaces, she now produces large-scale projects with NASA, museums, and collaborators worldwide. Devoted to building collaborative, interactive, and experimental frameworks – including spaces for immersive R&D – her practice weaves spatial sound, haptics, and narrative to spark empathy, foster collective resonance, and deepen connections with community, place, and environment.

**Miles Fulwider**
Miles is a producer, engineer, educator, and author. He is the President and CEO of ADK Microphones and the Associate Director of the Division of Creative Arts and the Program Director for the Music Technology, Digital Media, and Dance programs at the University of Saint Francis in Fort Wayne Indiana. Miles has produced and engineered for many Grammy- and Emmy Award–winning artists and produced/engineered several chart-topping and award-winning albums.

**Paul Geluso**

Paul Geluso works at the intersection of the theoretical, practical, and artistic dimensions of immersive sound recording and reproduction. He has contributed as a recording and mastering engineer, musician, and producer on hundreds of commercially released recordings and sound installations, including Grammy-nominated albums and award-winning films. He is currently Assistant Professor and Director of the Music Technology and Tonmeister programs at New York University and serves as the Chief Sound Engineer at Harvestworks Digital Media Arts and PAG Sound Services Inc.

**John Glass**

John Glass is a software engineer at Microsoft with research experience in spatial audio localization. He previously served as Director of Research Involvement for the Computing Student Union and co-founded the Computing Research Methods course at the University of Florida. He is passionate about computing research and development.

**Piet Goddaer**

Piet Goddaer is an author, composer, musician, and producer, working under pseudonyms: Ozark Henry, Studio Ozark Henry, Sunzoo Manley, and August Parker. Since David Bowie praised his debut album in 1996, he has successfully produced music ranging from avant-garde, electronica to pop and classical music. He is an artist with synesthesia, and his universe is defined by a blending of craft with experiment, audio with visual, heritage with technology. In 2017 he envisioned a 3D Creative Studio to explore and unlock the potential of Immersive Sound in the arts, spatial architecture, and life science.

**Marta Gospodarek**

Marta Gospodarek, New York University, specializes in spatial audio, sound design, and psychoacoustics with a focus on augmented reality (AR) and virtual reality (VR). She earned her PhD at the Music and Audio Research Lab at New York University and conducted research at IRCAM in Paris.

**Justin Gray**

Justin Gray is an award-winning producer, engineer, composer, and bassist who specializes in spatial and immersive audio. He collaborates with artists worldwide to bring music to life in stereo and immersive formats. A dedicated educator, Justin is a faculty member at Humber College in Toronto, where he teaches audio production, composition, and music performance.

**Kimio Hamasaki**

Kimio Hamasaki is a spatial audio researcher, an audio engineering and acoustics educator, a music recording producer and balance engineer, and an audio engineering consultant. He developed digital audio recording equipment, microphone techniques for natural immersive sound recording, and advanced spatial audio systems. He has many years of experience in international standardization related to audio technology. Kimio is president at ARTSRIDGE LLC and an AES fellow.

**Gregory Hanson**

Greg is a New York City–based theatrical sound designer and audiovisual consultant. With a love of synthesis, music, and all things sound, Greg spends his free time designing unique, site-specific installations and theatrical activations. Greg is an associate with the international theater, arts, and culture consultancy, Theatre Projects, and leads the performance sound, video, and communications team designing audiovisual systems around the world.

**Leslie Ann Jones**

A recording and mixing engineer and record producer for over 40 years, Leslie has held staff positions at ABC Recording Studios in Los Angeles, the Automatt Recording Studios in San Francisco, Capitol Studios in Hollywood, and now Skywalker Sound where she continues her career recording and mixing music for records, films, video games, and television and producing records primarily in the Classical genre. She is a past Chair of The Recording Academy's Board of Trustees and is the recipient of seven Grammy Awards, including four for Best Engineered Album–Classical and one for Best Immersive Audio Album. She serves on the Advisory Board of Institute for the Musical Arts, the Library of Congress's National Recording Preservation Board and is an Artistic Advisor to the Technology and Applied Composition degree program at the San Francisco Conservatory of Music. Leslie was also inducted into the NAMM TEC Hall of Fame in 2019 and is a Heyser lecturer. Leslie also chaired the committee that wrote 'Recommendations for Hi-Resolution Music Production', published by the Producers and Engineers Wing of the Recording Academy.

**Jean-Marc Jot**

Dr. Jean-Marc Jot is an audio technology innovator in consumer electronics and professional markets. He has driven the development of novel sound and music processing technologies, platforms, and standards for virtual reality, gaming, broadcast, cinema, and music creation. Before relocating to California in the late 1990s, he conducted applied research in spatial audio processing and perception at IRCAM in Paris.

**Brian F.G. Katz**

Brian F.G. Katz is a CNRS Research Director at Sorbonne Université's d'Alembert Institute, specializing in spatial 3D audio, room acoustics, HCI, and VR. With a physics-philosophy background, he earned a PhD in acoustics from Penn State (1998) and an HDR in engineering sciences from UMPC (2011). He previously consulted at Artec, ARUP, and Kahle Acoustics, and worked at UPMC, IRCAM, and LIMSI-CNRS.

**Gavin Kearney**

Gavin Kearney is Professor of Audio Engineering at the University of York, UK, specializing in immersive audio and interactive digital technologies. A leader in UK digital creativity research, he directs the AHRC-funded CoSTAR LiveLab and collaborates extensively with industry partners, including Google and the BBC. His research explores the intersections of sound, technology, and human experience across disciplines.

**Ralph Kessler**

Dipl.-Ing. Ralph Kessler has operated the pro audio company 'PINGUIN™ Ing.Büro' since 1989. His professional activities include developing and publishing of pro audio software (metering and 3D room simulation) as well as mastering in his own 'acoustic holodeck' construction in Hamburg. His inventions found their way into international patents or were published in scientific papers. In 2016 he received the AES Fellow Award.

**Hyunkook Lee**

Dr. Hyunkook Lee is Professor of Audio and Psychoacoustic Engineering and the founder/director of the Applied Psychoacoustics Lab (APL) at the University of Huddersfield, UK. His research covers a broad range of topics related to the perception, recording, and reproduction of spatial and immersive audio, as well as virtual acoustic rendering for binaural audio. He is a Fellow and a former Governor of the Audio Engineering Society (AES).

**Morten Lindberg**

Recording producer and balance engineer with 46 GRAMMY nominations, 38 of these in craft categories Best Engineered Album, Best Surround Sound Album, Best Immersive Audio Album, and Producer of the Year. Founder and CEO of the record label 2L. Grammy Award winner 2020.

**Nuria Lloret Romero**

Full Professor of Electronic Administration at the Polytechnic University of Valencia UPV, Nuria Lloret Romero has been director of the DCADHA department and has held various management positions at the UPV. She is currently deputy director of the Institute of Design and Manufacturing (IDF) of the (UPV) and Director of Institutional Relations of the IDF Association (https://institutoidf.com/). She has been Visiting Professor at New York University several times and was Visiting Professor at TEC de Monterrey and in 2022 at Stanford University. She is currently the Coordinator of the Berkeley School of Music as an affiliated center belonging to the UPV. She is director of the research project 'Atenea: Women in Art and Stem' (http://atenea.in/).

**Chenshen Lu**

Chenshen Lu received his MSc in Mechanical Engineering from the University of Florida in 2019 and is currently pursuing a PhD in Human-Centered Computing at the same institution. His research interests focus on efficient individualization of Head-Related Transfer Functions (HRTFs) and the study of spatial audio perception within virtual environments.

**Andres Mayo**

Andres Mayo is a Grammy-winning audio engineer, music and film producer, and runs his own company: Andres Mayo Immersive Audio. A Past President of the Audio Engineering Society with credits in over 4,000 titles released in all formats since 1991, as well as an immersive audio producer and certified Dolby Atmos mixer Neumann–Sennheiser Ambassador worldwide.

**Kyla McMullen**

Dr. Kyla McMullen, Associate Professor at the University of Florida, specializes in 3D audio technology, developing spatial sound systems to aid situational awareness and improve user interaction. Her research explores realistic sound rendering, rapid detection, and interference handling, making her a leader in advancing immersive audio experiences for real-world applications.

**Charlie Morrow**

Charlie Morrow is a composer, sound artist, producer, podcaster (Helsinki, Finland; and Barton, Vermont, USA). He is known for bringing experimental music to audiences through concerts, advertisements, installations, and public events. A student of avant-garde composer Stefan Wolpe, Morrow has created works ranging from sound portraits to orchestrated boat horn pieces and film soundtracks, while also producing famous jingles for brands like Hefty and Diet Coke. He founded the New Wilderness Foundation in 1974 and organized summer solstice celebrations across New York City. Charlie developed innovative immersive sound technologies featured in hospitals, museums, and corporate spaces worldwide, and patented immersive sound methods in physical and virtual space. Throughout his 50-year career, he has positioned himself as a 'framemaker', extending compositional skills beyond concert halls into everyday environments for healing and creativity.

**Daniel Neumann**

Daniel Neumann is a sound artist, organizer, and audio engineer. In his artistic practice, he is working in hybrid installation-performance formats, in which a main focus is how sound interacts with space and how space and spatial perception can be shaped by sound. His work is presented internationally and represented by Fridman Gallery, New York. Curatorially he runs an event series (New Ear:: Spatial) that engages in spatial sound works and focused listening.

**Lasse Nipkow**

Since 2010, Lasse Nipkow has been a renowned speaker on 3D audio music production. As a European expert, he simplifies complex topics. He has significantly contributed to music production, with over 500 recordings, including works with George Gruntz and the Stradivari Quartet. In 2022, he produced *Sonic Leaks* by Silent Work in Dolby Atmos/Auro-3D and began a PhD at the University of Huddersfield.

**Rolf Nordahl**

Rolf Nordahl joined the faculty of the Danish Technical University in 2026. Prior to that, he was Associate Professor at Aalborg University in Copenhagen and the leader of the multisensory experience lab together with Stefania Serafin. His research focuses on VR, presence, and the use of VR to help populations in need.

**Nathaniel Reichman**

Nathaniel Reichman, immersive mixer and producer, crafts three-dimensional audio experiences in music and film. He's produced Grammy-winning albums with artists ranging from John Luther Adams to Deadmau5, and is committed to advancing both the art and science of recorded music.

**Agnieszka Roginska**

Agnieszka Roginska is a Professor of Music Technology at New York University. She conducts research in the simulation and applications of immersive and 3D audio including the capture, analysis, and synthesis of auditory environments. Applications of her work include AR/VR/XR, gaming, mission critical, and augmented acoustic sensing. She is the author of numerous publications on the topics of acoustics and psychoacoustics of immersive audio. Agnieszka is a Fellow of the Audio Engineering Society (AES), and a Past President of the AES. She is the faculty sponsor of the Society for Women in TeCHnology (SWiTCH) at NYU.

**Michael Romanowski**

Michael Romanowski is the chief mastering engineer of Coast Mastering in Berkeley, California. He is one of the very first Immersive Audio Mastering engineers on the planet. As a sought-after consultant and advisor for companies such as Dolby, Sony, Fraunhofer, and others, his contributions have helped build and refine tools for creativity while pushing the boundaries of immersive audio production. Michael designed and built his first surround studio in 2001, added height channels in 2015, and created his first ATMOS music releases before the launch of Dolby Music. Five-time Grammy® Award winner, Ten-time Grammy® nominee. As an in-demand speaker he has presented at AES, NAMM, SXSW, Tape Op, Potluck Con, AES Mexico, AES Europe and has written articles for *Mix Magazine* and *Electronic Musician* and many more in his career of more than 30 years.

**Jorge Sastre Martinez**

Professor Jorge Sastre from the Universitat Politecnica de Valencia specializes in collaborative audiovisual technologies and computing. A former Fulbright Research Scholar at NYU, he is a member of the Computer Music Project at Carnegie Mellon University, where he was a visiting researcher. He directs the award-winning Soundcool.org project, with his work focusing on applications in neurocognitive therapies and education.

**Steve Schatz**

Steve is a 25-year veteran in sound for film. He has been an integral part of some of the earliest films mixed in immersive audio. He has worked on countless blockbuster studio films at Universal, Disney, and Sony. His knowledge and expertise have contributed to many colleagues' Academy Award nominations and wins.

**Michael J. Schumacher**

Michael J. Schumacher has innovated in the area of spatialized sound and algorithmic composition since the 1980s, creating multi-channel, generative *Room Pieces* presented in galleries, museums, concert halls, public and private spaces. His interest in the relationship of musical form and architecture led to the founding of Diapason Sound Art, a gallery devoted to the presentation of multi-channel sound installations, long-duration performances and intermedia artworks.

**Ulrike Schwarz**

Born and raised in Germany, Schwarz discovered she had perfect pitch at an early age and, as a teenager, was accepted into the exclusive Tonmeister program at the Hochschule der Künste Berlin. After earning her degree, Schwarz worked with German public radio and spent more than ten years serving as the senior sound engineer for large orchestral and jazz productions at the famed Bayerischer Rundfunk in Munich. Ulrike garnered numerous honors for her recordings and broadcasts, including multiple Echo

Klassik, Diamant D'Opera, Preis der Deutschen Schallplattenindustrie, and Diapason d'Or awards. In 2015, Schwarz moved to New York full time to team up with her now-husband and production partner Jim Anderson, with whom she's created a groundbreaking body of work in the field of immersive audio. Ulrike has received three Grammy nominations in the Best Immersive Audio Album category and was the President of the Verband Deutscher Tonmeister 2023–2028.

**Stefania Serafin**

Stefania Serafin joined the faculty of the Danish Technical University in 2026. Prior to that, she was a professor of Sonic interaction design at Aalborg University in Copenhagen and the leader of the multisensory experience lab together with Rolf Nordahl. Stefania received her PhD entitled 'The Sound of Friction: Computer Models, Playability and Musical Applications' from Stanford University in 2004, supervised by Professor Julius Smith III. Her research on sonic interaction design, sound for virtual and augmented reality with applications in health and culture, can be found here: HYPERLINK "http://tinyurl.com/35wjk3jn"tinyurl.com/35wjk3jn

**Wilfried Van Baelen**

Wilfried Van Baelen, founder of Auro Technologies and CEO of Galaxy Studios Group, is a pioneer of immersive sound – the term he introduced to replace 'Surround Sound + Height'. He revolutionized audio with Auro-3D, the first true 3D audio format offering an end-to-end solution across all audio markets. Trained at the Royal Conservatory of Antwerp, where he earned five first prizes including classical organ, Van Baelen opened his first studio at 18 before founding the world-renowned Galaxy Studios in 1995. As a producer and engineer, he has worked with international artists on more than 20 platinum albums and mixed multiple award-winning films. In September 2025, he launched AuroMasters, a groundbreaking streaming platform that delivers the full artistic intent – from lossy 200 kb/s to lossless 192 kHz/24-bit with adaptive bitrate – in Immersive Audio.

**Olivier Warusfel**

Olivier Warusfel studied acoustics at Université Paris 6 where he received his PhD degree in 1985. He has been lecturing in Musical Acoustics at University Paris 8 and Room Acoustics at University Paris 6. Since 1992, he has been leading the Room Acoustics group at IRCAM. His main interests focus on audio signal processing, sound spatialization, room acoustics, auditory virtual reality, spatial auditory perception and cognition.

**Wieslaw Woszczyk**

Wiesław Woszczyk's distinguished career as educator, record producer, and researcher spans over four decades. After working with such renowned artists as Philip Glass, Brian Eno, Gloria Gaynor, and Harry Belafonte, he established Canada's first and most successful academic program in music recording (McGill). Founding Director of the Centre for Interdisciplinary Research in Music Media and Technology (Montreal), his research lies on the cutting edge of virtual acoustics and immersive sound.

**Tomasz Żernicki**

Tomasz Żernicki, PhD, is a sound engineer, inventor, and entrepreneur specializing in 3D sound and immersive audio technologies. He is the co-founder of Zylia, a company that developed a renowned Ambisonics microphone and a suite of Ambisonics-based tools for 3D audio production. As a key inventor and pioneer in the field of Ambisonics, he invented the 6DoF Ambisonic interpolation technology. He also co-authored key sound coding technologies that became part of international ISO/MPEG standards, such as MPEG-D USAC and MPEG-H 3D Audio. As an expert, he actively participates in the work of the MPEG standardization group and is a member of the Audio Engineering Society. He is the author of over 100 scientific publications and more than ten patents. In 2017, he was honored as a New Europe 100 challenger by Respublica, Google, the Visegrad Fund, and the Financial Times.

**Gregor Zielinsky**

Gregor Zielinsky studied Tonmeister at the Music Academy in Detmold, Germany. He has worked with many famous artists, such as L. Bernstein, C. Abbado, and Count Basie Big Band. Since 2010 he has been an early adopter of 3D audio, working in that field with a lot of experience and excitement. He is giving worldwide training and lectures. Gregor is awarded with one Grammy, several nominations, and other awards.

# Introduction

## Artistic intention and technical considerations

*Paul Geluso and Agnieszka Roginska*

Our first book, titled *Immersive Sound: The Art and Science of Binaural and Multi-Channel Audio*, is intended for audio engineers, researchers, educators, sound designers, and composers who wish to gain in-depth knowledge about the methods and techniques of reproducing immersive sound. Our goal was to create a comprehensive and accessible guide to the rapidly evolving field of spatial audio technologies. The book focused on the theoretical and scientific foundations of spatial audio and methods to capture and reproduce immersive sound. We assembled contributions from leading researchers and practitioners across disciplines to address the scientific basis of the various methods of capturing and reproducing immersive audio, including stereo, binaural over headphones, binaural over loudspeakers, surround sound with height, object-based audio, sound field, and wavefield synthesis. We described the important psychoacoustic factors in the perception of immersive experience, as well as extended applications of immersive technique.

This current book, *Immersive Sound Volume II: The Design and Practice of Binaural and Multi-Channel Experiences*, will serve as a practical companion to our first book. This volume delves deeply into the conceptual ideas, philosophies, creative processes, implementation, and workflows that use those scientific and technical principles and turn them into compelling experiences. This book includes a collection of essays, interviews, and case studies that span projects in recording and production, installation art, live performance, museums, system design, and more. Readers will gain access to the processes and creative problem-solving approaches employed by industry pioneers. In the first section of the book, historical perspectives will trace the evolution of immersive audio practices from the early times. The section on *Guiding Principles and System Design* will delve into describing the philosophies and design principles of immersive experiences and advanced production techniques and methods. The second part of the book is devoted to *Production and Creative Practice* where we explore practical approaches, implementation, and tips and tricks to developing immersive experiences.

While our first volume gives readers the critical understanding of why spatial audio works the way it does and the methods and techniques to recreate immersive audio, this book focuses on how to effectively harness these principles in real-world scenarios, and the kind of experiential wisdom typically gained only through years of professional practice. It is a practitioner-centered approach, offering the philosophies of designing experience, and concrete methods and practices. It is designed for sound designers, recording engineers, producers, installation artists, educators, and composers.

When beginning to create an immersive experience, one might start by envisioning the most compelling sonic impression capable of evoking a specific feeling and emotional

DOI: 10.4324/9780429297526-1

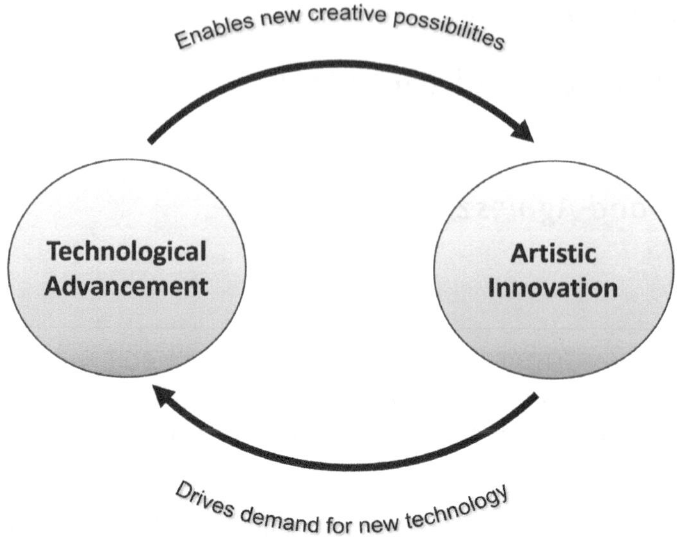


*Figure I.1* Continuous cycle of mutual advancement

response, one that is not only powerful but also deeply enjoyable. The pursuit of sound that captivates its listener, leaves a lasting impression, and stirs the emotions can be considered a purely aesthetic endeavor, serving as a guiding force for the creator's artistic intent.

This intent may take shape as the artist's conception, interpretation, or rendition – an idea that exists in thought but has yet to manifest in tangible form. It is within this conceptual space that the creative process begins, allowing the artist to explore, refine, and ultimately bring forth an experience that aligns with their vision.

Once the aesthetic goal is established, the creator must then consider and navigate the technical constraints. In the realm of immersive experience design, technological limitations are constantly being challenged and overcome through innovation. As new and enhanced technologies emerge, they push the boundaries of what is possible, allowing the art form to evolve. However, this relationship is not unidirectional. Just as technological breakthroughs expand artistic possibilities, creative visions often drive technological innovation in return, thus creating a powerful cycle (see Figure I.1). Artistic work can inspire, inform, and require new technologies to be developed. This symbiotic relationship accelerates progress in both domains: artists are inspired and acquire new tools to realize their visions, while technologists are driven by real-world applications that guide their innovations. This continuous advancement makes immersive world-building a dynamic and exhilarating field, where creative possibilities expand alongside technological progress.

Our authors discuss the use of sophisticated technical tools to create immersive experiences: binaural and multi-channel loudspeaker systems combined with visual media, carefully designed physical acoustics, and strategic location design – often blending multiple approaches simultaneously. These technologies unlock creative possibilities while imposing constraints that, if ignored, can shatter the illusion of immersion and destroy plausibility.

Plausibility stands as the cornerstone of successful immersive experiences. It means creating something that feels genuinely possible – real, believable, or, at minimum, undistracting to the audience. This principle may ultimately determine whether a production succeeds or fails, making it essential for both creatives and engineers (often the same person juggling multiple roles) to remain vigilant about this critical aspect of immersive media work.

Importantly, plausibility doesn't demand strict adherence to reality. Flying dragons, underwater cities, and other fantastical elements can feel entirely believable within the world you construct, as long as they remain consistent with the internal logic and sensory rules of that created environment.

## Design building blocks

Designing an immersive sound experience requires a holistic approach that goes above and beyond sound production, to encompass the complete physical and perceptual ecosystem. We can think of this multidimensional framework as an entire acoustic journey: from the spectral and temporal characteristics of sound sources, through the complex interactions of the sound waves as they propagate through and interact with the physical environment, to the psychoacoustic mechanisms by which these events are received and interpreted by the listener's auditory system. As Blesser and Salter (2007) articulate in their work on aural architecture, immersive sound design operates at the intersection of physics, acoustics, and psychoacoustics – creating environments where elements of sound are bound together into compelling perceptual experiences. Therefore, when we think of the basic design building blocks of such experiences, there are a few that are most prominent, including immersion, envelopment, directional sound, and the environment.

In the context of sound, **immersion** refers to the multidimensional auditory experience that envelops listeners in a complete sound environment and creates a sense of presence and spatial awareness that goes above and beyond conventional listening. This phenomenon is facilitated through various methods and techniques – from binaural recording and synthesis techniques that simulate natural hearing processes to wave field synthesis, Ambisonics that mathematically recreate sound fields, and many others (Roginska & Geluso, 2018). Through these techniques, immersive audio has the ability to transport a listener to another realm. Such experiences have the power of disrupting the physical reality of a listener's actual environment and instead create a sense of being surrounded by another soundscape.

**Envelopment** is an important psychoacoustic parameter that describes the subjective sensation of being surrounded by sound rather than perceiving sound as emanating primarily from specific directional sources. Beranek (1996) was one of the first researchers who systematically defined and measured listener envelopment (LEV) in his studies of concert hall acoustics. LEV represents the degree to which listeners perceive sound arriving from all directions around them, creating a sense of being 'enveloped' by the sound in the environment.

In more technical terms, **listener envelopment** is primarily created by lateral reflections and diffuse sound energy arriving at the listener's position from multiple angles, particularly from the sides and rear. These are the reflections that are created with appropriate time delays relative to the direct sound. According to Bradley and Soulodre (1995), the perception of envelopment correlates strongly with the lateral energy fraction and late lateral sound level measurements in acoustic spaces. Spatial audio reproduction systems typically recreate these conditions by using multiple speakers, through speaker placement, phase manipulation, and artificial reverberation to simulate natural acoustic environments.

The perceptual importance of listener envelopment goes beyond technical measurements – it serves as an important quality indicator in concert hall design, immersive audio production, virtual and augmented auditory environments, and immersive experiences. As Rumsey (2002) notes in his work on spatial audio perception, strong envelopment contributes significantly to listener preference, emotional engagement, and the overall sense of 'being there' that characterizes immersive sonic experiences.

**Directional sound** serves as another building block in creating immersive sound environments. Directional sound provides important spatial information that helps listeners construct mental maps of acoustic spaces and the sound sources within them. By controlling the perceived location of individual sound sources, designers can create auditory scenes with distinct foreground and background elements. This directionality enables several perceptual factors essential to immersion: (1) identifying the location of a sound source, which allows listeners to orient toward points of interest; (2) tracking the motion of a source, which creates the impression of movement through space; and (3) spatial separation reduces perceptual masking between competing sound sources, enhancing clarity and intelligibility (Bregman, 1990). The human auditory system has evolved to be sensitive to directional cues – Interaural Time Differences (ITD), Interaural Level Differences (ILD), and spectral coloration – which are all contained in the Head-Related Transfer Function (HRTF). Our modern immersive audio systems can now replicate the directionality cues to create a realistic and compelling experience.

Beyond technical reproduction, thoughtfully implemented directional sound can become a powerful narrative and emotional tool within immersive experiences. Chion's (1994) concept of 'spatial magnetization' explores how directional audio can (re)direct attention and create meaningful associations between visual and auditory elements in multimedia experiences. Collins (2008) describes how game audio designers use dynamic directional cues to generate tension and provide navigational guidance. Perhaps most importantly, directional sound enables designers to create contrasting acoustic zones including focused ones that command attention to diffuse and enveloping atmospheres that can induce emotional states.

The **environmental sound** characteristics and reverberation are an important part of immersive audio design. Reverberation serves as a spatial sonic signature that communicates information about an environment's physical dimensions, and the absorptive and reflective properties of walls and objects within the space. The specific pattern of early reflections (which arrive within 100 ms of the direct sound) provides listeners with the spatial orientation cues and information about where a sound source is located within the space; while the reverberant tail (including decay time, frequency response, and density of reflections) provides information about the space. Through various reverb technologies, designers can now recreate the acoustic 'fingerprints' of real spaces or create imagined sonic environments, allowing listeners to be transported between concert halls, cathedrals, small rooms, or imaginary landscapes. All of these building blocks, together, give designers a spatial vocabulary, and an assorted palette, with which they can paint captivating experiences.

## References

Beranek, L. L. (1996). *Concert and Opera Halls: How They Sound*. Acoustical Society of America.

Blesser, B., & Salter, L. R. (2007). *Spaces Speak, Are You Listening? Experiencing Aural Architecture*. MIT Press.

Bradley, J. S., & Soulodre, G. A. (1995). 'Objective Measures of Listener Envelopment.' *The Journal of the Acoustical Society of America*, 98(5), 2590–2597.

Bregman, A. S. (1990). *Auditory Scene Analysis: The Perceptual Organization of Sound*. MIT Press.

Chion, M. (1994). *Audio-Vision: Sound on Screen*. Columbia University Press.

Collins, K. (2008). *Game Sound: An Introduction to the History, Theory, and Practice of Video Game Music and Sound Design*. MIT Press.

Roginska, A., & Geluso, P. (Eds) (2018). *Immersive Sound: The Art and Science of Binaural and Multi-Channel Audio*. Routledge, 1st edition, October 2018.

Rumsey, F. (2002). 'Spatial Quality Evaluation for Reproduced Sound: Terminology, Meaning, and a Scene-Based Paradigm.' *Journal of the Audio Engineering Society*, 50, 651–666.

Chapter 1

# History of immersive sound experiences

*Braxton Boren*

The field known variously as 'spatial sound' or '3D audio' developed rather recently as a sub-discipline in laboratory settings, yet in some sense it has existed for far longer than it has been given a specific name. Tracing its historical path is complicated by the need to parse the names assigned to it at different times: the adjective 'spatial' creates an association with physical spaces (which will sometimes be helpful but is often misleading), while '3D' seemingly removes 1D or 2D spatial technologies from consideration; meanwhile the noun 'audio' gives a distinctively modern bias to the time periods in question. In fact, it is precisely because the field of audio shrank the perceived size of recorded sound that anyone bothered to think of what it meant for a sound to be 'spatial' at all: before Edison reduced sound to a tiny point, nearly all sounds would have been big or at least have carried a distinct spatial impression upon the listener. Some composers refuse to use the term 'spatialization' at all because it subtly conveys that the sound had no spatial component until one was added at a late stage (Nyström 2018).

Because of the possible confusion resulting from these terms, here we use the name 'immersive sound' to denote the sum of all sound events for whom a spatial or immersive component is important, if not the primary dimension along which the sound might be analyzed. While the previous volume followed the history of the field in general terms (Boren 2017b), this work focuses on the aesthetic 'experience' of those who wrote, sang, played, sculpted, chiseled, and – above all – *heard* immersive sounds throughout human history. This lens of analysis necessitates a smaller focus on individual technologies and more attention paid to the subjective, perceptual, and conceptual aspects of immersive sound. In short, in asking about an 'experience' we are in essence raising three questions:

1. What were the creators of these sounds trying to do?
2. What did they accomplish, accidentally or purposefully?
3. How was it heard and understood by their audience?

This is quite a different task from merely reporting the notable attempts and breakthroughs. To do so, we will have to define some bigger ideas and trace their development through a detailed analysis of a smaller number of cases.

## Types of immersive sounds through history

With the caveat that nearly all sounds had a spatial component before the advent of recorded audio, it is equally important to note that the average listener would have

DOI: 10.4324/9780429297526-2

distinguished between a 'space' and the source of sound itself. A cough is among the simplest of sounds, similar to a unit impulse, produced by exhaling against the closed vocal folds, followed by a sudden opening of the glottis. In an outdoor environment, the cough is short in duration, easily localized, and relatively unremarkable. However, a cough in a reverberant cathedral hangs in the air for several seconds and seems to come from everywhere and nowhere. In the latter case, it is clearly the physical interior space which creates the immersive sound experience of the listener, despite the initial impulse contributed by the cougher. Furthermore, while it may be tempting to see the two cases as the same acoustic phenomenon on different scales (a resonant enclosed volume, either the vocalist's chest cavity or the cathedral), this is a false equivalence from a perceptual framework: the vocalist's chest is heard primarily as a set of amplified frequencies due to the wavelengths that 'fit' inside the cavity, while the cathedral is primarily heard as an extension in space and time due to the ratio of the average distance between surfaces relative to the speed of sound. While the chest cavity does have a small but measurable reverberation time and the cathedral likewise affects different frequencies' intensities, from a listener's perspective the difference in degree becomes a difference in kind. Keeping this in mind, we would do well to separate 'Spaces' and 'Sounds' in our treatment.

But as our narrative continues, we will find cases where this distinction does not hold. In the past century, it has become possible to treat the sound sources and the enveloping room as part of a coherent whole in which generation and diffusion intersect and overlap. For these more holistic experiences, we will apply the label 'Sounds in Spaces'.

## *Spaces*

While the acoustic free field and diffuse field have always existed in some fashion, the oldest immersive sound fields we have direct evidence for comes from the archaeological record of the spaces themselves. Acknowledgement of spatial position or movement as an aesthetic consideration comes later, and full manipulation of both comes later still. So to understand the history in chronological order, we must go back to the beginning – or at least, as close as the evidence will take us – of human acoustic spaces.

The emerging field of Archaeoacoustics combines methods from archaeology and acoustics to examine the role of sound in the distant past (Scarre & Lawson 2006). Since sound is inherently transient, the field often uses spaces as a lens through which to examine the role of sound in human culture and how it would have been experienced by its listeners. Clapping one's hands at the base of the ancient Mayan temple of Kukulkan at Chichen Itza produces a chirped echo effect at the top of the temple, similar in quality to the call of the quetzal, a bird thought to be a messenger of the gods (Lubman 2008). Various studies have examined notable acoustic effects at Stonehenge (Till 2010) as well as other ancient European stone and timber circles (Reijs & Mussik 2017). In each of these cases, it is possible to construct 'weak' and 'strong' archaeo-acoustic arguments about the role of sound: the weak argument is that these spaces have unique acoustic effects which would have been perceptible by the listeners inside, and that this may have affected their use of sound within the space. The strong archaeo-acoustic argument is that aural experience was more important than other considerations (Reijs & Mussik 2017) or, in the strongest form, that the spaces were constructed primarily for acoustic reasons above all else. One does not have to fully endorse the 'strong' argument (indeed, it is rare enough today, even in concert halls, for aural design to trump all other considerations!) to acknowledge that the weak argument is almost certainly true.

Early human paintings date as far back as 40,000 years ago (Alvarez-Morales et al. 2023) and intact musical instruments have been dated as far back as 35,000 years ago to the Upper Paleolithic (Conard et al. 2009). An 18,000-year-old conch shell horn was carried 150 miles to the Marsoulas cave in the Pyrenees, where it may have been played in ceremonies connected to cave paintings (Kolar et al. 2022). Analysis of acoustic impulse responses suggests there may be some correlation between low-frequency audible modes of vibration and cave art in many cases, possibly with moderate amounts of reverberation as well (Jahn et al. 1996; Fazenda et al. 2017). Analysis of modern-day worship spaces also suggests that low-frequency reverberation and prominent modal resonances are an important component of listeners' emotional responses to these spaces (Algargoosh 2024). Thus in addition to a sensation of sound hanging in the air, such long-wavelength resonances between stone surfaces may have distinguished themselves via an emphasis of bass frequencies, which was rare in an early music culture dominated by flutes, percussion, and the human voice.

Of course, early freestanding human structures were constructed primarily for civic, military, or religious reasons, and spaces like caves were often not constructed at all. In the case of India, early caves became associated with local deities and were inhabited by hermits and ascetics, which in time led to the Hindu tradition of stone-cut monolithic caves as early as the fourth century BCE (Brancaccio 2014). As early Vedic traditions of open-air animal sacrifice around a fire transitioned to something more like modern Hinduism, the setting of such rituals likewise transitioned from the open air into cave temples and later to freestanding structures. These early stone temples, which often form a rectangular shape open on one side, generally have low reverberation times for most frequencies but may feature prominent modal resonances at particular frequencies within the human vocal range (Aswathanarayana & Boren 2025). The tradition of chanting the Vedas, starting at one tone, then transitioning to a tone slightly higher and another slightly lower than the base tone, could have been formed by the process of incrementally seeking out such resonances by Vedic priests. This connection, hard to prove definitively (like many archaeoacoustic hypotheses), nonetheless shows the possibility of 'space' itself forming the 'sounds' which would later take on a life of their own.

For ancient humans, whose sense of hearing had evolved primarily for localizing predators or prey in relatively free field settings, these spaces would have conveyed a variety of auditory cues that one had entered an environment which was categorically different from the normal spaces of life. Perhaps in settings so specific as Chichen Itza there was a narrative imposed on the space, with the sound coming from above representing the response of the gods to the worshipers present. But in many of these spaces, the feedback would have been much more general: a strange amplification of certain frequencies or a spreading out of sounds in time such that the natural world appeared to behave differently in that acoustic space. No obvious story need be imposed on such an experience, but in the most general sense it would have given these spaces a sense of otherness, being set apart, or (in the precise sense of the word) 'holy'. This tendency – for humans to be drawn to such spaces or to seek to emulate them virtually – will inform our understanding of all the spaces and spatial sounds that followed them.

As rulers and early proto-architects jointly embarked on the project of moving from natural caves to cutting into stone to make their own, doubtless some of the most observant stonecutters and masons may have noticed relationships between physical measurements and the sounds produced by the space. But the written record, already scant at such early periods of human history, leaves few examples of precise references to sound. More generally we might guess that, as such spaces proliferated and gained regular attendees, certain spaces became favored for a variety of reasons, one of which may have been their acoustic response. In the absence of a comprehensive understanding of the statistical analysis of the sound field in such spaces, new constructions often followed a pattern of basing their plans on existing spaces that were already preferred for whatever reason. Over time, this may have led to correlations between acoustic behavior and other physical measurements which temple planners used in their design process.

In ancient Egypt, temple construction began as simple shrines constructed of mud-bricks, which would have provided greater acoustic absorption than stone caves and consequently reduced the acoustic effect of the space relative to outdoor worship spaces. Yet these temples were obviously a cultural center, often elevated on a mound (either natural or human-made) and containing statues of gods, inside or outside the structure. Some such shrines were later reconstructed wholly from stone during the Middle Kingdom (second millennium BCE), as their importance to the larger economy merited such an expensive material (Kemp et al. 2000). This transformation would also have brought longer reverberation times and a greater auditory sense of the space's importance to the culture and its religious rites.

Around the same period, Bronze Age Europe began the construction of other spaces intended for a different type of sound: the ancient Minoan palaces on Crete contained large rectangular halls whose total purpose is unclear, but many have hypothesized a theatrical use for them. Commenting on the palace complexes at Knossos, Malia, and Phaistos, archaeologists Letesson and Vansteenhuyse (2006) observe that 'the large majority of the space in their Central Courts was used as a scene for very public performances by employing a loud voice, large gestures and a performer singled out'. Another large natural rock shelter in Spain has been hypothesized to have functioned something like a theater for small audiences (Alvarez-Morales et al. 2023). Whether these spaces were primarily theatrical or religious, they certainly establish a template of seating for spectators paying attention to a single individual, whose voice is strengthened and widened by the spatial parameters of the room in which it is located.

Whatever the precise purpose of the Minoan theaters, later spaces constructed purely for theater would follow through the famous Greek outdoor amphitheaters and later Roman theaters as well. The earliest Greek amphitheaters were constructed into the sides of hills as early as the fifth century BCE, and their main acoustic features were being somewhat isolated from wind noise and providing a strong early reflection from the stage. The Roman theaters, based on Greek designs, added walls around the audience area and formed an early example of the modern proscenium theater (Chourmouziadou & Kang 2008). These enclosed structures, largely built of stone, provided greater reinforcement of music and a longer reverberation time than that found in outdoor structures. Such theatrical spaces provided a secular alternative to temple worship but were likewise spaces set apart from everyday life, where drama unfolded on a cosmic level, and sound (both narrative and musical) was used to support the stories enacted on the stage.

The Jewish synagogue and its worship service developed based on the first and second temples at Jerusalem (originally constructed primarily from wood) and before that on the tabernacle, a fabric tent. Wood is acoustically absorbent, while fabric would act as an absorber via sound transmission to the outside. Both material cases create a relatively 'dead' acoustic response relative to stone structures, whether caves or freestanding temples. This would in turn skew the acoustics of such spaces towards clarity and away from envelopment, which would affect both the Jewish and later Christian liturgies (Lubman & Kiser 2001; Boren 2017a, 2021b). Across different religious traditions, the specific frequency spectrum and directivity of each liturgical element may affect the acoustic appropriateness of the corresponding worship spaces (Prasad & Rajavel 2013; Soeta et al. 2013). Individual composers who were accustomed to the response of a particular sacred space could also tune their music to be intelligible for that setting (Boren 2021a).

This pattern of the sound shifting to fit the space would also continue as the dominant cultural sonic events shifted from sacred to secular: as chamber music halls, concert halls, and opera houses propagated across the industrialized world, the subtle differences between different performance spaces would continue to shape the ensembles that called them home. The Philadelphia Orchestra's home in the Philadelphia Academy of Music, originally built as an opera house, is an acoustically dry space with a mid-frequency reverberation time of about 1.4 s. Under conductors Leopold Stokowski and Eugene Ormandy, the orchestra would develop its signature 'Philadelphia Sound' featuring unsynchronized bowing in the string section, giving a strong legato feeling almost akin to a reverberation within the ensemble itself to compensate for the lack of reverberation in the hall (Holland 2003). But when the orchestra went on tour, its internal 'washiness' could be a poor fit for more reverberant spaces. Thus the principle of being 'set apart' applies not only to the space itself but also to the sounds that will inhabit it. To paraphrase the 24th Psalm, only that which is holy may enter the holy place.

## *Sounds*

Skipping many examples featured in previous history (Boren 2017b), by the 19th century the ambitious artistic horizon of composers included some attempts to incorporate spatialization into their works. This generally manifested either through the purposeful placement of individual instruments or sections outside their normal orchestral positions, or through physical movement of sound sources. These two approaches might be generally categorized as 'static' versus 'dynamic' spatialization.

### *Static spatialization*

Composers from Berlioz to Mahler often experimented with offstage ensembles or spreading out instrumental sections around the concert hall. While such techniques add to the overall spatial impression of the room, they also reinforce the idea of the 'space' as an external and unchanging reality to which the music must accommodate itself. Since the musical elements cannot easily move, their position becomes another element of the performance space which may help or hinder the overall acoustic and musical impression of the performance. Even the changing the time delay of the suggested echo effect in Haydn's '*Das Echo*' (Hob. II:39) is merely the exception that proves the rule that rooms are generally fixed and unchanging over time. This static quality makes the space within which music takes place an interesting artistic choice, akin to the choice of soloist or orchestration, but space is not in this case an artistic variable over which musical expression can be projected.

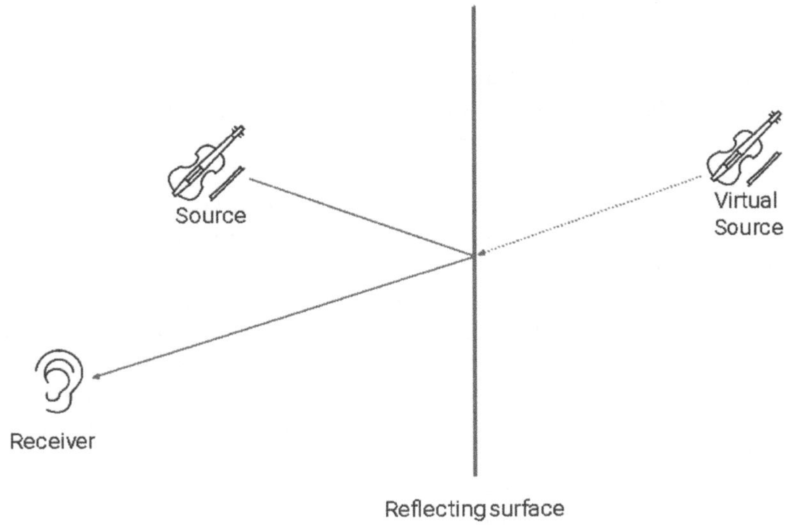


*Figure 1.1* Equivalence of space and virtual static spatialization

In a larger sense, the projection of an acoustic 'space' and the static spatialization of instruments within that space are in the end perceptually equivalent: each sound reflection off a physical surface yields a virtual source in a virtual room on the side of the reflecting surface (Figure 1.1). The direction and time delay are equivalent, while the spectral components of the sound are affected by the absorption and scattering properties of the surface itself (Allen & Berkeley 1970). Given infinite performing forces and infinite control of instruments' dynamics and articulation, a composer could theoretically suggest a physical space merely through the static positioning of performers in a free field. Note that each arrangement of virtual sources only holds for a single receiver position, with the consequence that the geometry of the virtual room being invoked changes for each listener in the audience.

The second approach, in which sounds themselves are moved over time, is difficult to do with acoustic instruments, though some have tried: a notable example is Carl Orff's concept of *Theatrum Mundi*, which sought to unite music, movement, and dance into a single artistic achievement. Orff's famous 'O Fortuna' from *Carmina Burana*, calls for unspecified motion by musicians and dancers to signify the whims of the goddess Fortune; however, in practice, because of the large musical forces required for the piece, these spatial directions are often omitted in performances (Babcock 2006). A more choreographed approach would be employed by Charles Ives's father George Ives, who once directed two marching bands to cross each other while playing different pieces in different keys, to the consternation of nearby residents (Zvonar 2006). But such an experiment would be nearly impossible to choreograph in the limited space of a concert hall, and technological innovation would be needed before such a trajectory could be realized before a traditional audience.

A more appropriate attempt would be made by the younger Ives as he tried to move past the mere cacophony of his father's experiment and instead incorporated a static separation between ensembles in his piece *The Unanswered Question* (1908). However, whereas previous static spatialization had served to envision a larger consonance of a single volume, Ives's ethereal and dissonant music was better heard in a dryer acoustic space, where the audience's sense of spatial location might help them make sense of the highly contrasting material. Many years later, in his memorial to the 9/11 attacks, *On the Transmigration of Souls*, John Adams would adopt Ives's spatial separation in an attempt to add clarity to the 'Ivesian' chaos his piece invoked, though in the end he felt the piece had failed to bridge the gap between the composer's intent and the victims' trauma (Adams 2023).

The most well-known 'spatial' composer of acoustic music, Henry Brant, was highly influenced by Ives and worked to develop this principle in more detail. The trend of 19th-century music had been to develop more complex harmonic relationships which, while shocking to contemporary audiences, nonetheless could be resolved through more distant relationships back to a tonal center. But in the 20th century, the atonal tradition ceased with the idea of a rooted tonal center in favor of more abstract serial relationships between notes which were not always perceptible to lay audiences. Brant realized that spatial separation between contrasting notes could help audiences more clearly make sense of the greater dissonance of 20th-century music (Brant 1967). Yet in treating spatial arrangement in this way, Brant only served to use space as a means to greater harmonic coherence. Space had still not yet been used as a primary dimension of musical composition itself (Boren & Musick 2018).

### *Dynamic spatialization*

By way of example, to see the difference between primary and secondary dimensions, we may examine the difference between the relationships of melodic ideas and rhythmic motives in Western and Indian music. In the West, it is not uncommon to find a melodic instrumentalist playing or improvising over drum loops. The rhythmic or temporal dimension acts as a frame over which to paint melodic ideas. Conversely, a Hindustani (North Indian) performance will often feature a percussionist (such as a tabla player) playing different rhythmic motives over a simple repeating melody on a tambura. In the second case, the melody is the frame for temporal exploration.

As the more interesting examples of spatialization moved away from both physical spaces and acoustic instrumentation and toward electronic realization, composers had more freedom to explore spatial motives, in which the dimensions of frequency and time were no longer primary. The exact principles which would serve to organize this spatial exploration, however, would vary somewhat from one composer to another.

Because the inclusion of physical spatialization connects to our everyday experience of the world in a way that 'frequency space' does not, physical principles are the most obvious organizational basis for dynamic spatial movement. Just as 'coherence' or 'consonance' in the frequency domain provided a sort of safe home state for the auditory system, and deviations from it signaled the autonomic nervous system to be in a state of heightened alert, so too Natasha Barrett has suggested that spatial music might be able to signal a spatial coherence with our default mental model of the world around us, consonant with our expectations. In this system spatial dissonance could likewise be achieved by defying our expectations about what sound sources ought to do in the physical world, drawing greater attention to themselves in the process. Finally, a return to spatial consonance after a period of dissonance could be considered a spatial cadence of sorts. One example of such dissonance could be the separate manipulation of physical velocity without Doppler shift, akin to how the phase vocoder allowed temporal manipulation of audio signals without 'chipmunk'-like spectral distortion (Barrett 2002).

One of the most basic physical principles which might be expected when we have control over a sound's 3D position is our expectation of gravitation (what goes up must come down). While previous analysis in electroacoustic music focused on *spectromorphology* (the change of an isolated sound's frequency spectrum over time), spatial positioning allowed the focus on the perceived spatial location of a sound event in motion. Just as higher frequency components tend to be drawn downward in a 'gravitational archetype' (Roy 1998), so too spatial sound might have a sense of gravity within it: Dennis Smalley (2007) writes

> A descent, depending on its energy and velocity, invites an analogy with downward drift or fall, seemingly attracted towards a lower region. If it disappears from view we might imagine it continuing out of sight or earshot: there is no need for it to reach a grounded position for us to recognise its gravitational tendency, and no need for a plane to be explicitly expressed. An ascending, disappearing spectral motion may embody similar implications, but there is a significant difference, because a continuing ascent can also appear to be distancing itself, moving into spatial infinity.

Smalley believed that, whereas physical systems require (and thus we tend to expect) the expenditure of energy to reach greater heights, in purely sound-based art, it is not clear that this holds perceptually since the 'energy' required is no less for low positions versus high. A possible rejoinder would be that just as the 'gravitational archetype' evolved over time in frequency space because composers often followed an upward jump with falling stepwise progressions, so too spatial archetypes might still evolve as best practice as spatial motives become more achievable with available technology.

In addition to simple up-down relationships, many early electroacoustic composers were drawn to the possibility of spatial rendering to trace larger geometric curves over time, evoking a 'space' without the acoustic reflections that such a space would bring with it. Among the more notable examples of this type of composition include Edgard Varèse, Iannis Xenakis, and John Chowning, among other examples still being explored today (Kendall 2006; Besada 2014; Chowning 1971; Chowning 1977). This geometrical approach has the perceptual advantage of a small number of sources which can move slowly enough for our auditory system to update their apparent position gradually.

Another physical principle evoked by complete control of the spatial environment, beyond the simple evocation of velocity or momentum, is a more emergent sense of chaos versus equilibrium. Sources moving in tension with one another can cause displacement and alter each other's trajectories in ways which may be difficult to intuitively predict. Much like physical examples of chaos theory, these sources themselves may begin in similar positions but trace very different paths according to slight variations in their initial conditions. However, such creations, which often contain hundreds or thousands of elements in the visual domain, are necessarily limited by human auditory perception and stream segregation, which vastly limits the number of auditory sources that can be simultaneously localized (McDermott 2009).

### *Sounds in spaces*

Beyond all the composers and creators mentioned so far, perhaps the most famous is Karlheinz Stockhausen, who engaged in spatial exploration throughout his career. Yet Stockhausen often pursued space as another subsidiary dimension (along with tonal pitch) to his primary goal of serialism (Boren & Musick 2018). Going beyond early serialist composers, Stockhausen desired to subject many other parameters to temporal order, with spatial position as one such parameter. While Brant used space to enhance frequency, Stockhausen used space to enhance temporal sequence.

But (assuming we do not experience an artistic revival of serialism anytime soon) Stockhausen's greater and more lasting contribution is the fusion of 'sounds' and 'space' together through his masterpiece *Gesang der Jünglinge* (1955–1956): Stockhausen, a devout Catholic, had originally planned to perform the piece within the iconic space of the Cologne cathedral. When church officials denied him permission, he was forced to alter his original plan to create sounds situated in a physical space, instead creating an electronic composition that consisted of both a synthetic acoustic space and the sound events that took place within it (J. Smalley 2000; Chait 2014). The piece's *Jünglinge* (youths) refers to the story of Shadrach, Meshach, and Abednego from the book of Daniel, who sing praise to God even after being cast into a fiery furnace. Perhaps because of the constraints of the framing narrative, the piece required a blend of recognizable semantic source material and an environment from which the *Gesang* (song) emerged.

This fusion of sound generation and spatial perception goes beyond Orff's *Theatrum Mundi* and even beyond Wagner's more comprehensive *Gesamptkunstwerk* – Wagner could design his Bayreuth *Festspielhaus* but could not significantly change the acoustic space throughout the course of an operatic performance. The ability to affect both 'sound' and 'space' simultaneously might be thought of as musical 'worlding'(Rickert & Salvo 2006), or we may also more simply call it 'sounds in spaces'.

The ability to generate sounds virtually and place them in virtual rooms has, of course, spread widely in the age of digital studio technology. However, for the 50 years or so after *Gesang*'s premiere in 1956, most such experiments were still in stereo (or, alas, mono), limiting their immersive possibilities. But recently disparate fields of musical creation and spatial synthesis have converged in tools, such as binaural panners and sophisticated game engines, that go beyond a standard digital audio workstation (DAW) and allow non-expert composers to create spatial compositions without significant investment in loudspeaker-based infrastructure. Object-based audio systems encode spatial trajectories in metadata, realizing the objects differently based on the playback capacity of different audio systems (Tsingos 2017). This space-first compositional environment allows 'space' to be manipulated freely over time as simply and naturally as frequency/pitch/melody.

This diffusion of technological capacity mirrors the early periods in Western music when pitch notation was becoming standardized and more church composers had the resources to experiment with different harmonic progressions and resolutions: many 'rules' of music theory emerged out of practical experience of what worked and what didn't. Human ears, while allowing some differences between listeners, are similar enough that there are broad principles that allow composers to predict an audience's attention, expectation, and satisfaction.

Given the meteoric rise of spatial audio applications in the early part of the 21st century, one is left to wonder where this technology is headed in the future. Spatial audio might become another fad that periodically pops up and is abandoned, akin to Hollywood's on-again/off-again relationship with 3D films. Spatial audio might be left to idiosyncratic electroacoustic composers and experimental sound artists, dissatisfied with the status quo, while the majority of musicians continue in stereo. Spatial audio might become a sort of high-end reverb plug-in, in which artists basically adopt the 'static spatialization' paradigm but without any constraints on their physical room or human ensemble. Or perhaps the confluence of intuitive tools and human imagination might lead to a renaissance in musical exploration and a new language of spatial motives, expectation, and innovation.

The first humans to discover enclosed cave environments probably could not imagine the acoustic potential of such reverberant spaces for human music making. Likewise, the first musicians to use a drone tone against a simple melody could not conceive of the coming flourishing of complex polyphony or the effect such compositions would have on future generations. Considering our position in this flowing channel of history, we too cannot fully envision the spatial possibilities ahead of us.

Unlike Captain Kirk, we cannot state categorically that space is sound's 'final frontier' (D. Smalley 2007). Perhaps advances in brain–computer interfaces will one day artificially stimulate our auditory cortex without the generation of any physical pressure wave. But if history is any guide, we know there remain significant unexplored frontiers which may yet overturn all our preconceived notions about sonic experiences.

## Acknowledgments

Many thanks to my friend and colleague Shashank Aswathanarayana for many wide-ranging conversations on long drives across rural India, which spurred some of the thoughts in this essay.

## References

Adams, J. (2023, December 11). The sound of memory. *The New Yorker*, 60–63.

Algargoosh, A. (2024). The link between the acoustic characteristics of worship spaces and their emotional impact. *Journal of the Acoustical Society of America*, *155*(5), 3027–3036. https://doi.org/10.1121/10.0025923

Allen, J., & Berkeley, D. (1970). Image method for efficiently simulating small room acoustics. *Journal of the Acoustical Society of America*, *65*, 943–950.

Alvarez-Morales, L., Rosa, N. S. da, Benítez-Aragón, D., Fernández Macías, L., Lazarich, M., & Díaz-Andreu, M. (2023). The Bacinete Main Shelter: A prehistoric theatre? *Acoustics*, *5*(1), 299–319. https://doi.org/10.3390/acoustics5010018

Aswathanarayana, S., & Boren, B. (2025). Acoustic analysis of two Hindu temples in Southern India. *JASA Express Letters*, *5*(3). https://doi.org/10.1121/10.0036033

Babcock, J. (2006). Carl Orff's 'Carmina Burana': A fresh approach to the work's performance practice. *The Choral Journal*, *46*(11), 26–40.

Barrett, N. (2002). Spatio-musical composition strategies. *Organised Sound*, *7*(3), 313–323. https://doi.org/10.1017/S1355771802003114

Besada, J. (2014). Computer, formalisms, intuition and metaphors: A Xenakian and post-Xenakian approach. In *Proceedings of the international computer music conference*. https://doi.org/10.13140/2.1.4610.0484

Boren, B. (2017a). A generalized version of the Lubman-Kiser theory of historical acoustics and worship spaces. *Journal of the Acoustical Society of America*, *141*, 3774.

Boren, B. (2017b). History of 3D sound. In A. Roginska & P. Geluso (Eds.), *Immersive sound: The art and science of binaural and multi-channel audio*. Focal Press.

Boren, B. (2021a). Acoustic simulation of J.S. Bach's Thomaskirche in 1723 and 1539. *Acta Acustica*, *5*(14).

Boren, B. (2021b). Word and mystery: The acoustics of cultural transmission during the Protestant Reformation. *Frontiers in Psychology*, *12*(564542). https://doi.org/10.3389/fpsyg.2021.564542

Boren, B., & Musick, M. (2018). Spatial organization in musical form. In *Proceedings of the 2018 international computer music conference* (pp. 33–38). Daegu, Korea.

Brancaccio, P. (2014). Cave architecture of India. In *Encyclopedia of the history of science, technology, and medicine in non-Western cultures*. Springer. https://doi.org/10.1007/978-94-007-3934-5

Brant, H. (1967). Space as an essential aspect of musical composition. In E. Schwartz & B. Childs (Eds.), *Contemporary composers on contemporary music*. Da Capo Press.

Chait, R. (2014). Gesang dream: Functions of faith in Stockhausen's electric mass. *Perspectives of New Music*, *52*(3), 185–196.

Chourmouziadou, K., & Kang, J. (2008). Acoustic evolution of ancient Greek and Roman theatres. *Applied Acoustics*, *69*(6), 514–529. https://doi.org/10.1016/j.apacoust.2006.12.009

Chowning, J. (1971). The simulation of moving sound sources. *Journal of the Audio Engineering Society*, *19*(1), 2–6. https://doi.org/10.2307/3679609

Chowning, J. (1977). Simulation of moving sound sources. *Computer Music Journal*, *1*(3), 48–52. https://doi.org/10.2307/3679609

Conard, N. J., Malina, M., & Münzel, S. C. (2009). New flutes document the earliest musical tradition in southwestern Germany. *Nature*, *460*(7256), 737–740.

Fazenda, B., Scarre, C., Till, R., Pasalodos, R. J., Guerra, M. R., Tejedor, C., Peredo, R. O., Watson, A., Wyatt, S., Benito, C. G., Drinkall, H., & Foulds, F. (2017). Cave acoustics in prehistory: Exploring the association of Palaeolithic visual motifs and acoustic response. *Journal of the Acoustical Society of America*, *142*(3), 1332–1349. https://doi.org/10.1121/1.4998721

Holland, B. (2003). Listening to walls and ceilings instead of music. *The New York Times*.

Jahn, R. G., Devereux, P., & Ibison, M. (1996). Acoustical resonances of assorted ancient structures. *Journal of the Acoustical Society of America*, *99*(2), 649–658. https://doi.org/10.1121/1.414642

Kemp, B. J., Boyce, A., & Harrell, J. A. (2000). The colossi from the early shrine at Coptos in Egypt. *Cambridge Archaeological Journal*, *10*(2), 211–242. https://doi.org/10.1017/S0959774300000081

Kendall, G. S. (2006). Juxtaposition and non-motion: Varèse bridges early modernism to electroacoustic music. *Organised Sound*, *11*(2), 159–171. https://doi.org/10.1017/S1355771806001440

Kolar, M. A., Fritz, C., & Tosello, G. (2022). Acoustics in music archaeology: Re-sounding the Marsoulas conch and its cave. *Acoustics Today*, *18*(2), 52–61. https://doi.org/10.1121/AT.2022.18.2.52

Letesson, Q., & Vansteenhuyse, K. (2006). Towards an archaeology of perception: 'Looking' at the Minoan palaces. *Journal of Mediterranean Archaeology*, *19*(1), 91–119. https://doi.org/10.1558/jmea.2006.19.1.91

Lubman, D. (2008). Convolution-scattering model for staircase echoes at the temple of Kukulkan. In *Acoustics '08* (pp. 4161–4166). Paris, France. https://doi.org/10.1121/1.2934779

Lubman, D., & Kiser, B. H. (2001). The history of Western civilization told through the acoustics of its worship spaces. In *Proceedings of the 17th international congress on acoustics*, published by SAGE Journals.

McDermott, J. H. (2009). The cocktail party problem. *Current Biology*, *19*(22), R1024–R1027. https://doi.org/10.1016/j.cub.2009.09.005

Nyström, E. (2018). Topographic synthesis: Parameter distribution in spatial texture. In *Proceedings of the 2018 international computer music conference*. Daegu, Korea.

Prasad, M. G., & Rajavel, B. (2013). Acoustics of chants, conch-shells, bells and gongs in Hindu worship spaces. *Acoustics*, *2013*, 137–152.

Reijs, V., & Mussik, R. (2017). *Acoustic and aural research at timber and stone circles*. Academia.edu. https://www.academia.edu/33279912/Acoustic_and_aural_research_at_timber_and_stone_circles

Rickert, T., & Salvo, M. (2006). The distributed Gesamtkunstwerk: Sound, worlding, and new media culture. *Computers and Composition*, *23*(3), 296–316. https://doi.org/10.1016/j.compcom.2006.05.001

Roy, S. (1998). Functional and implicative analysis of Ombres Blanches. *Journal of New Music Research*, *27*(1–2), 165–184. https://doi.org/10.1080/09298219808570743

Scarre, C., & Lawson, G. (Eds.). (2006). *Archaeoacoustics*. McDonald Institute for Archaeological Research.

Smalley, D. (2007). Space-form and the acousmatic image. *Organised Sound*, *12*(1), 35–58. https://doi.org/10.1017/S1355771807001665

Smalley, J. (2000). Gesang der Jünglinge: History and analysis. In *Masterpieces of 20th-century electronic music: A multimedia perspective*. http://sites.music.columbia.edu/masterpieces/notes/stockhausen/GesangHistoryandAnalysis.pdf

Soeta, Y., Shimokura, R., Kim, Y. H., Ohsawa, T., & Ito, K. (2013). Measurement of acoustic characteristics of Japanese Buddhist temples in relation to sound source location and direction. *Journal of the Acoustical Society of America*, *133*(5), 2699–2710. https://doi.org/10.1121/1.4796116

Till, R. (2010). Songs of the stones: An investigation into the acoustic history and culture of Stonehenge. *Journal of the International Association for the Study of Popular Music*, *1*(2), 1–18. https://doi.org/10.5429/2079-3871(2010)v1i2.10en

Tsingos, N. (2017). Object-based audio. In A. Roginska & P. Geluso (Eds.), *Immersive sound: The art and science of binaural and multi-channel audio*. Focal Press.

Zvonar, R. (2006). A history of spatial music. *eContact!*, *7*(4). http://cec.sonus.ca/econtact/7_4/zvonar_spatialmusic.html

Part I

# Guiding principles and system design

Chapter 2

# Hamasaki-Square, Hamasaki-Cube, and the development of the 22.2 system

*Kimio Hamasaki*

## History and philosophy

I started my professional recording career in classical music for the live broadcasting of concerts by the NHK (Nippon Hoso Kyokai) Symphony Orchestra in 1982, where we faced many constraints to get the best sound. There were hard limitations to setting the microphones at the preferred positions and placing the many microphones I wanted to use because it was a live concert with an audience and also a live broadcast with TV cameras. The sound in the concert was dramatically changed from the rehearsal without the audience compared to the live concert with the audience. At that same time, I also made many recordings of orchestral and chamber music in the recording studios where I could try out different microphone techniques to achieve the best sound. These different situations served me quite well as I developed microphone techniques aimed at capturing natural sound.

I always try to achieve the best natural sound for my recordings. What is the natural sound of classical music recording? It should have the very natural timbre of each musical instrument, very stable localization of each musical instrument, and a very natural spatial impression created by the musical performance.

This chapter discusses my contribution to developing the Hamasaki-Square and Hamasaki-Cube systems, as well as the history and development of the 22.2 multi-channel sound system.

## Hamasaki-Square and Hamasaki-Cube

When I started my music recording career, all the recordings were done in mono or two-channel stereo. I used the 'main + ambience' microphone technique at that time for classical music recording. This technique allowed me to control the distance of musical instruments and the feeling of reverberation. MS stereo microphone, AB technique with two omnidirectional microphones, and XY techniques with two directional microphones were often used for the main microphones. Two omnidirectional microphones were often used for ambience microphones.

In the late 1980s, I started recording for 3-1 and 3-2 multi-channel sound systems. The 3-1 multi-channel sound format (or 3-1 quadraphonic sound) could be transmitted using the high-definition TV broadcasting in Japan. These two approaches were used in my early surround sound recordings, as shown in Figures 2.1 and 2.2. These concepts were developed based on two-channel recording. Unfortunately, the 3-1 quadraphonic sound

DOI: 10.4324/9780429297526-4

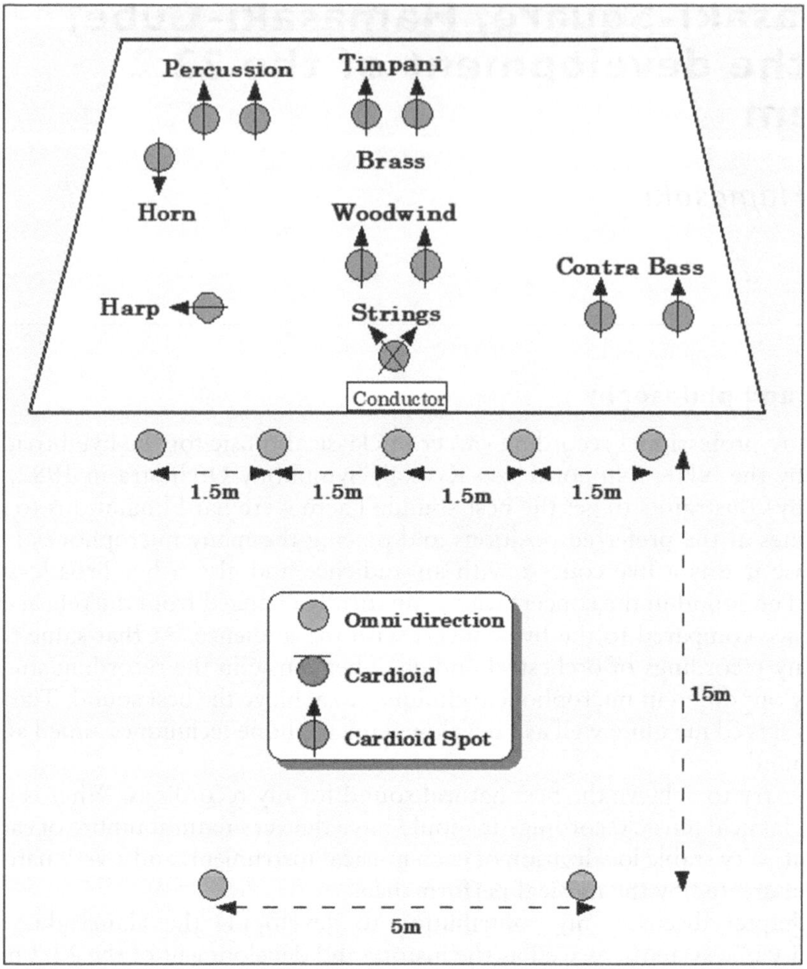


*Figure 2.1* Microphone technique using omnidirectional microphones for 3-2 multi-channel sound recording of an orchestra [2]

system was inappropriate for music, especially classical music. The mono surround channel couldn't reproduce the natural acoustic field like the concert hall. Therefore, I started to experiment with 5.1 surround sound. With 5.1 surround sound, I could create a more natural music sound field than the 3-1 quadraphonic sound could deliver. I focused on developing microphone techniques for 5.1 surround sound during this period. However, I noticed some issues again with recording and reproducing the natural sound field for classical music, even with 5.1 surround sound. The most challenging issue was how to enlarge the appropriate listening area, namely the sweet spot. The main reason was that no reflections could be reproduced from the higher positions using the 5.1 surround sound system [1, 2].

The omnidirectional microphone technique for recording 3-2 multi-channel sound for an orchestra shown in Figure 2.1 improved the spatial impression but also caused

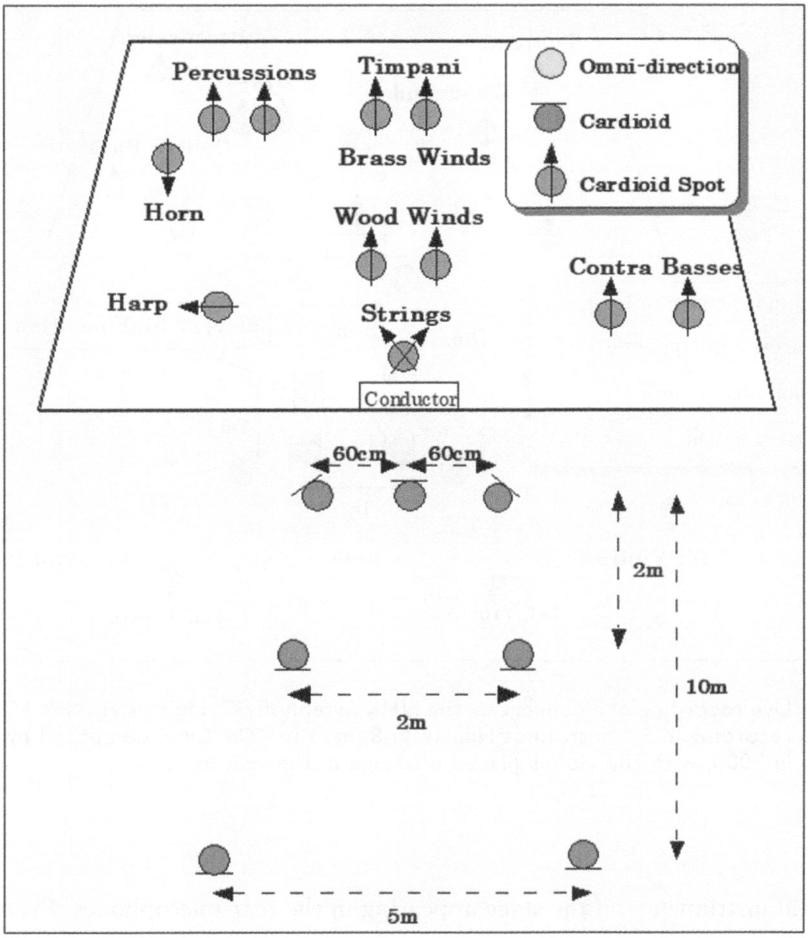


*Figure 2.2* Microphone technique using cardioid microphones for 3-2 multi-channel sound recording of an orchestra [2]

artifacts. When main microphone signals and ambience microphone signals are mixed electronically in two-channel stereo, the time delay between each microphone creates a good spatial impression. It does not cause any issues in listening. But when main microphone signals are reproduced from frontal loudspeakers and ambience microphone signals are reproduced from rear loudspeakers, the time delay can be heard significantly and causes some annoyances, particularly for impulsive percussion. It is necessary to place the ambience microphone near the main microphone to reduce these effects caused by the time delay. Therefore, the omnidirectional microphone should be replaced by a directional microphone because an omnidirectional ambience microphone must be set at a certain distance from the main microphone to get the proper reflections and reverberation.

With the microphone arrays using cardioid microphones shown in Figure 2.2, the problems due to the time delay just described were solved; however, this raised another issue of the unstable localization of musical instruments caused by the direct sound from

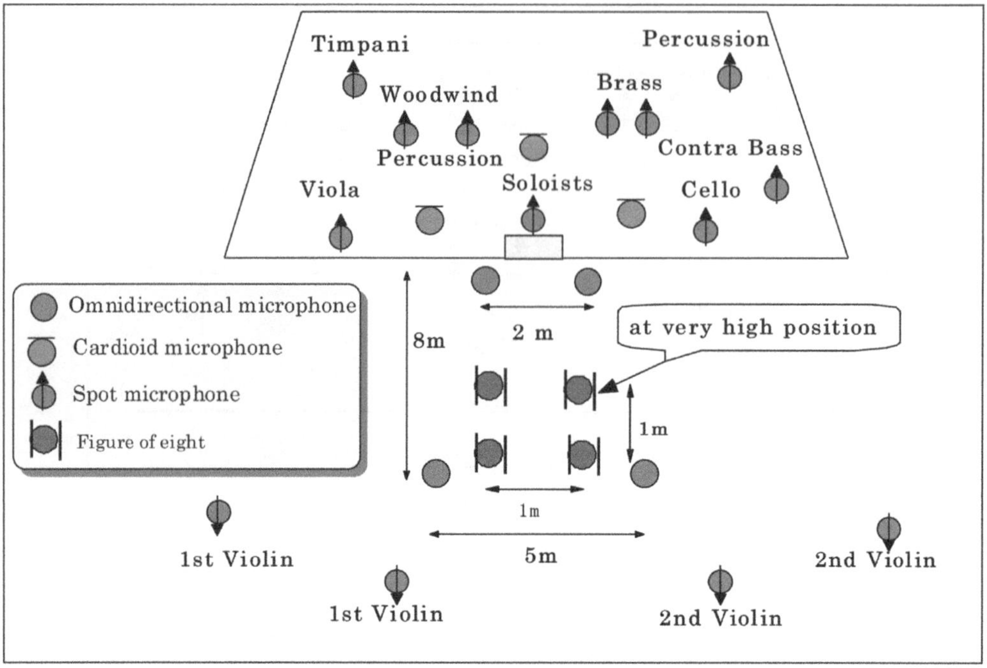


*Figure 2.3* Live recording of a concert by the NHK Symphony Orchestra at NHK Hall [3]: First recording in 5.1 surrounds Hamasaki-Square for *The Gate*, composed by Tan Dunin in 2000, with the violins placed off-stage in the seating area

the musical instruments on the stage appearing in the rear microphones. Even if a high-quality directional microphone was used for the rear microphone, I found that it was not ideal for capturing indirect sound. Since the rear microphone caught some direct sound of the orchestra, the sound image localization in the frontal sound field deteriorated as the level of the rear microphone increased. Loud sounds, such as brass wind instruments, were particularly affected.

In 2000, I developed a microphone technique for the ambience microphone array consisting of four figure-of-eight microphones to solve all these issues and achieve a natural orchestra recording in the concert hall. Florian Camerer named this technique the Hamasaki-Square. Figure 2.3 shows my first recording in 5.1 surround with Hamasaki-Square that I made in 2000. It was the orchestral piece *The Gate* composed by Tan Dun, which had a unique orchestra placement with violins situated in the seating area. This recording won the Audio Award of CATs (Creativity and Technology) Awards at IBC in 2001.

### *Concept of recording and mixing for natural sound*

My philosophy on music recording is getting the best balance between direct and indirect sound because we always hear both in any listening environment. An actual acoustic field gives us the natural sound of music performance, as shown in Figure 2.4. Indirect sound

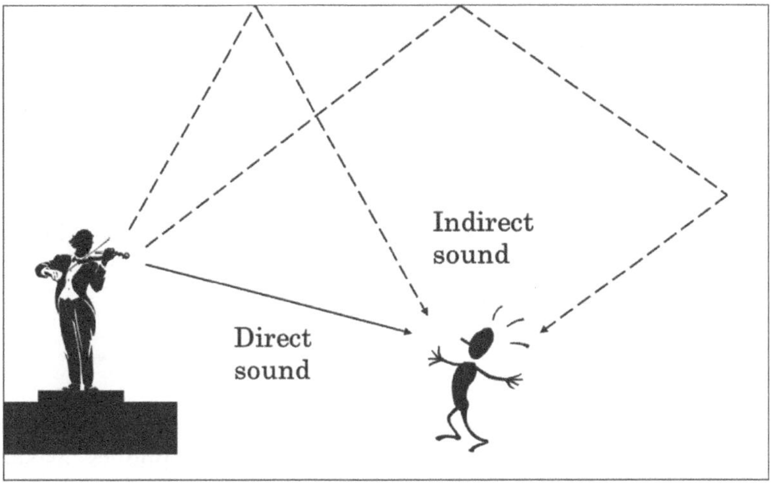


*Figure 2.4* A real acoustic field gives us the natural sound of music performance [2]

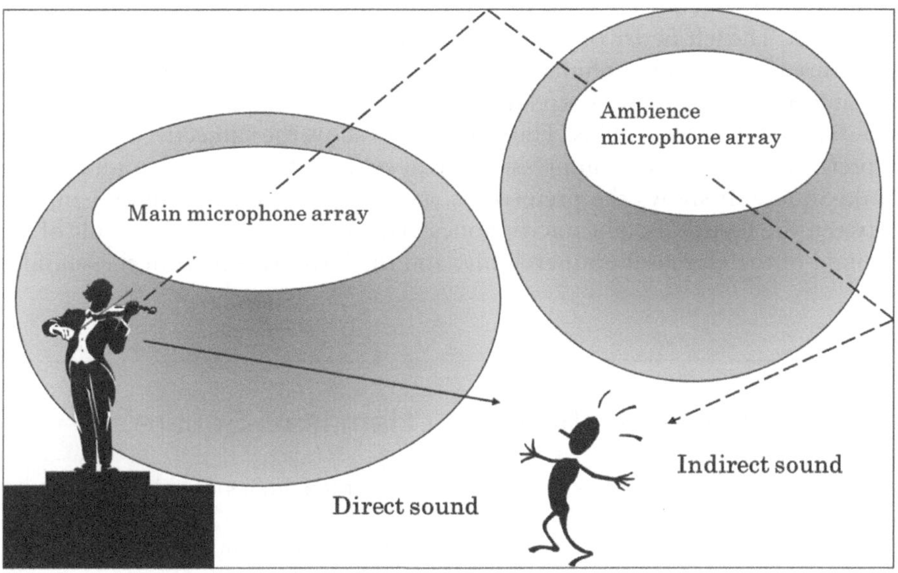


*Figure 2.5* Basic concept of capturing the direct sound and indirect sound by a main microphone array and an ambience microphone array [2]

consists of early reflections and late reverberation, reflected from the recording venue's walls, floors, and ceilings.

Suppose we can capture the direct and indirect sounds independently as much as possible, as shown in Figure 2.5. In this case, we can adjust the direct and indirect sound balance to achieve the most natural sound by using two microphone arrays, namely a main microphone array and an ambience microphone array. In practice, the main microphone array can't capture only the direct sound from musical instruments, and the ambience microphone array can't capture only indirect sound. Therefore, it is necessary to use the most appropriate microphone technique for the job and carefully adjust the positions of two microphone arrays. The main microphone array must capture each musical instrument's natural timbre and produce a sound image with stable localization, and the ambience microphone array must capture the natural spatial impression. The balance of direct to indirect sound can dramatically change between rehearsal and recording; therefore, it is essential to get the appropriate balance between direct and indirect sound during or after the recording, for which this method is the best choice.

### *Hamasaki-Square*

Hamasaki-Square was developed for the ambience array for the 5.1 channel surround sound recording, which consists of four figure-of-eight microphones, as shown in Figure 2.6. The null direction of each figure-of-eight microphone must face the stage, namely the musical instruments, which are the sources of direct sound. This enables microphones to capture very little direct sound but much of the concert hall's early reflections and late reverberation.

The distance between each figure-of-eight microphone of Hamasaki-Square must be more than 2 m. The left figure of Figure 2.7 shows the cross-correlation coefficient of the sound captured by two microphones, which are calculated according to the distances. It shows that 2 m gives the lesser correlation for the lowest frequency in three-dimensional and two-dimensional sound fields. The right figures show the subjective evaluation results on the preference for a spatial impression when subjects listen to the sound captured by Hamasaki-Square. It shows the preferences according to the distance of 1 m, 2 m, and 3 m between the figure-of-eight microphones. The upper figure is the result of listening to the sound of the Hamasaki-Square only, and the lower one is from the sound mixing

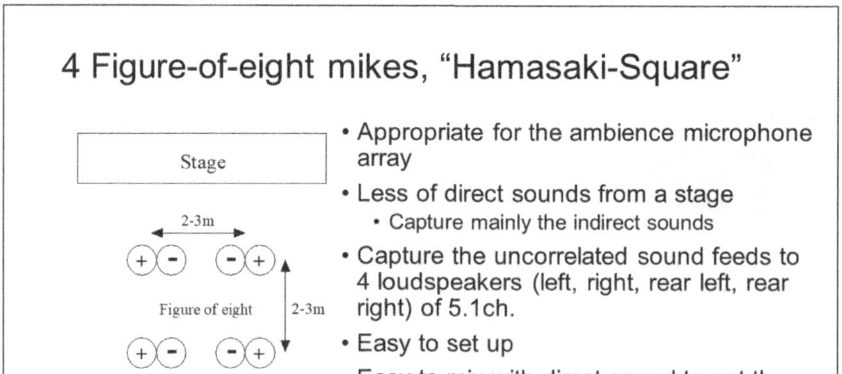


*Figure 2.6* Hamasaki-Square consists of four figure-of-eight microphones [4, 5]

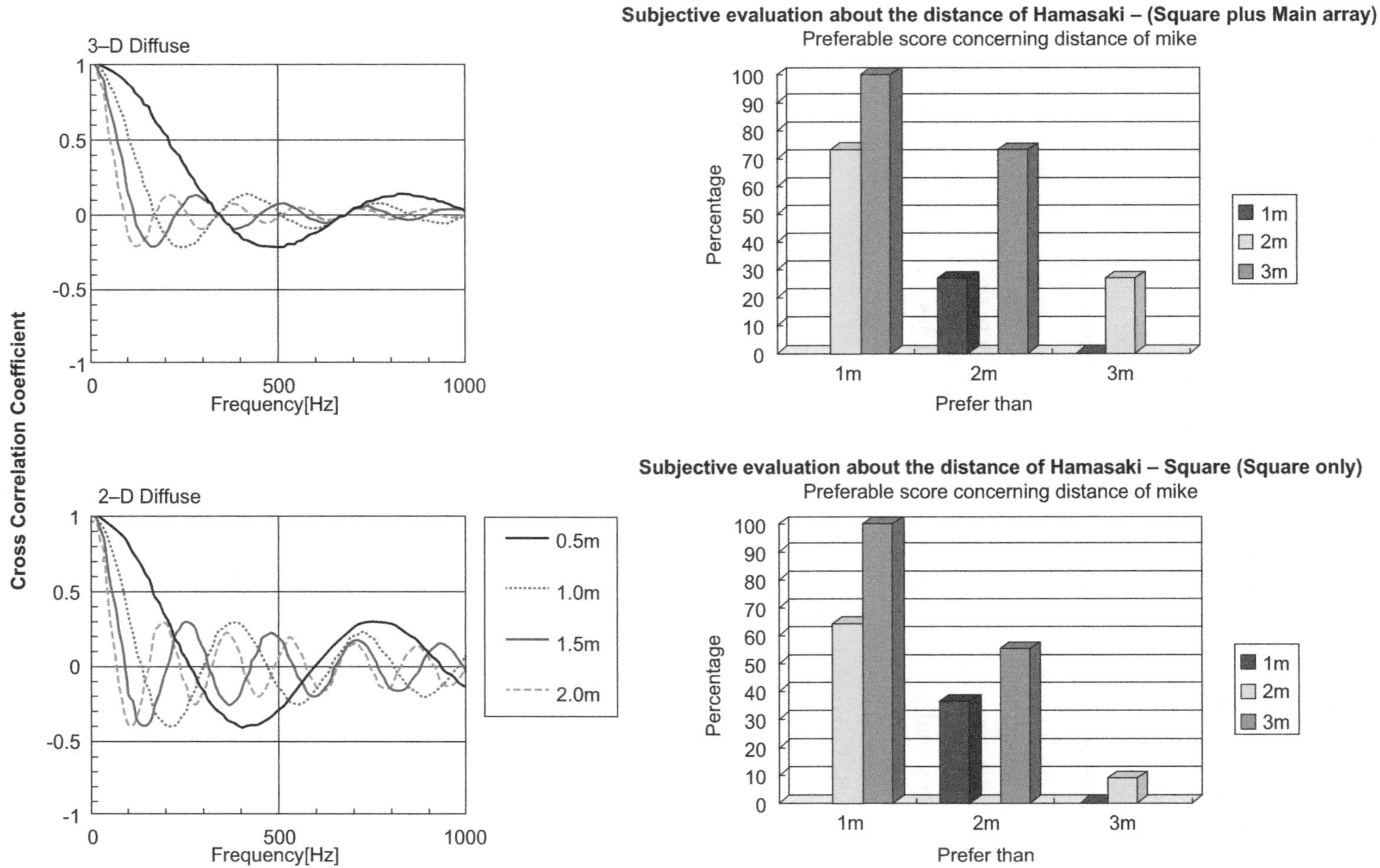


*Figure 2.7* Left: Cross-correlation coefficient of the sound captured by two microphones according to the distance of each figure-of-eight microphone. Right: Results of subjective evaluation on the preferences of spatial impression according to the distance of 1 m, 2 m, and 3 m. The results show that the distance between each figure-of-eight microphone of more than 2 m archives an improved spatial impression [4, 5]

of the Hamasaki-Square and the main microphone array mixed together. These results show that figure-of-eight microphones spaced more than 2 m achieved a better spatial impression.

It is mandatory to feed each individual audio signal captured by the Hamasaki-Square and the main microphone array to only one playback channel or loudspeaker of the multi-channel sound system to obtain a natural spatial impression. This means that it is not suggested to pan the audio signal from each microphone between playback multiple channels or loudspeakers).

### *Microphone technique for the immersive sound recording*

Two types of microphone techniques can be applied to the immersive sound recording of an orchestra in a concert hall, as shown in Figure 2.8. Type A is a microphone array consisting of omnidirectional or directional microphones, where the numbers and positions of the microphones can be changed according to the given channel or loudspeaker count of the immersive sound system. The Type B system combines a main microphone array used to catch mainly direct sound from the stage and an ambience microphone array to catch mainly indirect sound in the hall, which is placed at *x* m distance. The figure shows an example of a 5+5+0 sound system.

In a live recording situation, such as a broadcast or a concert recording, it is challenging to get a proper spatial impression using the Type A microphone technique because the presence or absence of an audience will dramatically change the live acoustic in the concert hall. And there is not enough time to strictly adjust the position of microphones

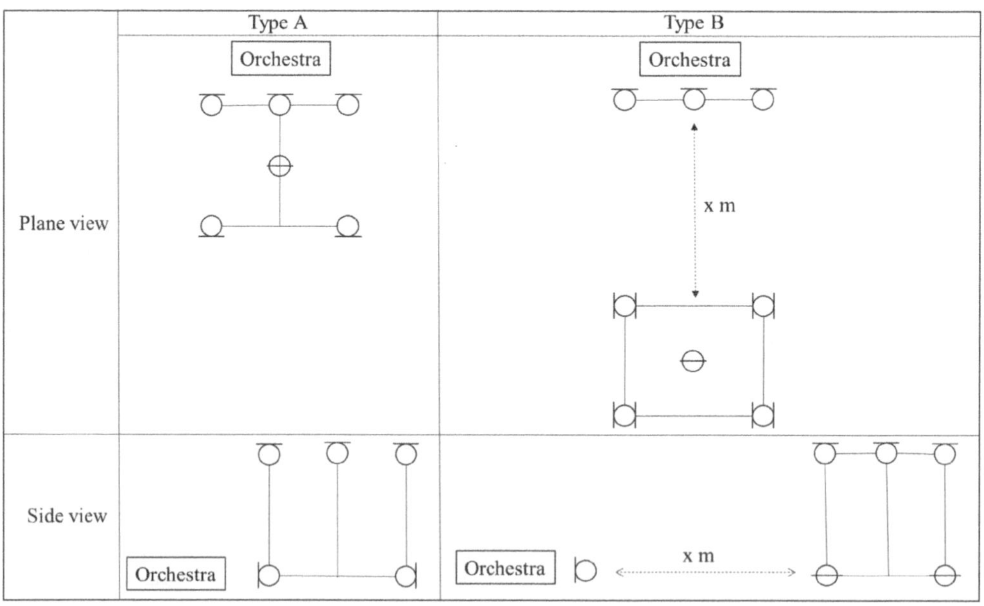


*Figure 2.8* Two types of microphone techniques, Types A and B, can be applied to the immersive sound recording of an orchestra in a concert hall [6]

*Table 2.1* Influence of acoustical components on the features of the immersive sound recording [6]

| | *Direct sound* | *Early reflections* | *Late reverberation* |
|---|---|---|---|
| Localization of each sound image in the frontal sound field | Large | Medium | Small |
| **Spatial impression** | **Small** | **Large** | **Large** |
| Timbre | Large | Medium | Medium |
| Localization of each sound image in the lateral, rear, and upper sound field | Large | Medium | Small |

in such productions. Therefore, the Type B plan can be a more practical choice than the Type A, which can easily control spatial impressions and realize stable sound source localization in the frontal sound field.

Table 2.1 shows the rate of influence of acoustical components, including direct sound, early reflections, and late reverberation. The table shows their effect on localization for each sound image in the frontal sound field, spatial impression, timbre, and localization in the lateral, rear, and upper sound field on the immersive sound recording.

It is necessary to use the appropriate ambience microphone array to get a better spatial impression of immersive sound. It must capture the reflections from behind and above an audience as much as possible and must not capture the direct sound from the frontal stage as much as possible. Audio signals captured by the ambience microphone array must be uncorrelated and reproduce the fluctuation in the low-frequency range of the concert hall. It is essential to understand that leakage of direct sound from the frontal sound field to the rear and upper channels (loudspeakers) sometimes breaks the natural frontal sound images created by frontal loudspeakers [6].

### *Hamasaki-Cube*

Under careful consideration of these conditions, I use the microphone technique for the immersive sound recording, as shown in Figure 2.9. It shows the example for 4+5+0 and 5+5+0 sound systems, which consists of the main microphone array for Left, Center, Right (L, C, R), Height left, Height center, and Height right (HL, HC, HR), and the ambience microphone array.

The Hamasaki-Cube was developed by extending the philosophy and concepts of Hamasaki-Square to adapt immersive sound recording. Hamasaki-Cube consists of figure-of-eight microphones for the middle layer and super-cardioid microphones for the upper layer.

As previously described regarding the 5.1-channel surround sound recording with Hamasaki-Square, it is mandatory to feed each audio signal captured by each microphone of the Hamasaki-Cube and the main microphone array to only one channel (loudspeaker) of the multi-channel sound system to obtain the natural spatial impression. It means it is never suggested to pan the audio signal from each microphone among multiple channels or loudspeakers.

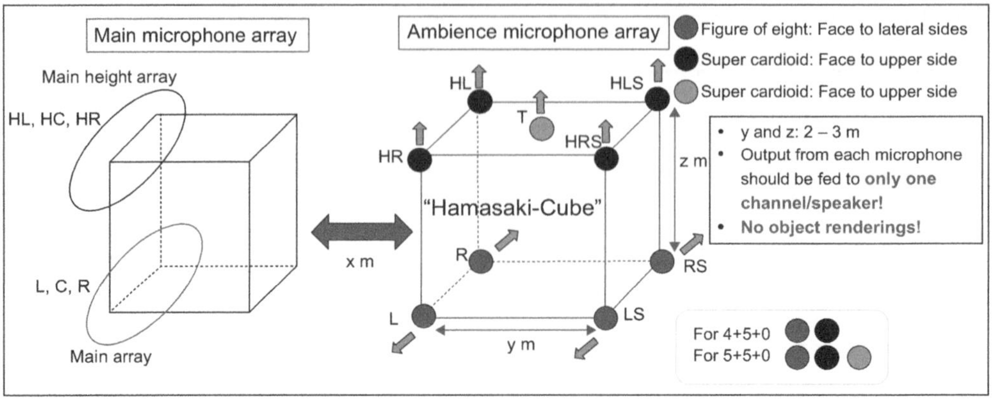


*Figure 2.9* Microphone technique for the immersive sound recording using Hamasaki-Cube [5]

Using a channel-based sound system to achieve the most natural immersive sound is essential. Object-based rendering usually reproduces the audio signal of a channel or the audio signal of a microphone of the ambience array in multiple loudspeakers that impairs the decorrelation between channels. This impacts the natural spatial impression of the original mix. The object-based format breaks the important rule of 'one microphone to one channel (loudspeaker)'. Therefore, using only the channel-based format for mixing, distributing, and reproducing the immersive sound is strongly recommended to achieve the best natural music recording and reproduction.

### *Recording examples using Hamasaki-Cube*

Figures 2.10 and 2.11 show the microphone arrangement using Hamasaki-Cube for my recordings of *Pini di Roma* and *Belkis, Queen of Sheeba suite* composed by Ottorino Respighi in immersive sound.

## 22.2 multi-channel sound system

In 2000, I was transferred from the NHK Broadcast Center to the NHK Science & Technical Research Laboratories, where I was expected to play two roles: to rebuild the declining acoustic research activities and to develop a sound system for the Ultrahigh-Definition Television (UHDTV). Therefore, I started the research for the sound system for the UHDTV in 2000, and this was a good chance for me to develop a new and ideal sound system from scratch for recording and reproducing the very natural sound field for music.

The UHDTV system with 4,000 scanning lines was developed to achieve this 100° horizontal viewing angle. The system has 16 times the resolution of HDTV and provides a viewing distance of 0.75 H (H: picture height). Table 2.2 compares the specifications of the UHDTV with 4K × 8K pixels with those of the HDTV system [7].

The UHDTV was developed to achieve the highest possible sense of reality, so the sound system used with UHDTV should also contribute to this impression of presence. The presence is the feeling of 'being there'. The sound field reproduced by conventional

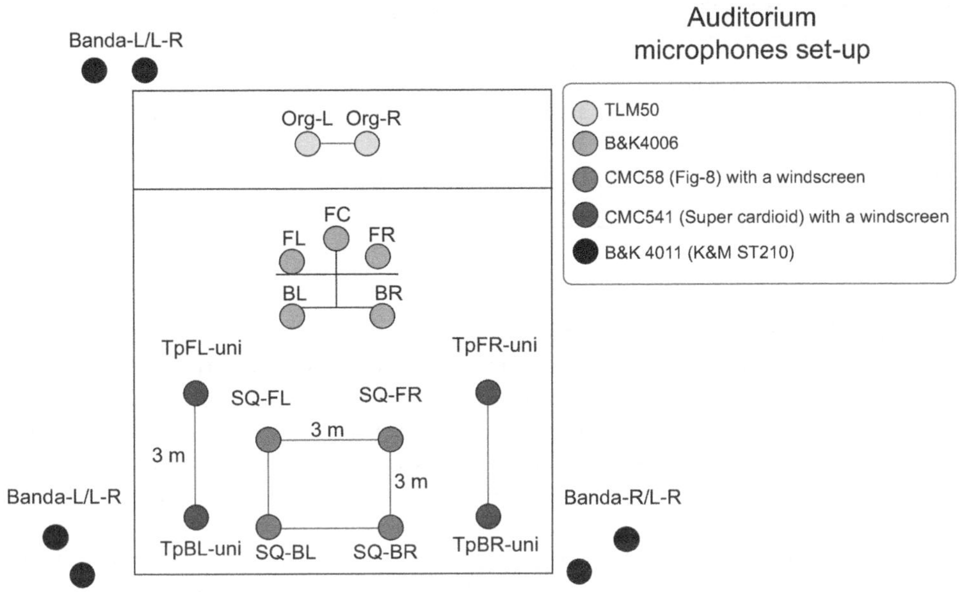


*Figure 2.10* *Pini di Roma*: Microphone arrangement for main and ambience microphone arrays using Hamasaki-Cube for my latest recording example of an orchestra in immersive sound [5]

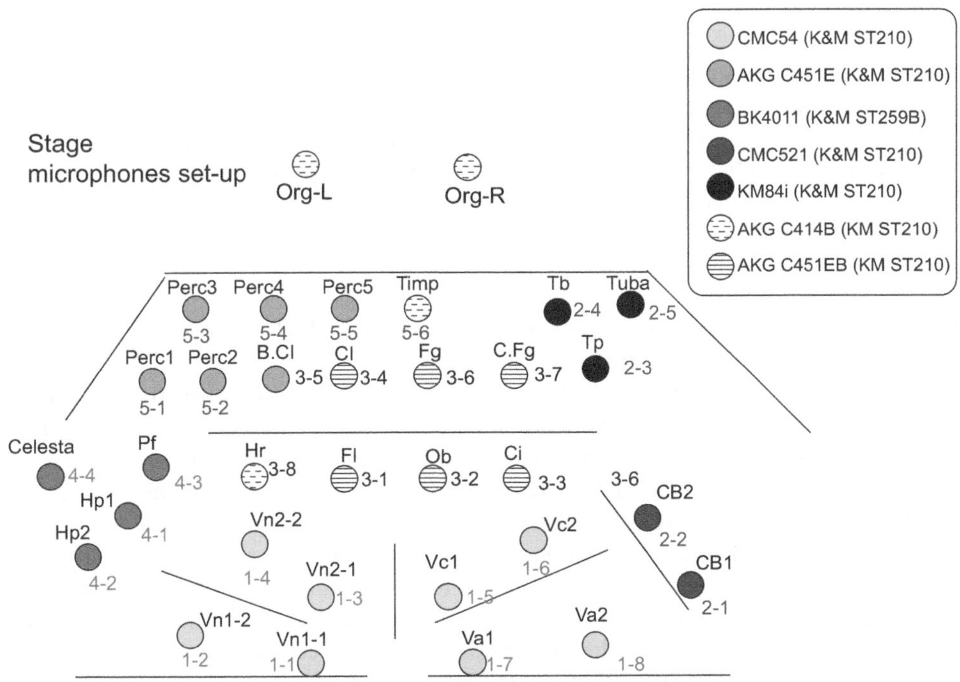


*Figure 2.11* *Belkis, Queen of Sheeba suite*: Microphone arrangement on the stage for my latest recording example of an orchestra in immersive sound [5]

*Table 2.2* HDTV and UHDTV system specifications [8, 9]

| | *HDTV* | *UHDTV* |
|---|---|---|
| Number of pixels | 1080 × 1920 | 4320 × 7680 |
| Viewing angle (pixel invisible) | 30° horizontally | More than 100° horizontally |
| Relative resolution | Equivalent to 35-mm motion film | More than twice 70-mm motion film |

multi-channel sound systems for movies, such as the 5.1 surround sound system, is strongly related to the visual image on the screen. Therefore, multi-channel sound systems for conventional movie theaters prioritize frontal sound reproduction over rear sound reproduction. However, the UHDTV system provides an immersive sensation due to the wide viewing angle associated with an extra-large screen. Therefore, the multi-channel sound system must provide a sound field that surrounds viewers with various sound sources.

The horizontal viewing angle of a UHDTV is approximately 100° in the front row and 40° in the back in a typical UHDTV theater. Therefore, the horizontal and vertical angles of movements in the visual image are extremely wide in UHDTV compared to those in conventional movie theaters. Consequently, the ability to localize sound objects over the screen images should be improved so that the horizontal and vertical directions of the visual image match the sound image. Based on these requirements, I made the conditions for the sound system of a UHDTV as follows [8, 9]:

- Stabilize frontal sounds over the entire screen area.
- Reproduce sound images in all directions around a viewer, including elevation.
- Reproduce a 3D spatial impression that augments a sense of reality.
- Maintain a wide listening area with exceptional sound quality.
- Be compatible with existing multi-channel sound systems.
- Be possible to make live recording and mixing to reproduce natural sound fields of music.

I also determined the concept of the sound system for UHDT, as shown in Figure 2.12, when I started the research in 2000.

The first sound system for UHDTV was introduced and demonstrated at the NHK STRL Open House in 2003, as shown in Figure 2.13. The three-dimensional sound system was constructed by adding an upper loudspeaker layer system. It is necessary to place at least eight loudspeakers equally spaced on concentric circles with the listener at the center for reproducing sound image localization and a sense of sound expansion in a two-dimensional horizontal plane 360°. Ideally, eight loudspeakers should be equally spaced at the height of the viewer's ears and above. However, due to the fact that the screen is not acoustically transparent and loudspeakers cannot be placed behind it, six loudspeakers were placed at the height of the audience's ears, eight concentrically above the audience, and one in the center of the upper layer. In addition, I placed one loudspeaker on the left, center, and right of the upper part of the screen and one to the center of the lower part of the screen to reproduce the up-and-down feeling of the sound image following

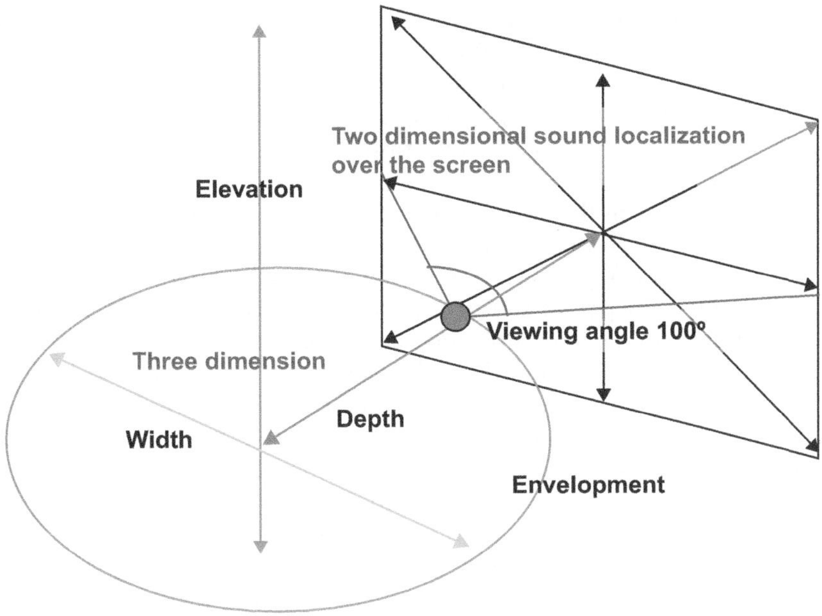


*Figure 2.12* Concept of the sound system for UHTDV

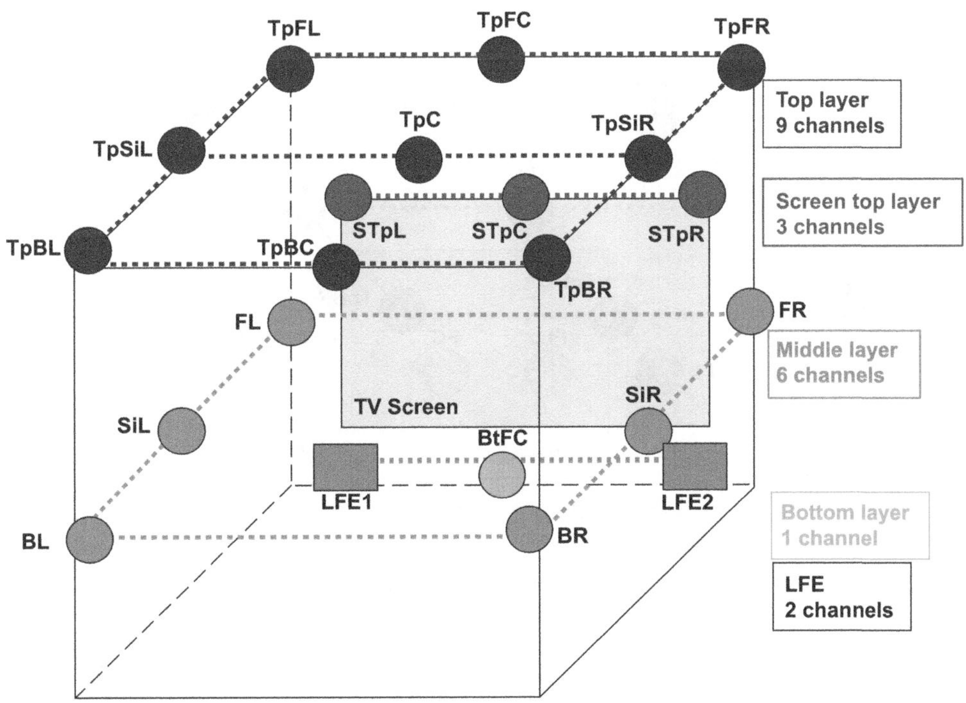


*Figure 2.13* 19.2 multi-channel sound system for UHDTV in 2003

the visual image on the large screen to the front of the listener. In addition to the 19 full-frequency-range loudspeakers, two subwoofers (150 Hz or lower) were placed on the left and right sides of the lower part of the screen to reproduce powerful bass sound.

It was confirmed that the three-dimensional sound system constructed for this open house in 2003 was superior to the conventional two-dimensional system in reproducing a more natural sound field and the localization of a wider sound image. However, the lack of loudspeakers in the center of the screen caused instability in the front sound image localization. Therefore, I thought the 20.2 channel system (nine for the upper layer, eight for the middle layer, three for the lower layer, and two dedicated subwoofers) with an acoustically transparent screen and loudspeakers also placed at the rear of the screen is appropriate for such an ultra-high definition large screen. After several recording trials and subjective evaluation, I finally decided to add two channels, one between the left and center and one between the center and right, on the middle layer as shown in Figures 2.14 and 2.15 in 2004. It was the birth of the 22.2 multi-channel sound system. The 22.2 multi-channel sound system was demonstrated the first time in the world at World Expo 2005, held in Aichi, Japan. I used 73 loudspeakers, as shown in Table 2.3 and Figure 2.16, for the 22.2 multi-channel sound system in the UHDTV theater with a 600-inch screen at World EXPO 2005.

The 22.2 multi-channel sound system was standardized at SMPTE in 2008 as SMPTE 2036-2-2008, Ultra High Definition Television1 – Audio Characteristics and Audio Channel Mapping for Program Production [11]. The channel maps and labels of the 22.2

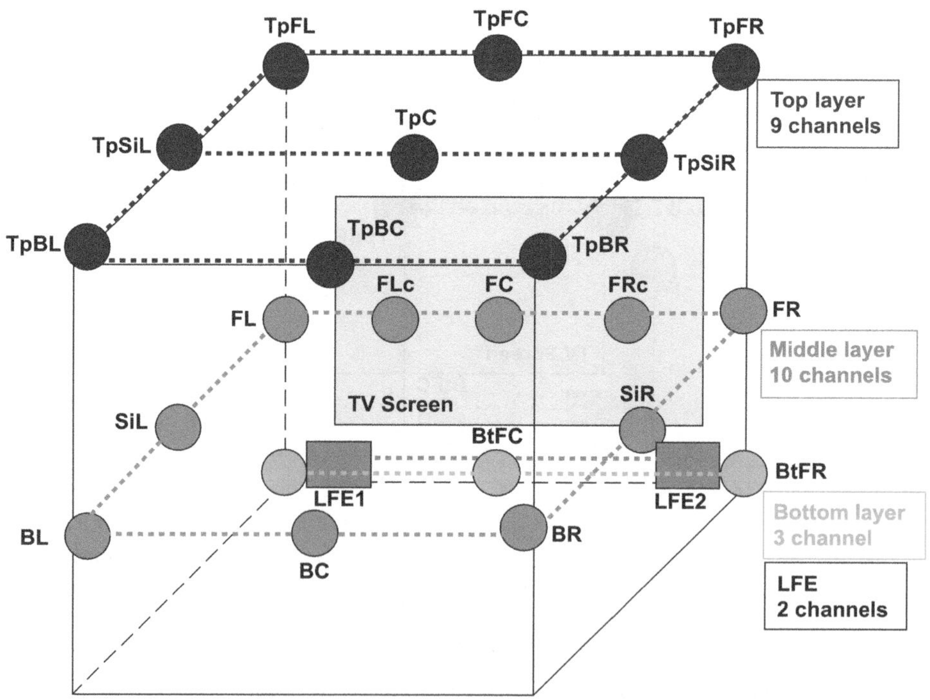


*Figure 2.14* 22.2 multi-channel sound system for UHDTV in 2005 [10, 11]

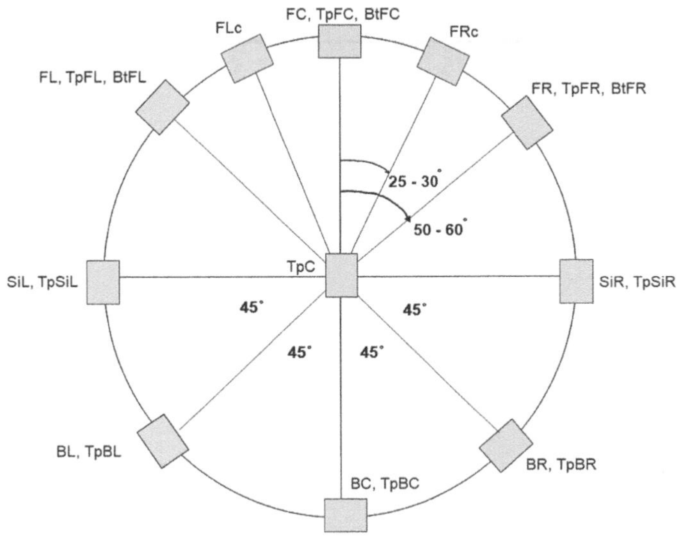


*Figure 2.15* Horizontal placement of each channel in the 22.2 multi-channel sound system in 2005 [12]

*Table 2.3* Number of loudspeakers used for the UHDTV theater in World Expo 2005

| *Channel* | *Number of loudspeakers* |
|---|---|
| FL | 3 |
| FR | 3 |
| FC | 4 |
| LFE1 | 6 |
| BL | 1 |
| BR | 1 |
| FLc | 4 |
| FRc | 4 |
| BC | 1 |
| LFE2 | 6 |
| SiL | 4 |
| SiR | 4 |
| TpFL | 2 |
| TpFR | 2 |
| TpFC | 2 |
| TpC | 2 |
| TpBL | 2 |
| TpBR | 2 |
| TpSiL | 2 |
| TpSiR | 2 |
| TpBC | 2 |
| BtFC | 2 |
| BtFL | 2 |
| BtFR | 2 |
| Extra subwoofers | 8 |

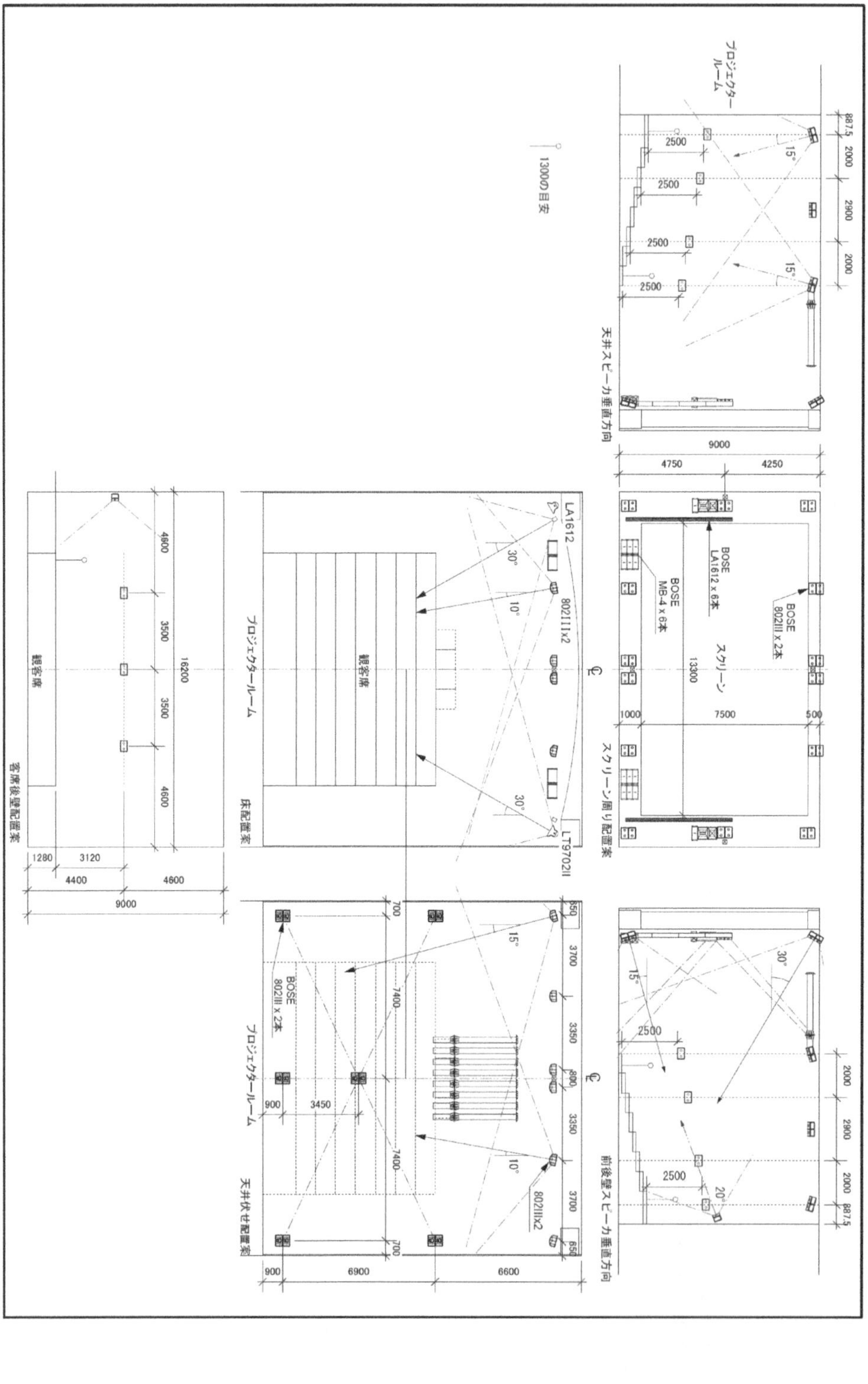


*Figure 2.16* Arrangement of 73 loudspeakers for the 22.2 multi-channel sound system with a 600-inch screen at UHDTV theater in World Expo 2005

*Table 2.4* Channel maps and labels of 22.2 multi-channel audio [10, 11]

| *AES pair no./ch. no.* | *Channel no.* | *Label* | *Name* |
|---|---|---|---|
| 1/1 | 1 | FL | Front left |
| 1/2 | 2 | FR | Front right |
| 2/1 | 3 | FC | Front center |
| 2/2 | 4 | LFE1 | LFE-1 |
| 3/1 | 5 | BL | Back left |
| 3/2 | 6 | BR | Back right |
| 4/1 | 7 | FLc | Front left center |
| 4/2 | 8 | FRc | Front right center |
| 5/1 | 9 | BC | Back center |
| 5/2 | 10 | LFE2 | LFE-2 |
| 6/1 | 11 | SiL | Side left |
| 6/2 | 12 | SiR | Side right |
| 7/1 | 13 | TpFL | Top front left |
| 7/2 | 14 | TpFR | Top front right |
| 8/1 | 15 | TpFC | Top front center |
| 8/2 | 16 | TpC | Top center |
| 9/1 | 17 | TpBL | Top back left |
| 9/2 | 18 | TpBR | Top back right |
| 10/1 | 19 | TpSiL | Top side left |
| 10/2 | 20 | TpSiR | Top side right |
| 11/1 | 21 | TpBC | Top back center |
| 11/2 | 22 | BtFC | Bottom front center |
| 12/1 | 23 | BtFL | Bottom front left |
| 12/2 | 24 | BtFR | Bottom front right |

multi-channel audio were specified in this standard, as shown in Table 2.4. And, finally, UHDTV broadcasting with 22.2 multi-channel sound was a world-first in Japan in 2018.

### *Subjective evaluations of 22.2 multi-channel sound system*

Subjective evaluations assessed the performance of three different multi-channel audio systems: two-channel stereo, 5.1 surround sound, and 22.2 multi-channel sound. The stimuli for the subjective evaluations were selected from the World Expo 2005 programs. The sounds and video were screened in a UHDTV theater (a 450-inch screen). The subjects were asked to report their impressions of the sound provided that the different sound systems reproduced while shown the UHD images on the screen. They were asked to sit in different positions so that differences in impressions based on position with regard to the screen could be discussed. A description and full details of the subjective evaluation methodology are described and first published in the SMPTE *Motion Imaging Journal*, 'A 22.2 Multi-channel Sound System for Ultrahigh-Definition TV (UHDTV)', 117(3) 40–49, April 2008 (Copyright © SMPTE).

*Summary of the evaluation*

Two programs, each about 8 minutes long and divided into four scenes, were produced for World Expo 2005. Table 2.5 lists the sound content of each scene (A and B). The two programs were originally produced using the 22.2 multi-channel sound system, so the 5.1 surround sound and two-channel stereo sound were produced by down-mixing from the original 22.2 multi-channel sound, based on the down-mixing equations shown in Figure 2.17. The coefficients ($k$, $m$, $n$) in the down-mixing equations adjust the total sound level and balance the sound levels in the upper, middle, and lower layers. The $k$, $m$, and $n$ during the subjective evaluations were set to 1.0 to maintain the same total loudness in the different multi-channel sound systems.

*Table 2.5* Sound content in each program of UHDTV for World Expo 2005 [10, 12–14]

| | *Program A* | *Program B* |
|---|---|---|
| Scene 1 | Short sound clip of three-dimensional movement without video followed by music with female singers with video | Three-dimensional movement of sound based on CG followed by symphony orchestra music |
| Scene 2 | Music and surrounding sounds of shooting location | Symphony orchestra music followed by three-dimensional sound of water gushing from a huge dam |
| Scene 3 | Typical three-dimensional sound scene for Japanese festival followed by music | Three-dimensional sound creation of sound effects with symphony orchestra music |
| Scene 4 | Music with children singing | Rocket launch |

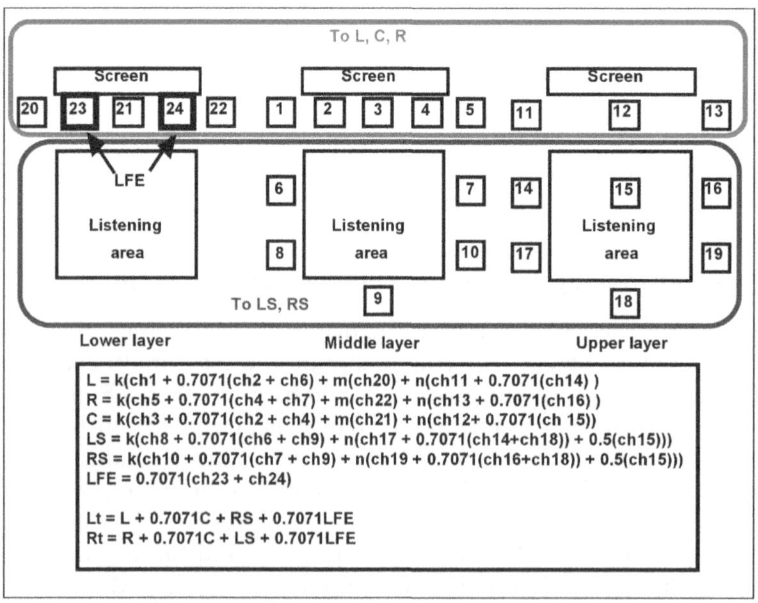


*Figure 2.17* Down-mixing sound from 22.2 ch to 5.1 and 2.0 ch [10, 12–14]

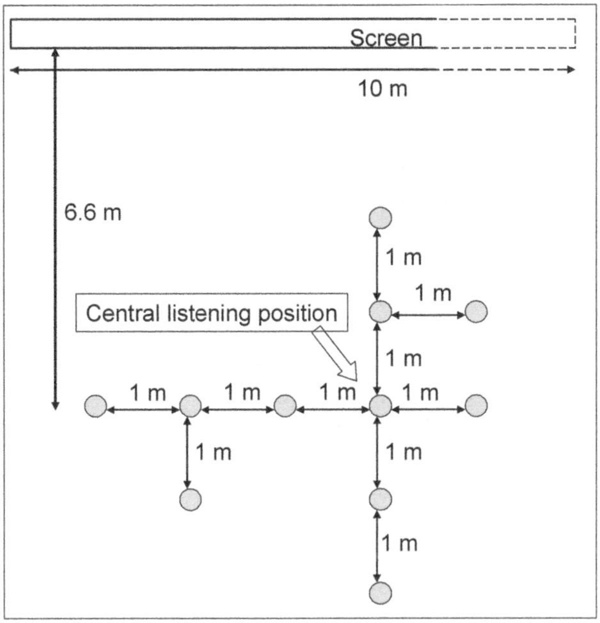


*Figure 2.18* Listening positions in the experiment [10]

The experiment was done in the UHDTV theater.

Between three to ten subjects listened to the stimuli simultaneously while sitting in positions spaced at 1-m intervals around the central listening position during screenings in the UHDTV theater. The layout of these positions is shown in Figure 2.18.

### *Evaluation method*

The Semantic Difference (SD) evaluation method was used in this experiment. Subjects were asked to rate their impressions on a 7-point scale for the pairs of evaluation terms listed in Table 2.6. Each pair contained two terms with opposite meanings, such as 'dynamic' and 'static'. Subjects were asked to select a score from 1 to 7, in which 1 means very dynamic, 2 means fairly dynamic, 3 means slightly dynamic, 4 means neither dynamic nor static, 5 means slightly static, 6 means fairly static, and 7 means very static. Subjects rated each stimulus (segment of content) using the 24 evaluation pairs listed in Table 2.6. Fifty-three subjects participated in the experiment.

### *Results of experiment*

Figure 2.19(a) shows the total mean values of all results at all listening positions for each pair of evaluation terms for each sound system to compare the basic features of the three different multi-channel sound systems. Each mean value calculated from all results obtained in the experiment is marked with a 95% confidence interval for different terms and for different sound systems. The horizontal axis represents the scale listed in Table 2.6, and the vertical axis contains each pair of evaluation terms.

*Table 2.6* Pairs of evaluation terms [10]

| *Term* | *Scale* | *Term* | *Short description in figures* |
|---|---|---|---|
| Dynamic | 1-2-3-4-5-6-7 | Static | Dynamic |
| Gaudy | 1-2-3-4-5-6-7 | Plain | Gaudy |
| Powerful | 1-2-3-4-5-6-7 | Weak | Powerful |
| Interesting | 1-2-3-4-5-6-7 | Not interesting | Interesting |
| Delightful | 1-2-3-4-5-6-7 | Dull | Delightful |
| Impressive | 1-2-3-4-5-6-7 | Not impressive | Impressive |
| Distinct | 1-2-3-4-5-6-7 | Indistinct | Distinct |
| Beautiful | 1-2-3-4-5-6-7 | Dingy | Beautiful |
| Transparent | 1-2-3-4-5-6-7 | Not transparent | Transparent |
| Clear | 1-2-3-4-5-6-7 | Muddy | Clear |
| Bright | 1-2-3-4-5-6-7 | Dark | Bright |
| Loud | 1-2-3-4-5-6-7 | Soft | Loud |
| Unique | 1-2-3-4-5-6-7 | Ordinary | Unique |
| Real | 1-2-3-4-5-6-7 | Unreal | Real |
| Natural | 1-2-3-4-5-6-7 | Unnatural | Natural |
| Presence | 1-2-3-4-5-6-7 | No presence | Presence |
| Deep | 1-2-3-4-5-6-7 | Flat | Depth |
| Wide in left and right | 1-2-3-4-5-6-7 | Narrow in left and right | Width |
| Wide in front and rear | 1-2-3-4-5-6-7 | Narrow in front and rear | Front/rear |
| Wide in above and below | 1-2-3-4-5-6-7 | Narrow in above and below | Up/down |
| Clear movement of sound | 1-2-3-4-5-6-7 | Obscure movement of sound | Movement |
| Sound come from every direction | 1-2-3-4-5-6-7 | Sound come from lopsided direction | Direction |
| Rich reverberant | 1-2-3-4-5-6-7 | Poor reverberant | Reverberant |
| Rich envelopment | 1-2-3-4-5-6-7 | Poor envelopment | Envelopment |

Figure 2.19(a) shows that the 22.2 multi-channel sound system was rated significantly better (smaller value) than the two-channel stereo system for every evaluation term except 'loud' and was rated significantly better than the 5.1 surround sound system for the terms 'gaudy', 'distinct', 'front/rear', 'up/down', 'movement', 'direction', 'reverberant', and 'envelopment' (short descriptions of terms are listed in Table 2.6). The results also show that the 5.1 surround sound system was rated significantly better than the two-channel stereo system for every term except 'loud'. The results suggest that there was no difference in the loudness between each system. They also suggest that the 22.2 multi-channel sound system provided a better three-dimensional spatial sound quality than either the two-channel stereo system or the 5.1 surround sound system.

Figure 2.19(b) shows the total mean values of all results at listening positions within 1 m distance from the central listening position for each pair of evaluation terms. Figure 2.19(a) shows that the 22.2 multi-channel sound system was rated significantly better (smaller value) than the two-channel stereo system for every evaluation term except 'bright', 'natural', and 'loud', and was rated significantly better than 5.1 surround sound system for the terms 'up/down', 'movement', and 'direction'. The results also show that

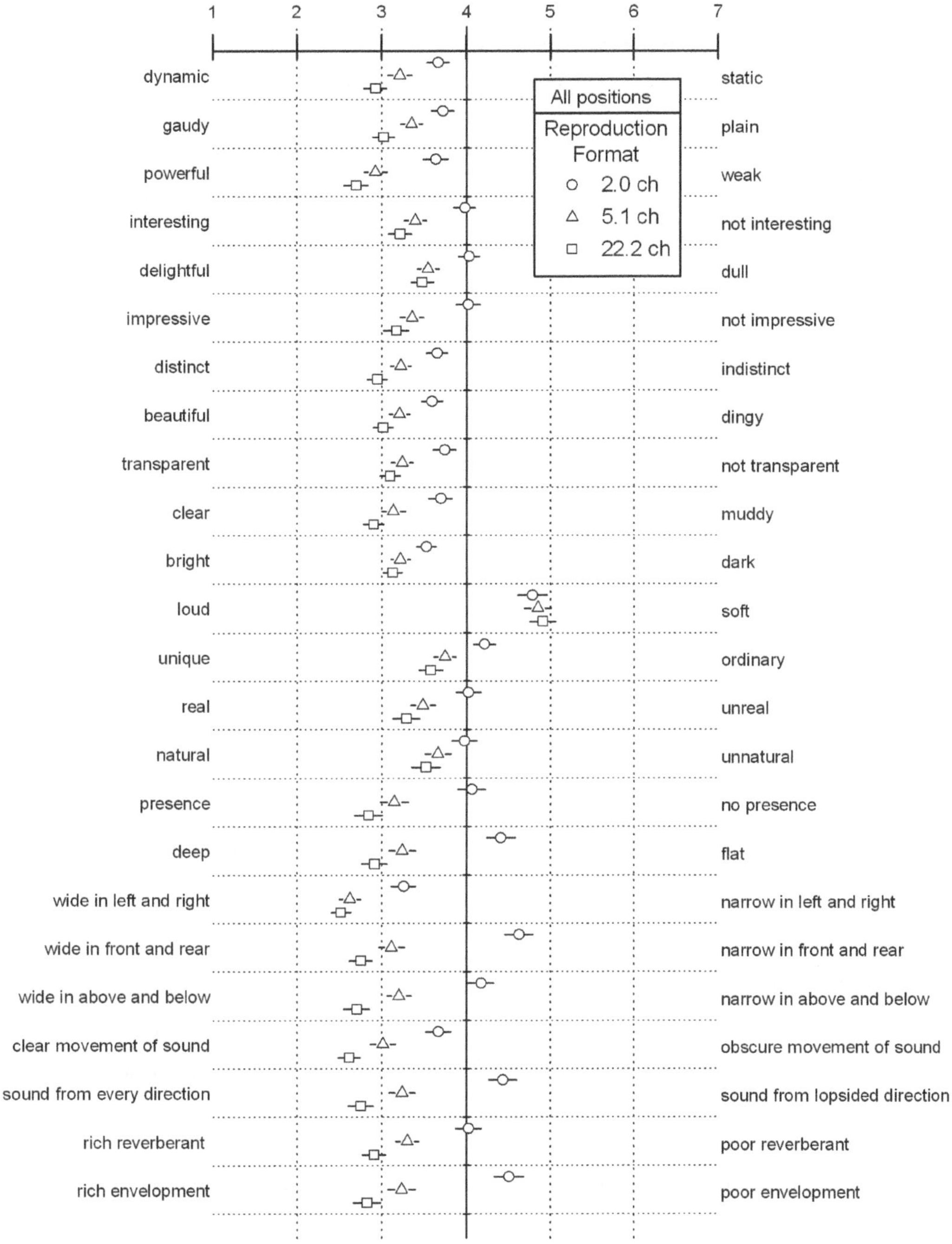


*Figure 2.19(a)* Results of 2.0, 5.1, and 22.2 channel sound systems using a large screen at all listening positions [10]

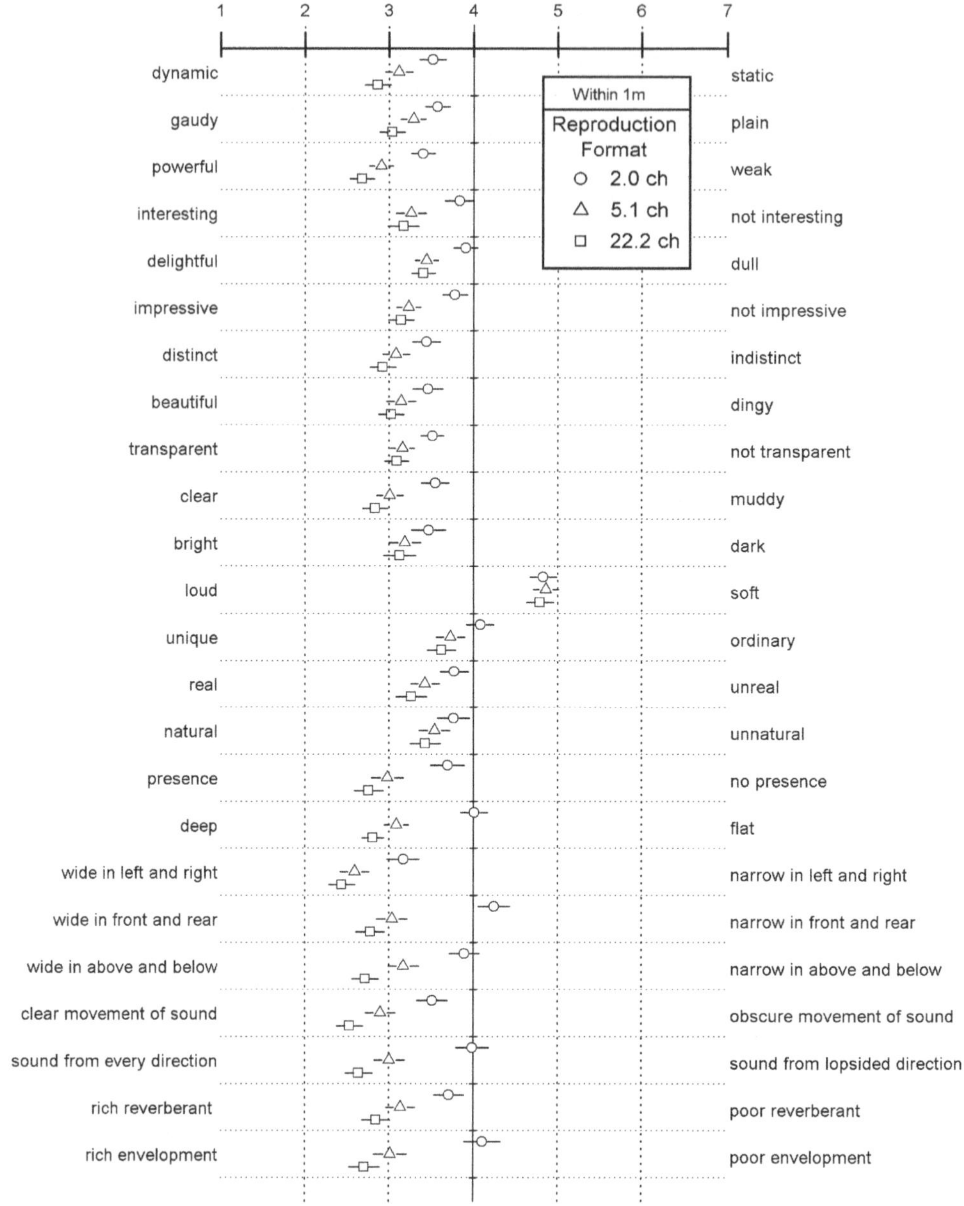


*Figure 2.19(b)* Evaluation of 2.0, 5.1, and 22.2 channel sound system using a large screen at listening positions within 1 m distance from the central position [10]

the 5.1 surround sound system was rated significantly better than the two-channel stereo system for every term except 'gaudy', 'bright', 'loud', 'unique', 'real', and 'natural'. They suggest that the 22.2 multi-channel sound system provided a better perception of the three-dimensional direction of sound images than the 5.1 surround sound system.

However, a significant difference could not be observed in the sensation of reverberant and envelopment between 22.2 and 5.1.

Figure 2.19(c) shows the total mean values of all results at listening positions outside of 1 m distance from the central listening position for each pair of evaluation terms for each sound system. It also shows that the 22.2 multi-channel sound system was rated

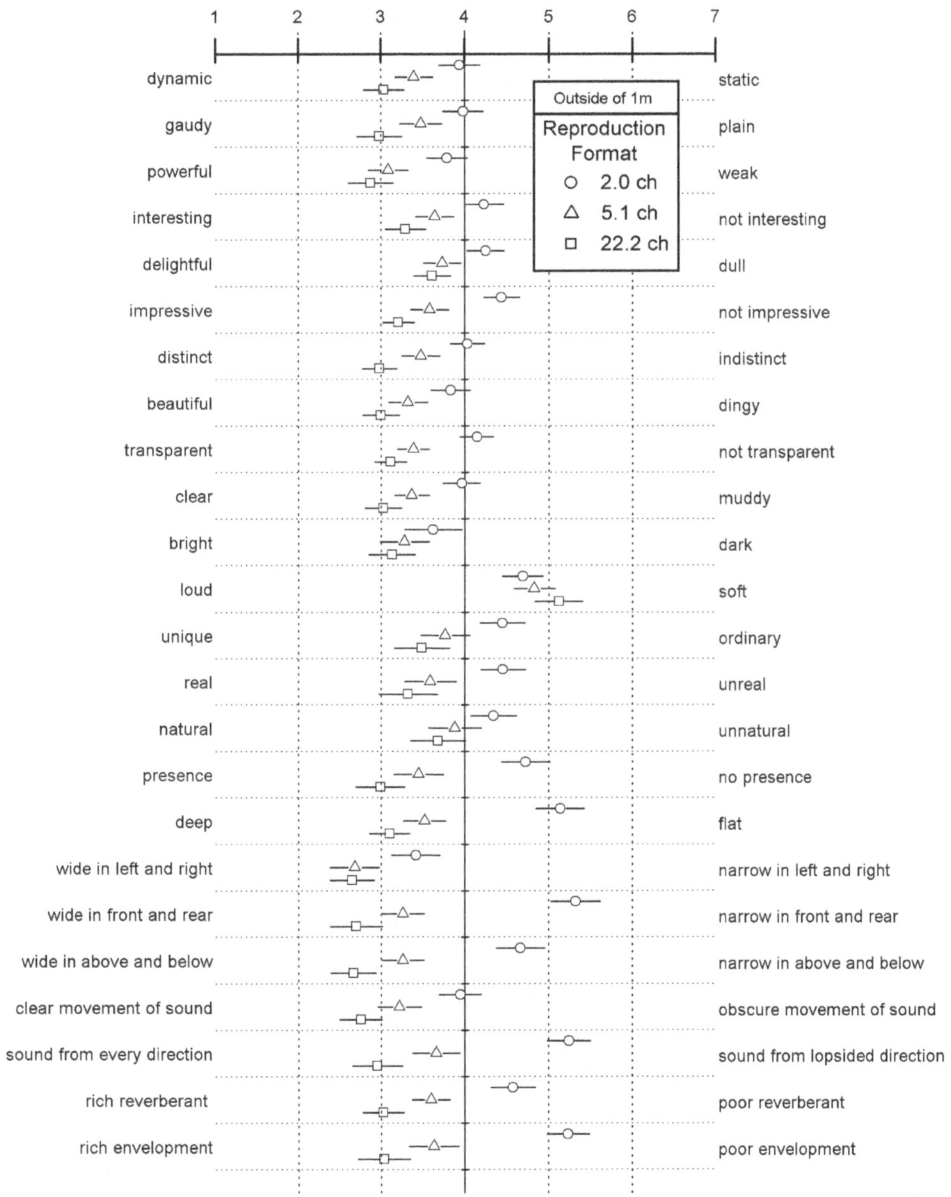


*Figure 2.19(c)* Evaluation of 2.0, 5.1, and 22.2 channel sound systems using a large screen at the listening position outside of 1 m distance from the central position [10]

significantly better (smaller value) than the two-channel stereo system for every evaluation term except 'bright' and 'loud' and was rated significantly better than 5.1 surround sound system for the terms 'distinct', 'front/rear', 'up/down', 'movement', 'direction', and 'reverberant'. The results also show that the 5.1 surround sound system was rated significantly better than the two-channel stereo system for every term except 'gaudy', 'bright', 'loud', and 'natural'. They suggest that naturalness, reality, presence, and three-dimensional spatial sound quality got extremely worse on the two-channel stereo and that three-dimensional spatial quality got worse on the 5.1 surround sound system. The 22.2 multi-channel sound system still maintained good three-dimensional spatial sound quality, and the results for most evaluation terms looked similar regardless of listening positions.

### *Discussion*

Subjective evaluations using the semantic differential method showed that the 22.2 multi-channel and the 5.1 surround sound were better at a 5% significance level for various sensations than the two-channel stereo in the theatrical environment with a 450-in UHDTV screen. Also, the 22.2 multi-channel sound was better at a 5% significance level for 3D spatial sound quality than the 5.1 surround sound. The differences between the 5.1 surround sound and the two-channel stereo are more significant than the difference between 22.2 and 5.1.

The evaluated results in Figures 2.19(a), 2.19(b), and 2.19(c) demonstrate that viewers have better impressions of various spatial attributes in the broader listening area by the 22.2 multi-channel sound system than the two-channel sound system and the 5.1 surround sound system. It also shows that 5.1 provides better impressions of various spatial attributes in the broader listening area than the two-channel stereo. This means that the 22.2 multi-channel sound system has a wider 'sweet spot' than the two-channel stereo sound system and the 5.1 surround sound system and that the 5.1 has a wider 'sweet spot' than the two-channel stereo.

The other attributes, such as reality and transparency, had no differences in ratings between the 22.2 multi-channel sound and the 5.1 surround sound in the theatrical environment. This might be due to the influence of the picture on the audio impression because the UHDTV theater had a much larger screen than that in the conventional home environment.

#### *Sound design and mixing approach in the three-dimensional sound reproduction system*

Magnitude estimation (ME) was used to investigate how the recording engineer and sound designer used each channel of the 22.2 multi-channel sound system in their mixing. Six engineers and one sound designer with much experience in mixing and designing sound in 22.2 multi-channel systems were asked to rate each channel's necessity for mixing the 22.2 multi-channel sound. They were asked to rate how much each channel contributed to overall mixing by rating each channel, compared to channel #3 (Front center), which was given a reference value of 100 points. The engineers and designers were free to score as they liked. If they felt a channel was twice as important as channel #3 (Front center), they could give it 200 points.

Figure 2.20 shows the results of assessing a total mix of various programs. Each channel's score is a geometrical means of the scores given by all subjects. Total mix means the final mixing of a program in which sound engineers and designers decide the final balance of the three stems (sound effects, music, and speech). Every channel except three channels (Top side left, Top side right, and Top back center) scored higher than 70 points. These results show that sound engineers and designers used every channel effectively.

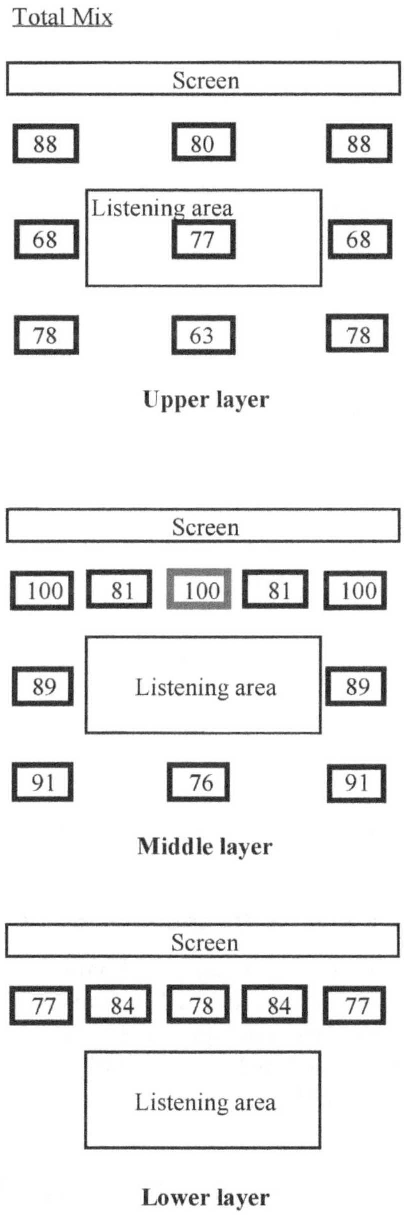


*Figure 2.20* Results of assessment of a total mix of various programs [10]

*Three-dimensional sound mixing*

The principal applications of each layer for mixing 22.2 multi-channel sound in the 2000s are enumerated here [10, 15].

### Upper layer

- Reverberation and ambiance
- Sound localized above, such as loudspeakers hung in gymnasiums, airplanes, and fireworks
- Special sounds, such as meaningless sounds

### Middle layer

- Forming the basic sound field
- Reproducing listener envelopment

### Lower layer

- Sounds of water, such as the sea, rivers, and drops of water
- Sound on the ground in scenes with bird's-eye views

In the early production stage using 22.2 multi-channel sound, sound engineers pointed out several issues in three-dimensional sound mixing:

- Effective use of the upper and lower layers
- Three-dimensional movement of sound sources
- Creating an elevation of sound
- Interaction between immersive audio and visual cues

The basic features and advantages of 3D sound with the 22.2 multi-channel system are well understood and have been shown by several evaluations and various productions. However, further research and production are still necessary to develop 3D sound production know-how. Three-dimensional sound production tools must also be developed.

## *Recording the Gruppen in 22.2 multi-channel sound*

I had a recording project for the Karlheinz Stockhausen's Instrumentation Orchesterwerke *Gruppen* at Suntory Hall in Tokyo in 2008. This work needs the three orchestras placed around the audience, as shown in Figure 2.21 (Orchestra A, B, and C). Therefore, only the recording in 22.2 multi-channel sound could reproduce the listening experience as the same as the audience sitting there. This recording project was a unique opportunity to test the value of the 22.2 multi-channel sound system I began researching in 2000. I decided to conduct the recording in 22.2 multi-channel sound.

The basic concept for this recording was using my recording technique consisting of the main microphone arrays and ambience array (Hamasaki-Cube). I put the four main microphone arrays and one Hamasaki-Cube, as shown in Figure 2.21. The results of this

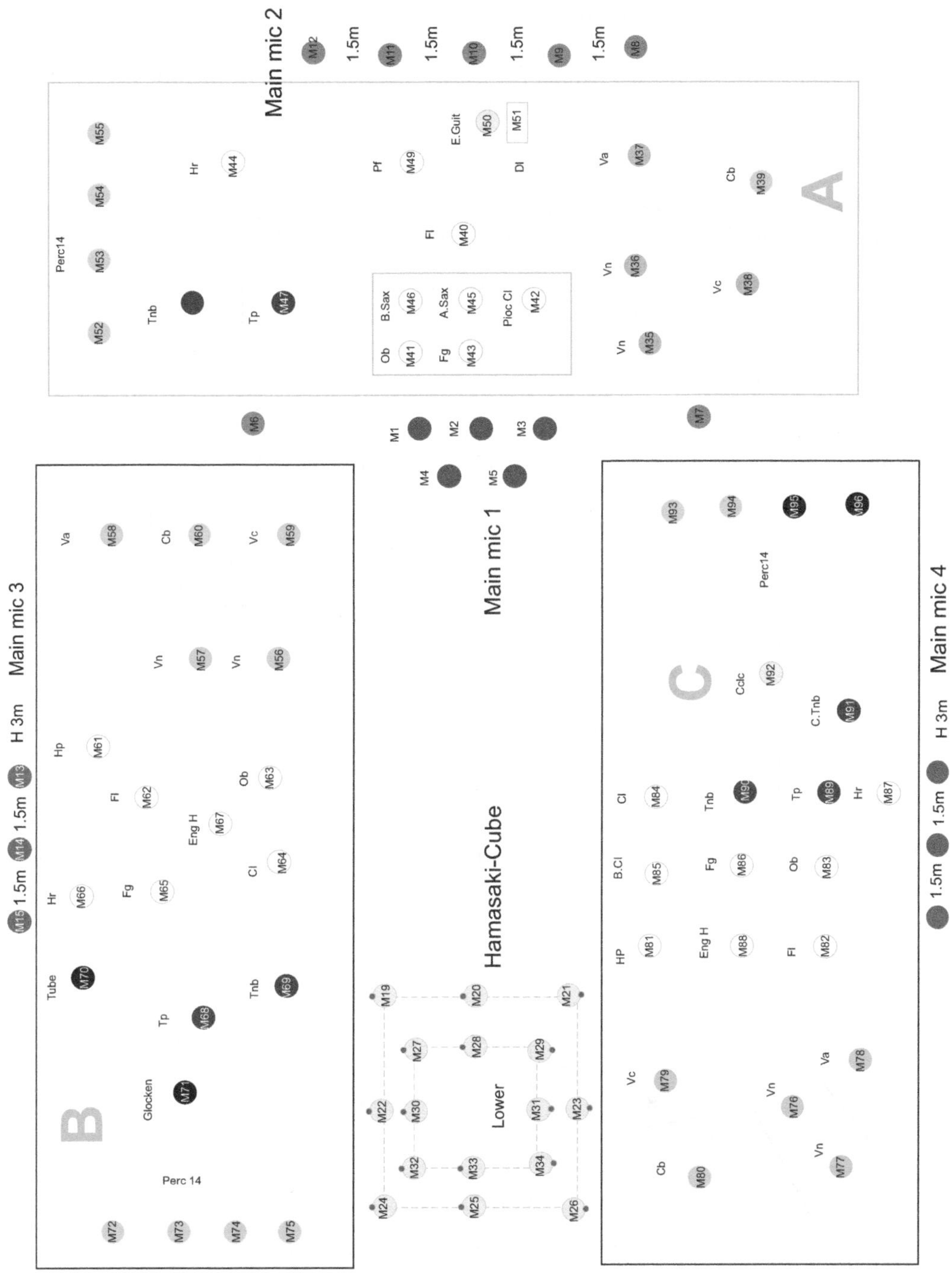


*Figure 2.21* Microphone arrangement for recording the Gruppen

recording were more than I expected, and in the listening room of the 22.2 multi-channel sound system, we could experience this work as Karlheinz Stockhausen created it. After the recording, I held listening sessions for music experts who had attended the concert. They said, 'It was natural and great, and this recording provided a clearer understanding of the composer's intentions than what we heard in the concert hall'.

### *World-first immersive sound footage recording and transmission using 22.2 multi-channel sound*

The London Olympic Games 2012 became the world's first showcase of immersive sound footage recording and transmission of UHDTV with 22.2 multi-channel sound. The production tools for the 22.2 multi-channel sound system were developed for this recording and transmission as shown in Figure 2.22.

The 22.2ch one-point microphone is a spherical microphone that uses acoustic baffles to partition a sphere into angular segments, onto each of which an omnidirectional microphone is mounted. These baffles produce the directivity. It was partitioned into a sphere of 45-cm diameter into eight parts horizontally and three parts vertically, corresponding to the three-layer setup in 22.2 multi-channel sound. This microphone was used for the Olympic Games 2012 to record 16 channels (8 channels of the middle and top layers, respectively) to capture the sound of the heat and cheers in the Olympic stadiums.

The live mixing and post-production systems used a newly developed three-dimensional sound panning algorithm developed by Akio Ando [16]. The 3D reverberator used the 'Space Builder' concept developed by Wieslaw Woszczyk, Brett Leonard, and Doyuen Ko from McGill University [17].

Footage recording and transmission of UHDTV in 22.2 multi-channel sound were carried out in the Olympic Stadium (including the opening and closing ceremonies), the Aquatics Center, the Basketball Arena, and the Velo Park, as shown in Figure 2.23.

A specific OB van was used for UHD video and 22.2 multi-channel sound recording, respectively, and the video and audio were transferred to the BBC TV Center in London

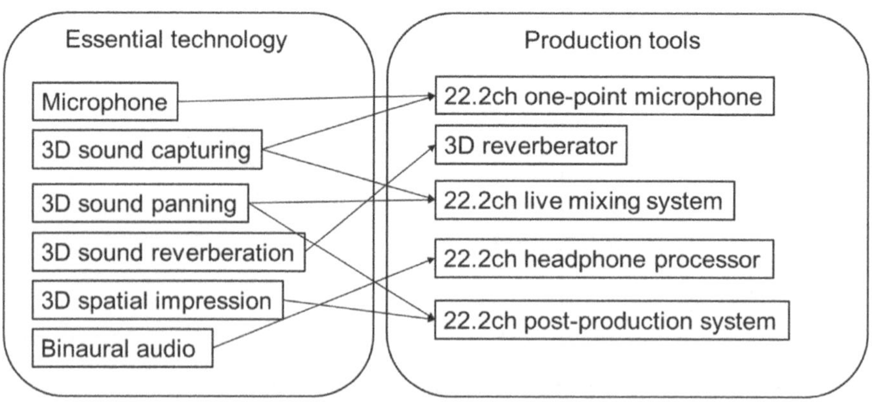


*Figure 2.22* Production tools for 22.2 multi-channel sound developed for the London Olympics 2012

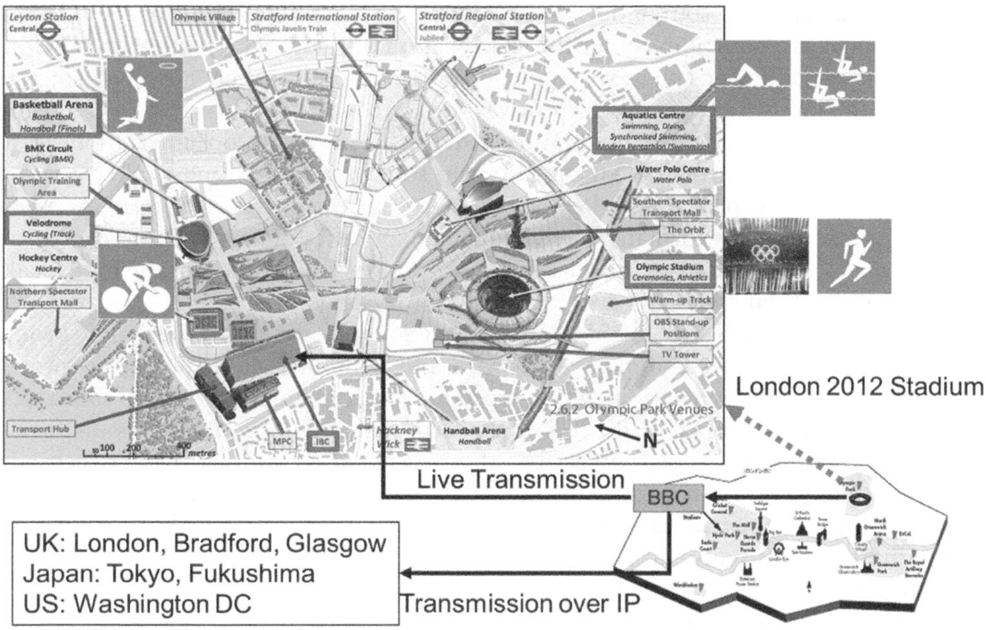


*Figure 2.23* London Olympics 2012 footage recording, live viewing, and public viewing using UHDTV and the 22.2 multi-channel sound

for post-production and transmission. The live viewing was done at IBC in the Olympic Parks. The public viewing was done in London, Bradford, and Glasgow in the UK, Tokyo and Fukushima in Japan, and Washington, DC in the USA.

## *22.2 multi-channel sound system state of the art*

The 22.2 multi-channel sound system has been used for 8K broadcasting in Japan since 2018. Several 8K TV programs have been mixed in 22.2 multi-channel sound, such as NHK Symphony Orchestra's concerts. The 22.2 multi-channel sound system has also been used as the reference three-dimensional sound system for the research on spatial sound and the development of immersive sound production systems worldwide. Since the birth of the 22.2 multi-channel sound, several international standards have been established regarding the 22.2 multi-channel sound system.

SMPTE ST 2036-2:2008, 'Ultra High Definition Television – Audio Characteristics and Audio Channel Mapping for Program Production', specifies the characteristics of digital audio for UHDTV program production and distribution. This document also defines the mapping and labeling of 22.2 multi-channel audio for UHDTV program production [11].

IEC 62574:2020 specifies the general channel assignment, including the 22.2 multi-channel sound system and the metadata for the multi-channel audio format. This document, with its channel mapping and labeling content, provides the unified usage of channel assignments for source devices, digital audio interfaces, and sink devices. The general channel assignment excludes the specification of the exact position of each loudspeaker. The metadata has the exact position of each loudspeaker; however, since the position of the loudspeaker is different in various surround sound formats, the position information of the metadata is not standardized. It is aimed at consumer applications; it is not targeted at theatrical environments. Up to 65 labels for loudspeaker positions and 15 labels for loudspeakers on the display are specified, which can be used for all current multi-channel audio formats. The channel assignment concept is room-centric and listener-centric. This general channel assignment specifies the position of the loudspeaker as seen from the listener [18].

Recommendation ITU-R BS.2051-3, 'Advanced Sound System for Programme Production', specifies an advanced sound system with and without accompanying picture. An advanced sound system is a system with a reproduction configuration beyond those specified in Recommendation ITU-R BS.775 or a system with any reproduction configuration that can support channel-based, object-based, or scene-based input signals or their combination with metadata. An advanced sound system uses audio data in combination with an appropriate set of metadata to specify a sound scene to be delivered/broadcasted. The specifications include requirements for signaling the properties of advanced sound content and loudspeaker layout to be used in content production for advanced sound systems. The advanced sound system can apply to the sound component of television and expanded large-screen digital imagery (LSDI) programs, as well as for sound-only programs. The 22.2 multi-channel sound system is specified as Sound System H (9+10+3) [19].

Recommendation ITU-R BS.1770-5 (11/2023), 'Algorithms to Measure Audio Programme Loudness and True-Peak Audio Level', specifies audio measurement algorithms for the purpose of determining subjective program loudness, and true-peak signal level. It also specifies the objective loudness measurement algorithm for arbitrarily placed loudspeaker configurations of the advanced sound system, including 22.2 multi-channel sound system [20].

Recommendation ITU-R BS.2076-2, 'Audio Definition Model', describes the structure of a metadata model that allows the format and content of audio files to be reliably described. This model, called the Audio Definition Model (ADM), specifies how XML metadata can be generated to provide the definitions of tracks in an audio file. Annex 2 contains a selection of examples of metadata that uses the ADM, including a 22.2 multichannel program with an alternative dialogue example [21].

Recommendation ITU-R BS.2127-1 (11/2023), 'Audio Definition Model Renderer for Advanced Sound Systems', specifies the reference renderer for use, including for program exchange, with the advanced sound systems including 22.2 multi-channel sound system specified in Recommendation ITU-R BS.2051-2, and the audio-related metadata specified by the Audio Definition Model (ADM) in Recommendation ITU-R BS.2076-1. The audio renderer converts a set of audio signals with associated metadata to a different configuration of audio signals and metadata based on the provided content metadata and local environmental metadata [22].

ISO/IEC 23008-3:2022, 'Information Technology – High Efficiency Coding and Media Delivery in Heterogeneous Environments – Part 3: 3D Audio (2019)', specifies technology that supports the efficient transmission of immersive audio signals and flexible rendering for the playback of immersive audio in a wide variety of listening scenarios. These include home theater setups with 3D loudspeaker configurations, 22.2 loudspeaker systems, automotive entertainment systems, and playback over headphones connected to a tablet or smartphone [23].

Japanese Association Radio Industries and Businesses (ARIB) STD-B59: 'Three-Dimensional Multi-channel Stereophonic Sound System for Programme Production', specifies basic parameters for the three-dimensional multi-channel stereophonic sound system (22.2 multi-channel sound system) with more than 5.1 channels stereophonic sound, sound signals used in the studio and so on. This standard intends to produce programs and exchange programs smoothly by this system and is applied to the sound system of UHDTV [24]. No ARIB Technical Report (TR-B44) specifies the technical guidelines for production using a three-dimensional multi-channel sound system, actually regarding the 22.2 multi-channel sound system [25].

## References


1. K. Hamasaki, 'Multichannel Sounds in Television – Technical and Aesthetic Approach,' *IMAGE TECHNOLOGY – Journal of the British Kinematograph, Sound and Television Society* 83(2), April 2001.
2. K. Hamasaki, T. Shinmura, S. Akita, K. Hiyama, 'Approach and Mixing Technique for Natural Sound Recording of Multichannel Audio,' Presented at the 19th AES International Conference: Surround Sound – Techniques, Technology, and Perception, Elmau, Germany, Paper #1878, June 2001.
3. K. Hamasaki, 'Multichannel Sound Recording of Music: Practical and Technical Aspects,' Presented at the AES 19th International Conference, Elmau, Germany, June 2001.
4. K. Hamasaki, K. Hiyama, 'Reproducing Spatial Impression With Multichannel Audio,' Presented at the AES 24th International Conference: Multichannel Audio, The New Reality, Banff, Canada, Paper #19, June 2003.
5. K. Hamasaki, 'Microphones Techniques for Natural Music Recording,' Presented at the AES 152th Convention, May 2022.
6. K. Hamasaki, W. Van Baelen, 'Natural Sound Recording of an Orchestra with Three-dimensional Sound,' Presented at the AES 138th Convention, Warsaw, Poland, Paper #9348, May 2015.
7. F. Okano, M. Kanazawa, K. Mitani, K. Hamasaki, M. Sugawara, M. Seino, A. Mochimaru, K. Doi, 'Ultrahigh-Definition Television System with 4000 Scanning Lines,' Presented at NAB 2004, Las Vegas, NV, April 2004.
8. K. Hamasaki, K. Hiyama, T. Nishiguchi, K. Ono, 'Advanced Multichannel Audio Systems with Superior Impression of Presence and Reality,' Presented at the AES 116th Convention, Berlin, Germany, Paper #6053, May 2004.
9. K. Hamasaki, K. Hiyama, R. Okumura, 'The 22.2 Multichannel Sound System and Its Application,' Presented at the AES 118th Convention, Barcelona, Spain, Paper #6406, May 2005.
10. K. Hamasaki, et al., 'A 22.2 Multichannel Sound System for Ultrahigh-Definition TV (UHDTV),' *SMPTE Motion Imaging Journal* 117(3), 40–49, 2008.
11. SMPTE 2036-2-2008, 'Ultra High Definition Television1 – Audio Characteristics and Audio Channel Mapping for Program Production,' Published by the Society of Motion Picture and Television Engineers (SMPTE).
12. K. Hamasaki, Y. Nakayama, T. Nishiguchi, R. Okumura, 'Wide Listening Area with Exceptional Spatial Sound Quality of a 22.2 Multichannel Sound System,' Presented at the AES 122nd Convention, Vienna, Austria, Paper #7037, May 2007.
13. K. Hamasaki, T. Nishiguchi, K. Hiyama, R. Okumura, 'Effectiveness of Height Information for Reproducing Presence and Reality in Multichannel Audio System,' Presented at the AES 120th Convention, Paris, France, Paper #6679, May 2006.
14. K. Hamasaki, T. Nishiguchi, K. Hiyama, R. Okumura, 'Influence of Picture on Impression of Three-Dimensional Multichannel Sound,' Presented at the AES 28th International Conference, Piteå, Sweden, June 30–July 2, 2006.

15. T. Nishiguchi, K. Hamasaki, R. Okumura, Y. Nakayama, M. Iwaki, A. Imai, A. Ando, S. Shimaoka, Y. Otuka, S. Kitajima, 'Production and Live Transmission of 22.2 Multichannel Sound with Ultra-High Definition TV,' Presented at the AES 122nd Convention, Vienna, Austria, Paper #7137, May 2007.
16. A. Ando, K. Hamasaki, 'Sound Intensity-Based Three-Dimensional Panning,' Presented at the AES 126th Convention, Munich, Germany, Paper #6679, May 2009.
17. W. Woszczyk, B. Leonard, D. Ko, 'Space Builder: An Impulse Response-Based Tool for Immersive 22.2 Channel Ambiance Design,' Presented at the AES 40th International Conference, Tokyo, Japan, October 2010.
18. IEC, 'IEC 62574:2020 Audio, Video and Multimedia Systems – General Channel Assignment of Multichannel Audio,' International Electrotechnical Commission, Geneva, Switzerland, 2020.
19. ITU-R, 'Recommendation ITU-R BS.2051-3: Advanced Sound System for Programme Production,' International Telecommunication Union, Geneva, Switzerland, 2023.
20. ITU-R, 'Recommendation ITU-R BS.1770-5: Algorithms to Measure Audio Programme Loudness and True-Peak Audio Level,' International Telecommunication Union, Geneva, Switzerland, Nov. 2023.
21. ITU-R, 'Recommendation ITU-R BS.2076-2: Audio Definition Model,' International Telecommunication Union, Geneva, Switzerland, 2019.
22. ITU-R, 'Recommendation ITU-R BS.2127-1: Audio Definition Model Renderer for Advanced Sound Systems,' International Telecommunication Union, Geneva, Switzerland, Nov. 2023.
23. ISO/IEC, 'ISO/IEC 23008-3:2022 Information Technology – High Efficiency Coding and Media Delivery in Heterogeneous Environments – Part 3: 3D Audio,' International Organization for Standardization/International Electrotechnical Commission, Geneva, Switzerland, 2022.
24. ARIB, 'ARIB STD-B59: Three-Dimensional Multichannel Stereophonic Sound System for Programme Production,' Association of Radio Industries and Businesses, Tokyo, Japan, 2016.
25. ARIB, 'ARIB Technical Report TR-B44,' Association of Radio Industries and Businesses, Tokyo, Japan, 2017.

Chapter 3

# Early immersive sound and the development of Auro-3D

*Wilfried Van Baelen*

## Origins of immersive sound

Some people call me the 'Father of Immersive Sound'(see Figure 3.1). I coined this term to be used as the accepted and industry standard term for true 3D experiences. I also pioneered a streamlined 3D audio format with end-to-end solutions for all audio markets. In only five years, Auro-3D was integrated and established in the markets of cinema, home, cars, mobile, games and music.

Actually, I was not the first guy thinking about 3D audio at all. Michael Gerzon was already working on Ambisonics in the 1970s. Ambisonics is a genius theoretical concept, but it is less scalable than it looks at first glance, especially because of the audible artifacts in each of the orders. Two decades later, Tomlinson Holman (THX) presented in July 1999 a 10.2 format for cinema with two height speakers behind the screen. Kimio Hamasaki started his research in 2000 and presented in 2005 the NHK 22.2 system. But all those systems did not have the technical end-to-end solutions needed to make it a standard in all audio markets. And this is what everyone was looking for: one solution that could work for every market and that reproduces the sound experience as intended by the creators of the content. This is something that mono, stereo, or surround sound had achieved in an evolution of about 100 years of sound recording and reproduction (see Figure 3.2). Note that it took more than two decades for surround sound to become established in cinema in the 1970s and in homes in the mid-1990s via the DVD format in compressed audio. It took another decade to bring it finally to the consumers in uncompressed audio quality (Blu-ray in 2003). That's a total of about 30 years to bring the surround sound experience from cinema into the home with a similar audio quality.

I was very excited that I was able to come up with end-to-end solutions that enabled a breakthrough in the addition of the missing dimension 'height' in less than one decade to all audio markets. I wanted to invent a three-dimensional audio format that is fully scalable, very flexible, and feasible from production up to the reproduction without any compromise in delivering everywhere the same sound experience as intended by the creators. Recognizing that nothing at that moment came close to developing that goal, this was a serious challenge. But the results of my research and developments and inventions inspired many companies and engineers in the audio industry who jumped on this train. The hundreds of Auro-3D demos I performed all over the world were an eye-opener. Everyone was immediately greatly impressed with how this very efficient system was able to deliver heightened emotional experiences with many new creative possibilities for film, music, games, sports, etc. Quickly, this three-dimensional sound experience was

DOI: 10.4324/9780429297526-5

*Figure 3.1* Wilfried van Baelen, 'Inventor of Immersive Sound'

poised to become the next big thing. 'Auro-3D is the major leap in the evolution of sound reproduction', according to A.R. Rahman, 2011 (Oscar winner composer of 'Slumdog Millionaire', and operator of the first Auro-3D certified studio in India). In the same year, Skywalker Sound decided to install the Auro-3D system. George Lucas of Lucasfilm announced in November 2011 that *Red Tails* will be mixed in Auro-3D and will be released in January 2012 as the first ever Hollywood movie in 3D audio. The team from Skywalker Sound came to Galaxy Studios for the mix in Auro-3D. This first release of a Hollywood movie in a 3D audio format from a legendary film producer as George Lucas was a game changer.

## The need for a new generic term

For decades, surround sound was marketed as a three-dimensional sound reproduction format, which is wrong. It is only a two-dimensional reproduction format as the speakers are arranged in a horizontal plane around the audience. I was annoyed that there was not a common definition for 3D audio because, when I talked about 3D audio, everyone thought it was about surround sound. When I went to the Hollywood studios to integrate this sound format in movies, their reaction was: 'I understand the visual 3D effect in a movie, but how about sound in 3D?' And that's the world I started in. You cannot believe how much pioneering work it took to evangelize the idea that surround sound is only a two-dimensional reproduction format. I often had public discussions with audio professionals who stated that surround sound creates more depth in the sound and since depth is the third dimension, surround sound is a three-dimensional reproduction format. I replied to them, 'There is also spatial depth in stereo; does it mean that stereo is then also a three-dimensional reproduction format? We should not mix up the psychoacoustical spatial depth created by our brain versus a physical location of loudspeakers enabling the reproduction of sounds coming from above, which is not possible with the surround sound speaker layouts located in a two-dimensional horizontal plane around the audience and missing the height dimension, the third dimension in sound. Moreover, it is especially by adding the height layer (as I defined it in the Auro-3D specs) that much more depth is experienced'. This reply typically killed such discussions. I received many apologies from those people. It shows that there was really a lot of pioneering work to do to settle the common misunderstanding that surround sound is a 3D audio format, which it is not.

Since 2006, I have tried to find a good brand name for my three-dimensional audio concept and thought about Aura-phonic sound. But the moment I went on the internet, I saw hundreds of names and products related to Aura and sound. My friend, Wolfram Burgert, had the idea to change just one letter and make it Auro as a combination of three Latin words: *audio* (sound), *audire* (listen), *aurum* (gold). Aurophony was born, which we updated two years later to the brand name Auro-3D. When we proposed the industry to use Aurophonic sound as the new generic term to replace 'Surround Sound + Height', we often heard the comment that this name is too close to the brand name of Auro-3D to be used as a generic term.

I remember when I was at the spatial conference in Tokyo in October 2010, where I launched the Auro-3D format, I saw the use of many different terms like 'periphony', 'spatial audio', 'aurophonic sound', '3D audio', '3D surround', 'surround sound with height', 'true 3D sound', 'sound in 3D', etc. while each trying to describe the same thing. It was as chaotic as the Babylonian confusion of tongues. It is at that moment that I realized that we really need a new generic term to be used by everyone, so the idea came to me to use the term 'immersive sound'.

Because . . . 'immersive' is exactly what this whole experience is all about when adding the third dimension in sound if it is done in the right way. It creates a heightened emotional impact by being totally *immersed* in sound from all around and above the listener. People in the industry all liked that term because it could be easily understood by everyone and described very well the new experience it is all about. In 2011, we launched together with our partner Barco (market leader in digital projectors for Cinema) immersive sound into the market. The release of *Red Tails* by Lucasfilm in AURO 11.1 by Barco (January 2012) immediately validated its game-changing impact on the film industry. Within one year, most major Hollywood Studios had installed one or two mixing studios with the Auro-3D system and other leading post-production film studios all over the globe followed (China Film Group, Bollywood studios, etc.).

So this new generic term spread very fast in the cinema industry. The definition of 'immersive sound' and 'immersive audio' became integrated in the 2098 SMPTE Immersive Audio Standards for Cinema Sound. The use of the term 'immersive audio' is more related to the technical components of the three-dimensional sound reproduction, whereas 'immersive sound' is more or less related to our hearing, the listening experience.

So that's where it all started and probably one of the reasons why they call me 'Father of Immersive Sound' as I pushed this term into the market with the sole intention to create transparency in communications. But I think the main reason why people often

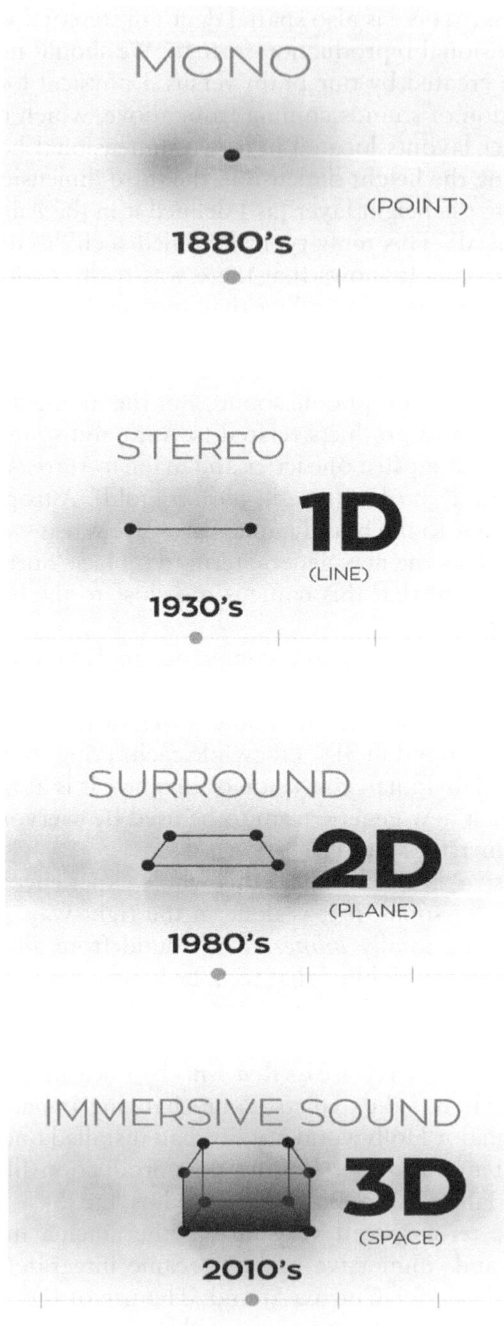


*Figure 3.2* Evolution of sound reproductionillustrating the transition from mono(a) to stereo (b), surround (c), and immersive audio formats(d).

give me that compliment is mainly due to my inventions and many contributions, which enabled me to bring the Auro-3D format to the world with groundbreaking end-to-end solutions for all 'Immersive Sound' audio markets.

My goal was to design a very scalable system that can reproduce in the most efficient way the same Immersive Sound experience as intended by the creators in each kind of content (film, music, sports, games, etc.). To achieve this goal, I combined scalable speaker layouts (for small up to large rooms maintaining everywhere the intended Immersive Sound experience) with a groundbreaking audio-codec that could bring up to 24 channels of uncompressed audio quality on existing carriers like DCP or even on Blu-ray, which was limited to only eight channels of uncompressed audio. Due to this Auro-codec, there was no need for a new delivery format to bring this Immersive Sound experience to all these markets. Auro-3D is designed to use those existing delivery formats without the need for any change in the specs and without any compromise in the sound quality. This means that the first ever Blu-ray player on the market could be used to playback Auro-3D, something that other formats later were not able to achieve without the need to change the specs. Moreover, this innovative technology enables 'single inventory', which means that the same audio file is used for the reproduction of the 5.1 or 7.1 surround mix as for the Auro-3D mix (see further on in the chapter) and this without any compromise in the sound quality of each audio format.

It was mainly due to my previous experiences with 5.1 surround that gave me this insight and vision (as will be explained). I conceived this Auro-3D codec in early 2006 and asked Guido Van den Berghe to write the code. He became my co-inventor of that patent, which I gave the name 'Octopus' (later it became the AURO-Codec) and was filed by Galaxy Studios. But there were so many other things to solve in each of those audio markets as part of the end-to-end solutions integrated in the Auro-3D format. I will give you some highlights, but let me start with how I came to this amazing journey.

## Pioneer in surround sound formats

When I heard in 1980 for the first-time quadraphonic sound as a teenager, I was so intrigued and impressed that I couldn't wait for that format to become the next listening standard after stereo. But unfortunately, quadraphonic sound didn't make it to the mass market. So, when the DVD was announced mid the 1990s with the ability to play back 5.1 surround, I wanted to produce music in that format in analogue and digital format. The sound quality of the DVD reproduction was not great because 5.1 surround could only be delivered via lossy compressed audio formats (AC-3 or DTS), which means that audible information gets lost and even creates audible artifacts (distortion). But at least, it was a step forward in enhancing the spatial experience compared to stereo.

My younger brother Guy and I (the founders of Galaxy Studios) decided to construct within our studio complex a new dedicated mixing room for 5.1 surround productions built upon the same concept as concrete bunkers on springs. We asked SSL to produce a custom made 9000J console with 5.1 monitoring and routing, which became their first music console with surround sound features, installed at Galaxy Studios in 1997. This was one of the first, if not the first ever studio with a large format console dedicated for 5.1 surround productions for the music industry. The popular engineer Ronald Prent at Wisseloord Studios (close to Amsterdam) was so impressed by it that he decided to move to Belgium to help introduce surround sound to the music market.

It was also in the same period that we were asked by Philips and Sony to make mixes for their upcoming Super Audio CD (SACD) format, using DSD technology. This allowed for high-resolution audio above 100 kHz. However, it was at that moment only a stereo format. We pushed Philips and Sony to expand SACD to 5.1 surround because that spatial component was, in my opinion, a much better selling point for SACD than just a better audio quality in stereo. They were already looking into that, but they could not deliver the same length as a normal CD on this new SACD format. But they saw the opportunity and worked on finding solutions. Shortly after, we were doing our first mixes for SACD in 5.1 surround. We produced many albums for the SACD format sponsored by Philips and Sony.

But there was the need for a good mastering studio that could handle 5.1 surround productions with high-resolution digital audio including DSD. The magic that many mastering engineers are doing is very often in the analogue domain. However, since mixing engineers tend to use the available headroom in the digital domain with 0 VU analog level set at –18 dB DBFS, converting digital sound that goes to 0 dB into analogue was creating a signal that is too strong in the analogue domain, causing audible distortion. So how can we solve that? We were looking into solutions to optimize the mastering process and created, together with Ronald Prent at Galaxy Studios, our Reference Mastering Studio. We asked the CEO of SPL to develop a custom-made mastering console that can handle an internal audio chain up to 120 V in order to have enough headroom to handle transfers from digital to analogue without causing audible distortion. In 2002, we installed the first 7.1 MMC mastering console from SPL, which became a new reference in the music mastering industry. (See Figure 3.3.) This attracted the Grammy-winning

*Figure 3.3* SPL MMC-1

mastering engineer Darcy Proper,[1] who moved from New York to Galaxy Studios located in Mol, Belgium. As of 2024, this console is still active in our mastering studio and is still a top reference mastering studio for stereo and surround sound.

Galaxy Studios became a reference studio to remix and remaster 5.1 surround productions. Many of those bands and artists reacted so enthusiastically that they wanted their next album to also be in 5.1 surround. But after the SACD was launched with success in the music market, Philips and Sony stopped further their financial support of mixing albums in 5.1 surround. At the time, producers responsible for each album stated that they needed all their budgeted time for the stereo mixes, so there was not enough time left to mix in 5.1 surround, although artists still wanted it. This experience taught me a very important insight for the success of a new audio format: it has to be possible to create it within the same budget as provided for the main stereo audio format. This means that any extra cost should be small enough that it is negligible in the overall budget. So the question is: how do we solve and achieve that?

## Developing an efficient workflow

The more I was confronted with that issue, the more I realized that we will not be successful in bringing 5.1 surround to the music market unless we find a workflow where you can create stereo and surround sound simultaneously without any compromise in the quality or the artistic intentions in each audio format, stereo and 5.1 surround. So I was wondering how that issue can be solved and started to do many tests and research the compatibility of stereo and 5.1 surround mixes. After a few months of many tests, I discovered that to make a stereo mix and 5.1 surround mix compatible with each other, I should not have to change the EQ settings when toggling between the stereo and the 5.1 surround versions, although panning and volume required different settings in each audio format.

This means that for most music content a stereo mix cannot be created as the engineer intended to do it just by making a simple down-mix from the 5.1 surround mix, as we used to do that for a movie in which a down-mix from the surround mix was intended to create the Dolby stereo mix, an analogue stereo carrier that allowed the decoder to recreate a surround mix, but this technology had cross-talk issues between the channels, and the decoded mix didn't sound like the original. Later technologies like Dolby Pro Logic and Dolby Surround impacted the sound quality with a bandwidth limit of 7 kHz in the surround channels. There was no consistently good solution at the time to create a stereo mix from a surround mix for music. It might work in some exceptional cases, but usually you need a different panning and volume control for each individual sound source as well as effects like reverbs etc. for the stereo and surround versions. It intrigued me to find a solution because no music console had such capabilities. At that time, we had a Capricorn from NEVE. I was the very first buyer of that console, which was just before the merger of AMS and Neve. Although the first one was installed at Abbey Road Studios – for practical reasons during the final testing of this new impressive digital console, the second one was installed at Galaxy Studios in 1993. I disclosed with the AMS-Neve team my idea about the requirements needed to mix stereo and surround sound simultaneously and asked them to make a software update for the Capricorn console that allows this independent routing and panning for a stereo and surround bus. The R&D team liked the idea. A few months later, I got a software update that allowed me to mix the stereo and surround version starting from the same mix setup on the console but with total control over an independent panning and volume changes in each of those audio formats.

I started experimenting to find my most efficient workflow because it was not clear to me in which format (stereo or surround) I should finalize first. I concluded that the overall best way was to toggle between the stereo mix and surround mix already during the setup of the mix to achieve a very good result in both formats before starting the mix automation. So the volume and panning were already set differently in each of those audio formats. Then I added automation and finalized first the stereo mix. Except for panning, all automations were done in the same way for both formats. To finalize the 5.1 surround master, I only had to tweak just a few minor things which took very little time. For example, what works well in stereo as left-right automated panning could be as nice as front versus back channel panning automated in the 5.1 surround mix. The small tweaks in panning or volume automation in the other format could be done in just a fraction of the time.

I was very excited that I achieved my goal to be able to create a stereo and surround mix simultaneously without any compromise in the quality within the same time frame it takes to create just one of those formats. And over some time, I realized that my stereo mixes had more depth because I wanted to get closer to the spatial experience I had in the surround mix.

I used the auxiliary busses to trim volumes and even mute certain input channels that I used in surround but not in stereo. A good example of this are the reverbs. In the Lexicon 960, it was possible to have a routing of one reverb to a surround output while another engine with the same reverb in stereo routed to the stereo output. Note that a stereo reverb does not sound the same as a down-mix of a surround reverb and not the same as only the left and right channel of the surround reverb. Because of this, down-mixing the surround reverb into the stereo bus was not an option for me. I started to learn how the highest compatibility can be achieved in both main audio formats and how to get the best out of each simultaneously. The good news was that I could offer my clients the surround mix within the budget of the stereo mix without any extra cost while creating great mixes in both formats. This solution created so much work for us that in 2002 it inspired Ronald Prent, who preferred analogue consoles, to work together with Paul Wolff from API to design a new console with a channel module that was based on the same principle as the routing features I had created in the Capricorn. I gave that API console the name VISION because there was a clear future vision behind it, allowing the creation of mixes in the main audio formats without the need of any extra time. The very first API Vision console with 80 channel strips and 170 motorized faders was installed at Galaxy Studios in August 2003 (see Figure 3.4).

## First steps going beyond the main 5.1 surround sound format

There was a lot of discussion in the music industry about the position of the surround loudspeakers in a 5.1 setup because some engineers liked the surround speakers more in the back, but that caused a bigger gap between the front and the surround speakers. So in 2005, we at Galaxy Studios started experimenting with a 7.1 setup in which the side speakers fill the gap between the front stereo and back surround speakers. We experienced how those side surround speakers had a completely different effect on our hearing and listening experience compared to the rear loudspeakers, which are located out of our visual field.

*Figure 3.4* Founders of Galaxy Studios, brothers Wilfried and Guy van Baelen with first API Vision

Moreover, it is challenging or often not possible to install side surround speakers following the 7.1 specs, not only in a control room but also in the home because at a 90° angle at the sides, there is, mostly for practical reasons, no place to put those loudspeakers except in dedicated home cinemas which are designed from the start with that goal in mind.

## Addition of the missing dimension in sound reproduction – height

All those experiences in the 5.1 and 7.1 surround formats became very important in my development of the Auro-3D format, which I started after a call from a German producer Tom Hapke in 2005. He asked me to co-produce a pop music production in the 2+2+2 format. He tried to enhance the overall experience using the same number of six channels as a 5.1 surround setup but organized in a different way using a set of quadraphonic early-level layer speakers (almost like the corner speakers of a 5.1 setup) with the addition of 2 height speakers. There was no LFE because every channel was full spectrum.

I was very skeptical because Tom Hapke told me that, so far, only classical music was recorded in that format and was wondering what kind of sounds were going to come from those two front height channels. Before embarking on this journey, I requested a demo of the 2+2+2 format at Galaxy Studios and met the two owners of the related patents: Wolfram Burgert and Dr. Alfred Schefenacker. Both were associated with the recording label DIVOX, which has released award-winning classical music albums recognized for their excellence in both artistic performance and sound quality.

I remember that moment when I heard those classical recordings in the 2+2+2 format for the first time at Galaxy Studios – that was really an eye-opener. I experienced a much more natural sound with greater depth, thanks to the four front speakers (standard stereo plus two height speakers) installed as a vertical quadraphonic sound field. It intrigued me to understand why we as human beings experience so much more depth and natural sound from the front channels in such a simple way. It required only two additional microphones about 1 m above the main stereo pair of mics with the signals reproduced by two *height* loudspeakers located above the stereo pair of speakers with a vertical distance equal to half of the distance between the stereo pair of speakers. I was convinced and wanted to try this system for the first time in a pop production together with producer Tom Hapke and his artist Sylvia Dias. This time, the challenge was not just delivering a stereo and a 5.1 surround mix – I needed to deliver three formats: stereo, surround, and 2+2+2. Again, I wanted to have this achieved in one workflow and without the need for extra time or costs compared to what it would have cost only for the stereo version.

Due to the lessons I learned about easily streamlining compatibility between all those formats, I used my 5.1 surround setup and added the two height front speakers to have control over all those formats simultaneously, which was easy to do on my Capricorn console. I remember that period very well – especially because I thought that, after having produced and engineered more than 20 platinum albums in various genres in both stereo and surround sound, as well final mixing for movies for cinema, I didn't expect any more big surprises since I found a way to maximize the emotional experience for the listener. But the journey starting those recordings with the two extra front height channels above a 5.1 surround setup was that of one surprise to another.

## The vertical axis

The addition of that vertical axis was very new and very different compared to the experiences with the loudspeakers located in a two-dimensional horizontal plane with 5.1 or 7.1 surround sound. I discovered things that I never was aware of. For example, I wanted to pan a sound from left to right and then move that sound from the ear-level speakers up to the height speakers directly located above the ear-level speakers, but it surprised me that I could not experience the vertical movement of the sound, only the horizontal between left and right. I also tried to position sounds between two speakers vertically installed above each other with an angle of about 30° and tried to locate a sound as a phantom source somewhere in the middle of that vertical axis, but that didn't work either. I thought, probably the angle in the vertical axis is not big enough to experience the movement of the sound. So I expanded the angle from 45° to 50°, but it didn't solve the problem. Although I could technically position the sound exactly between the main layer surround speaker and its corresponding height speaker, I was not able to experience the sound as a phantom source in the middle of that axis as we are used to experience that in the horizontal axis between the speakers of a stereo setup in which you can hear by example a voice located in the middle of the sound field as a mono source at a place where no speaker is installed. It became clear to me that the same experience was not possible in a 'vertical stereo field'. Surprisingly enough, I could experience the position of a moving sound very well when I was panning it from the left front ear-level speaker to the right front height loudspeaker and vice versa, or from the left front height speaker down

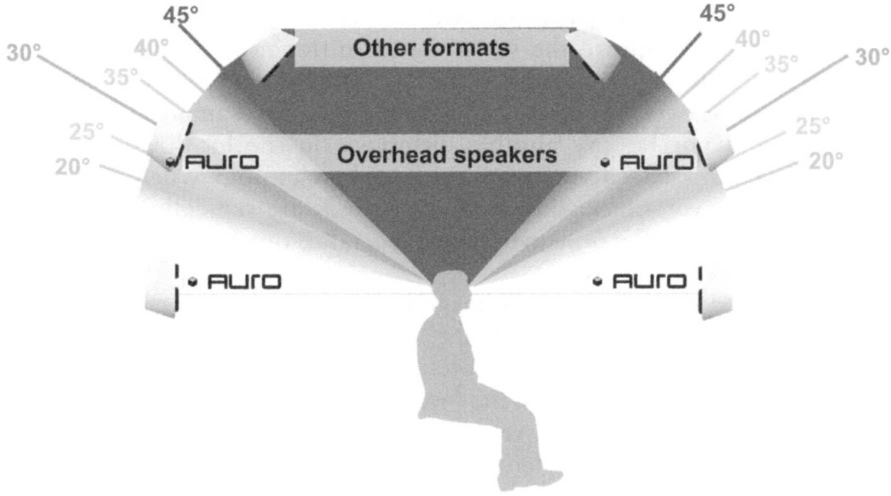


*Figure 3.5* Coherent vertical sound field



to the right front ear-level speaker and vice versa. I could also perfectly position a mono sound as a phantom mono source between the two front height channels.

The many tests I did in May 2025 gave me many surprising discoveries of things I never realized before. As such, I was intrigued about the vertical quadraphonic sound field in the front and was wondering if I would be able to create a phantom source in the middle of that cross – meaning positioning a sound above the center speaker – and I was excited to find out that it was possible.

As a result of those tests with this quadraphonic vertical sound field and vertical angle between those speakers, I discovered something very unexpected when I played back recordings for which I used a vertical quadraphonic microphone setup (two stereo pairs of microphones, of which the height microphones were about 1 m higher than the standard stereo pair microphones); by playing them back over a quadraphonic vertical setup of loudspeakers, the sound was very natural and very coherent up to the angle of about 35° to 40°. If the angle is greater, the coherence gets lost and it suddenly sounded artificial, like two different sources and not as a coherent sound field (see Figure 3.5).

Another surprising discovery was that panning a mono source between two speakers located vertically above each other did not have the same result as moving a sound with the same movement in front of the microphones (between the lower and height microphones). I experienced that if the angular distance between the vertical speakers and ear level speakers became greater than 40°, the coherence of the sound field as well as the natural depth disappeared. I did the same tests in a very dry recording booth (150 ms RT60); the movements in the vertical axis were still there but less obvious compared to the recordings I made in the Galaxy Hall using the same microphone setup.

All of this taught me how important the 3D reflections around moving objects are in order to have sounds located in the vertical axis.

I suddenly realized that it had to do with the fact that we don't have an ear on the top of our head to capture the vertical time differences. Further, object-based technology is not able to bypass that human characteristic as well, despite what some industry people were saying at the time. With the same 3D audio speaker setup, there should not be any audible difference in the localization of a mono sound between channel-based versus object-based technology when using the same panning laws. Moreover, object-based source sounds are typically mono and don't contain the original 3D reflections which are crucial to our brain to localize sounds, distance, size, etc.

Not experiencing a similar phantom image in the vertical axis as we are used to in the horizontal axis was really shocking to me. It would have been so normal to me that I could position sounds as a phantom source between two speakers vertically located above each other as we used to do in the horizontal axis in stereo or in the surround sound formats. So I realized that the vertical axis was completely different and wanted to know more about our hearing system, how it works with these vertical time and level differences as well as color differences, the differences that I heard when a loudspeaker was elevated while playing the same sound; some high frequencies got more pronounced. I was talking with a few top engineers about such things, but nobody could give any clear answer to my questions at the time. The feedback I mostly got was, 'Why do you want to know this? Consumers will never place height speakers in their home, so why would you bother?' I would explain that a much more natural sound with more depth in the front and improved timbre could be obtained, something that had to be experienced to truly understand. I felt that there was something very special in that experience, something that was really very interesting, and I didn't want to give up on my research because there was still something unnatural in the overall sound experience to be fixed; the two surround speakers at the back didn't match the natural sound coming from the front. The gap between the five speakers in the front with a vertical sound field became even more obvious compared to the 5.1 surround format. So I added two side surround speakers, creating a 7.1 surround ear-level layer, to fill that gap, but It didn't result in what I was hoping for. I tried to install the two front speakers above the side surround speakers (like a stereo pair of speakers directly overhead, like the 5.1.2 setup), but that made everything even worse because it created a much lesser overall sound experience and not natural at all.

I was wondering what the result would be by adding a pair of 'height surround' loudspeakers above the surround loudspeakers. Since I only had nine similar 8050 Genelec speakers at Galaxy Studios, I used my two side surround loudspeakers of the 7.1 surround setup and placed them above the 110° surround loudspeakers. The result was a 30° elevated quadraphonic height layer installed above the corner speakers of a 5.1 surround setup. I made some recordings of different instruments with also height surround microphones for those height surround channels.

And oh my God! . . . I will never forget the moment the first time I experienced those results because the emotional impact was not comparable with anything I heard before. It was as if I was standing in the recording room although I was listening in the control room. I never experienced the reproduction of sound so natural and so 'immersive', it was really magic for me and I got tears in my eyes from music that never touched me like that before. That was the moment I realized the addition of the third dimension executed in the right way could create a totally new emotional experience when listening to music.

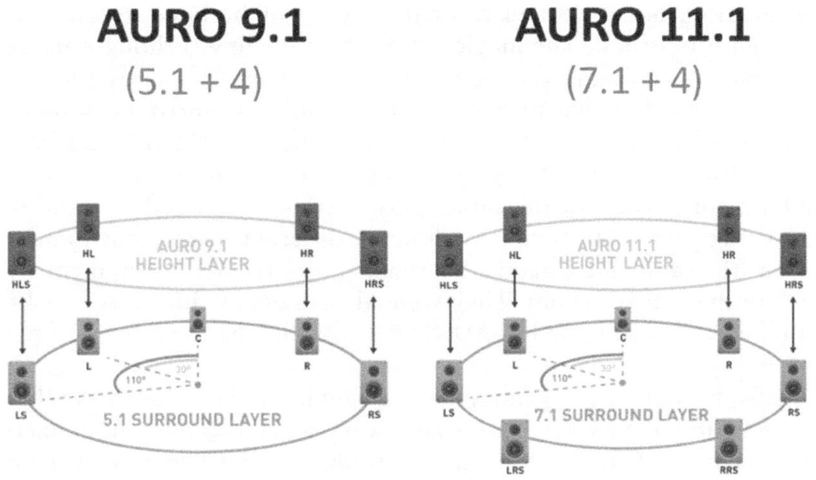


*Figure 3.6* AURO 9.1 and AURO 11.1 speaker layout

In fact, this is the moment that the Auro-3D format was born, during the summer of 2005 (see Figure 3.6).

But there was even more. The gap at the sides, present in surround sound, was suddenly not there anymore! There was no need to use the side surrounds to fill the gap between front and back surround loudspeakers. Another big surprise was a much larger sweet spot. The Auro-3D effect could be experienced even outside the inner circle of the loudspeakers as well. All these totally unexpected results in these experiments made me even more curious. How about using smaller speakers or different kinds of speakers? Perhaps we need another loudspeaker designed for the height layer. How about the results in a wet room versus one with dry acoustics? And what about adding more speakers to create a complete hemisphere? Would this improve the illusion of a natural sound reproduction even more? How about different microphone techniques that might improve the recording and reproduction of a more natural sound field. I had hundreds of questions and wanted to understand how I could create the next step in the history of sound in the most efficient way. I started to be interested in neuroscience and the speed of sound going to the cortex, which explained to me why human beings experience more spatial information when using sampling frequencies above 100 kHz.

## Higher emotional impact

I wanted to understand why the emotional impact is so much higher adding the third dimension in sound. It took me some years before I got answers to that question. A good example of that is tests I did with orchestral percussion instruments. Standing in front of a large gong being hit, you get goosebumps because the gong resonates with your body. If you use the best available microphones and a pair of top-end stereo loudspeakers to reproduce that sound, your brain recognizes the gong as a good reproduction, but it doesn't resonate with your body as if you were standing next to it when it is struck. The magic just isn't happening. I thought in the past that surround sound was going to reproduce that same magic resonating feeling as if you were standing in front of that instrument, but it did not. Surround sound made only a small improvement in the illusion of a real sound, but the magical impact – as if you were standing in front of the gong was not happening. And now, suddenly with the AURO 9.1 format, I had the same goose bumps compared to standing near the instrument being played. I tested it with many different instruments, and, each time, the result was the same, pure magic. Musicians, too, reacted spontaneously when listening back in the control room. They were always deeply impressed by how incredibly natural everything sounded in AURO 9.1 just like the experience of playing their instrument live.

Additionally, by adding the quadraphonic height layer above the ear-level 5.1 layer for the AURO 9.1 format, I experienced a noticeably more relaxing and immersive sensation when listening to Auro-3D compared to traditional 5.1 Surround. Curious whether this effect could be objectively measured, I invited professors and doctors from Austria to Galaxy Studios in 2008. They were conducting research on brain activity, and together we ran tests comparing Auro-3D with stereo and surround sound. Remarkably, the differences I had sensed subconsciously were clearly visible on their monitoring test equipment.

It was fascinating to see how this experiment required a completely different approach compared to typical audio tests, where we usually switch quickly between two examples due to the brain's short-term auditory memory, just a few seconds. In this case, however, we needed several minutes of continuous listening to measure the results. But once we did, it became clear: the brain showed signs of increased relaxation when listening in Auro-3D.

All of this made it clear to me that the addition of the third dimension in sound and its impact on human beings were totally underestimated. All those experiments were a true learning experience for me, and I realized that hardly anyone in the audio industry was aware of these fascinating findings. So I began sharing the results in workshops I presented at each AES convention in the USA and Europe from 2006 until the launch of the Auro-3D format in 2010. These sessions became real eye-openers for many audio engineers working in surround sound and exploring next-generation audio technologies. Naturally, there was also some skepticism, mainly because many couldn't yet envision how such a format could be successfully brought to market. There was indeed a missing link.

But the question I always got at the end of my workshops was how to bring that format to the market knowing that all music consoles were limited to 7.1 surround monitoring, with no distribution format for consumers above eight channels of audio like Blu-ray?' And what about the HDMI format that was used to connect all audiovisual devices at home? HDMI was also limited to eight channels of uncompressed audio. Moreover, there was not even an audio codec beyond 7.1 surround sound at the time. So I realized quickly the challenge bringing a high-quality three-dimensional audio format to market would require many new developments in order to produce, distribute, and reproduce more than eight channels of uncompressed audio. The more I searched for solutions, the greater the challenges became and started to feel as if I was on 'a mission impossible'.

## Development of the Voice-of-God channel – the Auro-3D third layer

Although I created Auro-3D for music production using only two layers (ear-level surround and height layer around 30° elevation), I realized that the audio formats which made it into all audio markets were established first in the cinema industry. But what about adding a height layer in a movie theater? As a dubbing engineer on more than 20 movies,[2] I was very familiar with film-sound in general and the compatibility issues between small and large places, which is partially solved with the X-curve. So, I was wondering how I could improve the movie sound by adding height channels. Soon I realized that in as a large space as a cinema theater, an extra overhead layer was needed to get the natural experience of fly-over effects. That's the moment I added the top-layer and called it the 'Voice-of-God' channel. This term had already been used by Imax, but Imax's speaker was located behind and on top of the screen while I found that the Voice-of-God should be experienced right above your head (see Figure 3.7).

There again, I was confronted with another new experience related to our hearing system. I was expecting a huge difference, and much more natural overall sound, by adding this Voice-of-God speaker right above my head which I first checked in the control room for music production. I was shocked that the difference was not big at all. I remember the first time I recorded the pipe organ (on which I gave many concerts) in the church and expanded my existing AURO 9.0 recording setup by adding a single microphone directed to the ceiling to record the reflections coming from above. When listening to that result in the studio where I put a top speaker directly above my head, I thought something must have been wrong with the calibration of my mics because I didn't hear the impact of that Voice-of-God channel in the way I imagined. I checked all calibrations and made new recordings, but the result was the same. That's the moment I learned that human beings are less sensitive to sounds coming from directly above us. All of this sounds pretty commonplace now, but at that time, it was groundbreaking at least for me and for most engineers I met except for Kimio Hamasaki who had similar experiences when he developed his 22.2 system, which also includes a top center speaker. Moreover, there are less source sound originating above our head, except for fly-overs and some special effects.

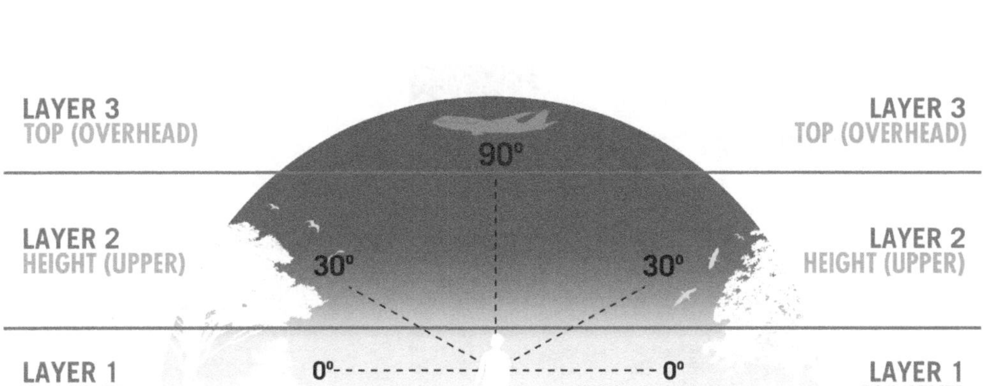


*Figure 3.7* Auro-coherence with 3 layers.

## Creating improved screen sound for cinema and home

I conducted many experiments with the height layer above the standard screen channels and was impressed how much more transparency and depth I could create having those height channels added behind the screen in a cinema theater or above a TV at home. It is that vertical quadraphonic field around the TV that does the magic. It became clear to me that defining the specifications for the Auro-3D loudspeaker layouts were the key to getting the most efficient and effective illusion of natural sound reproduction while maintaining full backward compatibility with the existing 5.1 main surround standard.

But in a cinema theater, I experienced a gap above the center loudspeaker and tried to fix that by moving the left height and right height speakers a bit more to the center loudspeaker and locate them above the left-center and the right-center loudspeaker. But then, I experienced immediately that the left and right upper corner speakers behind the screen were crucial to get natural motions of sounds from the upper left to the upper right corner of the screen, just as we experience with the standard left-center-right screen channels. So making an elevated copy of the three main screen channels became the best solution, while maintaining full backward compatibility with home installations. That's the way I came to the AURO 11.1 for Digital Cinema using three layers of sound; 5.1 surround main standard + 5.0 Height layer (including a Center height speaker beyond the screen) and the Voice-of-God channel for which I used two rows overhead to create a large sweet spot because at almost every seat, the audience has the same overhead experience.

The addition of that height center channel taught us something new: I could experience moving sounds from the left speaker to the height center speaker and down to the right speakers. Similarly for sounds starting from the upper left or right channels and moving to the center speaker. Having six screen loudspeakers divided over two layers (left, center, right, height-left, height-center, height-right) allowed for so much more triangular relationships between all those speakers behind the screen. This led to creating more depth, matching the visual information on the screen much better, and delivering much more transparency in the overall sound (see Figure 3.8).

Dubbing mixer Marti Humphrey, who first installed the Auro-3D system in the Hollywood area (The Dubstage at Burbank), stated publicly in interviews and at workshops in conventions how the Auro-3D screen sound is enhancing the intelligibility of the

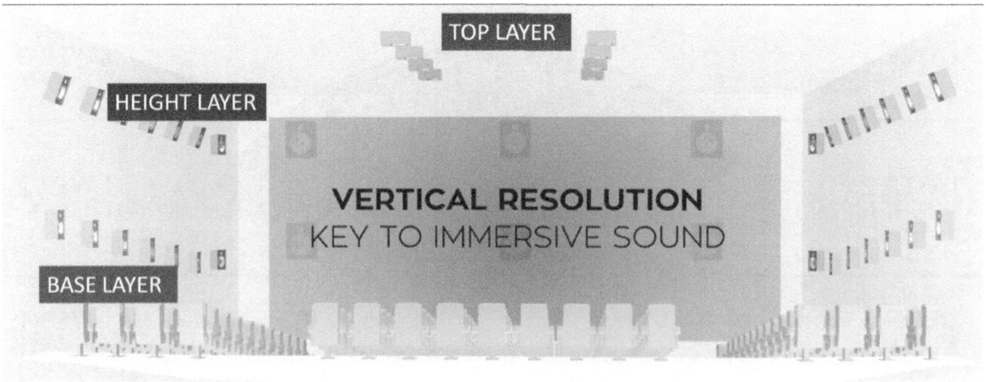


*Figure 3.8* Heightened cinema screen sound

dialogue, which is, of course, the backbone of a movie. Having the ability to spread hundreds of sounds over those six screen channels in two vertical layers is especially helpful in action scenes in which hundreds of sounds are coming together, creating much more clarity, detail, and depth. He explained that placing the music in the height layer created a more emotionally impactful experience. While the volume remained consistent with the 5.1 surround down-mix, the increased spatial clarity through the Auro-3D screen channels enhanced the perceived intensity of the music. The ability to spread the large amount of sound over more speakers in the vertical axis on screen results in less masking and less comb filtering, delivering a much higher transparency and making it easier for our brain to analyze the often complex sound field coming from the screen channels.

Knowing that about 80% of the movie sound is coming from the screen channels, it was very important to me that Auro-3D was able to deliver an optimized screen sound that was never achieved by other established immersive audio formats in digital cinema. Moreover, the screen sound of the AURO 11.1 system used in digital cinema has a natural compatibility with the 5.1 screen channels and is fully compatible with Auro-3D home systems. This is important because it allows the translation of the artistic intent everywhere – which has always been my goal. Also this speaker layout for digital cinema became integrated in the SMPTE 2098 cinema standard for Immersive Audio.

A very astonishing test was the following; we did blind tests with an audience comparing 2K and 4K projectors[3] and discovered that human beings experience more visual depth in the picture when Auro-3D was used versus 5.1 surround sound. We did a comparison with a 2K picture using Auro-3D versus a 4K picture with 5.1 surround sound. Everyone chose blindly for the version with 2K and Auro-3D, and when we asked why they had chosen that, many of them said that they experienced more depth in the picture which was of course not true, but that is how our brain creates more visual depth with the Auro-3D screen sound.

## Speaker layouts and patents

Naturally, I wanted to protect the intellectual property of all my speaker layout ideas by filing patents. The bad news was that I learned that a speaker layout could not be protected as an isolated invention. As such, I decided to share my experiences openly with the audio industry and prepared a workshop at the 120th AES Convention in Paris (May 2006). I then got a call in January 2006 from Kimio Hamasaki who invited me to the workshop on 'Surround Sound + Height' that he was chairing. I initially said that I want to do this workshop by myself, but he explained that he was working on it as well and recently presented a 22.2 multi-channel sound system with ultra-high vision (8K video) at World Expo 2005 held in Aichi, Japan, to which I asked whether the system has 24 discrete channels and a setup of 24 loudspeakers. Which he confirmed and was hoping that it would become an international standard for next generation ultra-high vision TV system. When I asked Kimio Hamasaki how he was going to distribute the 22.2 format to consumer markets, he replied that there was no solution yet, but they were working on some ideas with the NHK team.

I thought if AURO 9.1 already sounds immersive, what about 22.2? When I experienced it for the first time, I have to admit that it sounded impressive, although I was expecting a much bigger difference. If you see the evolution of sound after stereo was introduced at home, it took about one decade to add just one more channel. So I thought that I would be with my AURO 9.1 system at least one decade too early knowing that 7.1 surround wasn't even introduced in the consumer markets until 2012.

Kimio Hamasaki had a different approach. He started testing loudspeaker configurations in a research lab, where hundreds of speakers were installed in a dome around the listener. I had a chance to visit his place. It is really amazing. They used this setup to understand how to reduce the number of speakers till you start to experience gaps in the hemispherical sound field and that's the way they came to this 22.2 system.

I was always very concerned about backward compatibility with the main stereo and main surround standards. Therefore, I created the Auro-3D speaker layouts from this 5.1 main standard to ensure full backward compatibility to play back all stereo and surround content in the world as intended by the creators, which wasn't possible in the 22.2 system. But speaker layouts were only one part of the format. What about recording techniques, and what about technical solutions to enable the distribution and the reproduction of it in cinema, homes, cars, etc. . . . ?

Since object-based technology had already been discussed in broadcast for more than a decade, I was looking into the option to integrate object-based audio. However, I saw many problems integrating this technology in the workflow and reproduction. The technology offers interesting applications for broadcast, like multiple languages or volume control of the dialogue. This might be a solution for broadcast, where dialogues are typically mono tracks. But this is not the case in movies where the dialogue is mostly a multi-channel stem of at least 5.0, and this would become double when adding a Height layer that can represent the natural 3D reflections recorded at the film set. My vision had two phases: the first was to develop a 3D audio format that could immerse audiences in sound from all directions using traditional channel-based technology. The second phase would incorporate object-based technology, primarily for broadcast applications and cinema theaters. This advanced approach would leverage the extensive speaker systems in theaters to achieve better spatial resolution. With that intention, in 2011, I designed the AuroMax system, which has AURO 13.1 beds (the channel-based part with 7.1 + 5H + VOG) expanded with zones up to 27 discrete channels (26.1). This configuration makes sense for cinemas where at least 27 discrete channels could be installed. I never envisioned using AuroMax for the music market, in which I don't see the need for all those channels and don't see the need to use object-based technology for global music production. AuroMax became integrated in the SMPTE 2098 standard for immersive sound systems for digital cinema (see Figure 3.9).

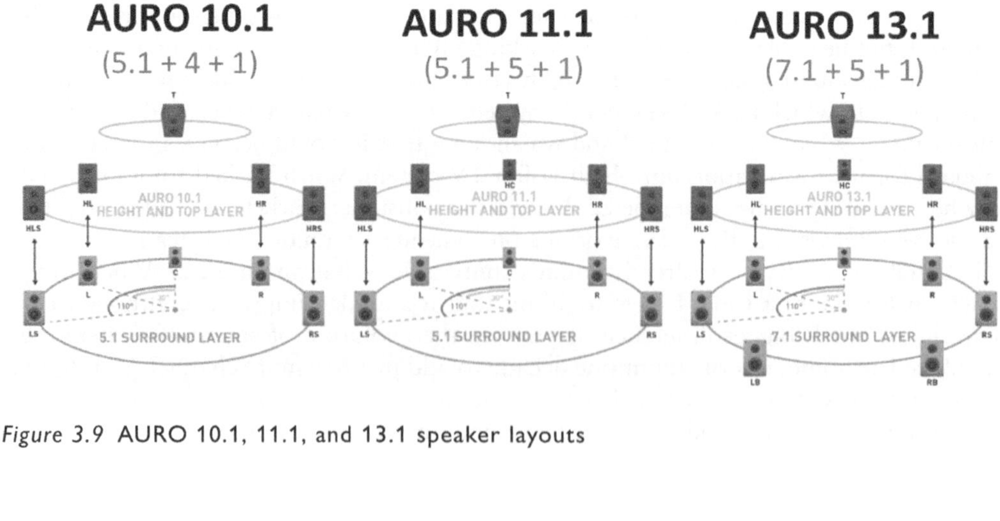


*Figure 3.9* AURO 10.1, 11.1, and 13.1 speaker layouts

Some people think that more loudspeakers result in a more immersive and more natural sound. The opposite is true, except for dedicated designs that don't require any compatibility with other formats. A larger number of loudspeakers can create issues with comb filtering and resulting in a smaller sweet spot. Research has shown that AuroMax is delivering a larger sweet spot than 5.1 or 7.1 surround sound versus Dolby Atmos. A smaller sweet spot means fewer people can experience the movie in the way it was intended by the creators.

If you are directing a channel-based surround mix in a theater and distributing the content over an array of loudspeakers, it may result in more comb filtering and increased phasing issues as compared to the traditional channel-based speaker arrays, which are designed to maximize the sweet spot area.

## Auro-Codec

The main question was how to bring the Auro-3D format with 10–14 channels to all those audio markets while everything in the chain from production up till reproduction was limited to eight channels uncompressed audio (PCM). I also didn't see any chance to start discussions with major companies to develop a new Blu-ray format that could allow up to 16 PCM channels or the hundred companies that worked on the HDMI standard for many years. So this was a huge hurdle to solve.

I was wondering if there wasn't a solution to hide information as metadata for a decoder in a PCM carrier. So I started to study the PCM bitstream and discovered that the least significant bits of a 24-bit signal were in fact not needed during playback. Most digital-to-analog convertors (DACs) in the market use 20-bit DACs, and amplifiers typically have a signal-to-noise ratio typically around 100 dB and very rarely better than 120 dB, which is about equal to the dynamic range of a 20-bit signal.[4] This means you would hear the noise of the amplifiers first and cannot detect a white noise produced by the least significant 4 bits (see Figure 3.10).

During the recording and mixing process, it is important to have 24 bits for headroom and rounding errors, but the moment the final mix is made, the master is only related to the dynamic range during playback. Considering that sound needs to be heard above

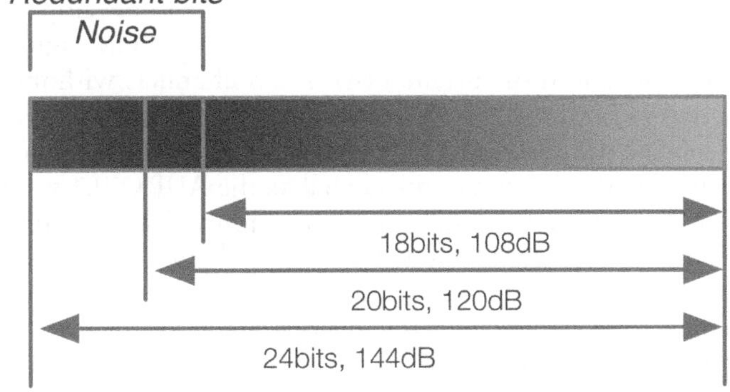


*Figure 3.10* Redundant bits in final 24-bit master for playback

the background noise of a room, and a very quiet room is about 25 dB. That would mean that we need more than 140 dB SPL to experience the full dynamic range in a silent room of a 20-bit audio track. Knowing that the best amplifier ever built has a signal-to-noise ratio (SNR) from 130 dB (CAT SL1 Legend Amplifier designed by Morris Kessler). Achieving an SNR above 120 dB is uncommon due to the complex circuitryand multifunctional design of amplifiers. High-end AV receivers typically exhibit SNR values ranging from approximately 100 dB to 110 dB which means that there is plenty of room in the least significant bits of a 24-bit playback format.

The moment that I realized that there are up to 4 bits in the signal that can't be reproduced by the best high-end home equipment, I started to think about the use of those non-audible four least significant bits to mix two PCM channels into one PCM channel and 'hide' metadata and mathematical solutions that a decoder will need to split the channels again so that the decoded signal has no audible differences with the original sound of each channel. This should be done without any audible artifacts and without the use of any psychoacoustic optimization. I wanted a pure mathematical solution (like lossless codecs) to ensure pure PCM audio quality in order to ensure a very fast decoding process within just a few samples, not a lossy codec such as MP3, AC-3, AAC, etc., which can result in audible artifacts and require one or two video frames to decode.

My goal was clear: I wanted to bring metadata in those inaudible least significant bits of a 24-bit signal in the encoding process while mixing multiple PCM channels in fewer PCM channels. The decoder should be able to reveal the original audio channels without any audible artifacts from this encoded PCM carrier. This means that two or three PCM channels of 24-bit can be down-mixed, not just matrixing but also changing volumes by the down-mix, into one PCM channel using an encoder during the down-mixing process. The bandwidth used for the encoded PCM channel would be the same as one of the PCM channels, which means by mixing three PCM channels into one PCM channel, the total bandwidth used was only one-third. I was wondering if this would be possible without any impact on the sound quality of that PCM carrier signal[5] or the decoded channels. The decoder needs to recognize the metadata in the PCM encoded carrier and is able to reveal from that one encoded PCM channel the three original PCM channels without any audible difference compared to the original three channels. Everyone told me that something like that would never be possible. Staying in the PCM digital domain while encoding and decoding a signal without any audible impact would be magic. But since I didn't see any other solution to bring Auro-3D to the market, I was keen to try this out and started to invest in it. I started a journey with Guido Van den Berghe to develop such a codec. After two years of code writing and many tests, we achieved two PCM signals mixed into one PCM channel and reproduced the original two PCM channels without any audible loss. We could do this in 48 kHz as well in 96 kHz due to the natural compatibility with PCM. So this allowed to bring at least AURO 9.1 into a 5.1 PCM surround carrier having two final masters in one file: the 5.1 surround as well as the AURO 9.1 version.

I saw in the past various articles in which I saw journalists explaining that Auro-3D is using 4 bits of the 24-bit signal to hide the height channels into the 5.1 PCM carrier. This is totally incorrect. The concept is that the 24-bit signals are dithered to a 20-bit signal to free up the 4 least significant bits. This process has no audible impact when done correctly. The Auro-decoder uses 4 bits from each sample of the Auro-encoded tracks as metadata. This metadata allows the decoder to extract the original 20-bit channels from the mixed channels without any audible quality loss. This groundbreaking result allowed mixing two vertical layers of sound into one 5.1 PCM layer. So this was already a perfect solution for almost every audio market in which only two layers were needed with AURO 9.1 or AURO 11.1 (7.1 surround + 4 height). But what about the three-layered system that I needed in cinema theaters or at high-end home theaters? I hoped it was also possible, without any compromise in the sound quality, to bring three PCM channels into one PCM channel. That would allow me to bring the three-layered system of Auro-3D that I envisioned to use for digital cinema (5.1 surround + five height channels + Voice-of-God overhead) into a single inventory master. This would mean that only one 5.1 PCM surround mix is the carrier out of the 5.1 surround mix and the 3-layered AURO 11.1 mix. About one year later (2009), that goal was achieved, and we started to create a plug-in suite to be used in Pro Tools so that engineers could create those encodings themselves.

There is another groundbreaking aspect to the Auro-3D codec. The decoder technically needs only one sample per channel to decode. For example, decoding two channels of audio would need two samples, resulting in a latency of only two samples. For three channels, the latency would be only three samples. This is one of the fastest digital decoders ever built.

To prove that the audible quality was the same as the original, I did a live demo for more than 100 audio engineers attending the AES Spatial Audio Conference in Tokyo in 2010 as part of the launch of Auro-3D. I had two mono tracks in a ProTools session and intentionally used two different songs to ensure that the codec would not use any kind of signals related to both channels. I did a live down-mix to a mono-track using the faders to dynamically change the volume of each track while creating the Auro-3D encoded mono PCM carrier track. This Auro-encoded mono PCM track was played back. Everyone witnessed the perfect PCM uncompressed audio quality and the dynamic movements I did, including one song that was almost completely inaudible at one moment. I did this to show the creative possibilities of this groundbreaking technology. Then I took the Auro-3D decoder and decoded in real-time the mono PCM track into the original two tracks with 100% channel split. I also zoomed in, and everyone could see that the latency between the encoded mono track and the two decoded tracks was only two samples at 48 kHz[6] (this would, of course, be the same amount in any other sample frequency up to 384 kHz). I synchronized the original two tracks with the two decoded tracks and reversed the phase of the decoded track to sum them with the original tracks. Then I recorded that result. Everyone could see that the delta was just a flat line in the fully zoomed-in ProTools session, meaning that there was no audible loss and no crosstalk occurred between the channels, both in the encoded as well as in the decoded tracks. A profound silence fell over the audience as I played back the delta signal. You could have heard a pin drop. The demonstration left everyone astonished, unable to comprehend how such a transformation was possible. What they witnessed seemed to defy their understanding of audio technology. A few years later in Japan, I received a special award for my contribution to the evolution in the history of sound recording and reproduction.

## AuroMatic up-mixing algorithm

While we started working in 2006 on the Auro-codec, I was wondering how I could bring such a system into people's homes without having the needed content. I've seen many new formats struggle in their early stages due to a lack of available content. This initial content gap has been a common challenge when introducing new technologies. I could imagine myself going to the film studios and then asking how many millions of devices would already be installed in the consumer market before they would invest in the delivery of Blu-rays with Auro-3D content. Similarly for the hardware manufacturer, would they integrate the Auro-3D decoder if there are not at least 1,000 movies available in the Auro-3D format? This classic *chicken or the egg* situation is very typical when introducing new audio or video formats. I wondered how I could solve that problem?

I imagined that it must be possible to create an algorithm that is able to render in real time a stereo or 5.1 surround mix into an Auro-3D experience. I had many ideas how such a thing could work, but I found it very important that the up-mixed result respect the original artistic intent. I often used up-mixing plug-ins to create a 5.1 surround version from a stereo mix, but the available solutions were changing the original mix too much. Stereo mixes that were pretty dry with no reverbs became much too wet because too much reverb was added in the up-mixed 5.1 surround experience. I didn't want that; I wanted a solution that sounds very natural while respecting the original intent of the creators.

I was wondering who had the knowledge to help me with such a challenging task. During the first quarter of 2006, I came in contact with Ralph Kessler who was the R&D engineer and project leader of the famous Quantec reverbs of which I was such a great fan, especially for their early reflections which sounded so natural. When I described my ideas, Ralph answered that it is already a big challenge to create an acceptable 5.1 surround version from stereo using up-mixing technology, and adding another four channels, while maintaining the intention of the original mix would make it even more challenging. But Ralph was incredibly ingenious and interested in this work. A few years later, we had excellent results as long as we kept it in the same music genre. But I wanted to have one algorithm that can be used not only for film but also in each kind of music genre and without needing to change the settings. My vision was that it should be fully 'automatic' – this word was the inspiration for calling the algorithm 'AuroMatic'. We continued to develop the algorithm to handle all kinds of content while maintaining the same end goal: an overall very natural immersive Auro-3D sound experience from each source (mono, stereo, surround) with respect to the original intent. My goal was to have this algorithm together with the Auro-3D codec in one 'Auro-3D Engine'. This engine would start within a few milliseconds of the AuroMatic up-mix algorithm if the original input source does not have the metadata of an Auro-encoded master. We worked on the AuroMatic developments for about seven years and both became the patent holders. AuroMatic became integrated as a separate plug-in in the Auro-3D Tool Suite (as will be discussed in this chapter).

## Mixing consoles for immersive audio

Having a codec to bring Auro-3D content to all audio markets alongside an up-mixing algorithm wasn't enough – it was crucial to cover all elements for an end-to-end workflow. This meant needing a large-format mixing console with monitoring systems supporting more than eight channels (ideally up to 16 channels) – not just digital mixing consoles for music and film but also analog consoles, which remain very popular in top-quality music production. At the time, everything was limited to eight-channel monitoring output without 3D panning or routing to vertical layers. So I created a console design that incorporated everything I envisioned for mixing Auro-3D, surround, and stereo simultaneously, providing complete control over panning and volume in each audio format. I created that design in 2007. It was a document of about 40 pages with all kinds of ideas and solutions. I went again to AMS-Neve asking if they could develop such a console for Galaxy Studios. The MD Marc Crabtree and his technical team were very intrigued about my vision and the potential solutions I proposed. He replied with two questions: 'Do you really believe that 3D audio is the future of sound for film and the music industry? And if we decide with AMS-Neve to create such a console, can you bring us at least ten clients?' Although I was not able to give Marc Crabtree the guarantees he asked for to start such an expensive development program, I told them that the next big step in the history of sound will be the addition of the missing dimension: height. However, it will not happen without having the right tools. I convinced them that AMS-Neve can be a part of history by developing such a large format digital console that can do at least 250 channels in 384 kHz (which is equal to 2,000 channels at 48 kHz). I told them that the solutions presented would not only be for the music industry. The same concept can be used for upgrading their DFC (Digital Film Console), which was already the leading digital console for film post-production in Hollywood major film studios as well in other leading audio post facilities worldwide. They decided to go for it and started with the development of the 88D music console which was based on my initial requirements. The first ever 88D console was installed at Galaxy Studios in 2010 (see Figure 3.11) and became a huge success. That console finally allowed me to do exactly what I had always dreamed of: creating stereo, surround, and Auro-3D mixes simultaneously, in the same amount of time it would normally take to do just a stereo mix, and without making any compromises in the quality of any of them.

One year later, AMS-Neve installed the first ever DFC-3D console for film at Galaxy Studios on which we mixed *Red Tails* from Lucasfilm, the first Hollywood movie in immersive audio to utilize Auro-3D 11.1 surround sound. (See Figure 3.12.) It was mixed

*Figure 3.11* First ever 88D console installed at Galaxy Studios in 2010

*Figure 3.12* AMS-Neve: First ever DFC-3D console at Galaxy Studios

with the team from Skywalker Sound at Galaxy Studios. This was announced by George Lucas (Lucasfilm) in November 2011 and released in January 2012 – historical moment in the evolution of film sound.

That same year, 2011, Skywalker Sound installed the Auro-3D system in their Studio A and upgraded their DFC Consoles to DFC-3D consoles. Within one year, almost every Hollywood major studio installed at least one Auro-3D system in their mixing stages. This was followed by the China Film Group and other leading audio post facilities in India, Japan, Korea, Russia, etc. In less than two years, with Barco as our partner, the Auro-3D format became established all over the planet.

The CEO of Dreamworks Animation was convinced after hearing the Auro-3D demo. He asked the head of audio at Skywalker Sound, to mix their upcoming film *Rise of the Guardians* in Auro-3D. He expressed concern about time constraints, believing that getting familiar with any new format would take additional time, and he couldn't imagine completing both a quality 5.1 surround mix and an Auro-3D mix within the time frame allocated just for the 5.1 mix alone. The head of post-production at Dreamworks Animation assured him that following the Auro-3D workflow wouldn't require extra time. The project was accepted, and the process went flawlessly – we actually finished both the 5.1 surround and Auro-3D mixes ahead of schedule. For many audio engineers, this proved that creating an Auro-3D mix simultaneously with traditional formats required no additional time.

## Creative Tool Suite

Digital audio workstations (DAWs) became very popular in music and film productions, but in 2009, most were limited to an eight-channel master output. We began developing a Creative Tool Suite that enabled routing up to AURO 13.1 (7.1 + 5 height channels + VOG), 3D panning, and mixing in the Auro-3D format on existing DAW stations – even though these were limited to 7.1 surround. The AuroMatic plug-in was also included in the Tool Suite, opening new creative possibilities. Importantly, this Auro-3D Tool Suite required no additional hardware. I presented the first public version at the 2010 AES Conference in Tokyo as part of the Auro-3D format launch.

Further developments of this Creative Tool Suite plug-in became one of the tasks of Bert Van Daele, my first employee when Auro Technologies was founded in 2010. The Creative Tool Suite facilitated the creation of content for film and music simultaneously in Auro-3D and a surround mix for film or the stereo mix for music. Suddenly, you could work easily with 12–14 discrete channels within Pro Tools, which was by design limited those days up to eight channels. I was thrilled to see that we had found so many solutions that made it possible to create Auro-3D films and music, something that had seemed impossible just a few years earlier.

## Microphone rigs

In the studio, we could easily position height surround microphones on stands, and we expanded our Decca-Tree rig to allow flexible vertical positioning of the front height microphones. For outdoor recordings, however, I wanted a microphone rig that was both flexible and mobile for film set usage. I began experimenting in 2006 with outdoor recordings on an Amsterdam film set, followed by various nature recordings.

Many engineers were intrigued by the Auro-3D experience and developed their own solutions based on my initial microphone rigs. The Italian engineer Paulo Carrer created an interesting compact rig that inspired manufacturers like Schoeps in developing their ORTF-3D microphone solutions. Florian Camerer designed his own 'Double UFIX' rig for mountain recordings and improved the microphone positioning. (See Figures 3.13

*Figure 3.13* Double UFIX by Florian Camerer (frame)

*Figure 3.14* Double UFIX by Florian Camerer (in the forest)

and 3.14.) His nature recordings are, to me, the most coherent natural sound recordings I've ever experienced with just nine audio channels using the AURO 9.0 speaker layout. Michael Williams, renowned for his amazing contributions in microphone techniques for stereo and surround, thoroughly explored various rig designs for Auro-3D recordings and beyond, as documented in great detail in his books alongside scientific research conducted at Galaxy Studios.

It was rewarding to see how many engineers and companies were inspired by the Auro-3D experience since its introduction at the 2006 AES Convention in Paris, each developing their own approaches and new products in response.

## Film sound and watermarking

Around the time that Barco decided to invest in Auro-3D in the year 2010, I learned that audio in digital cinema was 'watermarked' to prevent copying. By recording sound with a mobile phone in a cinema and putting the recorded file online, it's possible to determine in which theater it was recorded, including the exact time.

This raised questions about how the Auro-encoded 5.1 PCM surround carrier signal could survive such watermarking that disrupts the bitstream – another unexpected challenge when developing an audio format for all markets. As a solution, we integrated the Auro-3D decoder in the Doremi cinema server within the same chip generating the watermarking. This ensured all channels were decoded and watermarked exactly as required by the Digital Cinema Initiative (DCI) specs. Once again, we delivered a groundbreaking solution with just a simple software upgrade, requiring no other changes. We could offer an elegant solution because no extra equipment was needed in the B-chain (cinema sound system) to decode the Auro-encoded carrier, all while remaining fully compliant with DCI specs. The solution enabled a single inventory approach, meaning the same master could be used for both 5.1 surround and AURO 11.1 (3 layers: 5.1 + 5H + VOG). This efficiency reduced costs for the creation and worldwide distribution of movies. By example, the movie *Spiderman* from Sony Pictures was one of the first movies distributed worldwide with only the Auro-encoded 5.1 surround carrier, meaning theaters without Auro-3D installation experienced the 5.1 surround mix as intended, while Auro-3D theaters could experience the movie in Auro-3D exactly as intended by the creators.

Auro-3D proved to be a very straightforward immersive audio format for creating different language versions because it was based on familiar workflow processes. After examining other 3D systems, I became increasingly convinced that Auro-3D represented the best path forward as the most scalable system with superior sound quality to serve artistic intent universally. For this reason, I envisioned the importance of first reproducing 3D space with discrete channel-based technology, then adding object-based technology to enhance spatial resolution when sufficient speakers were available or to use it in broadcast for highly interactive features.

This led to the concept of the AuroMax system, which I developed in 2010. Lucasfilm announced in November 2011 the release of the movie *Red Tails* using the three-layered AURO 11.1 system introduced by Barco in March 2011 at Cinemacon in Las Vegas. This major international convention for professional cinema markets saw the first AURO 11.1 systems sold worldwide, with the first installation in Chengdu, China. Initially, Barco did not consider releasing AuroMax because it required additional hardware. The AuroMax system with its associated Barco renderer was eventually presented in April 2016 at Cinemacon in Las Vegas and later installed in various cinema theaters globally, delivering the Immersive Cinema Sound Systems within the SMPTE 2098 Immersive Audio cinema standard.

## AURIGA: an AV-receiver for home

It was always my vision to penetrate the home market once Auro-3D had been established in the professional cinema market. But there were no AV-receivers on the market that could reproduce more than eight channels in a single audio format. There were some receivers that had 12 output channels, but this was limited to 7.1 surround with two extra stereo outputs for two additional zones in a house. I was hoping that those two extra stereo outputs could be used to reproduce the Auro's quadraphonic height layer but that was technically not possible. A totally new design was needed. Unfortunately, no manufacturer was interested in developing this due to the lack of Auro-3D content for the home. We were not able to convince any manufacturer to develop an AV-receiver to support the Auro-3D format.

Together with my COO and CTO, we looked for a partnership with a manufacturer willing to develop a new AV receiver for us. We began development and, by late 2013, acquired that company, StormAudio. We launched at the CES Convention 2015 our

3D Audio AV-receiver for home use, which we named 'AURIGA'. It received numerous awards for its outstanding audio quality and for being the first ever AV receiver capable of reproducing immersive audio at home.

## Automotive

In order to reach all audio markets, the automotive market was a particularly important one because it is the place where many people have the time to experience music. We performed many tests in cars – first with Audi and later with Porsche. In 2016, Porsche integrated the Auro-3D experience in their Burmester premium audio car system for the new Panamera to be released early 2017. Porsche organized a worldwide four-day press event at Galaxy Studios where they invited more than 50 journalists from all over the world. The reception was extremely positive. Some journalists were describing a 'new driving experience' – which was stunning because it was only a new audio system, not a new engine or other mechanical innovation. This showed immediately the amazing impact of immersive audio in cars, followed by many others. A renowned Japanese journalist once said to me, 'This feels like a driving concert hall'. Due to its success, Porsche integrated the Auro-3D Burmester system in the Cayenne, Taycan, and Maccan. Also, the Chinese company Geely (owner of Volvo) integrated Auro-3D technology to enhance the sound in their cars. More than one million cars were sold with the Auro-3D experience on board (see Figure 3.15).

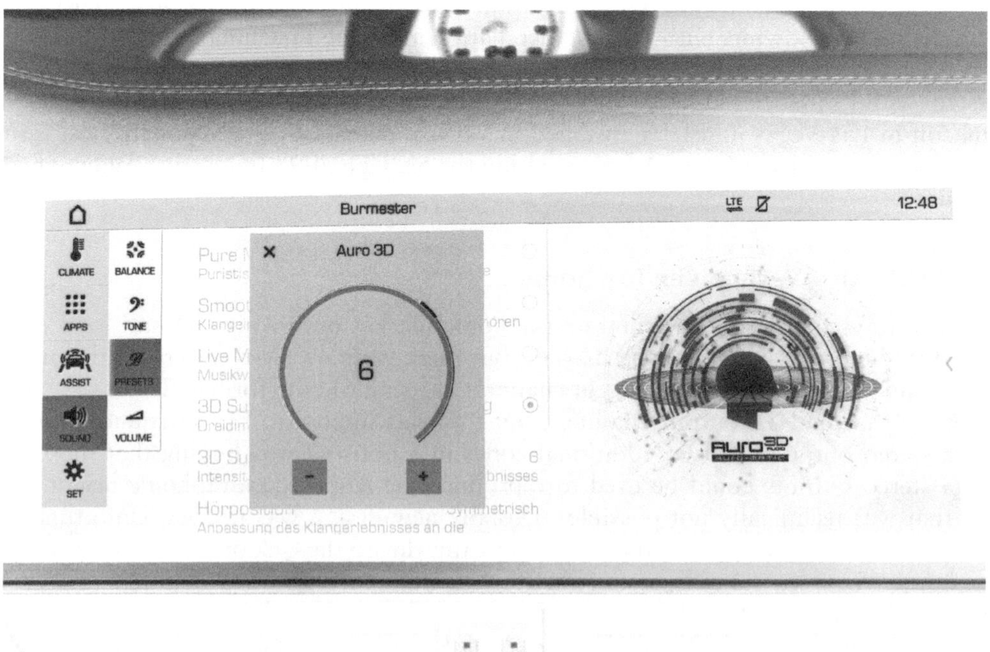


*Figure 3.15* Dashboard of Porsche

## Mobile and games with immersive sound

The company Tinno Mobile decided in 2015 to integrate the Auro-3D upmixing algorithm in their mobile phones. A few million phones were sold, but very soon after that moment, the company was acquired and stopped further integration. In the meantime, there was work on a new audio format being done that was released (called Vivid), which was based on the Auro-3D speaker layouts.

'Get Even' was released in 2016, with immersive audio, by the company Bandai Namco. This was at the start of many developments in the game engines to expand them with the third dimension. Our 3D panner was integrated in the Audio Kinetics gaming software; the company was acquired a few years later by Sony to integrate this technology in their game stations.

## The future

Immersive is certainly the present and the future – it's now a big step, but we will not go back again. There is too much traction by consumers and all major audio companies, including content creators in all kinds of audio markets. Finally, after 100 years of evolution, we have 3D Audio everywhere and delivering so much more joy in all kinds of content and entertainment. Very important is, of course, the educational part so that the Immersive Sound experience is not becoming a gimmick but rather is bringing to people the illusion of a natural sound experience. The same thing happened in the past with other new formats when mono became replaced by stereo, many productions had that kind of ping-pong stereo effect, while after some years, engineers started to discover much more creative possibilities and a more coherent sound field, creating more depth.

The same happened with surround sound in which everyone needed to find their way in using the center speaker. That led to many discussions, but it also opened many new creative possibilities. The same thing happened with immersive audio, with which many people in each audio industry needed to discover the creative possibilities and how they can use them to create a much higher emotional impact. Because finally, a higher emotional impact is what it is all about. Whether it's movies, music, games, or even sports, immersive audio creates a totally different level of emotional engagement. We did different tests with football and tennis competitions, and it is amazing how you feel much more of the tension of the game when experiencing these with immersive audio, as if you were there in the audience at the live event. Once you are used to this, you can't go back. This can be compared to HD video. If you could see what we were watching for so many years in standard resolution, it is hard to believe we did not complain. But there was no alternative.

With immersive audio, people are emotionally much more connected via the subconscious impact of the sound experience. One day, perhaps, it will be used to heal people – a domain that still needs much research. The good news is that immersive audio systems became very affordable and accessible to many consumers.

It is very clear to me that Immersive Sound is poised to stay here forever.

And now it is all about having more high-quality content, and a streaming platform that can bring this experience in the same quality to everyone and everywhere as was intended by the creators.

Looking back at the results of my devotion and passion over the past decades to bring Immersive Sound experiences to all audio markets, and seeing how it has inspired countless engineers across entertainment industries, brings me great satisfaction. Today, many professionals are involved and enjoying this technology – from content creators to millions of consumers worldwide who experience this heightened enjoyment daily. I feel fortunate to have executed this mission and to have witnessed how Immersive Sound has contributed to the evolution of entertainment industries. I want to express my gratitude to everyone who supported me throughout this remarkable journey.

## Notes

1 Darcy Proper was nominated more than ten times for a Grammy Award for best surround mastering. To date, she has won four Grammy Awards.
2 Including a movie I mixed in 2006 which received the Golden Lion of the Future award in the Venice Film Festival.
3 Galaxy Studios installed in 2006 one of the first prototypes of the Sony 4K digital projectors.
4 Each bit represents 6 dB of dynamic range.
5 The audible result of the PCM carrier needs to be exactly the same without using an encoder.
6 The latency is proportional to the number of channels. For two channels the latency is two samples, regardless of the sampling rate.

Chapter 4

# Immersive experience

## Height, acoustics and emotional impact

*Wieslaw Woszczyk*

As a kid, I remember listening to the grand piano we had at home. I would sit under the piano and listen while someone played. The sound was immersive; it filled the space between the floor and the piano, which was in a corner, creating a beautiful theater box of sound. My dad had a Philips mono tape recorder, and he requested that I record all of my piano recital pieces. In secondary school, I would perform my recital in front of the committee, and then I would have to come home and record it. That was done on a very heavy, old Philips tape machine.

When I started to listen to recorded sound in music, I was always interested and somewhat curious. Why is there this small painting-like representation of what the sound actually is? We experience sound in terms of dynamics, power, loudness, and extent. It becomes a bodily experience; you feel it in your body. Early on, listening to recordings on my modest system, it was like looking at a small miniature painting. I always thought I preferred paintings that are big, like Matejko or Rembrandt's *The Night Watch*. But of course, there are also artists like Van Gogh and Monet who have various techniques of representing nature or the image of a story. But I always thought there was an artistic aspect to this. It's about how we transform the vision of the world or a situation. It's also about how involving, or impressive, it is going to be for the senses and what kind of feeling you will get. Experiments in sound were always geared towards extending reality and making the images bigger. They were about being more involved in what the music or the sound originally provided. I always liked the hi-fi approach. They would have a big woofer, ribbon tweeters, and fully extended bandwidth. They would use double quad speakers and a stack of Quad ESL electrostatics, so you have a big surface pushing the air. Then there would be some ribbon tweeters and some Hartely 24-inch woofers that are super low in the audio spectrum.

### Stereo

The problem with stereo is that it only has two channels. A speaker system like this can fill the room with sound. The room then becomes an immersive chamber, which returns everything. You are then immersed.

In a good room, stereo can be enough. In fact, the room I am currently in is a loft. The loft has dimensions of a studio with tall ceilings where there is a small Bang & Olufsen Bluetooth speaker. It's a stereo system with three drivers. It sounds beautifully immersive. and, when you put it in the right place in the room, it sounds great everywhere. There's a feeling of sound presence when large speakers are emitting broad bandwidth over an

DOI: 10.4324/9780429297526-6

expansive surface area. This proximity creates a sense of presence and closeness to the sound. However, reflected sounds and the surrounding environment also play crucial roles, as they contribute energy from various angles that are incorporated into the image of sound.

## Surround sound experiences

Surround sound was present when I was studying in Warsaw. The Teldec (Telefunken-Decca Schallplatten) recording team came to do a recording in quadraphonic sound using a quadraphonic QM 69 microphone from Neumann. I must say, I wasn't that impressed by what I heard. It was still very preliminary. The balance between the presence of sound and the ambient envelopment or immersion was not satisfying. I thought maybe my expectations were too high. We saw them setting up, measuring everything at the Warsaw Chopin Academy. Then we came into the control room. The sound was thin and small, emanating from four speakers. The stage wasn't present. The room wasn't so present either. I felt no impact or engagement from the sound.

When we returned to the stereo, my impression of it needed to be expanded. Fortunately, when I worked in The Basement Recording Studio in New York, which was later the Big Apple recording studio, that studio had four overhead JBLs and a quad console. It had joysticks for mixing, KLH or Advents right in front to mix on, and there were also Auratones to check the balance.

When you really wanted to achieve an immersive sound for musicians, I would put their sounds into the overhead quad JBL speakers. It was like a party. You would put a lot of bass, bring the instruments away from the front into the center of the room, and bring everybody into the sound. You were in it. The room wasn't very big, so sound would saturate the control room, which was usually filled with people. Then we would make it very loud, immersive, and present. It had everything in it, close to the ears, present and personal

## Early digital

I didn't get further into immersive sound until surround sound came along. Although in my practice, I cultivated the idea of having the extended bandwidth, extending both top frequencies and bottom frequencies. When digital came, it was a disappointment because of the timbral coloration from the converters and the harshness of brick-wall filters within the audible band.

However, when Dr. Yamamoto from Pioneer brought the 96 kHz DAT recorder, which we could actually record at 96 kHz, he also brought some recordings from Japan. He brought two Pioneer speakers with ribbon tweeters on top that could reproduce 150 kHz. I suddenly realized when we had listening sessions with students at McGill University that we could hear the soundstage opening up and coming closer. This was unlike 48 kHz or 44.1 kHz, where the systems sounded harsh, suffering from steep anti-alias filtering and bad time domain response. But at 96 kHz, the sound started to open up. Early digital sound always started some distance away from you. It would never come forward. In analog, we didn't have this impression, and the sound would come forward. We made a lot of great recordings with the sound image starting from where you are, and extending all the way into the depth of the stereo image. But in the early days of digital, it was impossible to bring the sound closer. Then at 96 kHz, suddenly things came forward.

For me, the proximity of sound is an important aspect of auditory presence, of living in sound. If you can get close to it, it means you're really there. It's like standing in front of a musician, where everything is within reach. However, if the sound is behind a veil and only starts at some point away, it never gets wide enough. It always stays at a certain limited angle. To me, that's not immersive. That's not the future of sound.

### High-resolution

This realization drew my interest to high-resolution audio. I was always impressed by Super Audio CD and Direct Stream Digital (DSD) recordings, particularly those made with a very high sampling rate. In stereo, you can get the sense of sound coming close and also opening up to the sides and beyond the speakers. The auditory system just looks for all the tiny fluctuations, different modulations that happen independently in the right and left ears. If they're not there, the auditory system is not interested anymore. That tickling, from the excitement of sound, is gone. I could feel that there is that lightness, freshness, and airy transparency, openness that really lives in that sound with these formats.

### Center channel

Once surround sound came into play, there was a compromise due to the center channel. The center of the image, perhaps the most important thing, could become very small if everything is placed in a single speaker. There's a big difference between that and a virtual source, or a phantom source, being reproduced by multiple speakers in a proper relationship.

This is where wave field synthesis (WFS), for example, can create a much greater variety of wavefront shapes. Especially with the new possibilities, with the horizontal and vertical combination of WFS. However, in surround sound the center channel had to be used wisely. I thought that they had the right idea to basically keep most of the sound left and right. They used the center channel, and also the surround sound is there to include you in the story or the space of the screen.

We did some work with Dolby stereo-based matrixing techniques as well. We played with phase and polarity reversals and developed Dolby Surround encoding compatible microphone techniques. The goal was to have enough phase differences to channel that energy through the matrix decoder into the back speakers.

### Height

The real game changer began when we started to lift the speakers up. This was the revealing part because it turns out you can hear much more into the sound, a heightened sense of proximity. The sound comes to you, and you can hear much more inside it. If you broaden the front of the stage, the auditory system is extremely sensitive to additional information within the spectrum, temporal differences, and location of sound. By lifting the speakers in front and opening up the soundstage, we bring in something familiar to the ears. We always listen to the sound source integrated with early reflections that are around the source, which come from many different directions. We have that capability in the auditory system to integrate the reflected sound with the direct sound and create an image of the source in space.

After a while, within room reverberance, the recognition of delayed and incoherent sound comes. It's not just fused with the direct sound but is detached from the direct sound and reflected sound. This becomes something that paints the shape of the room and the kind of room you're in. There is a perceptual separation of what reflected sound belongs to a given source and what belongs to the room. This creates these two images, which our auditory system can track constantly and simultaneously because it's strategic. It's a kind of necessity to know, 'Where am I? Where are they?'

## Spatial sound and architecture

Reflecting on my impression of spatial sound, I've been sensitive to it since my experience under that piano as a child. My sensitivity was further heightened when I visited Hagia Sophia. This Byzantine church, built in Istanbul or Constantinople in 537 CE. It's a giant, magnificent structure commissioned by Emperor Constantine I. When the monks from Kyiv visited the church in the tenth century, they wrote, 'We knew not whether we were in heaven or on earth, for on earth, there is no such splendor or such beauty and we are at a loss how to describe it. We only know that God dwells there among men'.[1] The reverberation times in the church may be between 12 and 14 s. The space has a giant 31-m dome, two big semidomes, and six smaller semidomes. There is an all-marble balcony and very reflective surfaces.

You can imagine the spectacle of immersive sound that people experienced in the sixth century, where they could put the choir of castrati and singers at the top of the dome, amplifying the sound and bringing it like a shower down into the church, and then they had different places for the choir, the reader, and the cantor. You could bring 15,000 people, you could play with lit candles, and so on. Imagine the spectacle that people had experienced there.

## Sound exploration

In my practice, I recommend that everyone first explore, observe, and be sensitive to what they hear. Try to track that element of sound, expose it, get closer to it, understand it, and make it into something useful and strong enough to impress and be presented to others. The first recommendation for anyone working on sound is to keep your ears open. Listen and walk into an environment. You will get a sense of spaciousness, brightness, and layout, and anything that you could possibly reconstruct artificially is not the same. With that sensitivity and desire to explore the sound, you can go in and construct something. Experimentation is always part of finding the mystery of sound.

First, take control for creative augmentation, for creative exploration, and perhaps for exploitation of our auditory system. This is how to bring an artistic dimension into music and multisensory experiences. We are searching for that understanding of what really works. Every time, there should be a fresh approach. There's clearly a range of sceneries that one can create. The sceneries that are interesting and compelling always have some sort of a sound source and sound environment. There could be a break in between, meaning that you don't have to always continue. You can take that pause, like a musical pause, and create a new sound. Wait for the sound to die and another sound to be born. There's always a variety that drives the imagination.

Especially in music, it's a string of connected elements of information. When our perception decodes that string of information, we build a model of what is coming. We anticipate, understand, and have a feeling of what will be next, where it should be, whether this anticipation will be fulfilled or not. This is part of the beauty and the game of musical form or the musical interaction of players. Will they do it or not? Are they going to hit or not going to hit? What is the mystery that is coming ahead? We build an expectation based on our experience of what we like and don't like. We come with an attitude towards listening, as each individual has to a piece of music. Musicians, when they perform, when they create, when they discover in themselves what they can contribute and what will be interesting in the context of what everybody else is doing.

We help with sound recording and sound design to reveal as much as possible to the listener about the musician's intent and the construction of the piece of music. Where is the sense of what they are trying to communicate in terms of how to impart a feeling of that creation to the listener? Listeners are not just passive entities but active ones who seek out information. They focus on certain things and they want to hear certain things, like a dance. In a dance, you see a movement, and the movement ends. You can say, well, it ended. There's no more. It's finished. But then another movement comes from there and continues. In sound, it's also like that. Sound is created, then creates some sort of a reactive response, be it acoustical or an answer from another musician, but then it dies and then continues in some other way.

## Coherence and mystery

When designing an immersive sound experience, my first recommendation is to keep everything fresh and revealing. This means there has to be incredible spatial coherence and clarity in some ways, and a complete spatial mystery and blurring in others. This uncertainty, coupled with certainty, will constantly reveal and hide what is of interest to the listener. When listeners think that everything is always there as planned, they will stop listening. If things are fixed and not moving, they will become bored and uninterested. The mind will take over. You either stream and get completely caught up in what you're streaming, or you start thinking and become lost in your daily problems or other things you need to solve.

First, the compelling design of an immersive experience takes one completely into it by creating a focus on what is most important at the time. It considers the ongoing discussion, the forces at play, and how they are trying to have a discourse or perhaps some sort of struggle. Second, it always brings another piece of information to augment that view to make it refreshed.

## Focal points

We have the means of using sources to envelop or expand and enlarge the sources, creating a sense of depth. We decide what the focal point is now, what comes forward, what should be right next to you as a listener, and what is the most personal, private, and important message. We consider what are the sources that interact – are they assisting, are they enriching, are they not the primary focus of your observation as a listener? Then, ideally, we establish how the environment interacts with what's going on so that you, the listener, are included. If you hear the reaction of the environment, the room response, or the way the room changes shape or opens up because a new source comes in, and that source brings another room that works in tandem with the source, then you track that room and that source together. You can separate that image from another image. And you move through images, spaces, and scenes; you travel.

There's always a struggle, or, rather, there has to be a conscious plan of what to blend and what to fuse together, what to separate, and how to augment the impact, clarity, and the beauty of it.

## Observation and imagination

This process is a combination of observation and imagination. Observation stems from truly feeling what's happening. When I attend a concert or listen to live music, I always seek that ideal point of view, the ideal point of listening. I will search until I find a spot where I can sit down and say, 'Yes, I like this, I enjoy this view'. From this point, I see everything. I hear all the interactions between the players and the room, etc. I hear the right balance. You choose whether to sit close to the drum or closer to the piano. Ideally, the sound engineer chooses and observes where that point should be.

Then you add a bit of imagination to what you actually hear. What if something could be so magical that you don't hear the speakers? You don't hear the actual reproduction system, but you feel like you're in front of a band? What is there in the detail of this player's action or that instrument's character that makes it feel like it's in front of you? Do you just put two microphones where your head is, and is this supposed to capture everything that you hear? Well, it doesn't, because when you listen, your mind, as a recording engineer and the listener in that situation, can focus on a lot of things. Focusing will filter, boost, and change it, so you can hear it in a certain way.

Listeners of recorded music at home, for the most part, are blind. They rely only on what they hear. How do you bring them what you see, what you hear, and what you feel, and the impact of the music? Through the sound, through the design, through the capture. Where do you put that microphone? Where is the information in that situation that will reveal what is actually going on in the best way to the listener at home?

You may have to overemphasize because that listener doesn't have the multisensory advantage. You just realize, 'I'm listening, and maybe the speakers are not the best, and maybe this listener is away from the speakers'. It used to be that people sat between the left and right speaker and listened attentively, but it's not so anymore. People now walk around. So how do you reveal that complexity? Let's say you hear a certain response of the piano that is like the reflection of the floor and the expanding warmth of low frequencies. But it's subtle. It may be masked by all the other things if you just have two microphones and you don't take care of that particular attribute of sound separately and consciously.

## The engineer

The engineers and musicians are the hard workers here. Hopefully, the engineer helps them set up in a way that they perform comfortably, see each other, and hear everything for themselves without a struggle. It is actually a great conversation for them. It's like you have set a table with two cups of coffee, and people talk. They have a good conversation because they are comfortable.

Then the engineer comes in and explores once this great conversation is going on. They try to find the best viewpoint, and maybe the viewpoint changes depending on who is speaking, like cameras that switch from one face to another to show more expression, more dimension of conversation, the intensity of it, or the discord or the reactions. You have to observe all of it in a sonic context and sonic means.

## Environment

To me, sound does not exist without an environment. Most of the time, I would say that the source, to me, is less than 50% of what you hear. The environment is more of what you hear because we never hear sounds alone unless we're in an anechoic chamber or outdoors where it's not being supported. It just disappears too fast. But the room keeps it alive and allows us to have a greater chance to analyze it, to observe it, to enjoy it. It's almost like a dancer with a veil behind. The dancer moves and the veil follows. It continues in a way like the pattern of the dancer. It holds it for a bit longer, so we can notice and enjoy the beauty.

I always find that, once I have the source, a mold is found. I will do a lot of work trying to capture or to create the right environment so that the source has a place to live in, has a chance to shine or to sustain itself to live just a little longer in the light of the room. That means becoming more understood and more appreciated and maybe more separated from all this by the listener because the audition now has a better chance to be seen.

## Emotional impact

The listener absorbs these images slowly, not through a sensory bombardment, but in an systematic manner. We perceive music not as a sudden explosion but as a growing experience. It's the increasing awareness of the scene, the layout, the situation, and the relationships between the sources. The musical justifications for their harmony or discord, their consequential play, and the friction and resolution all become much clearer in time. The drama of music becomes more evident because you have the development in time, the layout, and the spatial follow-up from the room, which provides a wealth of information. You can fully comprehend this music to the point where it may move you to tears or make you leap for joy. It resonates with the emotional side of the listener, connecting with their experiences and their capacity to link music to the rest of their life.

The individual then feels understood, as if the music understands them and they understand the music, creating a connection. A significant part of this understanding is in the hands of the engineer. Their role is to clarify everything, to avoid building barriers between the performers, listeners, and the composer. They set the stage for this situation, allowing the musicians to interpret it and pour their heart, emotions, and experiences into it. Whether it's a sense of violence or softness, tranquility or rage, or any other emotion that justifies their performance, the musician can express it all. This connection and understanding can make the music resonate deeply with the listener.

The emotional impact from the composer, the performer, and the performance engages the room, prompting it to respond to the conversation. As an engineer, you capture all of this and infuse it with your own interpretation. There's an emphasis, a weight placed on certain elements and perhaps less on others, as you understand it. However, the fundamental goal of the conversation isn't altered, nor is it clarified to such an extent that the listener is physically affected because sound is physical. It does touch the skin and then resonates emotionally. I believe music isn't entirely rational or cerebral. It primarily exists on a surface of feelings, communicating its essence through these emotions.

Research exists, andThomas Lund has conducted an extensive investigation on this topic, linking music and reverberations with the sense of touch. This connection is often referred to as a 'gentle touch'. Thus music has the ability to touch you, to embrace you. Phil Ramone used to say that viola section tracks need to have a 'hugging' quality, they should envelop you. So there's this capacity for sound and music to touch you. Some people argue that reverberance brings pleasure, that it always feels more pleasant. And indeed it does.

There's nothing more beautiful than hearing the depth and perspective as well as intimacy, of something so close to your fellow singers, building it out, and then hearing the response somewhere. You feel heard, answered, and experience an incredible sensation. Apparently, the same areas of the brain are activated by this sound as they are by a gentle touch. When you touch somewhere, different senses, receptions, or feelings are generated by these areas, depending on how one is touched or touches someone else. So, yes, there's this incredible multisensory component in sound.

In music, because it has structure and we understand and follow it, we know where it's going. There is anticipation. The brain can focus on certain things to such an extent that in hearing and in all the other sensory peripheral devices, it can magnify the effect of a certain event in music. When the choir comes in, you're reaching out with your auditory, visual, and any other senses you know, tuning your attention for that particular channel of reception. We are very active in adapting as listeners, as humans.

A perceptual gain of 10 dB, 20 dB, 60 dB, or 100 dB in terms of extracting that feeling from music and sound is incredible, and it speaks to the economy of listening of the brain function. Now, apparently, human attention switches to maximum just to maximize that ability and your ability to transmit and focus on hearing its strokes. It puts peak attention to the left ear, then to the right ear, alternating six times per second, building an image, storing in memory, and comparing for the ultimate, not half, but a full analysis of the sound scene.

Of course, the brain interprets the scene even better because it has more information in its long-term memory that can put it into interpreting with the right feelings. The sound is spatially produced, spectrally diverse, spatially diverse, and dynamically diverse, and that sound is not masked. It is revealed in perception. Revealing is the unmasking of everything that it is trying to hide.

Our capacity, if we give it to the listeners with the right tools and the right appreciation of what they really need, is to uncover the emotion in sound all the way from the composer, from the player, to the first observer who is the recording engineer. The engineer is the one who has all those multisensory means of seeing it, being there, uncovering it, and then preserving it for everybody else to experience. That's what I love about this job and about teaching it. It's to teach young people to preserve the best of what they hear because they're the first ones, the first witnesses of that music being born into sound, and then providing it for others to enjoy and maximize that enjoyment.

## Vibro-sensory

We found the concept of vibro-sensory integration interesting. We had a vibration platform installed in our lab and were experimenting with adding a vibration sensation to low-frequency sound. Most of the time, it's challenging to use it coherently in a way that isn't disturbing because the vibration has to be just right. For instance, if you're recreating an earthquake, like the 7.0 earthquake demonstration in a San Francisco museum, it's different. But there was one moment where it worked perfectly for me in music.

During a recording session, I was sitting on a platform activated by four actuators, one in each corner. The platform was square, surrounded by six subwoofers, creating an immersive sound field with speakers in four layers and above. It was a surround recording of an organ from Redpath Hall at McGill University. Most of the time, you don't feel anything. But then comes the last note, the lowest one, and at that very moment, you feel the vibration. It's so justified that it feels as if the speakers have activated the platform, and the platform was not driven by something but was just a sympathetic vibration. It feels so natural and satisfying, so present, because it truly reflects the experience you would have if you were sitting on a bench in a church with a good organ. People are not usually open to these receptors, but if you put your hand on a pew during an organ concert, you feel the vibration all the time. It's just that we don't usually pay attention to it.

There were other experiments that demonstrated the difficulty of presenting vibro-sensory information. Achieving just the right subtlety of control was challenging. This technique is used in video games to create sensations, such as a bumpy ride or cornering. They have seats and other elements that use vibro-sensory information.

## Virtual acoustics

The area I'm most excited about now is recording with and in virtual acoustics. As we've discussed, indirect sound is a more important descriptor of sound than direct sound. Direct sound usually occupies a small space and comes from a specific direction, while indirect sound, which perceptually fuses with direct sound, can come from anywhere. The perceptual power of this sound is much greater, and we can incorporate a lot of indirect sound in music, which feels very satisfying.

To me, the best approach for future productions is to recreate this immersive sound in the recording studio with virtual acoustics and adjust it. We have two layers of recording engineering. The first is the virtual acoustics engineer, who creates the environment for the musician and player. This allows the musician to control the environment better, be aware of the environment, and play the room. Once a large immersive pedal on an instrument is satisfactorily used in this built environment, the recording engineer is brought in. They capture what's happening with high spatial resolution, transfer that experience to loudspeakers or headphones, and even augment it in the control room.

The transcript from the recording in the control room uses all the virtual acoustic components already in the room that the musicians are working with. It can also be further augmented in the control room during the remix. This process gives the recording engineer extra power to quantify and qualify the indirect sound, make it immersive, and directly influence the immersive quality of the players' and listeners' experience.

Of course, you can also work with a composer. The composer can imagine how certain things could sound in the studio, and you can incorporate it just as Berlioz would walk into his favorite church and compose for it. You can create a space for the composer to utilize the ambient forces in the best way.

To me, it's the greatest challenge because a great sound system is necessary in the studio room to produce believable rendered acoustics. The studio room, being a physical realm, interacts with what you're trying to create, so this must be acknowledged and worked with. However, all of this can be done without the need to travel and book rooms. You can even create rooms that don't exist, that couldn't be found. These could be constructions of science, freak accidents, or something interesting. Such environments will inspire both the performer and the composer.

This approach will immerse the listener in that prepared environment. They may not necessarily recognize it as synthetic, but it will provide them with a unique journey through the music. While such an experience might seem theoretically impossible, we can experiment with it on a more local basis rather than having to gather all these forces in a large church. It's easier to capture sound in a small room than in a large space because everything is contained in a smaller location and scale.

## Spatial awareness in sound

Of course, sound is the only medium that constantly provides for this spherical, all-around awareness. There's apparently even some data that tells us that auditory awareness is acting spatially while we sleep. It never disconnects, and it keeps us informed if something might be happening.

We're born with some aspects of spatial awareness. Eventually, kids will know how to distinguish the difference between correlated and decorrelated noise played through headphones. They can describe one as big and threatening, and the other one as nice, soft, and expanding. We have an amazing capacity as listeners. How do we couple that to making creative artwork that uses itself with other modalities as well? There are a lot of things we don't understand yet.

## Future of sound design

I'm very excited about bringing more of the beautiful experiences of sound alone, maybe with a touch of vibro-sensory but without the visual component. Vision creates a lot of demands, and then it requires coherence between sensors, which is difficult to spot. It can be done, but I'm thinking about the ability to create, in the future, the sense of a sound source that actually comes closer to you but sounds so real that you feel like you're almost in a wheelchair being moved around and without feeling any vibration.

This is different from just talking about the fullness of that impression, creating a sense of perspective, approaching the source, moving to the side or moving in the front, and all of it being so rich and true. It's a matter not just of energy panning but actually changing everything about the source, about the angle of the source, and about the coupling to the floor reflection and changing the response of the room in harmony and in great consonance with the source itself. This is to actually bring that realism to such an extent that it will allow you to float or to move and change your sense of where you want to be in the music effortlessly.

As a listener, you will just walk around without becoming confused because, at some level, you will have stable sound sources and the entire scene in there. But we want to be included and really present in the scene. We should be able to get just a little bit of proximity and sense of the realistic change, even to get so close that you can almost see or you can feel and even hear the arms of the drummer moving or the pianist's fingers touching the keys. It's that kind of sound. At the same time, you see that sound is not coming from the fingers but from the bottom, from up above, and from everywhere, in amazing unity of presence.

The sound source is not always considered as a unit where you say this sounds awesome or that sounds awesome. But the sound source that is free to evolve to any shape or size is not contained in any way. The recording should be able to capture and to produce it, and listeners should be able to adjust to it or to navigate through it. I thought this is like a dream to me, which means we have to get away from ever recording anything with a single microphone. You cannot extract space. Yes, you can synthesize it, but you start with a losing position if you just put one mic on the source. That's just like a postage stamp, and it's not a picture of reality.

My dream for the future is to release the sound sources and the rooms from being contained by technology, to really bring them alive around a listener.

## Note

1 From The Primary Chronicle St., Nestor the Chronicler (CE 1056–1114), Monastery of the Caves in Kyiv.

Chapter 5

# Natural listening, binauralization, and recording workflow

*Hyunkook Lee*

## Background

Since the age of 13, I've wanted to be a recording engineer. Growing up in Korea where formal university courses in sound recording were limited, I started my academic journey in 1996 as a student majoring in biochemistry. However, my passion for sound recording persisted, leading me to work as an assistant in a recording studio. The itch to delve deeper into sound recording prompted my decision to pursue formal education in the field.

In pursuit of this goal, I ventured to the UK to enroll in the Tonmeister course, an experience that ignited my fascination with acoustic recording and experimentation with stereo techniques. The turning point in my career occurred during a transformative three-month stint at the Aspen Music Festival in Colorado in 1999. Under the mentorship of distinguished recording engineers like John Eargle and Ron Stryker, I was introduced to the captivating world of surround recording, sparking a profound shift in my professional trajectory. Returning to the Aspen Music Festival in Colorado in 2000, I seized the opportunity to explore diverse miking techniques within the newly constructed music tent – a half-dome structure at the festival, providing an outdoor listening experience. Subsequent to this, I embarked on a one-year internship at the Metropolis in London. However, my deep dive into immersive sound occurred during my final year project with Dr. Francis Rumsey, my supervisor at the University of Surrey and one of the most prominent figures in surround sound at the time. Consulting his book, *Spatial Audio*, I found inspiration for my dissertation research.

Simultaneously, my work in psychoacoustics focused on surround microphone techniques, exploring how the localization accuracy is influenced by the transient or continuous nature of sound sources. This marked the inception of my journey into immersive sound research. For my PhD, I delved into the intricate realm of interchannel crosstalk and its impact on various sound attributes such as localization accuracy, locatedness, spaciousness, envelopment, and more. This research laid the foundation for my continued exploration of the nuanced and dynamic field of immersive sound.

While I was a PhD student, I also did some freelance work as a recording engineer, producing recordings and mixes for professional artists from both Korea and the UK. In my pursuit of expanding into surround sound, I encountered the challenge that, during that period, surround sound had yet to be mainstream. The norm was still stereo-centric.

During this transformative period, the profound influence of Francis Rumsey shaped my trajectory. Our collaborative research centered on the novel methodology for

DOI: 10.4324/9780429297526-7

subjectively evaluating surround sound – an emerging field at the time. In the 1990s, trailblazers like Francis Rumsey, Soren Bech, and a select few laid the groundwork for these methodologies, drawing inspiration from sensory evaluations such as food and wine tasting. A notable influence was the book titled *Perceptual Audio Evaluation – Theory, Method and Application* by Soren Bech and Nick Zacharov.

Under Francis's mentorship, I acquired a comprehensive understanding of these methodologies and imbibed the philosophy underpinning research in this domain. The 1990s marked a period of foundational work in sensory evaluation techniques, and I attribute my approach and philosophy to the systematic teachings of Francis. In homage to his impactful guidance, I refer to Francis as my 'research father', acknowledging the profound influence he had on shaping my research philosophy and methodology.

## What is immersion?

In my paper titled 'The Conceptual Model of Immersive Experience in Extended Reality', I conducted a comprehensive theoretical and practical review of all relevant concepts related to immersion. As the onset of the COVID-19 pandemic disrupted my subjective experiments, I took the opportunity to delve into extensive readings on immersion, aiming to discover effective ways to measure immersion. Conventional listening tests, I observed, were insufficient in capturing the essence of immersion, which led me to a deeper exploration of this topic.

During this research, I identified a significant level of confusion within the literature surrounding the discussion of immersion. Some scholars referred to *immersive systems*, while others discussed *immersive experiences*. To bring clarity to this, I organized my thoughts and synthesized the ideas into a coherent model, placing the immersive experience at the top of the hierarchy. Underneath were three supporting pillars: physical presence, social/self presence, and involvement. These were crucial components in creating this desired immersive experience.

Physical presence is characterized by the sense of 'being there' and is very important in both the real and virtual environments. This perception plays a vital role in transporting individuals into their chosen environment, shaping the context for a more immersive experience. Involvement, on the other hand, is centered around the content, emphasizing the need for engaging and captivating material. A higher-level immersive system more easily creates the sense of presence, while genuine involvement and engagement are based on the appeal of the content.

It is noteworthy that some researchers conflate presence with involvement, using the latter term when describing the concept of cognitive absorption. However, these two concepts are intrinsically linked and should be considered together. In the realm of immersive audio experiences, the aim is to transport the listener into a virtual auditory environment, whether realistic, synthetic, or artificially creative. The narrative of the content also plays a crucial role; an interesting and compelling story is essential for true immersion.

In the audio domain, achieving immersion can be challenging, particularly when dealing with different genres. While realistic expectations guide concert recordings, pop-rock music introduces complexities, questioning the spatial placement of instruments like drums and guitars. Object-based audio mixing raises meaningful questions about the significance and purpose behind spatial gestures. The current challenge lies in content creation for maximum immersive potential. When music and content are thoughtfully composed with immersion in mind, the result is a more effective and engaging immersive experience. Conversely, attempting to retrofit existing stereo content into an immersive framework proves challenging and often results in a lack of both presence and involvement when the original project did not prioritize immersive considerations from the outset.

## 'I am there' versus 'They are here'

During the process of recording and sound reproduction, I think about two distinct approaches to bring the sound scene and the listener together: either transporting the listener to the listening environment ('I am there') or bringing the listening environment to the listener ('They are here'). I think about this a lot due to the psychoacoustic considerations. At the AES Europe convention in 2022, Thomas Lund and I conducted a session delving into the concept of the 'Inception Dilemma', drawing inspiration from the movie *Inception*, which features dreams within dreams. In sound reproduction, this is similar to the 'room within a room' idea – recording in a concert hall and then playing it back through speakers in your living room. Essentially, it's akin to recreating a concert hall experience within your living room – a room within a room or an Inception pattern.

This dilemma is particularly pertinent in binaural reproduction, as our perception of sound in real life is inherently tied to an internal reference, influenced by the acoustics of the surrounding space. Despite objective physical qualities, these internal references shape what we perceive as the most natural acoustics. Consider, for instance, the mixing console reflection that engineers encounter as sound bounces from speakers onto the console, creating a comb filter in the spectrum. But nobody cares about it. Nobody thinks that sounds weird.

When we introduce an immersive recording from a concert hall into our living room, a unique concert hall listening experience emerges, blended with the room acoustics of our living space. However, we carry expectations and an internalized reference of how it should sound in our living room. Even with a high-quality concert hall recording played through preferred speakers, the visual bias introduced by the speaker placement and the surrounding environment can influence our auditory perception. For instance, a visit to a famous mixing studio may create the illusion of listening to the best mix in the world, despite objective acoustic measurements suggesting otherwise.

Delivering a virtual experience of 'being there' is a deeply personal endeavor, influenced by multiple factors – the acoustics of the environment, the visual setting, and the method of binaural and immersive reproduction. Externalization, or the ability to convincingly project the virtual experience, becomes challenging if the acoustics in the virtual playback do not align with the physical space where the listener is seated.

## Importance of HRTFs

I find that my exploration extends beyond one focus. It's crucial for people to understand the intricacies before delving into discussions about HRTF. How do we naturally perceive sound and our surroundings in real life? The essence lies not solely in the HRTF but in our movements – both physical and visual elements, acoustics, and spatial dimensions. While HRTFs hold significance, it's just a fragment of the bigger picture painted by the general HRTF.

Personal HRTFs hold value, but it's incomplete without considering the broader element. Individuals tend to adapt and even appreciate certain HRTFs over time, irrespective of their personal profiles. Take, for instance, the universally liked KU-100 HRTF – it might not match everyone's HRTF, but it somehow works.

Height perception, too, transcends individual HRTFs. I possess three sets of personalized HRTFs, yet none suffice when applied as mere filters. Dry HRTFs fall short in conveying elevation. The key lies in the interplay of acoustics within a space, enabling precise localization. Realistic room simulations, when added, introduce a vertical dimension that enhances the overall experience. People are gradually recognizing the importance of this interconnection, aligning with the broader question of how to recreate the feeling of 'being there'.

Consider the scenario of recording in a concert hall and wanting to replicate that experience in your living room, even without VR. While VR provides the ultimate visual immersion, optimizing your space's acoustics can transport the auditory essence. This can be an excellent experience – I personally encounter it nightly, immersing myself in audio, whether it's films in 7.1 on Apple TV or other content using my binaural plug-in, Virtuoso. These creations may be originally designed for dubbing studios or large cinemas, but, through optimization, I bring the large extent of these experiences to my modest office room. It's an example that showcases the transformative power of audio in creating a sense of presence and realism.

## Virtual objects in a natural listening environment

The ability to seamlessly integrate virtual objects into one's physical environment is a fascinating topic. However, delving deeper into the concept of presence raises crucial questions. The essence of presence extends beyond mere physical coexistence; instead, it navigates the nuanced realms of self-presence and social presence. In my current focus on augmented reality (AR) audio and extended reality (XR), these dimensions take center stage.

Our recent work involves the development of a sophisticated six-degree-of-freedom (6DoF) audio renderer. This tool allows for the precise mapping of virtual objects to different positions in space. An important consideration in this process is the congruency between the virtual and physical environments. Achieving room congruency becomes instrumental in simulating accurate acoustic environments in both virtual and augmented reality. Without this alignment, discrepancies emerge, impacting perceived distances, and subsequently diminishing the overall sense of immersion.

In our recent experiments, we employed motion capture technology to record a string quartet's performance in real time. This data was then utilized to generate virtual representations of the string players within the Unreal Engine. Leveraging our audio renderer, we mapped spatially precise microphone recordings to the visual positions of the virtual performers. Subsequently, we enriched the experience by introducing virtual acoustics in binaural format, utilizing the HTC Vive Puck and motion tracker to create an immersive experience. The outcome is remarkably realistic, offering an experience akin to listening to a live performance by actual string players.

## Natural real-world listening

It is important that we first understand the essence of a natural concert hall experience or any real-life scenario where movement is inherent – both in the swaying of heads and the shifting of bodies. My emphasis in recording lies in valuing the significance of the sweet spot's size. I aim to craft a recording that emulates a wide, natural sweet spot, recognizing that listeners may not always be seated at its center. Horizontal spacing between microphones becomes crucial to achieve this.

The second aspect I prioritize is making sure the recording's presence aligns with how human ears naturally capture sound. I steer away from creating artificial sound effects. Take, for instance, the use of height channels; capturing sound sources with multiple omni microphones can introduce crosstalk, impacting localization and spaciousness. In my earlier research, this was a focal point – understanding the effects of crosstalk in 5.1 recording. While crosstalk can be detrimental, creating multiple phantom images, it can also contribute to pleasing spaciousness with the right balance.

However, when it comes to immersive experiences, especially with height channels, crosstalk poses challenges. Human ears are horizontally spaced, lacking another layer for capturing height information. If microphones are placed solely above, capturing an additional layer of direct sound, issues arise, such as image shifting and coloration due to comb filtering. In the vertical plane, comb filtering becomes more problematic than in the horizontal plane.

My philosophy revolves around capturing sound in a way that aligns with how humans naturally perceive real sound. While creativity can widen or envelop sound beyond reality, I advocate for maintaining realism, especially in height channels. I find that straying too far from a natural listening experience can result in a perceptual disconnection, sounding more like a sound effect than an authentic representation of natural sound.

In essence, my approach aims to balance creativity and realism, acknowledging the potential for widening and enveloping sounds while preserving the integrity of a natural and realistic listening experience. It's about creating an immersive experience that resonates authentically with the way we naturally perceive and enjoy sound in our daily lives.

## Evaluation criteria for real-world versus virtual immersive experiences

The goal of creating immersive experiences is not merely about replicating reality; it's about achieving plausibility. Creativity plays a crucial role in this, allowing for a certain level of artistic expression. There's room for enhancing aesthetics in recordings beyond the constraints of a real-life event. Merely copying and pasting a faithful reproduction can sometimes lead to a dull outcome. For instance, recording in a challenging acoustic environment with unpleasant reflections may not be an ideal scenario for replication. In the realm of reproduction, there's an opportunity to elevate the aesthetic quality, making it more pleasing to the ear. This can be achieved through careful attention to microphone techniques.

Even with these enhancements, the authenticity of the original space can still be preserved. The key is to focus on delivering the essential perceptual cues rather than attempting to replicate every detail. This approach is very important to me – it's not about duplicating everything; it's about accurately representing the critical cues that define the essence of the sound. Striking this balance makes sure that the listener experiences a plausible and aesthetically enhanced experience, transcending the limitations of a mere reproduction.

## 6DoF versus fixed perspective

The relevance of six degrees of freedom (DoF) or 3DoF with respect to binaural audio is contingent upon the content at hand. In the realm of music, I find that 6DoF lacks the necessity for normal listening purposes. In practical terms, it may not prove beneficial. The assumption in binaural listening is that the listener remains in the sweet spot, where optimal sound reproduction happens. Moving away from this spot might not align with the typical scenario for music enjoyment. While 3DoF head tracking can help resolve front-and-back confusion and potentially enhance externalization and localization, the practicality diminishes when one simply wishes to sit down and enjoy music. In the context of music and film, where fixed orientation is common, the need for constant head movement to resolve localization confusion doesn't align with real-life listening habits.

Personally, I find 3DoF tracking necessary for brief moments of confusion in binaural scenarios, especially with dry binaural sound lacking distance and height perception. Head tracking proves helpful in such instances. However, once adapted to the environment, I often choose to disable it.

For music, I don't anticipate that head tracking or 6DoF will yield a dramatic improvement. However, in the realm of gaming, 6DoF becomes crucial for an immersive experience, mirroring real-life movements. While games may have spatial limitations, 6DoF can still enhance the overall experience, particularly in interactive storytelling. Mapping sounds to physical objects, even without VR, can create a rich, immersive experience, particularly in educational settings, museums, and galleries. However, in the context of music, I remain unconvinced of the necessity of 6DoF for an appreciable improvement.

## 3D choir recording

I would like to share insights from a notable recording project – a 3D choir recording in Merton College Chapel at Oxford University. This historic venue holds significance as the site of Michael Gerzon's first Ambisonic recording. It is an excellent space for choral music. I produced a Dolby Atmos Auro-3D Blu-ray album for a choir named Siglo de Oro, utilizing my PCMA 3D microphone array. Interestingly, despite the Chapel's very good natural reverb, the record label opted for spot mics and artificial reverb in stereo mixing. In contrast, I chose an immersive approach, relying solely on the main microphone array to capture the ambiance. This decision involved the balancing of the choir's direct sounds and ambient reflections, considering the array's distance from the choir and understanding the chapel's acoustics. The result was highly successful, offering a broad sweet spot and unique perspectives, creating an immersive experience that resonated even from non-central listening positions.

This approach stands in contrast to traditional Ambisonics, where straying from the sweet spot can lead to a collapse of the audio image and coloration. This project shows how immersive choral recordings can be achieved with a minimal setup, emphasizing the importance of capturing natural acoustics and balancing spatial elements.

In another immersive choral recording with the Ebo Singers based in York, we recorded a project in York Minster's chapel called the Chapter House, the excessively reverberant space posed challenges (with a reverberation time of around 10 s). It's almost like a reverberation chamber where you don't understand what a person is saying. Five meters away is very reverberant. However, the decay of the reverb is so smooth and lush, without any strange resonances or echoes. We recorded the singers and did some experimentation, with the singers standing in an arc formation as a classic audience performer perspective or 180°. We found the recording overly reverberant. I used a microphone array without any spot mics. The microphone array was around 3.5 m from the singers. In a reasonably dry concert hall, this leads to a good balance and clarity. In this space, however, everything was muddy and blurry. We rearranged the signers in a 360° formation, placing the singers around the microphone array, still about 3.5 m away from the array. It was magical. The problem was solved naturally, resulting in excellent intelligibility, clarity of the singers, and the right balance of reverb. This experience mirrored the techniques employed by renowned sound engineer and producer Morten Lindberg, who utilizes circular arrangements for orchestral recordings in cathedrals. The lesson learned from these projects is that thoughtful spatial arrangements and an understanding of the acoustic environment contribute significantly to the success of immersive recordings'. In such reverberant environments, recording a choir in a traditional 180° formation can be counterproductive as it amplifies the presence of reverberation arriving from the back. Instead, a more effective approach involves strategically placing sound sources in a 360° formation. This not only mitigates spatial unmasking issues but also enhances the overall musicality, resulting in a profoundly immersive experience. During the recording session with the choir in York Minster, I experienced this firsthand. Upon returning to the university to review the recording, I chose to recreate the ambiance by sitting in a darkened room. The cathedral's dim lighting added to the mood, and as I immersed myself in the playback, the sensation was overwhelming. It was almost midnight, and I was emotionally moved, almost to tears. The recording transported me back to the cathedral, evoking a profound sense of presence as if the singers surrounded me. This transformative experience resonated deeply, akin to the concept of bringing virtual elements into one's immediate environment. It felt as though my lab had been transfigured into York Minster, creating a surreal and unforgettable moment.

If the musical genre were different, for instance, if it were modern classical music, this spatial arrangement in York Minster might not prove effective. The suitability of a particular acoustic environment is intricately tied to the type of music being performed. Traditional concert halls with extended reverb times are well-suited for later romantic music, while chamber halls with shorter reverb times are more fitting for certain genres. In the case of choral music, York Minster's acoustics align perfectly with the genre's requirements. However, it's essential to acknowledge that the success of this technique is contingent on the specific acoustics of the venue and the nature of the musical ensemble. What worked seamlessly in one space may not yield the same results in another, especially if the acoustics differ significantly or if the musical ensemble presents unique characteristics. Attempting this approach in a space characterized by intense, undesirable reflections and a short reverb would likely prove ineffective, as the technique is tailored to specific acoustic conditions and musical contexts.

## Immersive experience for audio books

Another project I worked on in collaboration with Michael Stewart and Stephen Ely, who specialize in creative writing and poetry. We worked on creating an immersive audio experience for audiobooks based on their poems 'The Dogs' and 'Orasaigh'. Recognizing the effectiveness of immersive audio, especially binaural audio, for audio books, we delved into recording dialogues and various soundscapes, both natural and urban, using a 3D microphone array.

Additionally, I collaborated with a talented sound designer and researcher Katarzyna Sochaczewska. She composed the entire score for the binaural audiobook of Michael's poem 'The Dogs', which consists of 30 short sections, totaling around 40 minutes. Together, we aimed to craft a binaural mix for this immersive audio book. The initial version incorporated Ambisonics, but after realizing that mixing Ambisonics and decoding it to loudspeakers alongside binaural processing compromised the resolution and externalization, we decided to rework our approach.

In the revised version, we streamlined the process by using a simple 3D Panner within Reaper, directly mapping it to virtual loudspeakers in my binaural plug-in Virtuoso. The results were remarkable. The sound became vividly externalized, offering a natural spatial experience. We intentionally kept Michael's voice dialogue inside the head in stereo, while binauralizing background music and 3D reverb to expand the spatial dimensions. This spatial contrast, where narration is intimate and up-close while other elements exist in the broader soundscape, proved highly effective.

Rather than homogenizing everything into a single spatial perspective, our approach enhanced the overall experience by leveraging relative contrasts. While complete externalization is often praised, introducing contrasts allows for a more nuanced and noticeable 3D content, providing a richer and more engaging auditory experience.

## 3D recording workflow

My approach to recording involves a thorough understanding of the acoustic characteristics of the space. I do this by going into a space and clapping to get a sense of the acoustics. The best approach is to listen to music being played in that space in order to get a sense of the acoustics and the performance. While some may initially focus on microphone techniques, I prioritize determining the right distance for microphone placement. Unlike fixed presets such as XY, Decca tree, or an AB pair at 50 cm, I consider distance as a variable that needs adjustment based on the unique acoustics of each recording environment.

Factors like reverberance or unwanted reflections influence my decisions. In overly reverberant spaces, I may opt for closer microphone placement. If there are disruptive first-order reflections, omnidirectional microphones are avoided to prevent capturing excessive energy that can't be removed later. Depending on acoustics, I might choose a smaller array within a 1-m radius, maintaining a vertically coincident flat array to capture height information akin to human perception. This is a concept derived from my 2013 JAES paper, which also provides a theoretical basis for Schoeps ORTF-3D microphone array

Horizontal spacing is flexible and tailored to the acoustic context. Evaluating factors like front-back distances and the strength of back reflections guides my decisions. I address issues like direct echoes from back walls by using supercardioid or figure-of-eight microphones pointing sideways. Balancing the desire to maintain the original impression with the need to minimize disruptive echoes is crucial. Adjusting microphone angles and placement in the hall enables me to find the optimal balance, effectively bringing the acoustic world closer to the stage. Ultimately, I adapt my microphone techniques based on a careful consideration of the acoustic nuances in each recording space.

I started out doing a lot of experimenting with techniques and microphone arrays. As you accumulate more experience, entering a space becomes almost intuitive. It's a matter of drawing from your wealth of experiences and recognizing the familiar nuances. Your approach takes on a seasoned quality, guided by the techniques you've honed over time. It's a process of starting with a foundation based on past experiences and then fine-tuning it according to the unique characteristics of each space.

## Main lessons learned

Some of the main lessons I learned relate to the spatial and tonal quality. Regardless of how impressive the spatial characteristics are, if the tonal quality is lacking, the overall result won't sound good. Spatial considerations, although often emphasized too much, should not overshadow the crucial element, which is tonal quality. From my research and practical experience, I've learned that spatial and tonal aspects are interconnected. Crosstalk, for instance, is one factor that can impact both spatial separation and tonal quality. The key is finding a balance between these elements.

I strongly advocate that enhancing localization, refining image distribution, ensuring linearity, and managing the direct-to-reverb ratio in a recording also elevate tonal quality. Even if a 3D recording is down-mixed to 2D, it should retain its tonal character. Carefully designing the microphone array and stereo mix is essential for maintaining a high-quality tonal experience.

In our evaluations, we observed a striking correlation between tonal quality and the sense of presence (i.e., being there). If the tonal aspect is compromised, the immersive experience suffers. This aligns with the idea that artificial or synthetic sounds, even when mathematically accurate, fail to evoke the genuine sense of being in a particular space. We also found a strong correlation between listener envelopment (LEV) and overall listening experience. This correlation reinforces the importance of capturing decorrelated ambient sounds in immersive recordings. LEV is often considered important for creating the sense of presence, but a greater magnitude of LEV does not necessarily mean a greater sense of being there. LEV experienced in real concert halls is often disappointing, but immersive recording and reproduction allow us to manipulate the environment creatively, enhancing the overall experience, but care must be taken not to sacrifice tonal quality and the sense of presence.

Ambisonics, while mathematically sound for spatial reconstruction, often encounters limitations in the realm of tonal quality. In essence, maintaining simplicity is key. Like circuit designers crafting high-end audio equipment with straightforward designs, we too can benefit from keeping our recording workflow simple. This approach allows us to focus on the essentials and achieve the artistic goals without being burdened by overly complex technological workflows.

I'll also touch upon the potential challenges of the Dolby Atmos workflow, where an abundance of objects can complicate the mix engineer's process. I argue that while technology is a tool to achieve artistic goals, sometimes being forced into specific workflows and technologies can distract us from our creative objectives. In the end, it's about finding the right balance between technology and creativity to produce recordings that stand out both spatially and tonally.

Another thing I have learned over the years relates to understanding the limitations of the auditory system and the differences with the visual system. For me, observing objects moving within the Dolby Atmos renderer can be quite distracting. While visually engaging, the representation of objects moving around in a virtual space may create a misleading impression that we can perceive their exact positions with our ears. In reality, it's a graphical representation, and our auditory system doesn't capture such precise details. People might focus on the visual spectacle without paying enough attention to the auditory experience.

Through my research, I've come to realize the inherent limitations of human hearing. Our perception is far from perfect, and it's crucial to acknowledge and work within these constraints. Attempting to overcome these limitations is futile. In real-life scenarios, we can't hear things with pinpoint accuracy, and this reality extends to sound reproduction. Panning sounds at precise positions in a virtual space lacks a solid perceptual basis because our ears don't precisely perceive such details. Despite this, there's often an inclination to strive for this precision, relying on visualization provided in the graphical user interface.

I've shifted my approach to prioritize the higher-level experience rather than fixating on micro-details. Instead of being overly concerned with local edges, accuracy, and sharpness, I focus on delivering a holistic and immersive experience. This shift allows me to concentrate on the overall emotional impact and essential attributes, steering away from getting bogged down in the minutiae of positional accuracy that our auditory system doesn't fully appreciate. Focusing on the artistic intent is key.

Chapter 6

# Immersive workflow

## Mixing in binaural and mixing for large space

*Tom Ammerman*

### Background

My journey into immersive audio started with surround sound. Back in 1989, I personally purchased a surround generic loudspeaker system and two System 6000 processors by TC Electronics. With these tools, I aimed to create an immersive surround sound system that could handle both dynamics and a binaural mix simultaneously. My goal was to minimize the need for redoing the work and get as close as possible to the final version right from the start.

As I delved deeper into the world of immersive audio, I began selling binaural productions to the film industry and crafting music mixes in surround sound. I consider myself self-taught, as I learned everything on my own. During that period, I had the opportunity to work on some of the first surround mixes for Sony and Blu-ray. I also ventured into offering binaural mixes and explored my first 3D mixes. Since we had two extra channels available, it seemed only natural to explore what could be done with them. It became evident to me that we would eventually need tools capable of sending objects to the upper channels. Consequently, I founded my own company to develop specialized software, and in 2011, we released our first set of audio tools.

During that time, the engineers from Skywalker Studios, renowned for their film mixing expertise, joined Galaxy Studios due to the availability of Wilfred's unique 3D loudspeaker system. Auro-3D, a three-dimensional audio format, proved to be effective for movie projects, and my plug-in became highly sought-after since nothing similar existed at that time.

It's truly fulfilling to witness how my passion for immersive audio led me to establish my own company, develop tailored software, and contribute to the advancement of the field.

For me, achieving an immersive mix is about striking a balance between being enjoyable and captivating, while still maintaining plausibility. We must remember that we are still at the early stages, especially when it comes to mixing music and the general understanding of 3D audio. In a way, we are pioneers in this field, just as we were when stereo mixing was still in its infancy. Over time, we gained a wealth of knowledge and expertise in stereo mixing, and the same will happen with immersive audio.

The development of immersive mixing involves not only how we reproduce and deliver the audio but also how we work with objects and sound sources. Object mixing is not a new concept; it is similar to what we did with stereo mixes, where we would move elements from a single track to the left or right for stereo monitoring. Similarly, with

DOI: 10.4324/9780429297526-8

Ambisonics or object-based or channel-based approaches, we work and monitor based on the number of channels and the spatial setup. The key is to work in a plausible manner that maximizes the enjoyment for both the mixer and the end-user.

While some engineers may still be anchored in a stereo mindset, I believe that most people outside the industry have no problem embracing immersive audio experiences. There is ample room for exploration and experimentation in the realm of immersive audio. For instance, one exciting effect is the ability to have a band encircle the listener, allowing them to move their head while the band remains in a fixed position. Such creative possibilities open up new avenues for captivating experiences.

In essence, as we continue to evolve in the world of immersive audio, it is important to push boundaries, try different approaches, and create mixes that are not only fun but also plausible. By doing so, we can unlock the full potential of immersive audio and provide truly captivating experiences for listeners

## Workflow

My workflow for immersive mixing is streamlined and strategic. I begin by placing panners on all the tracks, ensuring that everything is available and prepared for the mix. Next, I set up my monitoring system. From there, my approach becomes highly intuitive.

I employ certain strategies. For instance, when working with a bass drum or a snare, I typically position them in the middle of the front, but I split the signal into two parts: the low end and the high end. I place the high end mostly in the middle, where I want it, and diffuse the low end throughout the room. When I want something to be in the middle, I don't rely on a phantom center; instead, I use the center channel to ensure it sits precisely in the middle. There's no reason to do it differently. So, I put high-frequency elements like clicks on the bass in the middle, while moving the low end into the room and sometimes adding something to the low-frequency effects (LFE) channel. This technique helps create a sense of spaciousness and adds a different dimension to the bass.

To make the mix feel larger, I utilize all the speakers available to me. It's a form of decorrelation where not only is the bass louder, but it also creates a distinct spatial feeling. Similarly, I experiment with moving certain elements, including bass, around the room for specific effects. These are some of the main strategies I employ, but beyond that, my approach becomes more intuitive.

When working with keyboards, I often pan them hard left and right. However, depending on factors such as the number of keyboards and the specific role they play (pad, arpeggio, etc.), I sometimes introduce subtle movement by slowly shifting the arpeggio upwards and downwards or changing directions. This helps maintain interest and avoids monotony, ensuring there is a sense of development. I strive to maintain a certain level of homogeneity in the mix, but in my future mixes, I intend to explore more unconventional and experimental approaches. I might venture into more extreme techniques and assess their effectiveness later. For example, I might experiment with a mix that predominantly emanates from a single speaker, such as the left surround or heights, and then flows back out. This can create a captivating effect, but it needs to be compatible with the song itself. Choosing the right song for such experimentation is crucial.

I believe the time is ripe to work with these innovative and unconventional effects. The necessary tools are available, and while some panning tools in digital audio workstations (DAWs) can handle these ideas to some extent, I see room for improvement and enhancement. Most DAW panners simply extend their stereo and surround sound strategies to work in 3D immersive formats. However, I believe there are opportunities for panning tools to evolve and better support the execution of these new, wild, and exciting ideas more efficiently.

## Mixing in binaural

From the very beginning, my fascination with surround sound led me to explore the world of binaural audio. In 1999, during the AES event in Paris, I had a transformative experience when I encountered the Dolby headphones demo and Studer's head-tracking demonstration. It was mind-blowing to hear binaural audio with head tracking in a room where the measurements were captured from the loudspeakers. Studer collaborated with IRT on this project, and it left a lasting impression on me.

There are two primary approaches to binaural audio. The first approach aims to provide listeners with the ability to experience an immersive mix even without a physical loudspeaker system. In this case, the challenge lies in creating a virtual loudspeaker system that sounds as natural as possible. It's important to note that achieving complete neutrality is not the goal since every system imparts some coloration. The second approach is to embrace the possibilities offered by binaural audio and open the doors to an entirely new realm of audio experiences. With binaural, you can venture into unconventional territories, exploring scenarios and effects that would be impossible to replicate in reality. Why not experiment with working in different rooms simultaneously as an artistic effect? For certain projects, I create a unique binaural version alongside the main mix for Blu-ray or other delivery formats, adding an extra dimension to the overall experience.

Working with multi-channel audio provides a tactile experience, as you can physically feel the sound waves resonating through your body. Additionally, natural head tracking further enhances the immersion. However, binaural audio presents both exciting possibilities and challenges.

One of the difficulties lies in working with the low end. In my personal opinion, many binaural versions I've come across tend to have less low end than they could utilize. Surprisingly, even small headphones can reproduce low frequencies effectively, and this opens up opportunities for utilizing low end reproduction in various rooms as an enhancement or artistic effect.

Exploring the boundaries of binaural audio and pushing its potential is a constant endeavor. By harnessing the capabilities of binaural technology, we can create captivating and unique auditory experiences that go beyond the limitations of traditional audio reproduction.

The main strategy I employ for working with multiple rooms in an immersive mix is to place the groove-oriented elements, such as drums and bass, in a smaller room to maintain tightness. The other elements are distributed among one or two additional rooms to create layers and depth around the listener. I often assign choirs or effects to the second and third rooms to add further dimension. Additionally, I dedicate an extra room solely to handle the low end, sometimes even using multiple low-frequency effects (LFE) channels. The low end is crucial for creating a sense of depth, distance, and overall spatial experience. It extends beyond just the bass frequencies, encompassing everything from around 500–600 Hz down to the sub-bass range. I recall a technique I instinctively employed when up-mixing from stereo to 5.1 or 7.1, where I would boost a wide bell around 800 Hz by a couple of decibels on the side channels to open up the space. The effect was truly remarkable. It's important to strike a balance when incorporating reverb in the mix. Too much reverb can detract from the realism and believability of the mix.

Excluding certain objects from the binaural mix is something I dislike as it disrupts the immersive experience. I've encountered this issue many times, especially with renowned producers who transitioned from stereo mixing. They often mix the voices in the phantom center and neglect the center channel. I believe there is room for more development and inspiration in the way we approach working with the center channel and LFE. They are not inherently problematic elements. If the center channel is overly dominant and clear, it may indicate a mistake in the mix. Distance and delay panning are also important tools to consider.

## Domes, halls, theaters, and planetariums

When working in spaces like domes, larger halls, or theaters, the input formats can vary, ranging from mono sources to multi-channel setups such as 5.1 or stereo. Each venue presents unique challenges. Domes tend to have a controlled environment and a large sweet spot. The loudspeakers are generally uniform, minimizing issues with delay and making distance panning more feasible.

However, in conventional theaters, localization can pose a challenge, particularly when a performer wearing a body tracker moves to the back of the stage while the audience is seated in the second or third balcony. Three-dimensional audio doesn't work effectively in such situations, and placing sound sources on the roof isn't feasible. Speech intelligibility becomes paramount, ensuring that everyone can clearly hear what is happening on stage. Panning options may be limited in these cases. The people seated in the balcony won't be able to hear loudspeakers positioned on the opposite side. While some movement effects can be applied, it's crucial to acknowledge the technical constraints and to understand that certain approaches may not be practical. However, subtle movements of sound sources can still be effective for the audience in terms of creating an immersive experience.

In an arena setting, considerations must be made to avoid critical groove-related issues due to the longer runtime of events. If a sound is delayed from the back speakers and heard by someone in the front, it can cause unwanted echoes. Surround sound can help alleviate this mono situation by providing a more enveloping experience. In the early days when only two main channels were available at the front, being too far to the left or right would result in a mono experience. Immersive audio can address this limitation, allowing for a more consistent and engaging sonic environment regardless of the listener's position within the arena.

In a live situation, a common strategy involves deploying a large additional line array system at the front. Typically, side speakers, one or two intermediate speakers, and a center channel are employed to disperse the band's sound. Depending on the desired effect, panning the singer as they move on stage can be implemented, although caution is necessary to avoid creating issues for the crowd located close to the stage. To address this concern, a new strategy has emerged, which involves utilizing front fills. This allows for panning effects, such as when the guitar player or singer moves across the stage. In the main arena, everyone experiences the panning effect, but those situated in the front rows do not hear it. By creating speaker clusters and implementing panning within each cluster, better coverage is achieved, ensuring a positive experience for everyone. Effects play a crucial role in enhancing the overall immersion and envelopment, whether through reverbs, keyboard pads, or other elements that complement the show or music style.

On the other hand, the approach differs significantly in theme parks. Unlike a dome, arena, or fixed installation, theme parks feature a non-homogeneous loudspeaker system. With two speakers here, ten speakers there, and another speaker elsewhere, the task becomes entirely different. Immersive audio takes on a new level of complexity and interactivity in this context. The creation of content specifically designed for theme park experiences requires a fresh perspective and innovative approaches to working with audio and music. It presents exciting opportunities for audio professionals to craft unique and captivating soundscapes in this dynamic setting.

One of our recent projects involved a theater that also functioned as a planetarium. In the past, audio domes often featured loudspeakers arranged in non-homogeneous clusters. However, for this planetarium, we adopted a layered approach using three main layers and a central Voice-of-God (VOG) speaker positioned at the center height. Each layer consisted of 12 speakers with a 30°spacing between them. This spacing provides adequate separation without creating noticeable jumps in sound. Typically, the first layer starts around two meters above the audience, which can pose challenges for those seated in the last few rows. To address this, we developed a strategy that involved using fills placed slightly lower to cover the last two or three rows. This solution significantly improved the experience for those audience members without disturbing the overall immersion for others in the room.

An interesting aspect of our approach is that we use 12 speakers in each layer, allowing for increased resolution in the upper layers where the speakers are closer together. This setup compensates for the reduced localization perception accuracy at higher elevations. Placing the speakers closer together helps mitigate this issue. Moreover, by aligning the speakers in a vertical line for each layer, we ensure consistent coloration at every height. This consistency enhances localization and contributes to a more coherent and plausible audio image. Overall, this layered approach has proven to be highly effective in planetarium environments.

It's worth noting that planetariums may mistakenly believe that they require thousands of loudspeakers. However, we utilize full-range speakers, enabling us to reproduce bass frequencies, including those below 100 Hz, even from the upper layers. This capability presents its own set of challenges, but through experimentation and refinement, we have successfully implemented this strategy in multiple projects. This approach works remarkably well and is applicable to different formats, be it 5.1, stereo, or object-based content in the dome.

Currently, we are witnessing a shift in planetariums, with an increasing focus on hosting music streaming shows, other forms of entertainment, and shorter movies of around an hour or less. I am involved in another project for a dome that extends all the way to the floor, which offers a truly unique experience. Recently, I had the opportunity to experience a similar installation in Montreal, where they have two planetariums – one traditional and the other with a dome that extends to the floor. It was an incredible experience, and now I'm excited to work on a similar project.

Once the installation is complete, it is crucial to train the staff and creators on how to work within these immersive environments. While some fortunate venues may have an immersive loudspeaker system in place, the majority do not. In such cases, binaural monitoring becomes essential. We conduct measurements in planetariums and theaters and even go on-site with a laptop connected directly to the loudspeaker system for fine-tuning. The workflow typically involves working with binaural monitoring for the initial stages (around 60–70%) and then conducting the final adjustments on-site. The goal is to deliver the master that meets their specific requirements. Licensing is also a significant aspect of the business, as shows created for one planetarium often find their way to other venues with different configurations. Hence, careful consideration must be given to compatibility across various systems. An object-based mix, combined with our specialized processor, allows for flexibility and seamless adaptation to different loudspeaker configurations. Rendered masters can be exported and easily played back from media servers or other playback devices.

It's important to emphasize the value of creating mono exports, as the playback capabilities in different planetariums can vary significantly. Some venues may have multiple synchronized media servers, necessitating flexibility in the playback format. The music and sound producers and sound designers working in these environments must adapt and learn new techniques. These immersive audio environments are evolving, and we are just at the beginning of exploring their full potential.

## As a producer

As a producer, it's my role to help artists discover what they truly want and provide options and suggestions. Presenting ideas and offering emotional connections by saying, 'What do you think about this? Could this capture what you envision?' is crucial. I listen carefully to the artists and avoid imposing my own agenda. I try to understand their vision and what they truly have in mind. It was a similar situation when working with Kraftwerk in the beginning. Fritz Hilbert of Kraftwerk had experience with surround sound but not immersive sound. I created a platform where he could freely explore his ideas. It's about providing the freedom to work on any idea and inspiring the artist to experiment with objects, height, and movement. However, it's essential not to over-inspire or interfere too much. Eventually, I started stepping back and allowing him to follow his creative direction. If he had a good idea, he would still seek my opinion, and I would offer my thoughts. There were times when I heard something but didn't comment because I felt he was satisfied with his idea. It's also crucial to understand that critiquing decisions can be counterproductive. As the producer, my role is to create an environment that facilitates the artist's journey toward the best possible outcome, ensuring their happiness and creative satisfaction.

## Lessons learned

The first important lesson I've learned is to avoid predictability and push boundaries. It's essential to strive for development and growth while staying true to my beliefs. If 70% of people love the result while 30% dislike it, I consider it a great achievement because it means the work has character. This is a personal philosophy I try to adhere to. I also believe in crossing borders and experimenting, but at the same time, I respect the artists who have put in a lot of effort and attention into their work. While I may approach some of my old mixes differently today, there are others that I wouldn't change a thing about. It's important to keep an open mind and realize that everything I know is the result of a wealth of experience. Stay true to your beliefs and learn from the mistakes you've made, while also allowing others the freedom to make their own mistakes and explore different avenues. Find a good balance where you can work independently and confidently without seeking validation from others, but also be open to hearing their opinions and thoughts. Follow through on your ideas before seeking input.

### Looking to the future

Thinking about the future and looking ahead, my hope is to witness a shift in the mindset of producers and artists, where they become more open-minded about immersive ideas. It's important to break free from the constraints of a primarily stereo-based enhanced 3D mix. Simply adding an extra element doesn't necessarily make it better; we need to explore new possibilities. We should not get stuck in the stereo world. I am confident that the next generation will embrace this change, although it can be challenging for those of us who come from a stereo and surround sound background. On the reproduction side, we also need to be more open and adaptable. I have been involved in a project where every available device around you can be used to reproduce an object-based master, utilizing mobile devices, TVs, soundbars, or any other equipment you have. We should learn to work with fewer controlled environments, where perfection is not necessary. Whatever devices and equipment people have should be capable of delivering immersive experiences. We should develop new technologies that can recognize and work with various devices and content. Additionally, it would be exciting if users could have some control over the positioning of sound sources to a certain degree. For example, someone may prefer the bass on the floor or a more spread-out band. Why not make this kind of customization easy and accessible? In conclusion, I look forward to the future and the potential for further growth and innovation in immersive audio.

Chapter 7

# Immersive acoustics and reverberation

*Ralph Kessler*

## Background/history

I began my journey in the 1970s while studying as an electrical engineer. During that time, I also worked as a residential engineer in a local recording studio, where I interacted with artists exploring new music trends. One pivotal moment was when I was asked how we could integrate new technology into arts, particularly for multimedia performances. This led me to ponder the possibilities of utilizing new tech in a show that catered to around 200 people back in the 1970s.

During an internship at a studio in South Germany, I had the chance to work closely with a device from a German reverb company called Quantec. This device intrigued me with its four outputs, prompting me to delve deeper into the concept of decorrelated reverb. It was an eye opener for me to realize the potential of creating immersive soundscapes using multiple outputs. That was a moment of revelation, and I thought to myself that with four outputs you can have a voice, vocals, speech in the middle, and a decorrelated noise outside. You can create a sound field. That was moving. So I decided to make that a performance, and we toured in Austria with four speakers. However, this innovative approach also brought challenges, as there were limited resources like mixers designed for such setups. Despite the hurdles, I remained fascinated by the possibilities that new technology offered.

Later I had the opportunity to work at Quantec and with the company's owner, Wolfgang Schwarz, who was a genius. I gained invaluable knowledge from him as he had already pioneered a device utilizing a format that was not widely discussed – the 'quad format', which was a significant revelation for me. I learned about a number of design principles for electronic hardware from him, who possessed an exceptional understanding of electronic design. He had a deep grasp of chip compatibility and timing considerations, which was so important for creating algorithmic reverbs in an era where CPU limitations were a major challenge. The quality of reverb directly correlated with CPU power and frequency density, making natural reverbs achievable only with high computing power back then. This experience was instrumental in my learning journey, and I hold immense respect for the technological advancements of that time.

Later in the 1990s, I expanded my knowledge into surround sound technology, focusing on music production and CD mastering. Working with companies like Weiss, I delved into designing compressors, equalizers, and understanding hardware interfaces with computers. This period marked significant advancements in user experience, especially with remote-controlled hardware integrated into digital audio workstations (DAWs).

DOI: 10.4324/9780429297526-9

Reflecting on my journey, I recognize the importance of aligning technology with natural experiences, aiming for authenticity in sound reproduction. This principle guided me through designing algorithms and fine-tuning hardware for optimal performance. While technology has evolved significantly since the 1970s, those pioneering experiences laid the foundation for my continued exploration and innovation in the field of audio engineering.

Later, following a brief stint in Switzerland with Daniel Weiss, I constructed for his modular signal processor system (compressors and equalizers) a modern, PC-controlled GUI, adding also new measurement tools for audio quality. My focus shifted toward mastering stereo tracks, where I honed my skills in stereo mastering and also developed computer-based metering tools for Daniel's projects. This journey led me to venture into CD mastering on my own.

I aimed to understand the significant impact mastering has on music perception, heavily influenced by the monitoring systems used by listeners. Mastering isn't just about the finishing touches; it's about translating artistic intentions into a final product that resonates across various playback environments. Whether it's in a car with pronounced bass response or in a quieter setting like a restaurant, the mastering engineer's role is crucial in ensuring the music retains its intended essence and impact regardless of the listening conditions. This exploration during the 1990s provided me with valuable insights into the intricate nuances of mastering and its influence on the overall music experience.

By the end of the 1990s, there was a significant shift towards surround technology, particularly the 5.1 format for music. This development intrigued me greatly, given my previous experience with quadraphonic systems, which were ideal for simulating reverb and enhancing live experiences. I remember thinking how quadraphonic systems never gained widespread success, yet with the emergence of 5.1 surround sound, especially in cinema with Dolby 5.1 standards, there was potential for the music industry to catch up.

A pivotal moment was attending the AES conference at Schloss Elmau in 2001, where designers came together to exchange ideas and insights. This conference marked a shift in mindset, prompting my involvement in mastering audio for live DVDs, which quickly became a standard practice. While 5.1 surround didn't dominate the music industry's disk formats, it became a staple for live DVDs, enhancing the overall viewer experience.

The transition to 5.1 surround required learning new techniques, such as optimizing the center speaker and creating immersive sound fields. Despite its limited market share in the music industry, the journey was fascinating and filled with valuable lessons on adapting to evolving audio technologies.

I dedicated over a decade to mastering the art of creating high-quality surround mixes for music, a pursuit deeply intertwined with considerations of design aesthetics. One key aspect I've encountered is the temptation among some new producers to assign a separate speaker for each musician, assuming it will automatically enhance the listening experience. However, I've learned that more speakers don't always equate to better quality or artistic intent.

In my experience, particularly working with larger ensembles like the Armenian orchestra or symphony orchestras with choirs of up to 200 people, the challenge lies in balancing the intricacies of the surround system while honoring the composer or arranger's vision. I've had conversations with arrangers, such as the German orchestrator Bert Kaempfert from the 1960s, who employed clever techniques like blending violins and flutes to create a unified sound. The danger arises when inexperienced surround producers prioritize spatial distribution over artistic integrity. For instance, if a young producer decides to place flutes in the height speakers and violins in front without understanding the intentional blending of sounds, they risk stripping away the mystery and depth carefully crafted by the arranger.

This underscores the importance of not just technical prowess but also a deep understanding of musical nuances and artistic intent in creating compelling surround mixes for music. Balancing technology with creativity is key to preserving the magic and essence of the original composition.

## Intention of the music and the composer

I come from the technical side, and part of my role as a music producer and engineering professional is to bridge the gap between the technical aspects and the artistic vision of musicians and producers. It's important to understand that a producer or musician may have a specific sound direction in mind, and they prefer to keep the intricacies of their sound elements hidden. This can be challenging for newcomers who may be excited about the technical capabilities they have at their disposal but may not always align with the artistic intentions of the arranger or producer.

This disconnect highlights the importance of effective communication in the production process. For instance, a producer might be thrilled about the various technical dimensions they can manipulate, but if it goes against the intended sound or composition, it can lead to misunderstandings or compromises in artistic integrity. This issue isn't unique and has become more pronounced in recent times, especially with the shift to remote work and collaborations due to circumstances like the COVID-19 pandemic.

Navigating these challenges requires a balance of technical expertise and a deep understanding of artistic intent. It's about fostering open communication and collaboration between technical teams and artists to ensure that the final product resonates with the intended creative vision while leveraging the available technological advancements effectively.

In my experience, effective communication plays a crucial role in achieving the desired musical outcome. When people can sit together and discuss the musical intentions, it enhances the technical setup process. I always emphasize to my students and colleagues the importance of understanding the composition and the orchestrator's vision before diving into any technical setup. When dealing with a group of musicians, it's essential to grasp how the composition functions as a whole before mapping out instrument placements. Unlike a live stage where physical locations dictate instrument positions, recording setups have more flexibility.

Clear communication is key to avoiding misunderstandings and achieving accurate results. I believe that we often face challenges due to insufficient communication, leading to misinterpretations or suboptimal outcomes. Therefore, prioritizing communication and understanding the musical intentions should be the first step in any technical setup or recording process.

### *Philosophy*

My philosophy can be summarized as follows: in an A/B comparison between a processed and unprocessed audio signal, the best algorithm is one that enhances the sound subtly yet noticeably enough that you miss it when it's turned off. It should add value by simulating great acoustics with a high degree of frequency-dependent reverberation time and well-structured reflections. However, it's crucial to understand that reflections in nature don't always exponentially increase; they have a natural pattern that shouldn't result in excessive noise buildup. Achieving this balance requires a nuanced approach, as the optimal reflection pattern varies depending on the music genre and context. There's no one-size-fits-all formula for creating a generic room simulation that works universally across different musical styles and environments.

One of the mysteries and traditions in acoustics is the distinction between early reflections and the diffuse part of late reverberation. Conventionally, this distinction is made within a time frame of 50–150 ms, depending on room size, marking the transition from early to diffuse reflections. However, I have reservations about this fixed boundary because it's more of a continuum where one phase transitions into the other gradually.

A high-quality reverb should evolve dynamically, avoiding a static or artificial feel. While some theories, like Manfred Schroeder's ideas involving prime numbers, have been explored, they don't provide a complete solution. Nature itself presents complexities such as cancellations and constructive interference, leading to a nonlinear buildup of reverberation rather than a simple progression towards white noise.

This complexity inspired me to work on a project in the 2000s, creating an extensive library of authentic acoustic spaces in North Germany. Using advanced recording techniques and meticulous observation, I sought to understand why certain spaces are perceived as acoustically superior by thousands while others are not universally appreciated. Acoustics, like music, is subjective and depends on various factors beyond mere technical specifications.

Modern advancements in computing power have enabled me to design sophisticated plug-ins capable of handling complex convolution processes, like my plug-in with 14 input and 14 output capabilities. This level of convolution power was unimaginable in earlier decades, allowing for research into reverb characteristics and optimization methods.

One critical design criterion I've identified is the input source for the reverb engine – it should ideally be stereo or even multi-channel to capture the spatial characteristics of the original sound source. This approach significantly enhances the realism and richness of reverberation, reflecting the inherent spatiality of real-world sound sources.

However, integrating such advanced reverb systems into digital audio workstations (DAWs) can be challenging, requiring careful synchronization and routing. These technical details are vital for achieving a natural and immersive reproduction of acoustics in audio production.

### *Acoustics*

Numerous factors contribute to achieving good acoustics in a concert hall. One question that arises is why the perception of a Beethoven symphony differs between different rooms. This disparity not only is due to the acoustics but also relates to how musicians are trained to perform in specific spaces.

There is a large concert hall in Hamburg, Germany, called the Elbphilharmonie, built in 2017. From the beginning there were debates regarding the hall's acoustic quality. The concert hall is designed with the idea in mind that there is intimacy between the musicians and the audience and that audience members are seated in proximity to the stage. For instance, having the audience close to the conductor, with the maximum distance from the last row to the conductor being less than 30 m, creates a more intimate setting. However, accommodating such proximity while maintaining excellent acoustics is very challenging. There are areas within the Elbphilharmonie that I found particularly enjoyable for listening to new and fast-paced music. For fast-paced genres like jazz or rock music, I prefer minimal reverb time to avoid blending notes together. The acoustic environment tends to meld notes, turning arpeggios into a single indistinct sound. This issue is pivotal in assessing whether an acoustic space is deemed good or not, as listeners want clarity, especially when focusing on song lyrics. Different music styles require specific optimization targets; for instance, legato strings necessitate a different reverb time than music with spoken lyrics. So determining what constitutes good acoustics is complex and context-dependent.

An account was once reported to me about Sir Simon Rattle, a notable conductor, during a soundcheck at the Elbphilharmonie. Rattle was the conductor of the London Philharmonic at the time and visited the venue with the orchestra – one of the first world-class orchestras to do so. During their performance, Rattle carefully listened to the acoustics, noting the specialties compared to other venues. To adapt to this acoustic environment, Rattle instructed the orchestra to adapt their playing accordingly (e.g., length of the notes), allowing the audience to experience a unique musical quality. This showcases the artistry involved in adjusting to varying acoustic conditions. While shorter reverberation may suit some orchestras better, others may struggle to adapt due to their unique sound qualities and playing styles. This highlights the intricate nature of orchestral performance and the importance of acoustics in shaping musical experiences.

The criteria for achieving good **artificial** reverb is a complex question. In the 1980s, it was considered a mystery in terms of knowledge, almost like a secret sauce. One criterion is the down-mix capability of a reverb. Consider a scenario where you have nine or more channels that are down-mixed to binaural, stereo, or even a single channel. However, there may be instances where the music is not entirely decorrelated. For example, when something is played through the center speaker, it automatically becomes 100% correlated unless specifically mixed differently. This high correlation can affect the down-mix balance, creating a different experience from what is heard in an immersive monitor layout. This factor is crucial but often overlooked by those designing reverb generators, as they tend to focus on creating a spectacular effect.

Some plug-ins even include features like phase flipping by 180°. However, in a mono down-mix, this can cause the reverb to disappear completely, which is not the producer's intention. Therefore, in my opinion, the correlation between channels is a critical factor that should not be ignored when evaluating the quality of artificial reverb.

Another factor is the frequency dependency of reverberation time, where high frequencies decay faster than lower frequencies. Convolution is excellent if you have the right methods to capture the room and if the CPU power is fast enough. It can give you goosebumps, and when you close your eyes, you feel like you are there. However, it is crucial that the signal-to-noise ratio of the impulse response is good enough. Toward the end of the impulse, it always merges into the noise floor, which can sound artificial, almost like aliens. This noise can be turned into an effect, but it's not desirable when seeking a natural reverb. Therefore, it's important to avoid increasing the noise floor during mastering or using compressors, as it can distort the convolution and create unintended effects. These considerations are essential when working with reverberation effects.

The tuning process is crucial in any preset generation for artificial reverb. However, in the case of a virtual acoustic life system, it becomes far more complex. I recall a colleague and pioneer of virtual acoustic systems once mentioning that, given the time, he would prefer spending a week in the room to match the real room and its electroacoustical elements, as achieving perfection is an ongoing process full of compromises. Just as with a piece of art, one never reaches a point where it's deemed completely finished. The complexity increases with larger rooms and different seating areas. In a virtual acoustic environment, you encounter challenges different from those in a controlled studio setting. That's why I advise customers seeking advice on virtual acoustics to dampen the room as much as possible.

I had the pleasure of working with my Armenian orchestra for a full week in Lebanon in 2015 for a very large show. We were fortunate to have a setup time of two weeks in the venue for television purposes. Every evening, we had an audience of 2,000–3,000 people. I had the opportunity to specify that I wanted the room to be completely dry because the owner of the orchestra also owned a carpet company. I informed them that if they wanted me there, they needed to ensure that we had two layers of carpet in the hall. The room was 50 m by 50 m and 20 m high. We dampened that room to nearly 300 ms, similar to the control room of a studio. It was perfect.

Another important aspect was hiding the speakers from the audience's view. If the audience were to see loudspeakers everywhere upon entering, they would have been fearful, at least for this type of music, which was folklore. We hid 50 speakers in the ceiling using cloth, so the audience was unaware of their presence distributed around them. We had three line arrays, and then we utilized the 50 speakers to create a virtual acoustics system and a completely dry environment. With this setup, it was easy to create an acoustically beautiful experience similar to what you would achieve in a studio. We didn't have to struggle against poor acoustics with our virtual acoustic system. You can truly simulate the acoustics of a church and use the same presets you have in a studio reverb engine, but only in that specific environment.

Normally, if you don't have the luxury of having these black two layers of curtains, you end up fighting against the acoustics, and that is what makes the life of the tuning person very complicated. The tuner has to decide which area of the room they should optimize for. Any tuner has their tricks, secret sauce, and makes measurements. Personally, I try to subtract measured acoustics from my library so that I have something like acoustic one, acoustic two, and acoustic three. With mathematical methods, you can do some work to subtract and reach a target acoustics. In a way, you have to think about it as a 'thermostat', where you measure the temperature and only add energy in those frequency areas where you think something needs adjustment, without touching the others.

A person must consider the balance between relying on quantitative measurements and tuning reverberation by ear. As a manufacturer of measurement units, one might expect that measurements are relied upon 100%. However, I believe that the ear is much better than measurements in certain situations. In music, I sometimes use a spectrum meter but not excessively. The spectrum meter may not always show differences between 0.5–1 dB at certain frequencies, whereas my ear can clearly perceive these differences.

My approach involves listening to music while tuning a reverb. I always have a test playlist with critical material that I use for this purpose. For instance, I listen to jazz with a double bass to identify and work on the room's modes. It's not feasible to do this for every seat, so I ask the customer which seats are the most important, and we optimize for those specific seats.

### *Static versus interactive applications*

In an active configuration, it's much harder for the designer of a reverb plug-in to ensure that all parameter changes are free of artifacts. For example, when transitioning from a small room or corridor to a large hall, there are challenges to consider. What can be done during the transition period when you are halfway between one room and another? To achieve this transition correctly, you would need both algorithms running simultaneously – one for the small room at the back and another for the large room at the front. This presents a real challenge for designers.

So the devil is in the details, as always. I think no company really deals with multiple reverb systems in simulations because everybody wants to save CPU power. They are happy to use 80% of all the CPU power for visual effects, so for audio, they try to minimize everything as much as possible. However, we had a customer once who introduced a convolution plug-in at one point in time, which offers much nicer quality than their algorithmic implementation. They also had an algorithmic implementation, but I believe it must have been done by someone who studied and understood Manfred Schroeder's work. Convolution, of course, is more CPU intensive and has the drawback of not being as flexible regarding the parameters compared to algorithmic ones. The future lies in hybrid solutions, I guess.

### *Lessons learned*

During my lengthy career, I have learned many lessons, and I would like to highlight a few here. The first one is not to overdo any effect. Avoid falling into the trap of spreading too much around you, especially in music production. While in games, effects need to be spread and moved to make sense as you navigate through different spaces, in music, be cautious about excessive motion. For example, I appreciate effects like the guitar slide in Elton John's Rocket Man, where it creates the illusion of the rocket soaring behind you and reaching the ceiling. This effect is excellent, especially when the rest of the tune maintains a calm aesthetic. Like a good cook, don't over-spice your food.

The second lesson is to test your work on as many monitoring systems as possible. Unfortunately, with the invention of all these new formats, it has become more complicated to test all possible environments. This is nearly impossible now because some companies release a new product every few months. The mastering engineer cannot cover everything available on the market. However, they should have a good understanding of how it will sound in a consumer environment. What they hear in their calibrated monitoring environment in the studio may not translate perfectly to home systems, but they should be aware of how it will sound. So don't forget this last point in your production chain.

My final point is about the importance of understanding the other side. Musicians should understand technology, and technologists should understand musicians. As an artist working in media, the landscape is much more challenging now than it was in the 1970s – it's more diverse. Understanding both the science and the art, and being able to communicate effectively with both types of people, is crucial, especially for teamwork. Take the time to learn about other disciplines; it will enhance your skills and make you better at what you do.

Chapter 8

# Creating and distributing immersive audio experiences

*Jean-Marc Jot, Thibaut Carpentier, and Olivier Warusfel*

In recent history, professional and consumer audio and music reproduction technology has progressed along multiple axes – including sensory immersion, electronic transmission, content formats, and creation tools. The authoring and consumption or performance of immersive media experiences increasingly leverages a global set of connected frameworks and ecosystems. These experiences – traditionally restricted and categorized into separate content markets such as music, movies, video games, and virtual reality – are becoming interoperable, ubiquitous, navigable, and adaptable to listener conditions and personal preference or language. In this chapter,[1] we examine aspects of this evolution with particular attention to flexible immersive audio creation and reproduction. Following a comprehensive review of the design principles of IRCAM `Spat`, an immersive audio creation and performance tool, we discuss the evolution of object-based immersive audio technology toward the unification of broadcast content and embodied experience ecosystems. We introduce the notion of Acoustic Objects, providing a universal spatial audio scene representation model for the creation and distribution of versatile and navigable music, multimedia, and virtual or extended reality sound, in entertainment and business applications.

## Introduction: artist-driven and physics-driven immersive audio experiences

### *Audio reproduction technology: enhancing listener immersion*

As reviewed in [Roginska and Geluso, 2018], the creation and performance of immersive audio experiences is rooted in ancient theatrical and concert practices that exploit our natural hearing faculty to spatially parse multisource soundscapes and the subtle features of reflective or reverberant acoustic environments. In modern times, audio technology has enabled creators to realize, record, and reproduce both realistic and imaginary artificial immersive audio experiences.

Applications of immersive audio technology fall broadly into two categories, depending on whether the end user experience is linear or nonlinear. *Linear* application scenarios are those in which a temporal sequence of events composing the audio or audiovisual media specified by the author is prescribed and cannot be modified interactively by actions of the listener or spectator. They are typical in the distribution of recorded or cinematic media such as music, movies, and podcasts. *Nonlinear* applications, in contrast, trigger events in response to user actions, and include video games and virtual, augmented, or extended reality experiences (VR, AR, XR) [Llewellyn and Paterson, 2021].

DOI: 10.4324/9780429297526-10

In multimodal experiences, both linear and nonlinear, creators can enforce the coincidence between audio and visual spatial cues to further enhance the spectator's sense of presence [Schubert et al., 2001; André et al., 2014; Lee, 2025]. Importantly, in modern consumer systems, this is technically facilitated by the fact that the audio renderer can query and adapt automatically to the end-user's playback configuration (headphones or loudspeaker array), as discussed further in the section 'Flexible loudspeaker or binaural immersive audio distribution'.

### *Aesthetics-driven linear immersive audio for music and soundtracks*

Spatial audio recording and reproduction technology has evolved over several decades from two-channel stereophonic recording to flexible immersive audio reproduction, through extended channel-based formats, including surround and height channels [Roginska and Geluso, 2018]. Throughout this evolution, the mixing signal flow employed by creators to produce audio recordings has remained consistent with the generic architecture now common in modern software digital audio workstations (DAWs). This signal flow is represented in Figure 8.1, where various multi-channel positional audio *panning*

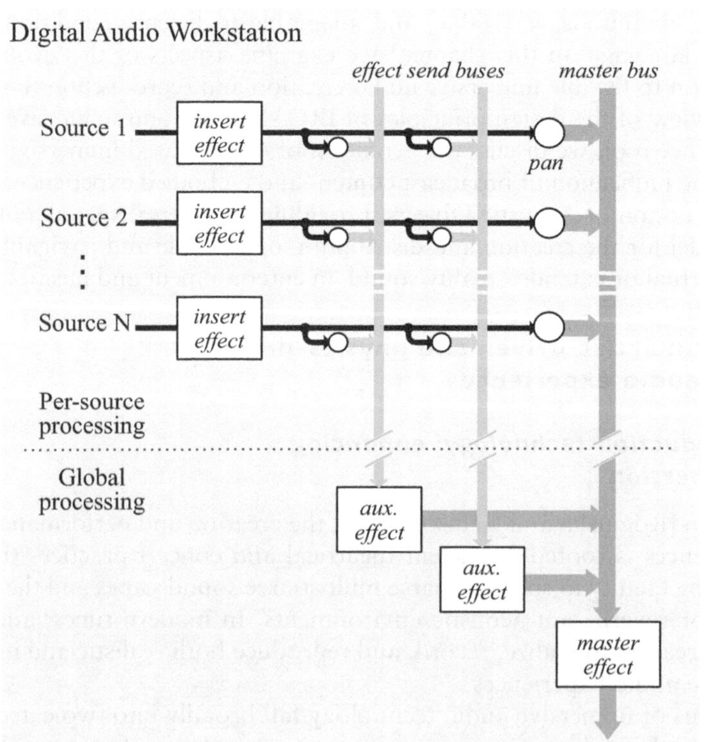


*Figure 8.1* Generic digital audio workstation (DAW) signal flow: The topology is consistent regardless of the master output format, from two-channel stereo to modern immersive sound formats. It allows panning and mixing of multiple source tracks, with optional per source insert effects, along with global processing including shared master and auxiliary effects.

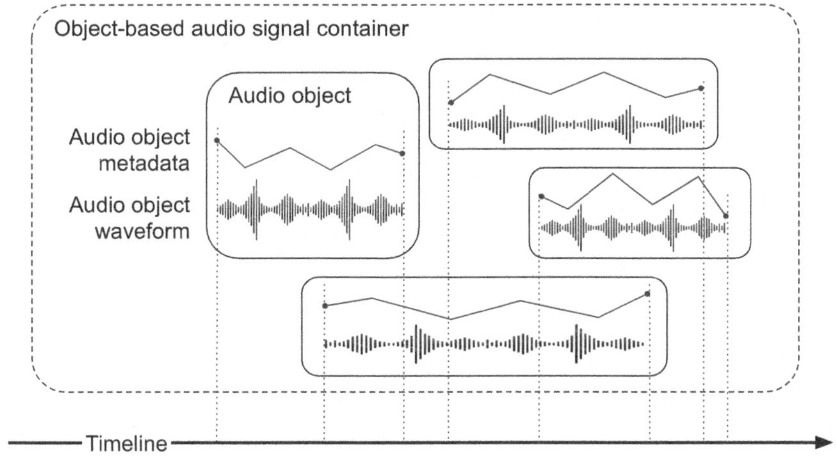


*Figure 8.2* Illustration of object-based immersive audio signal format for encoding multiple virtual sound sources as audio objects: typically, each audio object is represented by a single-channel audio waveform signal accompanied by metadata specifying its intended position and loudness.

techniques may be selected, depending on the immersive audio output signal format carried on the *master bus* [Pulkki and Karjalainen, 2008; Zhang et al., 2017].

In traditional *channel-based* formats and panning techniques, each of the audio output channels is destined to discretely feed an individual loudspeaker or cluster within an array facing or surrounding the listener, according to a predefined geometrical layout. Recent layout-agnostic formats and low-bit-rate coding specifications, reviewed in the section 'Flexible loudspeaker or binaural immersive audio distribution', are deployed in professional or consumer-grade media authoring software, distribution services, and playback equipment. These include *object-based* immersive audio formats, illustrated in Figure 8.2, and *scene-based* (a.k.a. Ambisonic) formats.

During the production and mixing of a new immersive audio recording using a DAW, the author is able to adjust the values of the parameters of all processing applied to each source signal (illustrated in Figure 8.1). These parameter settings and their variations over time are stored and recalled for execution by the DAW automation system. Their combined effects in the master output signal can be auditioned during the creation and reflected in the final exported file. When exporting to an object-based file format, the DAW software converts these parameter values to per-object metadata that encode the intended spatial position irrespective of the playback configuration. Accordingly, at the consumer playback end, a functionally equivalent rendering engine can reproduce the intended immersive audio scene over a flexible loudspeaker configuration or over headphones with optional head-orientation tracking.

### *Physics-driven interactive immersive audio for gaming and XR experiences*

In nonlinear application scenarios such as video games or VR, the end-user may be placed in the role of a virtual listener located in a virtual acoustic environment containing one or more virtual sound source(s), some of which may be the voices of characters or avatars of remote end users. Types of end-user interaction simulate six-degree-of-freedom (6DoF) movement of the virtual listener and/or sound sources, including both translation and three-degree-of-freedom (3DoF) rotations. The spatial audio rendering operations are performed in real time by the program and are comparable in nature to those performed by a DAW mixing engine (Figure 8.1). The output format is determined at playback time according to the end user's headphone or loudspeaker playback configuration.

Unlike linear audio reproduction systems, an interactive audio engine must obey a format-agnostic 6DoF rendering application programming interface (API) to enable runtime audio sample triggering and control of audio signal processing and mixing operations by the game or playback software application [Geronazzo and Seraphin, 2023]. In addition to per-object positional and loudness parameters, this API should include reflection or reverberation effect properties, along with source directivity and orientation parameters to support 6DoF navigation of the listener and/or motion of individual objects. Table 8.1 illustrates the capabilities of several generations of APIs for interactive immersive audio rendering. In the late 1990's, Personal Computer (PC) sound cards introduced the ability to simulate per-object distance, orientation, directivity, and sound muffling effects caused by intervening acoustic obstacles [IA-SIG, 1998; IA-SIG, 1999; Hiebert, 2005]. Later extensions enabled simulating room acoustic effects for increasingly complex reflective or reverberant environments [Peacock et al., 2006; Jot et al., 2021; Jot and Cohen, 2025].

In video game and VR or XR programs, the audio rendering parameters are automatically computed and dynamically updated by a numerical propagation modeling function that derives them from a physical and geometrical description of the virtual acoustic environment containing the listener and the sound sources [Vorländer, 2008; Savioja and Svensson, 2015; Pind et al., 2022]. By tying these control data to a virtual 3D world description, the application developer can manage and maintain the correspondence between the audio and visual presentations, while affording six degrees of positional freedom to each of the sound sources and to the listener. Examples of propagation modeling include the estimation of acoustic reverberation parameters and of acoustic diffraction effects depending on the position of the source and listener and on the geometries and

*Table 8.1* Examples of interactive object-based immersive audio rendering API capabilities

| *Audio rendering API features* | *(3DoF)* | *I3DL1, DSound* | *I3DL2, EAX2* | *OpenAL EFX* | *MSA* |
|---|---|---|---|---|---|
| Direction of arrival *(per-object)* | • | • | • | • | • |
| Distance/proximity *(per-object)* | | • | • | • | • |
| Source orientation *(per-object)* | | • | • | • | • |
| Source directivity *(per-object)* | | • | • | • | • |
| Filtering effects *(per-object)* | | | • | • | • |
| Listener's room *(global)* | | | • | • | • |
| Multiple rooms *(global)* | | | | • | • |
| Grouped reflections *(per-object)* | | | | | • |

Adapted from [Jot et al., 2021]

material properties of obstacles, reflectors, and room walls. Some of these calculations must be executed at runtime in response to the movements of the sources or the listener, while others may be precomputed for efficiency [Liu and Manocha, 2022].

### *Flexible loudspeaker or binaural immersive audio distribution*

Support for flexible multi-channel loudspeaker listening configurations was first introduced commercially in the early 2010s to render interactive audio in PC sound cards. Later, automatic adaptation to the consumer's loudspeaker count and layout was introduced in home theater systems supporting modern immersive linear audio formats. Today, creators and developers of linear or interactive immersive audio experiences can expect automatic adaptation of the spatial presentation to the listener's loudspeaker or headphone configuration.

For linear immersive audio production and archiving, the Audio Definition Model (ADM) [Füg et al., 2016] offers a widely adopted open file format that can serve as precursor to low-bit-rate content encoding for distribution to consumers. Audio elements can be represented by any of these methods:

- *Scene-based* coding, where the audio signal is encoded in a multi-channel Ambisonic format [Nicol, 2018; Zotter and Frank, 2019]. This method is implemented in MPEG-H [Herre et al., 2015] and IAMF [Hwang and Lim, 2024].
- *Object-based* coding [Riedmiller et al., 2015; Tsingos, 2018] carries a collection of *audio objects* each assigned a fixed or time-varying position (Figure 8.2) on a virtual surface centered on the listener (Figure 8.3) or surrounding the audience. This option is implemented in MPEG-H, Dolby Atmos [Robinson et al., 2012] or DTS:X [Jot and Fejzo, 2011].
- *Channel-based* coding, equivalent to a set of audio objects assigned fixed positions matching industry-standard layouts [ITU, 2022], is the most common immersive audio encoding paradigm, supported in all the modern formats just referenced.

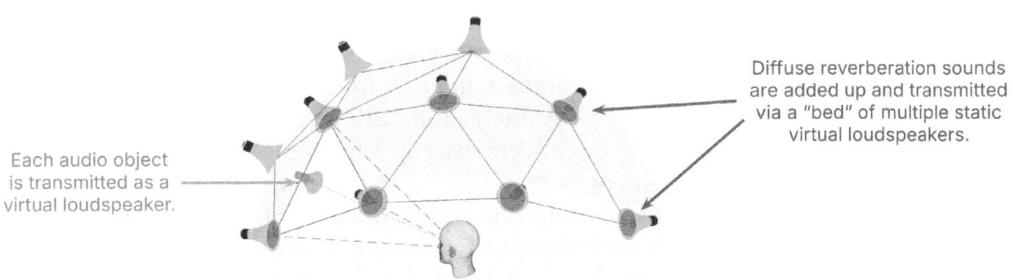


*Figure 8.3* In three-degree-of-freedom (3DoF) object-based immersive linear audio formats, each object represents a virtual loudspeaker assigned a fixed or time-varying angular position via per-object metadata. A fixed set of these objects, often referred to as the 'bed', is reserved for recording a mixture of the reverberation of multiple sound sources.

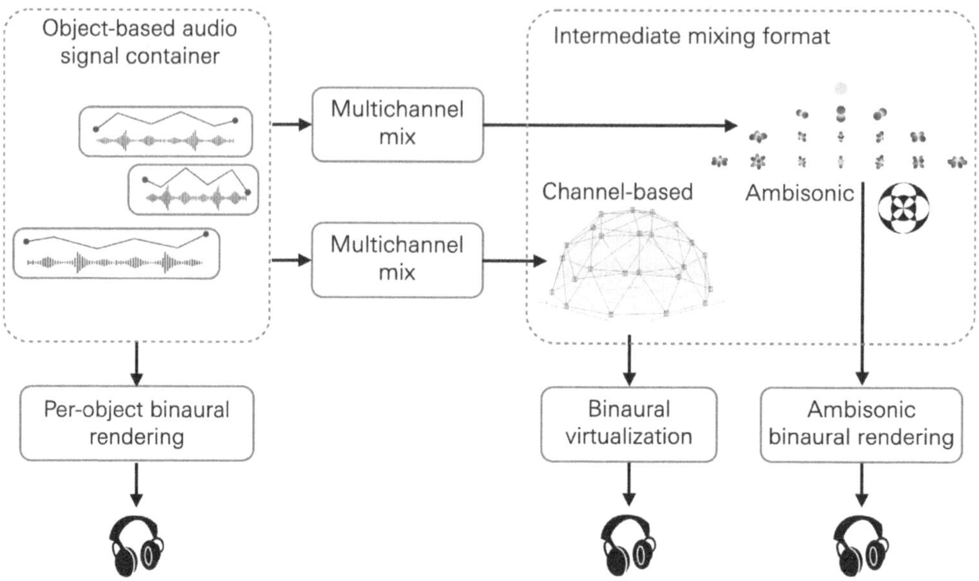


*Figure 8.4* Binaural output signal may be obtained by per-object binaural rendering or by binaural virtualization of a multi-channel mix.

Immersive audio reproduction over flexible loudspeaker configurations is needed in home theaters and in automotive environments. It is commonly realized via channel-based panning techniques (see e.g. [Pulkki, 1997; Tsingos, 2018]) or Ambisonic methods (see e.g. [Zotter and Frank, 2019]).

Many application scenarios require binaural reproduction over wearable devices such as earphones or smart glasses. In this case, a per-object positional rendering function can be realized by implementing a pair of dynamic head-related transfer function (HRTF) filters [Roginska, 2018; Jot et al., 1995; Zotkin et al., 2004]. Computationally efficient implementations of object-based binaural rendering are realized by computing and virtualizing an intermediate multi-channel mix (see Figure 8.4). The choice of intermediate encoding format may, however, undermine rendering fidelity. In the case of Ambisonic mixing, as in [Noisternig et al., 2003; Rafaely et al., 2022], this method is perceptually equivalent to the binaural rendering of a scene-based coded signal. Object-based binaural rendering methods advantageous in both fidelity and computational efficiency employ bilateral multi-channel mixing, where the left and right ear signals are mixed separately (as illustrated in the section 'Computationally efficient multi-room 6-DoF immersive audio renderer' and Figure 8.14).

By delivering a pair of audio signals individually to the two ears of the listener, binaural rendering methods aim to produce an immersive sound impression comparable to real-world listening or to reproduction over a multi-channel loudspeaker system. In headphone or wearable audio playback, factors that impact the success of this illusion include HRTF individualization, head-tracking, and listening room acoustic simulation [Roginska, 2018; Best et al., 2020; Frankenbach et al., 2025]. An approach to enhancing externalization and restoring front/back discrimination in headphone immersive audio is introduced in [Jot et al., 2022]. Immersive binaural rendering can also be achieved via loudspeaker arrays or soundbars, in desktop or near-field personal configurations, by combining cross talk cancellation processing, beamforming, and ear position tracking [Choueiri, 2018; Gálvez et al., 2019; Gardner, 1997].

## *Limitations and fragmentation challenges addressed in this chapter*

Despite remarkable advances in the ability to create and commercialize compelling immersive music recordings and cinema or game soundtracks, the democratization and success of immersive audio technology has been hindered by distribution ecosystem fragmentation and content creation constraints.

### *Creation: facilitating aesthetically motivated immersive sound design*

In authoring linear or nonlinear immersive audio experiences, with or without accompanying picture, creators face limitations affecting artistic expression and practical workflow.

- **Plausibility versus physical modeling:** Designing reverberation effects by specifying the geometry and wall materials of a virtual room can be artistically limiting, even with the support of accurate acoustic simulation software. In established music and soundtrack production practices, creators or sound designers may wish to apply aesthetically motivated sound effect processing, free of a constraint of prescriptive visual correspondence or simulation of an authentic acoustic environment [Filimowicz, 2019].
- **Democratization:** DAW software frameworks were originally designed for the mixing and mastering of recordings in channel-based formats. Producing object-based linear immersive audio often implies unfamiliar workflows or routing configurations to ensure compatibility with recently developed and evolving recording formats. Although object-based audio is in theory amenable to monitoring via headphones or low-cost loudspeaker arrays, it is often observed that the ability to critically monitor and qualify immersive audio creations involves substantial investment in studio equipment hardware, including multi-channel loudspeaker and amplification apparatus. There is a need to facilitate and democratize the creation of immersive audio content, toward making it as familiar and accessible as mixing and mastering in conventional stereo formats.

### *Deployment: unlocking immersive audio experience ecosystems*

In gaming or interactive XR applications, as previously mentioned, creation and playback systems must support the rendering of sound scenes in which the position and orientation of the listener or any sound source are determined at playback time, rather than prescribed at the authoring stage. Predictability of the end user quality of experience is facilitated by the fact that the different consumer platforms (developed by companies such as Sony, Meta, or HTC) are tightly associated with specialized authoring software supplied by the same vendors. In linear applications such as music or movies, on the other hand, consumers expect that any playback device or appliance, such as a TV set, should operate in a manner agnostic to the authoring tool employed to produce the soundtrack.

Ostensibly, as previously mentioned, both interactive and linear immersive audio technologies can leverage an object-based representation of the audio scene delivered to a listener. This notable technology advancement enables distributing immersive audio content that adapts automatically to the consumer's playback setup, including their particular loudspeaker layout. Despite this commonality, fundamental differences and limitations have hindered technology adoption and the dissemination of immersive audio content:

- **Interoperability:** Immersive audio content markets and ecosystems are fragmented between linear and nonlinear media and across multiple incompatible interactive media authoring and consumption platforms commercialized by competing companies (such as Sony, Apple, or Meta)
- **Personalization:** In object-based linear immersive audio recording formats, end-user interaction or customization (such as remixing, karaoke, or character language replacement) is inherently limited because, in standard practice, the reverberation components of multiple sound sources are summed up into a fixed set of additional recorded audio waveforms, commonly referred to as the immersive recording's "bed" [Riedmiller et al., 2015] (Figure 8.3).
- **Navigability:** In these linear formats, the audio objects composing the scene are assumed to be localized at a fixed distance from the listening position (see Figure 8.3) and are lacking sound source directivity and orientation properties. This contrasts with interactive audio experiences, which allow free 6DoF movement of the virtual sound sources and/or the virtual listener.

### *Organization of this chapter: from IRCAM `Spat` to Acoustic Objects*

In the next section, we describe an example of a perceptually based audio creation platform, IRCAM `Spat`, which enables plausible aesthetically motivated immersive music creation and performance, circumventing dependency on physical modeling of acoustic environments. The section entitled 'Acoustic Objects: bridging immersive audio creation and distribution systems' advocates for alleviating ecosystem fragmentation, by showing (a) how a universal device-agnostic immersive audio rendering model can support the creation and distribution of both physics-driven interactive audio applications and aesthetically motivated immersive audio content; (b) how object-based immersive linear audio content formats can be extended, via the notion of Acoustic Objects, to support reverberant object substitution (including language replacement), musical remix, or end user interaction (including 6DoF navigation in a recorded immersive audio scene). The section 'Unifying immersive audio experiences' concludes this chapter by drawing the evolution of immersive audio technology toward accelerating commercial deployment and enabling rich end-user personalization, in any linear or interactive entertainment or business application.

## IRCAM `Spat`: an aesthetics-driven immersive audio creation and performance platform

### *Designing a spatial reverberation system emphasizing plausibility*

At its inception (in the late 1990s), the *Spatialisateur* project aimed to develop a spatial reverberation processing system controlled directly via the set of perceptual attributes uncovered by the psycho-experimental study reported in Appendix A – which yields an efficient high-level parametrization of the perception of room acoustical effects in concert halls, opera houses, or auditoria. In this section, we review the original design and subsequent enhancements of the `Spat` library developed by IRCAM [Carpentier et al., 2015; Carpentier, 2024] in the Max/MSP programming environment [Max/MSP].

#### *Perceptually motivated high-level user interface*

Figure 8.5 displays the high-level user interface (referred to as `Spat_OPer`) implemented in the 1995 release of the `Spat` software library for a single sound source. It includes four main categories of control parameters, also shown in Figure 8.7:

- The `localization` of the sound source relative to the virtual position and orientation of the listener is represented via azimuth and elevation angles expressed in degrees, with distance expressed in meters.
- A set of nine `perceptual factors` (vertical sliders), each associated with a measurable `acoustical criterion` as described in Appendix A, jointly specify acoustic reverberation and reflection effects.
- The `radiation` properties of the sound source are specified, as detailed in Figure 8.10, via its orientation relative to the position of the listener, specified via `yaw`, `pitch`, and `roll` angles, as well as its directivity properties, specified by an aperture angle and two spectral correctors: `axis` (on-axis spectral correction) and `omni` (diffuse power spectrum average)
- The configuration of the loudspeaker or headphone system employed for auditioning/monitoring the immersive audio output signal.

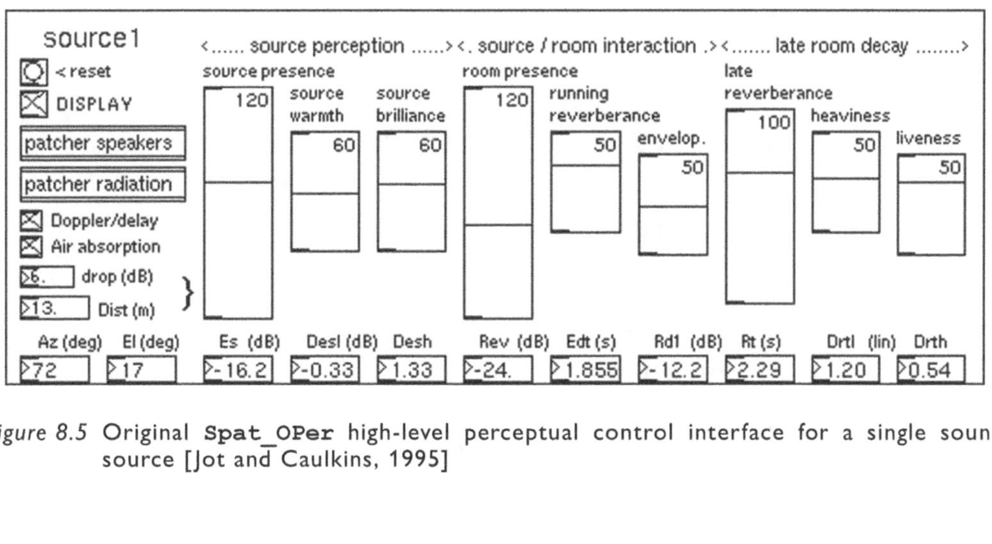


*Figure 8.5* Original **`Spat_OPer`** high-level perceptual control interface for a single sound source [Jot and Caulkins, 1995]

*Derivation of low-level reverberation rendering parameters*

Each one of the `acoustical criteria` which together form the high-level description of the reverberation effect can be expressed from energetic measures derived from a segmentation of the renderer's impulse response over four temporal sections and three frequency bands (Figure 8.6), assuming:

- Frontal incidence of the direct sound (R0)
- Random incidence of the early reflections (R1) within a frontal angular sector in the horizontal plane (60° wide)
- Isotropic diffuse distribution of late reflections (R2) and reverberation tail (R3)
- Time limits (l1, l2, l3) equal respectively to 20, 40, and 100 ms relative to the time of arrival of the direct sound
- Reference low and high frequencies (fl, fh) equal respectively to 250 Hz and 4 kHz.

The Appendix section 'Expression of the perceptual attributes' reviews the set of equations that relate the high-level `acoustical criteria` to these energetic measures, per [Kahle, 1995]. The implementation of the perceptual control paradigm requires performing the inverse conversion in order to recompute the time-frequency-energy distribution following an update in the settings of the `acoustical criteria`. This high-level to low-level control mapping function executes the calculations reviewed in Appendix A, section 'Derivation of the low-level rendering parameters' [Jot et al., 1996; ISO (Int. Org. for Standardization), 2005]. Each `acoustical criterion` is paired with a `perceptual factor` exposed in the `Spat_OPer` control panel. The scaling of each slider in the graphical user interface (Figure 8.5) takes into account the relative sensitivity of the `perceptual factors`, averaged across listening test participants and reported in Table 8.3.

Optionally, the set of calculated low-level processing parameter values may be further corrected by a *context compensation* operator (Figure 8.7), which automatically performs, in the time-frequency-energy domain, a deconvolution accounting for the acoustic response of a loudspeaker reproduction system's environment [Jot et al., 1996; Jot, 1997].

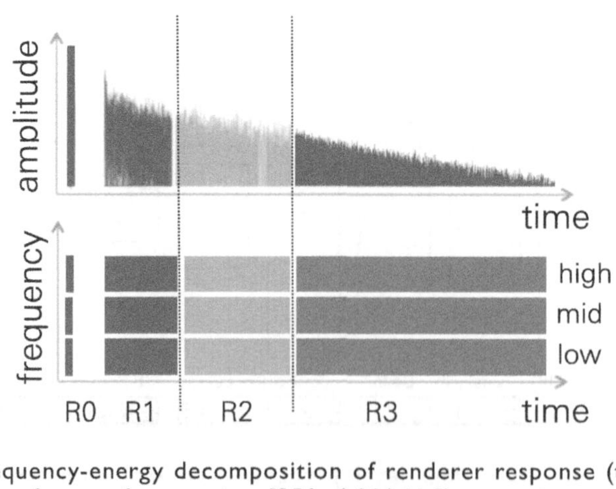


*Figure 8.6* Time-frequency-energy decomposition of renderer response (time limits: [20, 40, 100 ms]; reference frequencies: [250, 4,000 Hz])

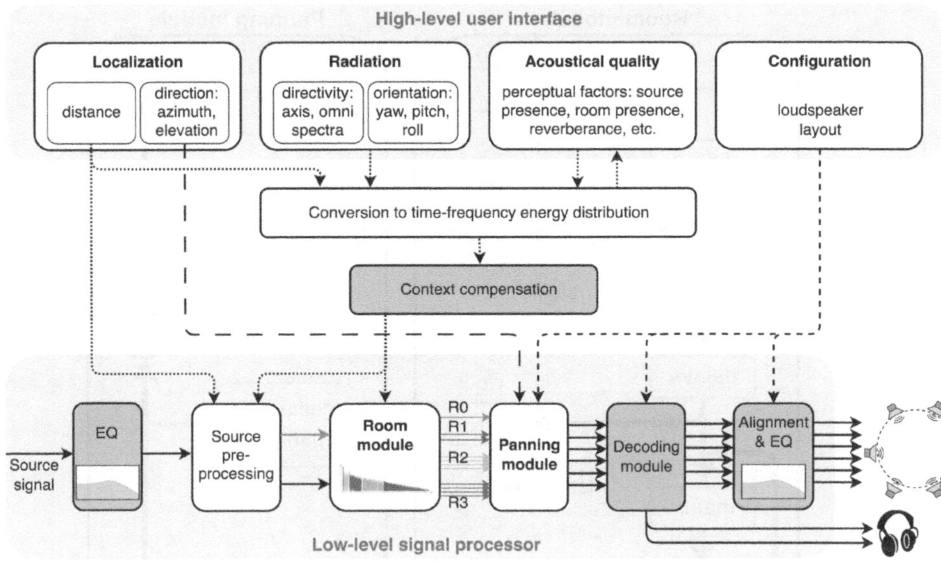


*Figure 8.7* Structure of a single-source Spat processing module, showing the control paths from application-level representation to low-level renderer parameters. Modules in gray are optional.

### *Spat renderer algorithm for a single sound source*

In addition to room acoustical descriptors, the high-level user interface `Spat_OPer` exposes controls for the `localization` and the `radiation` of the sound source. The internal topology of the core rendering functions, the *Room* and *Panning* modules, is shown in Figure 8.8 [Jot, 1997, 1999]. An archetypal example of impulse response produced by the system is displayed in Figure 8.9.

The four elementary filters (R0 . . . R3) are three-band spectral correctors that shape the time-frequency-energy distribution in the reverberator response following adjustments of the `perceptual factors`. Additionally, for a directional sound source, their gain settings are offset according to the source's `radiation` parameters: its orientation and its frequency-dependent directivity pattern, defined by its `axis` and `omni` spectral corrections and an '*inside cone*' `aperture` angle parameter (for details, see Jot et al. (2021), regarding parameter definitions, and Appendix for implementation in `Spat`).

The *Panning* module controls the perceived direction of sound arrival relative to the listener's head orientation, by adapting the output signals of the *Room* module to a selected positional audio rendering method. In its initial release, the `Spat` library supported several panning techniques (including stereo, pairwise, Ambisonic, and binaural) and various planar loudspeaker configurations [Jot and Caulkins, 1995]. A nonzero setting of the `azimuth` angle is equivalent to a rotation of the listener's head in the opposite direction, applied to both the direct sound (R0) and the synthetic early reflections (R1) by the elementary *Panner* modules (see Figure 8.8). This preserves the spatial relations between the components (R0 . . . R3), thus ensuring the conservation of the acoustical criterion `Es` and of its corresponding perceptual factor, `Source Presence`.

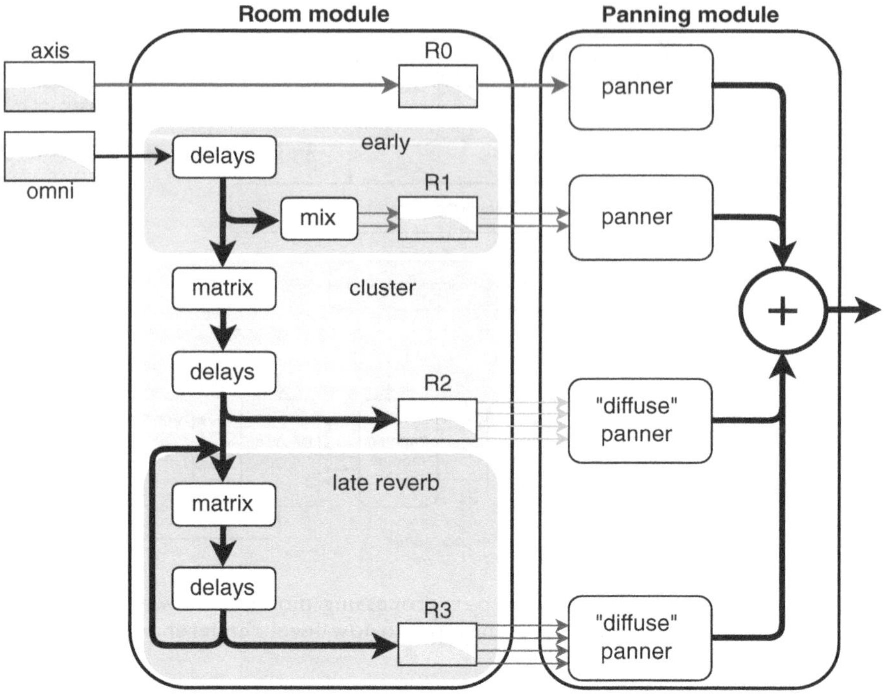


*Figure 8.8* Core `Spat` rendering algorithm architecture (*Room* and *Panning* modules) for a single sound source: Thicker lines indicate multi-channel audio signal paths.

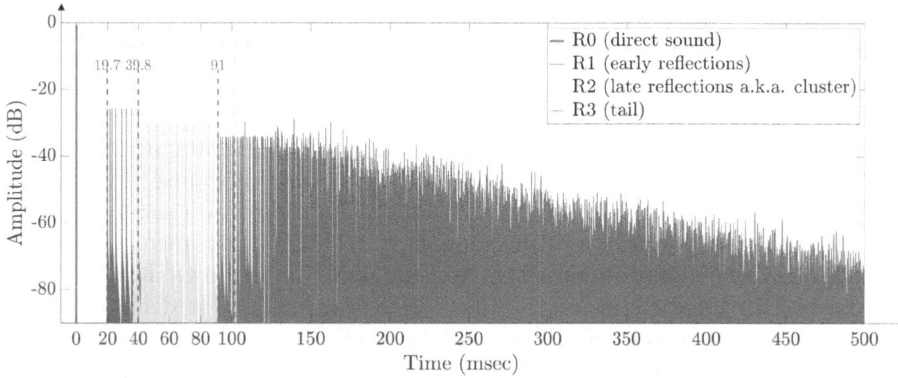


*Figure 8.9* `Spat` impulse response for a particular setting of the low-level temporal settings min, max, and distr of the early, cluster, and reverb segments (see user control interface in Figure 8.10(g))

Independently, as noted at the end of Appendix section 'Overview of psycho-experimental procedure and results', the perceived source-to-listener distance is controlled by offsetting `Es` so that increasing distance has the effect of reducing the energies R0 and R1, per equations (8.12), (8.13) of Appendix.

## Extensions from the initial renderer design and audio object model

Following its initial design, the Spat rendering framework and its `Spat_OPer` control interface were implemented in various forms – including *external* modules in the Max/MSP environment [Carpentier et al., 2015; Carpentier, 2021]; modules in the *Open Music* and *o7* computer-aided composition frameworks [Garcia et al., 2017; Bresson et al., 2017]; digital audio workstation (DAW) plug-ins [Spat Revolution]; devices for Ableton Live [Ableton Live; Spat for Live; XP4L]; standalone applications [Carpentier, 2016; Spat Revolution]; or hardware processor [Holophonix]. While the original design architecture (Figure 8.7) was retained, several key features were extended or altered.

### Multi-object, multi-room implementations

The system was extended to facilitate handling multiple sound objects and several rooms concurrently. The `Spat_OPer` user interface was revised to expose tabs presenting control panels for the different sources and rooms (Figure 8.10). When several sources are assigned to the same room, their contributions sum at the multi-channel input stage of the late reverberation computation module (gray box labeled 'R3' in Figure 8.8), thereby sharing its processing cost.

*Panoramix* is a standalone software application for immersive audio mixing and production. It relies on the same multi-object and multi-room audio engine and facilitates hybrid mixing (combining various panning techniques) along with concurrent distribution in multiple immersive audio output formats [Carpentier, 2016]. As shown in Figure 8.11, its graphic user interface simulates the mixing control panel of a software DAW or of a hardware mixing console.

### From 2D to 3D immersive audio

The *Panning* module supports recently developed multi-channel positional audio rendering techniques (see e.g. [Spors et al., 2013; Roginska and Geluso, 2018; Xie, 2022]), including binaural, VBAP, wavefield synthesis and high-order Ambisonics, enabling object-based immersive audio spatialization along both azimuth and elevation. This extension to 3D loudspeaker geometries required overcoming limitations of the original design, such as the directional distribution of the early reflections (R1) and the number of audio channels needed for reproducing the spatial impression of a diffuse reverberant sound field [Carpentier et al., 2015]. The latter is a topic of ongoing research [Hiyama et al., 2002; Walther and Faller, 2011; Agrawal and Braasch, 2018; Kirsch et al., 2021] both for channel-based reproduction (where the '*diffuse*' *panner* module is a diffusion matrix) and scene-based rendering (where it Ambisonically encodes the R2 and R3 output channels to simulate impinging plane waves).

### Beyond auditoria and concert halls

The psycho-experimental study presented in section 'Overview of psycho-experimental procedure and results' spanned a limited range of acoustical conditions typically

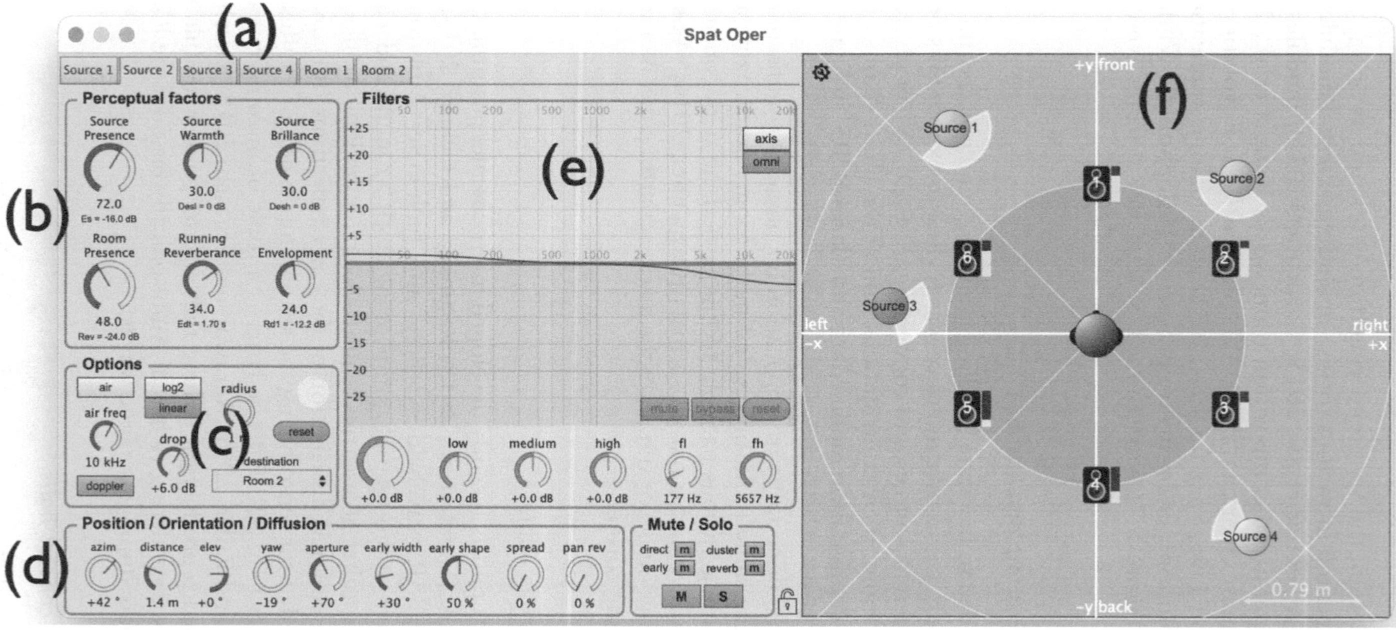


*Figure 8.10* Multi-object, multi-room `Spat_OPer` control panel: (a) Tabs for selecting sources and rooms; (b) `perceptual factors` for a selected source; (c) assignment of a source to a room; (d) additional source parameters; (e) `axis` and `omni` spectral settings for a selected source; (f) graphical representation of the scene (top-down view); (g) settings of a selected room; (h) detailed view of the `acoustical criteria` for a selected source.

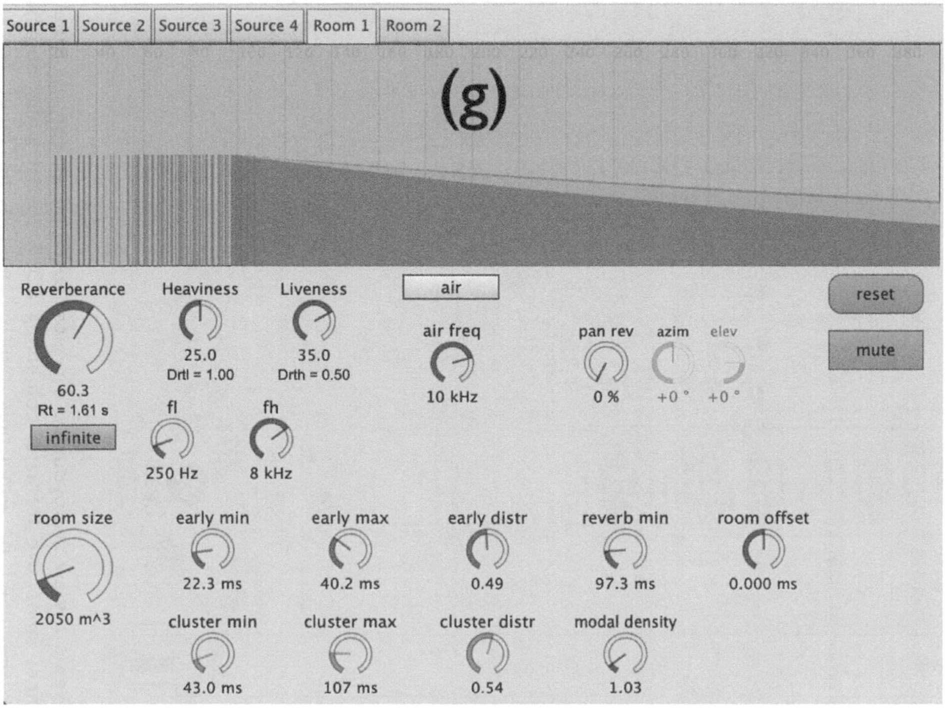


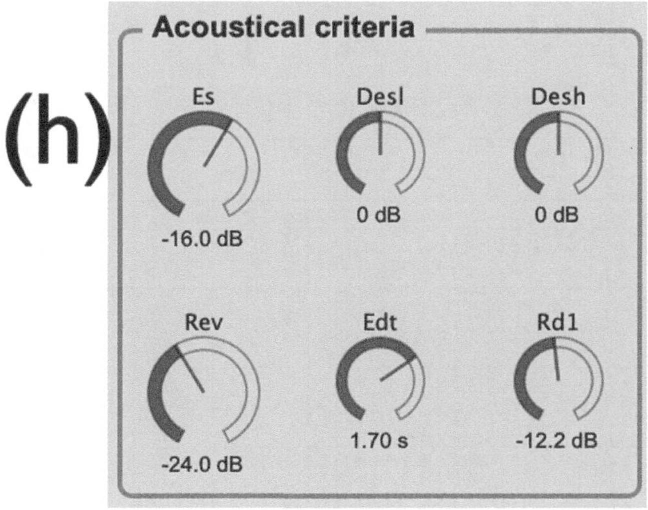


*Figure 8.10* (Continued)

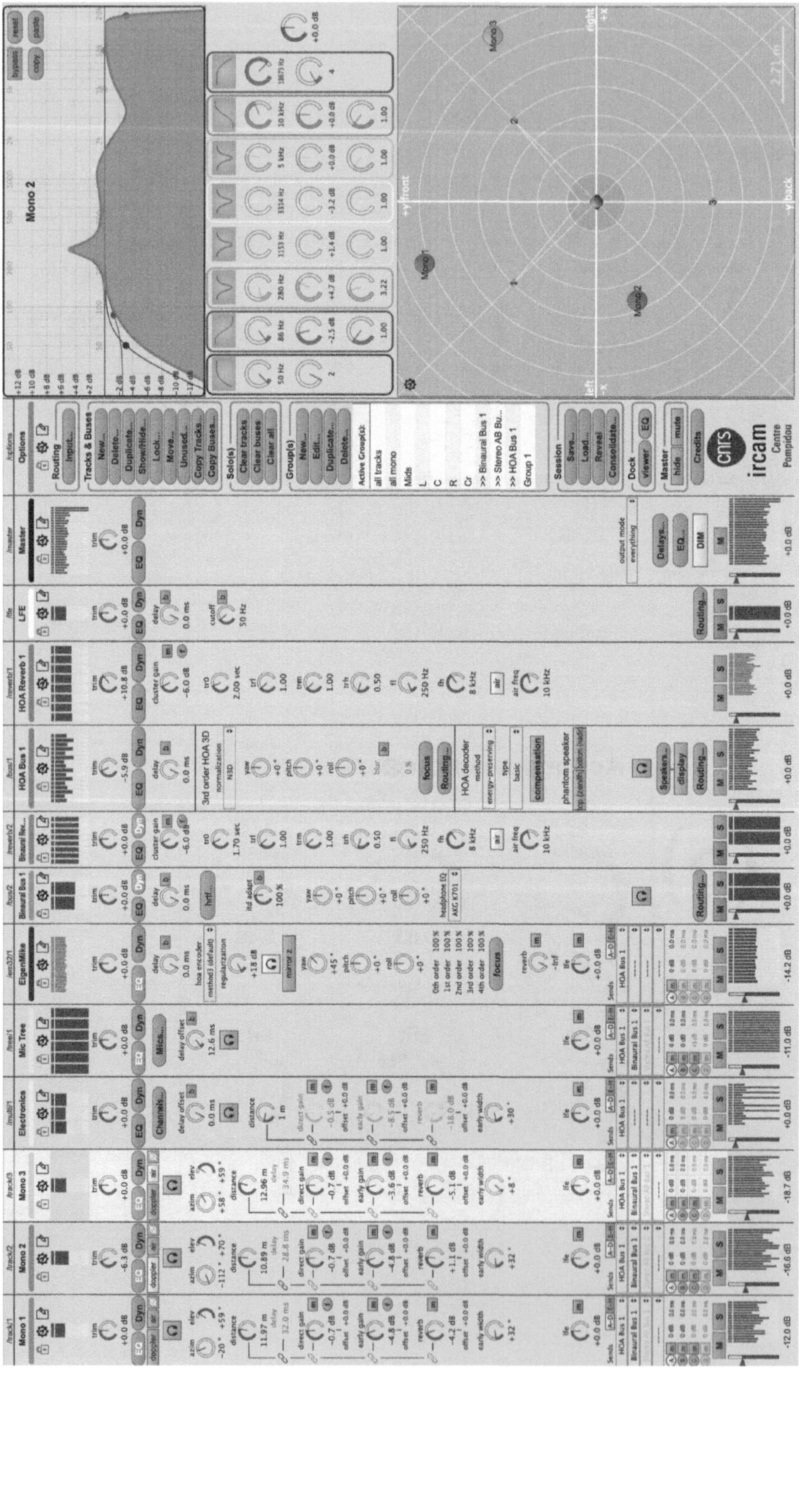


*Figure 8.11* User interface of the standalone application *Panoramix* presents a view analogous to a typical DAW mixer control panel. It allows for linear audio production in various multi-channel immersive formats, combining 6DoF spatial audio objects with microphone-captured or pre-recorded stereo, pairwise-panned or Ambisonic multi-channel audio input tracks. [Carpentier, 2016]

representative of music stimuli in midsize auditoria. For artistic or simulation applications, it is desirable to cover a larger class of room types (from bathrooms to cathedrals). The meta-parameter `room size` has the effect of stretching the time limits of the R1, R2, and R3 sections (see Figures 8.9 and 8.10[g]) according to the desired room volume. This simple approach has proven satisfactory for some artistic usages. In the current implementation, the calculation of the `acoustical criteria` is performed according to equations (8.2)–(8.11) provided in Appendix, even when the values of the time limits (l1, l2, l3) differ significantly from their default settings.

### *Hybrid reverberation and analysis/synthesis*

In [Jot et al., 1997], a procedure based on the notion of energy decay relief (EDR) was proposed for the time-frequency analysis/synthesis and denoising of room impulse responses, facilitating the estimation of the time-frequency-energy distribution and reverberation decay parameters that control the `Spat` reverberator. By applying Equations (8.2)–(8.11) provided in Appendix, the high-level perceptual control parameters exposed in `Spat_OPer` are calculated automatically to simulate an existing, measured, reverberation response. This mapping to the perceptual domain enables natural sounding alterations and interpolations between room acoustical conditions. In order to support the authentic simulation of a measured acoustic response while exposing a highly flexible perceptually based user control, an alternative realization of the *Room* module was developed [Carpentier et al., 2014], in which the early response (R1, R2) is rendered by a convolution-based algorithm, whereas the time-frequency envelope of the reverberation decay (R3) is accurately reproduced by a feedback delay network (FDN), provided that the measured reverberation response presents an exponentially decaying time-frequency envelope.

### *Anisotropic late reflections and reverberation*

The canonical reverberation response model presented in the section 'Derivation of low-level reverberation rendering parameters' assumes that late reflections (R2) and reverberation (R3) conform to an ideal isotropic sound field model agnostic to the listener's orientation. This assumption, theoretically valid inside a *mixing room* [Polack, 1993], does not necessarily generalize to semi-open spaces, environments presenting elongated or irregular shapes, or enclosures comprising acoustically coupled volumes [Kuttruff, 2016]. Recent studies investigate the perception of the anisotropy of a reverberant field, the development of objective metrics for the characterization of directional reverberation, the use of compact microphone arrays for its measurement and analysis, and techniques for its synthesis and reproduction [Romblom et al., 2016; Nolan et al., 2018; Alary and Politis, 2020; Massé et al., 2020; Alary et al., 2021; Hold et al., 2022].

In `Spat`, the "*diffuse*" *panner* module (Figure 8.8) implements a *divergence panning* algorithm that assigns uncorrelated input signals to an array of notional sources distributed along an arc having adjustable extent (Figure 8.12). The automatic hybrid reverberation analysis/synthesis procedure just described may be extended to the modeling of anisotropic reverberation decays.

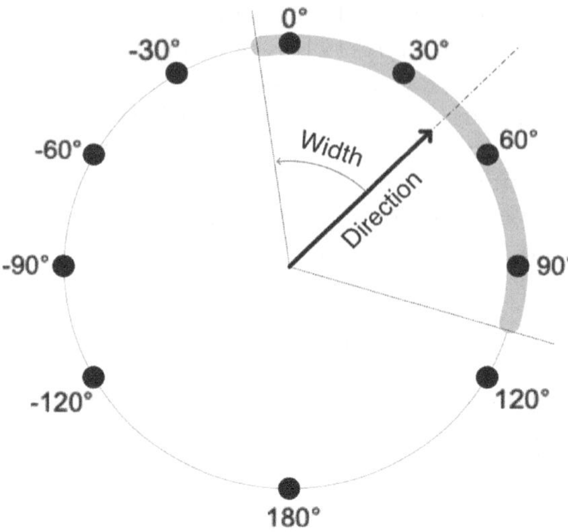


*Figure 8.12* Representation and rendering of the angular position and extent of a sound, employing a normalized *divergence panning vector* whose magnitude spans from 0.0 for a diffuse sound component to 1.0 for a point source [Jot et al., 2006]

## Acoustic Objects: bridging immersive audio creation and distribution systems

In this section, we review the technology evolution illustrated generally in Figure 8.13, prolonging the perceptual immersive sound design concepts developed in `Spat` to overcome the artistic creation, democratization, and ecosystem fragmentation obstacles previously outlined.

First, we review a general spatial audio rendering engine optimized for computationally efficient physics-driven VR or XR applications and show that it is compatible with immersive sound design according to the `Spat` model, reviewed previously and taken here as an archetype of perceptually motivated immersive audio creation framework. This yields a portable, application-agnostic spatial audio rendering engine that is equally adept at physics-driven acoustic simulation and at perceptually motivated sound design in applications such as:

- Immersive entertainment, including games, sports, music, and virtual concerts.
- Social copresence, including virtual meetings, remote collaboration, and virtual assistants.
- Virtual travel or tourism, including navigation and situational awareness.
- Assisted training, education, diagnosis, surgery, design, or architecture.

Then we introduce the notion of Acoustic Object, extending object-based immersive audio metadata specifications to enable application-agnostic scene representation and

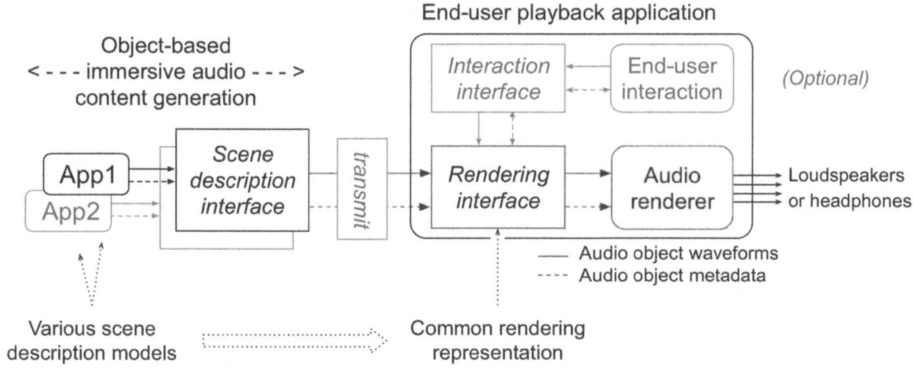


*Figure 8.13* Interoperable ecosystem synopsis: From application-level audio scene generation to low-level universal spatial audio rendering interface, with optional end user interaction. The content generation scenario may be linear or interactive, perceptually or physically driven. For any chosen application-level scene representation paradigm, a *scene description interface* translates its audio scene representation to an object-based rendering description compliant with the low-level *rendering interface*, thus ensuring faithful reproduction on any compliant device. Optionally, an *interaction interface* allows playback-end repositioning or substitution of audio objects received as Acoustic Objects. The intermediate *transmit* function allows for the juxtaposition of several concurrent audio scenes generated by independent applications.

optional playback-end interaction in recorded or broadcast immersive audio scenes. Applications include:

- In fixed- or free-viewpoint VR or TV presentation of live music or sport events: incorporating sound elements that coincide spatially with displayed visuals.
- In cinematic movie or video playback, facilitating dialogue translation with automatic matching of the original sound design qualities.
- In music or soundtrack production, exporting stems that are readily remixable, navigable, or amenable to karaoke-like performance, to the extent authorized by content rights owners.

### ***Universal 6DoF spatial audio rendering engine and interface***

In [Jot, 2012; Jot et al., 2021; Jot and Cohen, 2026], a generic object-based rendering model is proposed for interactive 6DoF spatial audio and multi-room reverberation, emphasizing auditory plausibility and computational efficiency. By favoring a perception-based construction of the rendered audio output, it aims for independence from application-level scene generation strategies.

#### *Object-based audio rendering interface for interactive XR and gaming*

The *rendering interface* encompasses the OpenAL EFX feature set [Peacock et al., 2006], summarized in Table 8.1, and extends it to enable efficient per-object parametric control of the spatialization of grouped early reflections. In order to meet the requirements of interactive gaming and XR scenarios, it allows for multiple virtual sound sources to be seamlessly instantiated by applications at rendering time and for assigned dynamic position and orientation within a navigable multi-room acoustic environment [Jot et al., 2021; Jot and Cohen, 2026]. Each sound source is represented as an audio object characterized by its frequency-dependent radiation pattern (a catalog of acoustic source directivity properties can be found, for instance, in [Ackermann et al., 2024]).

An update in the position or orientation of the virtual listener or of a sound source triggers updates of low-level renderer parameters. The acoustic environment may be managed dynamically as a navigable set of virtual rooms, each characterized by a 'reverberation preset' that encodes its response to an omnidirectional sound source located at a reference distance. This encoding incorporates the notion of the *reverberation fingerprint* of a room, which affords a data-efficient characterization of the perceptually relevant attributes of acoustic reverberation that are independent of source or receiver properties and positions in the room: the reverberation fingerprint is simply specified by the volume of the room and its frequency-dependent reverberation decay time [Jot and Lee, 2016].

In music creation and performance scenarios, the spatial audio scene composition conceived by the author or sound designer may be most effectively represented in the perceptual domain [Filimowicz, 2019]. A representative example of this approach is the perceptually based scene description paradigm implemented in `Spat`, described previously in the section 'IRCAM Spat: an aesthetics-driven immersive audio creation and performance platform'. In the following, we demonstrate the compatibility between this perceptually based paradigm and the generic spatial audio renderer considered here.

### *Computationally efficient multi-room 6DoF immersive audio renderer*

Figure 8.14 displays the architecture of a computationally efficient spatial audio rendering engine compatible with flexible loudspeaker-based or binaural playback configurations, previously introduced in [Jot, 2012; Jot et al., 2021]. It presents the following features:

- **Per-source processing:** Rendering of direct sound, acoustic reflections, and reverberation sound field for arbitrarily complex audio scenes comprising multiple audio objects, including both point sources and spatially extended sound sources, with minimal incremental computation cost per audio object
- **Effect processing:** A shared set of *effect slots* that can include multiple parallel reverberators to simulate complex acoustic environments, along with a spatialized reflections synthesis module (for discussions on the modeling and perceptually plausible rendering of early reflections, see for instance [Gospodarek et al., 2022; Jot et al., 1995; Meyer-Kahlen et al., 2024])
- **Anisotropic reverberation:** Reverberator output *pan* modules enable directionally weighted or spatially diffuse rendering of the acoustic reverberation emanating from the room in which the virtual listener is located, as well as several adjacent virtual rooms simulated by additional reverberators.

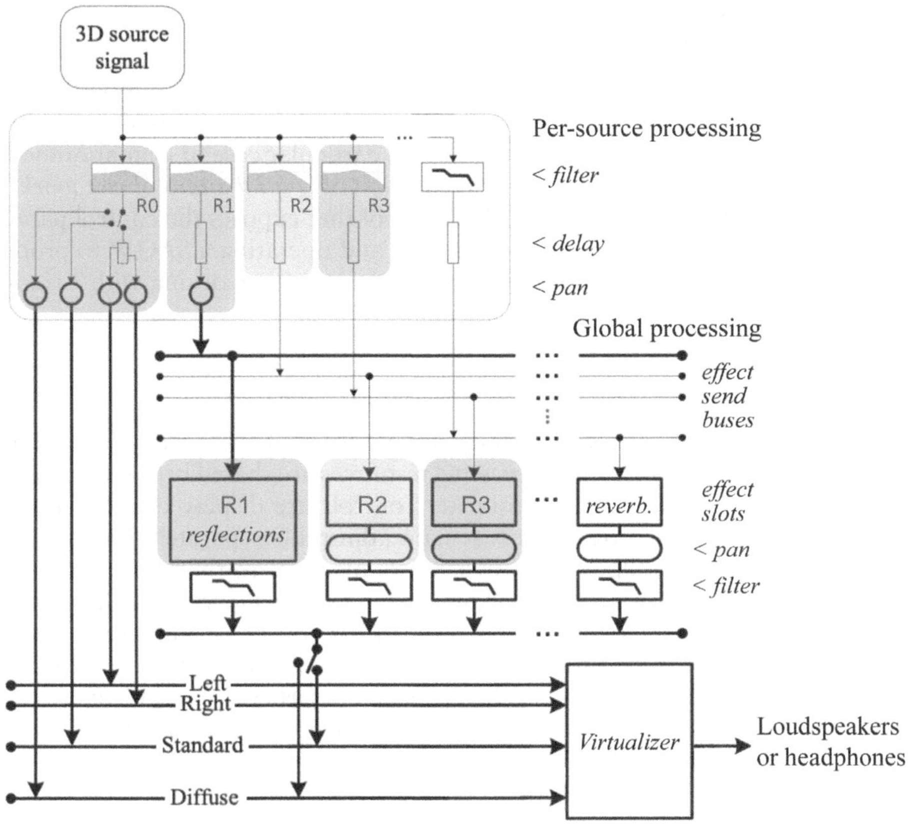


*Figure 8.14* Architecture of a generic immersive audio renderer for computationally efficient multisource/multi-room object-based spatial audio over loudspeakers or headphones. Thicker lines indicate multi-channel audio signal paths or processing functions. Labels (R0 . . . R3) illustrate the realization of a single-room renderer perceptually equivalent to `Spat's` low-level renderer algorithm, shown in Figure 8.8. Adapted from [Jot et al., 2023]

The shared *virtualizer* function can support various immersive loudspeaker or headphone playback modes, complementing the positional audio panning methods employed in the per-source direct-sound *pan* modules:

- *Standard* channel-based or scene-based (Ambisonic) methods [Tsingos, 2018; Spors et al., 2013]
- *Left/right* bilateral multi-channel methods [Larcher et al., 2000; Jot et al., 2006; Ben-Hur et al., 2021]
- *Diffuse* bus dedicated to the rendering of spatially extended sounds [Jot et al., 2006; Potard and Burnett, 2004].

## *Adaptation to aesthetically motivated immersive audio creation*

In this section, we demonstrate that the generic renderer of the section on 'A universal 6-DoF spatial audio rendering engine and interface' can be configured to emulate the Spat renderer of Figure 8.8 and can be viewed as a particular case of Digital Audio Workstation rendering engine. We hence define the notion of *spatial audio object workstation* (SAOW) as a linear immersive audio authoring tool that exposes the control parameters of the generic renderer. For a creator, configuring and operating a SAOW to produce an object-based immersive audio recording is as user-friendly as working with a traditional DAW to mix in conventional stereo or surround formats.

### *Perceptually based renderer configuration and control model*

Figure 8.14 illustrates a `Spat`-compatible configuration of the generic audio renderer, where reflections and reverberation processing resources are shared between audio objects assigned to the same room (whose parameter controls are displayed in Figure 8.10(g), while the other `Spat_OPer` parameters remain adjustable separately for each source). The per-source parametric filters (R0 . . . R3) are identical to those of Figure 8.8. The effect slots labeled R1, R2, and R3 each contain a reflections or reverberation processor configured to simulate the corresponding segment of the `Spat` impulse response (see Figures 8.8 and 8.9). The per-source processing block adjusts onset time and spectral correction for each of the early and later reflections and reverberation segments.

This configuration exercises the following renderer properties (detailed in [Jot et al., 2021]):

- Per-source processing parameters:
  - `Direct_pan` (azimuth and elevation angles)
  - `Reflections_pan` and `Reflections_focus`
  - Effect send delays, gains, and filters (R0 . . . R3)
- Global processing parameters:
  - Effect algorithm loaded in each of the active *effect slots* (reverberator or reflections processor)
  - Reference frequencies (fl = 250 Hz, fh = 4 kHz)
  - Reflections time span (for R1: 20 ms; for R2: 60 ms)
  - Reverberation `Decay_time`, `Decay_hf_ratio`, `Decay_lf_ratio`, and `Density` (for R3)

In order to support a perceptually based representation, the low-level rendering interface must be listener-centric: object position and orientation coordinates are defined relative to the listener's coordinates and head pose. Although the notion of 'room' may be exposed at the application level (as in the `Spat_OPer` interface of Figure 8.10), the renderer sees its task merely as a superposition of sound components (direct sound, reflections, reverberation), each specified via a minimal set of rendering commands. Similarly, the source's orientation and directivity parameters are not required in the low-level rendering interface.

## Spatial audio object workstation *for linear content authoring*

Immersive music and soundtracks can be created without deviating from the familiar practices adopted for mixing legacy channel-based stereo or surround recordings and exported in any legacy or modern immersive audio format. Examples of perceptually motivated high-level control interface are given by `Spat_OPer` (Figure 8.10) and *Panoramix* (Figure 8.11). The corresponding low-level audio renderer may be realized by the `Spat` renderer topology of Figure 8.7 or the generic renderer of Figure 8.14. As shown in Figure 8.15, this renderer can be thought of as a particular case of a Digital Audio Workstation.

Unlike traditional DAWs, the *spatial audio object workstation* of Figure 8.15 features a non-standard *effect send bus* format determined by the configuration of the *effect slots.*

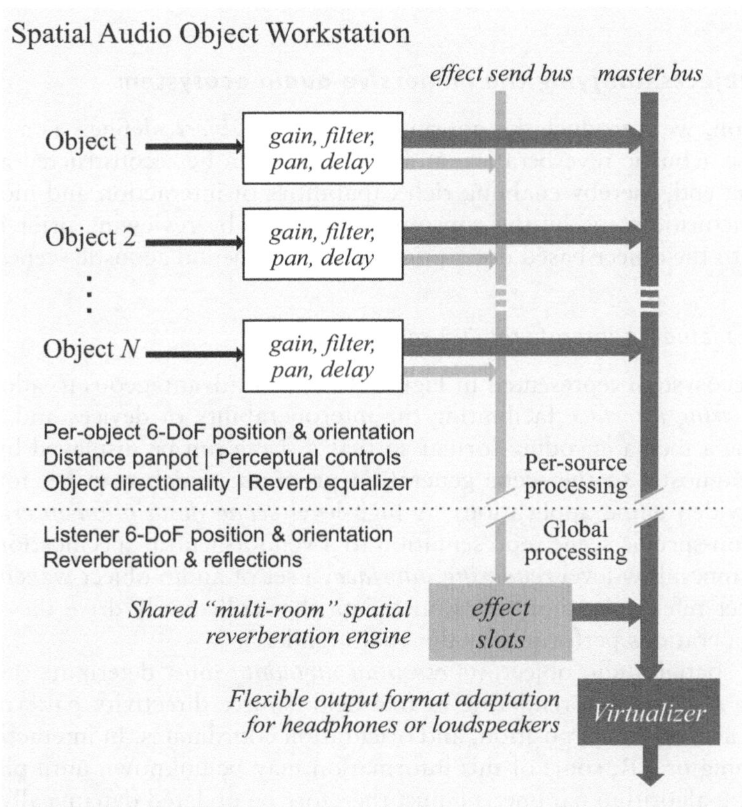


*Figure 8.15* Realizing a *spatial audio object workstation* by mapping the generic renderer of Figure 8.14 to a standard digital audio workstation signal flow (Figure 8.1). This exposes in the familiar DAW framework the functionality illustrated in Figures 8.10 and 8.11, including specific sound design features reviewed in the section '*Spatial Audio Object Workstation* for linear content authoring'. The per-object processing functions and the signal formats of the internal multi-channel summing buses (the *effect send bus* and the *master bus*) are apparent in Figure 8.14. Adapted from [Jot et al., 2023]

This variation enables all of the features described in Figures 8.10–8.11, including per-source sound design functions not offered in traditional audio production tools:

- A practical and effective per-object *distance* control, realized by simply mapping this parameter to a `perceptual factor`: the `Source Presence`
- Per-object control of the `perceptual factors` {`Source Presence, Source Brilliance, Source Warmth`} and {`Room Presence, Running Reverberance, Envelopment`}
- Per-object `axis` spectral correction, which affects the direct-sound frequency response according to the source's orientation and to its other radiation parameters, including:
- Per-object `omni` spectral correction, which, importantly, enables fine-tuning the spectral contour of the reverberation and reflections individually for each sound source (simulating the *diffuse-field transfer function* of this sound source, as defined in [Jot et al., 1997]).

## *Acoustic Objects: unifying the immersive audio ecosystem*

In this section, we introduce the notion of *Acoustic Object*, defined as a spatial audio object whose acoustic reverberation and reflections can be reconstructed at the decoding/rendering end, thereby enabling rich capabilities of interaction and modification of the virtual acoustic scene by the consumer. We begin by reviewing prior research and approaches to the object-based description of environmental acoustic scenes.

### *Audio scene metadata interoperability requirements*

Within the ecosystem represented in Figure 8.13, it is advantageous to adopt a vendor-neutral *rendering interface* facilitating the interoperability of devices and applications, and to define a media encoding format so that a scene can be displayed by a playback application agnostic to the scene generation program (which may be, for instance, a DAW or a video game application). A high-level *scene description interface* converts an application-specific scene representation to a vendor-neutral specification compatible with the common low-level *rendering interface*: a set of audio object waveform, stream, or synthesizer references, along with metadata that collectively drive the spatial audio processing operations performed by the audio renderer.

For each spatial audio object, its *essential metadata* must determine the parameters required for direct-sound rendering in free field: source directivity pattern (frequency-dependent), size or shape, position, and orientation coordinates. In interactive use cases, such as gaming or VR, some of this information may be unknown until playback time, and rendering algorithm parameters must therefore be updated dynamically.

As for the additional metadata that drive the rendering of reverberation and reflections, two approaches have been proposed:

- **Object-based:** Incorporating reverberation processing parameters in an object's metadata, as in:
    - *MPEG-4 Advanced AudioBIFS (Perceptual Approach)*, with a parametrization based on IRCAM `Spat` [Väänänen and Huopaniemi, 2004]

  - *RSAO (Reverberant Spatial Audio Object)*, which provides a set of discrete reflections combined with a parametric reverberation decay tail [Coleman et al., 2017].

- **Environment-based:** A separate metadata interface describes the acoustic environment, as in:

  - *MPEG-4 Advanced AudioBIFS (Physical Approach)*, which is based on the DIVA research project [Savioja et al., 1999; Väänänen and Huopaniemi, 2004]
  - *OpenAL Effects Extension*, an audio programming interface providing immersive reverberation as a shared parametric effect processing option [Peacock et al., 2006]
  - *MPEG-I*, which includes the constructs *Acoustic Environment* and *Geometry*, separate from *Sources:* [Herre and Disch, 2023].

In the latter approach, the environment may be described in architectural terms, including boundary and obstacle geometry, openings, and materials. From such geometrical and physical information, combined with each object's essential metadata, an acoustic propagation computing engine can derive the values of the rendering parameters (see e.g. [Savioja et al., 1999; Raghuvanshi and Snyder, 2018; Liu and Manocha, 2022; Pind et al., 2022]). A common method consists of partitioning the acoustic environment into a set of rooms, so that a playback-time update in the localization coordinates of an object will inform its assignment to one of these rooms and determine its contribution to effect send buses [Peacock et al., 2006; Jot et al., 2021].

### *End-user interaction with Acoustic Objects*

Optionally, the playback application of Figure 8.13 may enable end-user interaction with the presented immersive audio scene via the *rendering interface* and the *interaction interface*, which can support bidirectional functionality:

- Audio object generation or substitution (analogous to the *scene description interface* function shown at the left-hand side of Figure 8.13)
- Retrieving the values of the properties of an incoming audio object to reflect them in the end-user interaction interface (shown at the right-hand side of Figure 8.13)

In defining a universal Acoustic Object rendering interface, desirable objectives include:

- Allowing maximum flexibility in per-object sound design intent by the content creator;
- Ensuring faithful reproduction of spatial audio scenes at the playback end;
- Letting the end user take advantage of the interactivity and personalization benefits enabled by the Acoustic Object representation, within limits authorized by the content creator;
- Enabling content creators and playback application developers to select among several possible levels of end-user interaction and implementation complexity.

In the rendering interface, the metadata set associated with an Acoustic Object may include acoustic reverberation properties in addition to the per-object loudness and angular position properties already supplied in existing object-based immersive audio formats. In Table 8.2, we categorize the reverberation properties of an Acoustic Object into a hierarchy of *functionality profiles*, exposing increasing levels of end-user interaction. The *Basic* profile includes the minimum set of reverberation properties necessary for faithful immersive audio reproduction. The *Remix* profile enables object substitution such as language replacement, where both an audio object's waveform and its source radiation properties can be swapped, with sound design qualities preserved. The *Navig.* profile supplies additional per-object properties enabling 6DoF repositioning of an Acoustic Object and/or 6DoF navigation of the listener.

For example, end-user interaction may be performed via a graphical user interface equivalent to `Spat_OPer` (see Figure 8.10), with the initial settings of all perceptual attributes of an Acoustic Object automatically derived from incoming per-object metadata, by application of the equations provided in Appendix. Alternatively, end-user interaction may be realized via a first-person head-mounted VR display, or via the 2D graphic user interface of a virtual meeting or teleconferencing software application executed on the user's computer. The incoming Acoustic Objects may be generated by a remote instance of `Spat`, or by a spatial audio object workstation (SAOW), or by an interactive video game or VR server application (illustrated by 'App1' or 'App2' on the left-hand side of Figure 8.13).

*Table 8.2* Hierarchy of per-object metadata functionality profiles in Acoustic Object encoding specification. Adapted from [Jot et al., 2023] with a reverberation response decomposed into N segments (N is typically 2 or 3 and may be freely set for each Acoustic Object.

| | |
|---|---|
| *Basic* | Minimum Acoustic Object metadata set to ensure faithful rendering of this object (also allows gain scaling, removal, substitution, or 3DoF repositioning):<br>• *Gain* factor to be applied to this object.<br>• *Position* and *Extent* of the sound source may be jointly represented by a normalized *divergence panning vector* (p0) (see the section on 'Anisotropic late reflections and reverberation' and Figure 8.12)<br>• *Reverberation response*: for an Acoustic Object whose reverberation is decomposed into *N* response segments (denoted R1 . . . R*N*., see Figure 8.6), their energies and the late reverberation decay time (frequency dependent), their onset times (I1 . . . I*N*), and their divergence panning vectors (p1 . . . p*N*). |
| *Remix* | Additional metadata to enable and control spectral equalization for this object (useful, e.g., in playback-end remixing, mashup, or karaoke application scenarios):<br>• *Spectral corrections: on-axis* (affecting the object's direct sound) and *diffuse-field* (affecting its reverberation and reflections) |
| *Navig.* | Additional metadata to enable and control playback-end 6DoF navigation and repositioning of this object:<br>• *Orientation* coordinates and frequency-dependent off-axis directivity;<br>• *Distance* in 'world units', and optional parameters for controlling *distance-based attenuation factors* (see e.g. [Jot et al., 2021; Peacock et al., 2006]);<br>• Optional parameters for controlling features such as *air absorption*, *Doppler effect*, or *propagation delay*. |

*Dynamic Acoustic Object renderer and synchronized metadata*

Upon reception (or termination) of an incoming audio object, the *rendering interface* function of Figure 8.13 can instruct the Audio Renderer to automatically configure its audio processing signal flow (see Figure 8.14), based on interpreting per-object metadata:

- Instantiate (or deallocate) the *per-source processing* module associated with this audio object;
- If this audio object requires reverberation or reflections rendering:
  - Load (or free up) *effect slot(s)* necessary to render reflections and/or reverberation according to the parameters received via the incoming per-object metadata (see Table 8.2);
  - Load (or free up) the corresponding *effect send buses* as well as their connections with the *per-source processing* module associated with this audio object.

The end-user playback application of Figure 8.13 is compatible with the playback of legacy stereo, as well as channel-based or object-based immersive audio formats (see the section on 'Flexible loudspeaker or binaural immersive audio distribution'). The transmission of one or more Acoustic Objects is optional and involves extending the description of each of these audio objects with a synchronized metadata stream that specifies its reflections and reverberation effects. In practice, this solution can be implemented, for instance, as an extension of the Audio Definition Model (ADM) [Füg et al., 2016]. The additional per-object metadata may be transmitted in a low-bit-rate 'sidecar' file accompanying a 'main' object-based immersive audio file coded, e.g., in Dolby Atmos, DTS:X, or MPEG-H format (see the section on 'Flexible loudspeaker or binaural immersive audio distribution'), by ensuring consistent audio object identifier cross-reference between the two data files ('main' and 'sidecar') and that the 'main' file's immersive audio coding scheme preserves the separation of audio objects (see e.g. [Riedmiller et al., 2015]).

## Unifying immersive audio experiences

In the modern connected media ecosystem, it is essential to adopt interoperable transmission standards such that independent content generation applications can seamlessly deliver sounds presented to end-users employing generic spatial audio playback devices. By enabling the transmission and rendering of sounds in the form of Acoustic Objects, the end-to-end solution shown in Figure 8.13 addresses this need and overcomes the market fragmentation challenges reviewed in the section on 'Limitations and fragmentation challenges addressed in this chapter'. For linear object-based immersive audio content, the inclusion of Acoustic Objects implies a moderate additional computation to implement the effect slots and effect send buses (see the section on 'A universal 6-DoF spatial audio rendering engine and interface').

As illustrated in Figure 8.16, the unification of linear and nonlinear platforms facilitates the dissemination of immersive audio and music experiences, and the widespread adoption of the consumer devices employed to enjoy them:

- Consumer adoption is promoted by the consistently met expectation of a compelling and natural-sounding hearing experience and by a rapidly growing content and experience catalog.
- Content production and experimentation are encouraged by increasing consumer demand and confidence, along with the deployment of efficient creation workflows.

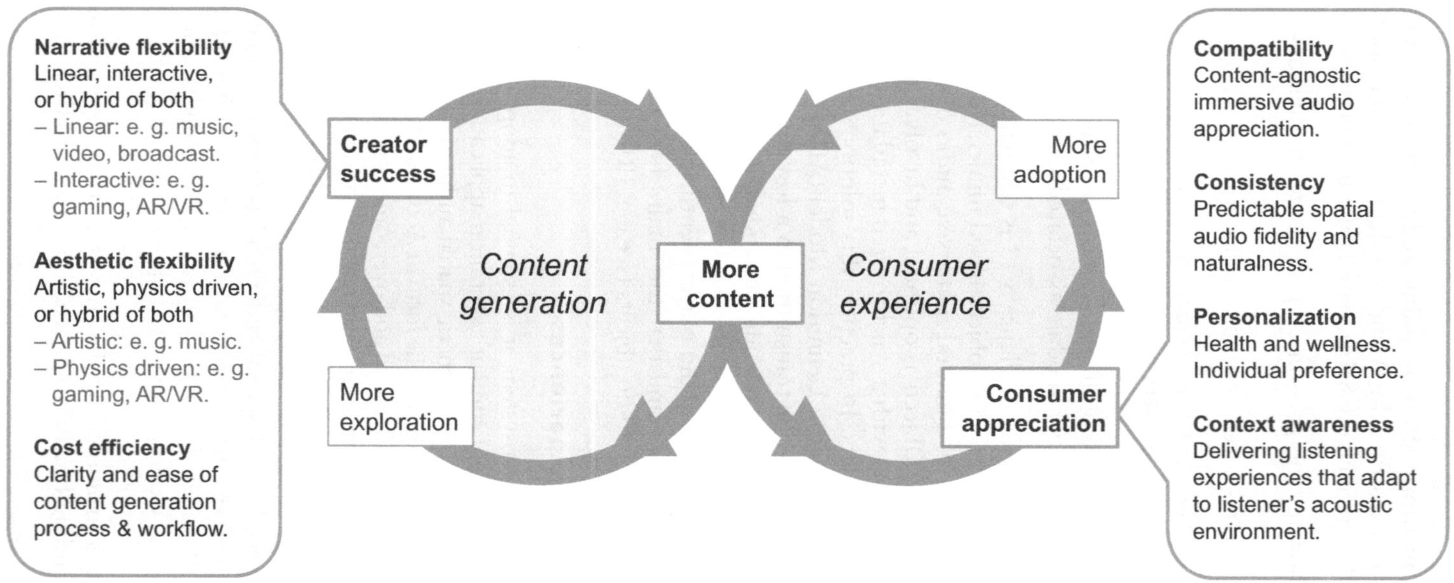


*Figure 8.16* Representation of the immersive audio ecosystem in the form of two interconnected 'value flywheels': technology adoption is accelerated by decoupling and democratizing content generation and augmenting consumer experience, and by allowing the two industries to develop independently while supporting each other, through leveraging an interoperable content format.

### *Universal model for flexible immersive audio generation*

At the content generation end, the encoding model introduced in the section on 'Acoustic Objects: unifying the immersive audio ecosystem' provides the option to transmit as Acoustic Objects one or more of the spatial audio objects composing an immersive scene, extending the capabilities of previous immersive audio coding formats. An Acoustic Object is a spatial audio object for which acoustic reflection or reverberation effects can be reconstructed at the receiving end, optionally augmented by user-end customization or interaction features such as remixing, karaoke, automatic language dubbing, or 6DoF navigation within the immersive audio scene (see the section on 'End-user interaction with Acoustic Objects').

At the receiving end, the incoming immersive audio content is handled identically whether the **narrative intent** of the author is linear (e.g., a cinematic soundtrack) or interactive (e.g., a VR game soundtrack). The choice of a linear or nonlinear narration is determined solely at the content origination end, as is the choice of **aesthetic design** (e.g., physics-driven or artist-driven). The *scene description interface* function associated with a content generation application (shown on the left-hand side of Figure 8.13) ensures that the application-specific audio scene representation will be automatically translated to standard renderer-compatible Acoustic Object metadata. In linear content creation, such as immersive music production, the *spatial audio object workstation* (SAOW) alleviates the 'immersive audio barrier to entry' evoked in the section 'Creation: facilitating aesthetically motivated immersive sound design', by enabling an intuitive extension to familiar digital audio workstation (DAW) workflows.

### *Unlocking the immersive audio experience ecosystem*

In immersive audio playback systems – including home, automotive, mobile, or wearable consumer devices – the capability to decode and render Acoustic Objects enables device optimizations and user experience enhancements:

- **Interoperability:** The playback application is agnostic to the type of incoming immersive audio content. It can be optimized to run permanently using dedicated computation resources and can render audio objects generated by several concurrent independent programs.
- **Consistency:** As a result, the timbral and spatial audio fidelity experienced by the consumer can be consistently personalized and predictable, irrespective of the type of content or experience (e.g., a motion picture soundtrack, an immersive music recording, or a VR meeting or concert).
- **Personalization:** For any sound received as an Acoustic Object, a rich set of customization or interaction features may be authorized by the content creator, ranging from hearing enhancements to musical remix, language replacement or navigation (as outlined in the section on 'End-user interaction with Acoustic Objects' and Table 8.2).
- **Context awareness:** In binaural reproduction scenarios, using, e.g., earphones or wearables, one important type of consumer-end personalization is the dynamic simulation of the user's local environment reverberation fingerprint, enabling natural-sounding augmented reality audio experiences. In loudspeaker reproduction, on the other hand, acoustic objects can be reproduced faithfully by accounting and compensating for the user's local environment reverberation (see context compensation operator in the section on 'Derivation of low-level reverberation rendering parameters' and Figure 8.7).

The development of multi-channel audio formats amenable to flexible headphone or loudspeaker configuration has played a critical role to enable the commercial dissemination of linear immersive audio content, including music and cinematic soundtracks. By further extending object-based immersive audio formats to support the optional transmission of Acoustic Objects, the audio industry can overcome a persistent fragmentation obstacle: ensure that any immersive audio experience, interactive or linear (including broadcast or streaming), aesthetics-driven or physics-driven, can be reproduced seamlessly on any consumer device – while allowing personalization options including hearing enhancement, automatic language dubbing, 6DoF navigation, or creative manipulation. As illustrated in Figure 8.16, this evolution opens up conditions for the accelerated growth of the immersive audio content and device markets.

## Acknowledgments

This chapter is partly derived from a 2023 conference paper by the same authors, entitled 'Perceptually Motivated Spatial Audio Scene Description and Rendering for 6-DoF Immersive Music Experiences' (Jot et al., 2023). Elements borrowed from that paper are reproduced here with the permission of its publisher. The authors would like to thank Agnieszka Roginska, Alan Silva, David Pearl, Earl Vickers, Luiz Zanardo, Paul Geluso, and Peter Otto for thoughtful comments and feedback on aspects covered in this chapter.

## Appendix: principles and implementation of perceptually based scene representation and control in IRCAM `Spat`

### *Perceptually based scene representation*

In this section, we summarize the background, methodology and results of a psycho-experimental study that informed the development of `Spat`, a spatial reverberation processor controlled by a perceptually based interface. The design of this tool and of the `Spat` library is reviewed in the section on 'IRCAM Spat: an aesthetics-driven immersive audio creation and performance platform'.

#### *Research on the perception of auditorium reverberation*

The interest in performance venues for the enjoyment of live orchestral music, opera, or theater has prompted extensive scientific research on the architectural design of auditoria and on the objective characterization and perceptual evaluation of their room acoustical quality (e.g. [Beranek, 2004; Barron, 2010; Kuttruff, 2016]). Following the distinction proposed in [Gade, 2007], one can differentiate several domains for describing the effect of room reverberation in the context of music listening, recording, or production (Figure 8.17):

- **Architectural:** Referring to geometrical and physical parameters, such as room shape, size, or wall materials
- **Objective:** In terms of metrics calculated from an impulse response or transfer function representative of the effect
- **Subjective:** Expressing sensations such as 'intimacy' or 'sweetness', much like one would describe the timbre of a sound or the taste of a food

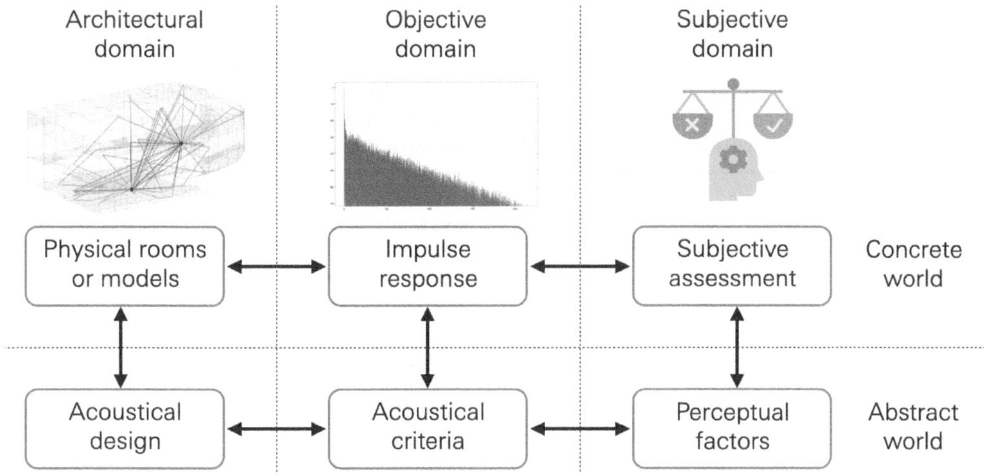


*Figure 8.17* Overview of concepts related to subjective room acoustics (Inspired by [Gade, 2007]).

Research on the objective characterization of the acoustics of performance spaces has led to the definition and standardization of a set of objective parameters widely adopted by acousticians, including reverberation decay time (*Rt*), early decay time (*Edt*), early-to-late energy ratio (*C50*, *C80*), strength (*G*), lateral energy fraction (*LF*), and interaural cross-correlation coefficient (*IACC*) [Barron and Lee, 1988; Beranek, 1992; Gade, 2007; Vorländer and Mechel, 2008; ISO (Int. Org. for Standardization), 2009].

Several investigations have sought to exhibit a minimal set of subjective parameters suitable for explaining or predicting listener sensations or preference judgments, or revealing correlations between objective metrics and subjective attributes [Schroeder et al., 1974; Ando, 1985; Barron, 2005; Lokki et al., 2012; Lokki, 2014; Ando, 2015]. In [Beranek, 1992], consolidating the results of multiple studies carried out in the 1950s–1980s, a concise set of subjective parameters is proposed to characterize the music concert audience experience: *Loudness*, *Reverberance*, *Running Liveness/Reverberance*, *Clarity*, *Intimacy*, *Warmth*, and *Spaciousness* (later split into *Apparent Source Width* (*ASW*) and *Listener Envelopment* (*LEV*) [Gade, 2007]).

### *Towards a structured perceptual model for controlling virtual acoustic systems*

The definition of a structured description model is a key step toward realizing a perceptually based control interface for virtual acoustic systems. Two essential properties should be fulfilled: its conciseness and its uniqueness. In such a model, a virtual acoustic condition is a point in the perceptual space described by a minimal set of mutually independent control dimensions (referred to in this chapter as *perceptual attributes*). This mutual independence (orthogonality) is important for the uniqueness of the description and, in practice, to ensure that modifying the setting of a given perceptual attribute will affect as little as possible the sensations controlled by adjusting other perceptual attributes in the set.

*Table 8.3* The set of nine independent acoustical criteria (left) and corresponding perceptual factors (right) implemented in the Spat_OPer control interface (see Figure 8.5). The first three represent the perception of the source itself; the next three characterize the interaction between the source and the room; the last three control the room's late reverberation decay. Some entries in the table are derived according to equations (8.19)–(8.24).

| Acoustical criterion | *Range* | *Unit* | Perceptual factor | *Range* | *Sensitivity* |
|---|---|---|---|---|---|
| *Es* | [−40 0] | dB | Source presence | [0 120] | 4/dB |
| *Desl* | [−10 10] | dB | Source warmth | [0 60] | 3/dB |
| *Desh* | [−10 10] | dB | Source brilliance | [0 60] | 3/dB |
| *Rev* | [−40 0] | dB | Room presence | [0 120] | 3/dB |
| *Edt* | (derived) | s | Running reverb. | [0 50] | (derived) |
| *Rd1* | (derived) | dB | Envelopment | [0 50] | (derived) |
| *Rt* | [0.1 10] | s | Reverberance | [0 100] | 5/dBs |
| *Drtl* | [0.1 10] | – | Heaviness | [0 50] | 2.5/dB |
| *Drth* | [0.1 1] | – | Liveness | [0 50] | 5/dB |

These objectives motivated a series of perceptual tests conducted in the late 1980s by IRCAM [Lavandier, 1989; Jullien et al., 1992; Jullien, 1995; Kahle, 1995]. This extensive psycho-experimental research initiative ultimately led to the design of the perceptual control interface module implemented in the Spat library, named Spat_OPer (see Figure 8.5 and Table 8.3). Each test consisted in collecting perceptual dissimilarity judgments between virtual acoustic conditions presented in pairs. Such a non-verbal approach was preferred over preestablished questionnaires, in order to avoid possible semantic bias.

From the dissimilarity judgments, the INDSCAL multidimensional analysis method was used to build a representation space whose dimensions reflect not only the variance expressed in the collected data but also the individual sensitivity of each participant along each dimension [Carroll and Chang, 1970]. The resulting representation space allows projecting the relative positions of the different objects under study (here the different virtual acoustic conditions) and quantifying the weight that each participant assigns to each dimension. In other words, the representation space of a given participant (that would reflect his/her individual dissimilarity judgments) may be obtained by stretching or compressing the common space along its principal dimensions.

### *Overview of psycho-experimental procedure and results*

In order to bound test duration and complexity, each perceptual experiment was based on a limited set of acoustical conditions (7–10), thus spanning only a small portion of the whole perceptual space. In total, 16 tests were conducted [Lavandier, 1989]. Virtual acoustic conditions were varied along different classical architectural acoustic criteria (e.g., *Rt*, *C80*, *G*, . . .) or low-level parameters (time and spatial distribution of the first and late reflections). They were applied to musical excerpts (solo instrument or singer) and were synthesized with a set of digital audio processors available at the time (delay units, gains, filters, reverberator). Tests concentrating on time and spectral distribution were generally conducted on headphones, whereas tests involving the manipulation of the spatial distribution of synthetic reflections were conducted in IRCAM's anechoic room equipped with a dome of nine loudspeakers (seven in the horizontal plane and two at elevated positions in the median plane [Lavandier, 1989]).

The analysis of each of the 16 tests provided a partial representation space driven by two to four axes, yielding a total of about 50 dimensions. The next step of the study was to classify these dimensions, trying both to detect which were common to several tests and to determine their optimal objective correlation over the full set of tests [Jullien, 1995]. The analysis led to the definition of ten perceptual dimensions whose objective definitions are reported in Appendix section 'Towards a structured perceptual model for controlling virtual acoustic systems', and whose labels were proposed by a small group of expert listeners. Of these ten dimensions, nine were retained in the practical implementation of the `Spat` processor (details in Appendix section 'Derivation of the low-level rendering parameters'). It should be noted that the expression of the objective quantities that drive the perceptual attributes are nontrivial. Some of them include conditional contributions of time segment energy (such as described in [Lochner and Burger, 1958]), energy masking effects and spatial dependence of late reflections. Masking effects proved critical, for instance, to formulate the loudness of the reverberation (or `Room Presence`) and to differentiate `running reverberance` (reverberance audible while music is playing) from `(Late) reverberance` (reverberance audible after music stops) – see Table 8.3 and [Griesinger, 1992; Haapaniemi and Lokki, 2014; Green et al., 2019].

As observed in Kahle [2013], a feature of the resulting perceptual model is an implicit 'stream segregation' within the perception of an audio object, whereby the early and late energies that shape the impulse response are associated with two independent percepts (we note that this emergent property is analogous to the subjective parsing of sound events into *foreground* and *background* components observed in Griesinger [1995]). The perceptual factors `Source Presence` and `Room Presence` (see Table 8.3 and Figure 8.5) correlate to the principal sensations that vary depending on the position or directivity of a sound source within a given room. In the development and applications of the `Spat` processor, reviewed in the next section, it was observed in particular that isolated control of the perceptual factor `Source Presence` emulates with remarkable effectiveness a variation in source-to-listener distance.

## `Spat_OPer` perceptual control equations

### *Expression of the perceptual attributes*

In this section, we review the set of equations resulting from the psycho-experimental study reported in Appendix section 'Towards a structured perceptual model for controlling virtual acoustic systems'. These equations express the `acoustical criteria` exposed in the high-level `Spat_OPer` control panel (Figure 8.5 and Table 8.3) as functions of the time-frequency-energy distribution in the impulse response (Figure 8.6).

The impulse response energy is decomposed into four time sections as defined in Table 8.3. The energy $Ri$ of the $i$th section (with $i \in \{0, 1, 2, 3\}$) is given by:

$$\mathrm{R}i = \int_{t=li}^{t=l(i+1)} h^2(t) \cdot dt \tag{8.1}$$

| *Section* | *Ri* | *Ii* | *I(i + 1)* |
|---|---|---|---|
| Direct sound | R0 | 0 ms | 20 ms |
| Early reflections | R1 | 20 ms | 40 ms |
| Late reflections | R2 | 40 ms | 100 ms |
| Reverberation tail | R3 | 100 ms | ∞ |

We denote $Ri_{low}$, $Ri_{high}$, and $Ri_{mid}$ the energies in the low, high, and mid-frequency bands, respectively. Similarly, the variation of the reverberation decay time with frequency is represented by the values $Rt_{low}$, $Rt_{high}$ and $Rt_{mid}$ (also denoted $Rt$). This defines a total of 15 low-level rendering parameters. The psycho-experimental study reported in Appendix section 'Overview of psycho-experimental procedure and results' and [Kahle, 1995] identified ten mutually independent acoustical criteria which uniquely characterize the acoustical sensation perceived by a listener, produced by a sound source, both located at given positions in a concert hall or auditorium:

- The perception of the sound source is characterized by three criteria: the early energy *Es* and its relative variations at low and high frequencies denoted *Desl* and *Desh*. Assuming that $R0_{mid} + R1_{mid} > R2_{mid}$,

$$Es = R0_{mid} + R1_{mid}; \tag{8.2}$$

$$Desl = \frac{R0_{low} + R1_{low}}{Es}; \tag{8.3}$$

$$Desh = \frac{R0_{high} + R1_{high}}{Es}. \tag{8.4}$$

- The perception of the room is characterized by three criteria: the reverberation time *Rt* and its relative variations at low and high frequencies, denoted *Drtl* and *Drth*.

$$Drtl = \frac{Rt_{low}}{Rt}. \tag{8.5}$$

$$Drth = \frac{Rt_{high}}{Rt}. \tag{8.6}$$

- The perception of source/room interaction is characterized by four acoustical criteria: *Rd1*, *Rd2*, *Edt*, *Rev*.

$$Rd1 = \frac{0.3 \cdot R1_{mid} + 0.05 \cdot R2_{mid}}{Es}. \tag{8.7}$$

Assuming that $R0_{mid} > R1_{mid}$,

$$Rd2 = \frac{0.5 \cdot R1_{mid} + 1.5 \cdot R2_{mid}}{R0_{mid}}. \tag{8.8}$$

Noting $C = 10^{\frac{-1.2}{Rt}}$, $D = Es/R3_{mid}$ and $D_{max} = 10^{\left(-\frac{3}{2}\right)} - 1$.

If $Es + R2_{mid} \leq D_{max} \cdot R3_{mid}$, then:

$$Edt = 0.4 + Rt \cdot \left(1 - \frac{2}{3} \cdot \left(1 + \frac{Es + R2_{mid}}{R3_{mid}}\right)\right); \tag{8.9}$$

Otherwise:

$$Edt = \frac{0.6}{\left(1 + \frac{Es + R2_{mid}}{R3_{mid}}\right)}. \tag{8.10}$$

Lastly, if $(1 - C) \cdot R3_{mid} \leq 4 \cdot Es$, then:

$$Rev = C \cdot R3_{mid} + \frac{\left((1 - C) \cdot R3_{mid}\right)^2}{8 \cdot Es}; \tag{8.11}$$

otherwise: $Rev = R3_{mid} - 2 \cdot Es$.

### *Derivation of the low-level rendering parameters*

In this section, we review a set of equations previously reported in [Jot et al., 1996; ISO (Int. Org. for Standardization), 2005], solving the inversion of these relations in order to compute reverberator parameters given the settings of the `acoustical criteria` in the `Spat_OPer` control panel. For an omnidirectional sound source having flat `axis` and `omni` spectral corrections, the 12 energy values $Ri_{mid}$, $Ri_{low}$ and $Ri_{high}$ (with $i \in \{0, 1, 2, 3\}$) are obtained by:

$$\begin{aligned} R0_{low} &= Desl \cdot R0_{mid}; \\ R0_{mid} &= Es - R1_{mid}; \\ R0_{high} &= Desh \cdot R0_{mid}. \end{aligned} \tag{8.12}$$

$$\begin{aligned} R1_{low} &= Desl \cdot R1_{mid}; \\ R1_{mid} &= \frac{\left(Es \cdot Rd1 - 0.05 \cdot R2_{mid}\right)}{0.3}; \\ R1_{high} &= Desh \cdot R1_{mid}. \end{aligned} \tag{8.13}$$

$$R2_{low} = R2_{high} = R2_{mid}. \tag{8.14}$$

$$R3_{low} = R3_{high} = R3_{mid}. \tag{8.15}$$

If $Edt > 0.4$, then:

$$R2_{mid} = R3_{mid} \cdot \left(10^{\frac{3}{2}\left(1 + \frac{0.4 - Edt}{Rt}\right)} - 1\right) - Es; \tag{8.16}$$

Otherwise:

$$R2_{mid} = R3_{mid} \cdot \left( 10^{\frac{0.6}{Edt}} - 1 \right) - Es. \tag{8.17}$$

If $\frac{Rev}{Es} \leq 2 \cdot \frac{(1+C)}{(1-C)}$, then:

$$R3_{mid} = \left( -C + \sqrt{C^2 + \frac{Rev}{2 \cdot Es} \cdot (1-C)^2} \right) \cdot \frac{4 \cdot Es}{(1-C)^2}; \tag{8.18}$$

Otherwise:

$R3_{mid} = Rev + 2 \cdot Es$ .

The `acoustical criteria` *Rd*1 and *Rd*2 are related respectively to the percepts of *apparent source width* (*ASW*) and *listener envelopment* (*LEV*) previously mentioned. In order to control these two effects independently, it would be necessary to afford an additional degree of freedom in the low-level reverberation rendering model, enabling finer control of the spatial distribution of reflections. Instead, it was decided to discard *Rd*2 from the high-level control interface. The percept of *listener envelopment* (*LEV*) is therefore governed indirectly by adjustments of any of the criteria *Es*, *Rd*1, *Edt*, *Rev* or *Rt*, per Eq. (8.8) and Eqs. (8.12)–(8.18).

In the implementation of the Room module, frequency-dependent gains are realized by the spectral correctors denoted *Ri* in Figure 8.8, realized by proportional dual-shelving filters [Jot, 2015]. For a directional sound source, the `omni` spectral correction is applied by offsetting accordingly the low/mid/high dB gains in the spectral correctors *R*1, *R*2 and *R*3. The direct sound component *R*0 is corrected similarly, according to the sound source's `axis` and `omni` spectral correction settings, its orientation, and its directivity pattern.

In the realization of this high-level control scheme, it is necessary to implement constraints ensuring that the values of *R*0, *R*1, and *R*2 remain always positive. Additionally, the maximum value of *Rd*1 is limited in order to prevent *R*0 from vanishing (since the direct sound arrival time provides the temporal reference on which the definition of all the `acoustical criteria` relies). Consequently, the allowed setting ranges of *Rd*1 and *Edt* depend on the settings of the other acoustical criteria:

$$Rd1_{min} = 0.05 \cdot \frac{R2_{mid}}{Es}. \tag{8.19}$$

$$Rd1_{max} = 0.27 + Rd1_{min}. \tag{8.20}$$

If $2D \leq D_{max}$, then:

$$Edt_{min} = 0.4 + Rt \cdot \left( 1 - \frac{2}{3} \cdot (1 + 2D) \right); \tag{8.21}$$

Otherwise:

$$Edt_{min} = \frac{0.6}{(1+2D)}. \tag{8.22}$$

If $D \leq D_{max}$, then:

$$Edt_{max} = 0.4 + Rt \cdot \left(1 - \frac{2}{3} \cdot (1 + D)\right); \tag{8.23}$$

otherwise:

$$Edt_{max} = \frac{0.6}{(1 + D)}. \tag{8.24}$$

## Note

1 A portion of this chapter was previously published and is available in open access: Jean-Marc Jot, Thibaut Carpentier, and Olivier Warusfel, 'Acoustic Objects: Bridging Immersive Audio Creation and Distribution Systems', in *Proceedings of the 158th Convention of the Audio Engineering Society (AES)*, Warsaw, Poland, May 2025. https://aes2.org/publications/elibrary-page/?id=22909

## References

[Ableton Live] Ableton, 'Ableton Live'', www.ableton.com

[Ackermann et al., 2024] David Ackermann, Fabian Brinkmann, and Stefan Weinzierl, 'A Database with Directivities of Musical Instruments', *Journal of the Audio Engineering Society*, volume 72(3) pp. 170–179, 2024. https://doi.org/10.17743/jaes.2022.0128

[Agrawal and Braasch, 2018] Sarvesh Agrawal and Jonas Braasch, 'Impression of Spatially Distributed Reverberation in Multichannel Audio Reproduction', in *Proceedings of the 145th Convention of the Audio Engineering Society (AES)*, New York, October 2018. www.aes.org/e-lib/browse.cfm?elib=19802

[Alary et al., 2021] Benoit Alary, Pierre Massé, Sebastian J. Schlecht, Markus Noisternig, and Vesa Välimäki, 'Perceptual Analysis of Directional Late Reverberation', *Journal of the Acoustical Society of America*, volume 149(5) pp. 3189–3199, 2021. https://doi.org/10.1121/10.0004770

[Alary and Politis, 2020] Benoit Alary and Archontis Politis, 'Frequency-Dependent Directional Feedback Delay Network', in *Proceedings of the IEEE International Conference on Acoustics, Speech and Signal Processing (ICASSP)*, pp. 176–180, Barcelona, Spain, May 2020. https://doi.org/10.1109/icassp40776.2020.9054717

[Ando, 1985] Yoichi Ando, *Concert Hall Acoustics*, Springer, 1985. https://doi.org/10.1007/978-3-642-69810-1

[Ando, 2015] Yoichi Ando, *Opera House Acoustics Based on Subjective Preference Theory*, Springer, 2015. https://doi.org/10.1007/978-4-431-55423-3

[André et al., 2014] Cédric R. André, Étienne Corteel, Jean-Jacques Embrechts, Jacques G. Verly, and Brian F.G. Katz, 'Subjective Evaluation of the Audiovisual Spatial Congruence in the Case of Stereoscopic-3D Video and Wave Field Synthesis', *International Journal of Human-Computer Studies*, volume 72(1) pp. 23–32, January 2014, ISSN 1071–5819. https://doi.org/10.1016/j.ijhcs.2013.09.004

[Barron, 2005] Mike Barron, 'Using the Standard on Objective Measures for Concert Auditoria, ISO 3382, to Give Reliable Results', *Acoustical Science and Technology*, volume 26(2) pp. 162–169, 2005. https://doi.org/10.1250/ast.26.162

[Barron, 2010] Michael Barron, *Auditorium Acoustics and Architectural Design*, Spon Press, New York, 2nd edition, 2010. https://doi.org/10.4324/9780203874226

[Barron and Lee, 1988] M. Barron and L.-J. Lee, 'Energy Relations in Concert Auditoriums. I', *Journal of the Acoustical Society of America*, volume 84(2) pp. 618–628, August 1988. https://doi.org/10.1121/1.396840

[Ben-Hur et al., 2021] Zamir Ben-Hur, David Lou Alon, Ravish Mehra, and Boaz Rafaely, 'Binaural Reproduction Based on Bilateral Ambisonics and Ear-Aligned HRTFs', *IEEE Transactions on Audio, Speech, and Language Processing*, volume 29, January 2021. https://doi.org/10.1109/taslp.2021.3055038

[Beranek, 1992] Leo L. Beranek, 'Concert Hall Acoustics – 1992', *Journal of the Acoustical Society of America*, volume 92(1) pp. 1–39, July 1992, ISSN 0001–4966. https://doi.org/10.1121/1.404283

[Beranek, 2004] Leo L. Beranek, *Concert Halls and Opera Houses: Music, Acoustics, and Architecture*, Springer, 2nd edition, 2004. https://doi.org/10.1007/978-0-387-21636-2

[Best et al., 2020] Virginia Best, Robert Baumgartner, Mathieu Lavandier, Piotr Majdak, and Norbert Kopčo, 'Sound Externalization: A Review of Recent Research', *Trends in Hearing*, volume 24 pp. 1–14, 2020. https://doi.org/10.1177/2331216520948390

[Bresson et al., 2017] Jean Bresson, Dimitri Bouche, Thibaut Carpentier, Diemo Schwarz, and Jérémie Garcia, 'Next-generation Computer-aided Composition Environment: A New Implementation of OpenMusic', in *Proceedings of the International Computer Music Conference (ICMC)*, Shanghai, China, October 2017. http://hdl.handle.net/2027/spo.bbp2372.2017.042

[Carpentier, 2016] Thibaut Carpentier, 'Panoramix: 3D Mixing and Post-Production Workstation', in *Proceedings of the 42nd International Computer Music Conference (ICMC)*, pp. 122–127, Utrecht, Netherlands, September 2016. http://hdl.handle.net/2027/spo.bbp2372.2016.023

[Carpentier, 2021] Thibaut Carpentier, 'Spat: A Comprehensive Toolbox for Sound Spatialization in Max', *Ideas Sónicas*, volume 13(24) pp. 12–23, June 2021. https://hal.science/hal-03356292

[Carpentier, 2024] Thibaut Carpentier, *Orchestrating a Software Ecosystem for the Creation, Manipulation, and Distribution of Spatial Audio*, Ph.D. thesis, Sorbonne Université, Paris, December 2024. https://theses.hal.science/tel-05018624

[Carpentier et al., 2014] Thibaut Carpentier, Markus Noisternig, and Olivier Warusfel, 'Hybrid Reverberation Processor with Perceptual Control', in *Proceedings of the 17th International Conference on Digital Audio Effects (DAFx-14)*, pp. 93–100, Erlangen, Germany, September 2014. https://hal.science/hal-01107075/document

[Carpentier et al., 2015] Thibaut Carpentier, Markus Noisternig, and Olivier Warusfel, 'Twenty Years of Ircam Spat: Looking Back, Looking Forward', in *Proceedings of the 41st International Computer Music Conference (ICMC)*, pp. 270–277, Denton, TX, September 2015. http://hdl.handle.net/2027/spo.bbp2372.2015.056

[Carroll and Chang, 1970] J. Douglas Carroll and Jih-Jie Chang, 'Analysis of Individual Differences in Multidimensional Scaling Via an N-Way Generalization of "Eckart-Young" Decomposition', *Psychometrika*, volume 35 pp. 283–319, September 1970. https://doi.org/10.1007/BF02310791

[Choueiri, 2018] Edgar Choueiri, 'Binaural Audio Through Loudspeakers', in Agnieszka Roginska and Paul Geluso (editors), *Immersive Sound – The Art and Science of Binaural and Multi-Channel Audio*, Chapter 5, pp. 124–179, Routledge, New York, 1st edition, October 2018. https://doi.org/10.4324/9781315707525

[Coleman et al., 2017] Philip Coleman, Andreas Franck, Philip J.B. Jackson, Richard J. Hudges, Luca Remaggi, and Frank Melchior, 'Object-Based Reverberation for Spatial Audio', *Journal of the Audio Engineering Society*, volume 65(1/2) pp. 66–77, January/February 2017. https://doi.org/10.17743/jaes.2016.0059

[Filimowicz, 2019] Michael Filimowicz, *Foundations in Sound Design for Linear Media – A Multidisciplinary Approach*, Routledge, 2019. https://doi.org/10.4324/9781315106335

[Frankenbach et al., 2025] Cornelius Frankenbach, Carsten Herzog, Julia Seitz, and Janina Fels, 'Evaluating Localization Accuracy of Dynamic Binaural Synthesis: Differences Between Individual, Estimated, and Generic HRTFs', *Journal of the Audio Engineering Society*, volume 73 pp. 05–23, March 2025. https://aes2.org/publications/elibrary-page/?id=21939

[Füg et al., 2016] Simone Füg, David Marston, and Scott Norcross, 'The Audio Definition Model – A Flexible Standardised Representation for Next Generation Audio Content in Broadcasting and Beyond', in *Proceedings of the 141st Convention of the Audio Engineering Society (AES)*, Los Angeles, CA, October 2016. https://www.aes.org/e-lib/browse.cfm?elib=18430

[Gade, 2007] Anders Christian Gade, 'Acoustics in Halls for Speech and Music', in Thomas D. Rossing (editor), *Handbook of Acoustics*, pp. 301–350, Springer, 2007. https://doi.org/10.1007/978-0-387-30425-0

[Gálvez et al., 2019] Marcos F. Simón Gálvez, Dylan Menzies, and Filippo Maria Fazi, 'Dynamic Audio Reproduction with Linear Loudspeaker Arrays', *Journal of the Audio Engineering Society*, volume 67(4) pp. 190–200, April 2019, ISSN 1549-4950. https://doi.org/10.17743/jaes.2019.0007

[Garcia et al., 2017] Jérémie Garcia, Thibaut Carpentier, and Jean Bresson, 'Interactive-Compositional Authoring of Sound Spatialization', *Journal of New Music Research – Special Issue on Interactive Composition*, volume 46(1) pp. 74–86, 2017. https://doi.org/10.1080/09298215.2016.1230632

[Gardner, 1997] William Gardner, *3-D Audio Using Loudspeakers*, Ph.D. thesis, Massachusetts Institute of Technology, September 1997.

[Geronazzo and Seraphin, 2023] Michele Geronazzo and Stephania Seraphin (editors), *Sonic Interactions in Virtual Environments*, Springer International Publishing, 2023. https://doi.org/10.1007/978-3-031-04021-4

[Gospodarek et al., 2022] Marta Gospodarek, Olivier Warusfel, Pablo Ripollés, and Agnieszka Roginska, 'Methodology for Perceptual Evaluation of Plausibility with Self-Translation of the Listener', in *Proceedings of the AES International Conference on Audio for Virtual and Augmented Reality*, Redmond, WA, August 2022. https://www.aes.org/e-lib/browse.cfm?elib=21874

[Green et al., 2019] Evan Green, Eckhard Kahle, Vincent Berrier, and Emilie Carayol, 'Beyond 80ms: The Subjective Effects of Sound Energy Arriving Shortly After the "Early" Sound Period', in *Proceedings International Symposium on Room Acoustic (ISRA)*, pp. 409–416, Amsterdam, Netherlands, September 2019. https://doi.org/10.18154/RWTH-CONV-240133

[Griesinger, 1992] David Griesinger, 'IALF-Binaural Measures of Spatial Impression and Running Reverberance', in *Proceedings of the 92nd Convention of the Audio Engineering Society (AES)*, Vienna, Austria, March 1992. www.aes.org/e-lib/browse.cfm?elib=6841

[Griesinger, 1995] David Griesinger, 'How Loud Is My Reverberation?', in *Proceedings of the 98th Audio Engineering Society Convention*, Paris, France, February 1995. www.aes.org/e-lib/browse.cfm?elib=7823

[Haapaniemi and Lokki, 2014] Aki Haapaniemi and Tapio Lokki, 'Identifying Concert Halls from Source Presence vs Room Presence', *Journal of the Acoustical Society of America*, volume 135(6) pp. EL311–EL317, May 2014. https://doi.org/10.1121/1.4879671

[Herre and Disch, 2023] Jürgen Herre and Sascha Disch, 'MPEG-I Immersive Audio – Reference Model for the Virtual/Augmented Reality Audio Standard', *Journal of the Audio Engineering Society*, volume 71(5) pp. 229–240, May 2023. https://doi.org/10.17743/jaes.2022.0074

[Herre et al., 2015] Jürgen Herre, Johannes Hilpert, Achim Kuntz, and Jan Plogsties, 'MPEG-H Audio – The New Standard for Universal Spatial/3D Audio Coding', *Journal of the Audio Engineering Society*, volume 62(12) pp. 821–830, January 2015, ISSN 1549-4950. https://doi.org/10.17743/jaes.2014.0049

[Hiebert, 2005] G. Hiebert, *OpenAL 1.1 Specification and Reference*, Creative Labs, June 2005. openal.org/documentation/

[Hiyama et al., 2002] Koichiro Hiyama, Setsu Komiyama, and Kimio Hamasaki, 'The Minimum Number of Loudspeakers and Its Arrangement for Reproducing the Spatial Impression of Diffuse Sound Field', in *Proceedings of the 113th Convention of the Audio Engineering Society (AES)*, Los Angeles, CA, October 2002. www.aes.org/e-lib/browse.cfm?elib=11272

[Hold et al., 2022] Christoph Hold, Thomas McKenzie, Georg Götz, Sebastian J. Schlecht, and Ville Pulkki, 'Resynthesis of Spatial Room Impulse Response Tails with Anisotropic Multi-Slope Decays', *Journal of the Audio Engineering Society*, volume 70(6) pp. 526–538, June 2022. https://doi.org/10.17743/jaes.2022.0017

[Holophonix] Holophonix, 'Holophonix by Amadeus'', https://holophonix.xyz

[Hwang and Lim, 2024] SungHee Hwang and Felicia Lim, 'Immersive Audio Model and Formats, v1.1.0', 2024. https://aomediacodec.github.io/iamf/

[IA-SIG, 1998] IA-SIG, *3D Audio Rendering and Evaluation Guidelines*, MIDI Manufacturers Association, June 1998.

[IA-SIG, 1999] IA-SIG, *Interactive 3D Audio Rendering Guidelines – Level 2.0*, MIDI Manufacturers Association, September 1999.

[ISO (Int. Org. for Standardization), 2005] ISO (Int. Org. for Standardization), 'ISO/IEC 14496–11:2005: MPEG-4 – Part 11: Scene Description And Application Engine', ISO, December 2005. www.iso.org/standard/38560.html

[ISO (Int. Org. for Standardization), 2009] ISO (Int. Org. for Standardization), 'ISO 3382–1:2009: Acoustics – Measurement of Room Acoustic Parameters – Part 1: Performance Spaces', ISO, June 2009. www.iso.org/standard/40979.html

[ITU, 2022] ITU, 'Recommendation ITU-R BS.2051–3 – Advanced Sound System for Programme Production', 2022. https://www.itu.int/rec/R-REC-BS.2051/en

[Jot, 1997] Jean-Marc Jot, 'Efficient Models for Reverberation and Distance Rendering in Computer Music and Virtual Audio Reality', in *Proceedings of the International Computer Music Conference (ICMC)*, Thessaloniki, Greece, September 1997. http://hdl.handle.net/2027/spo.bbp2372.1997.064

[Jot, 1999] Jean-Marc Jot, 'Real-Time Spatial Processing of Sounds for Music, Multimedia and Interactive Human-Computer Interfaces', *ACM Multimedia Systems Journal*, volume 7(1) pp. 55–69, January 1999. https://doi.org/10.1007/s005300050111

[Jot, 2012] Jean-Marc Jot, 'Interactive 3D Audio Rendering in Flexible Playback Configurations', in *Proceedings of the Asia Pacific Signal and Information Processing Association Annual Summit and Conference*, Hollywood, CA, December 2012. http://www.apsipa.org/proceedings_2012/papers/401.pdf

[Jot, 2015] Jean-Marc Jot, 'Proportional Parametric Equalizers – Application to Digital Reverberation and Environmental Audio Processing', in *Proceedings of the 139th Convention of the Audio Engineering Society (AES)*, New York, NY, October 2015. www.aes.org/e-lib/browse.cfm?elib=17916

[Jot et al., 2021] Jean.-Marc Jot, Rémi Audfray, Mark Hertensteiner, and Brian Schmidt, 'Rendering Spatial Sound for Interoperable Experiences in the Audio Metaverse', in *Proceedings of the International Conference on Immersive and 3D Audio (I3DA)*, Bologna, Italy, September 2021. https://doi.org/10.1109/I3DA48870.2021.9610971

[Jot et al., 2023] Jean-Marc Jot, Thibaut Carpentier, and Olivier Warusfel, 'Perceptually Motivated Spatial Audio Scene Description and Rendering for 6-DoF Immersive Music Experiences', in *Proceedings of the 2nd International Conference on Immersive and 3D Audio (I3DA)*, Bologna, Italy, September 2023. https://doi.org/10.1109/I3DA57090.2023.10289196

[Jot and Caulkins, 1995] Jean-Marc Jot and Terence Caulkins, *Spat Reference Manual*, IRCAM, 1995. https://support.ircam.fr/docs/spat/3.0/spat-3-ref/

[Jot et al., 1997] Jean-Marc Jot, Laurent Cerveau, and Olivier Warusfel, 'Analysis and Synthesis of Room Reverberation Based on a Statistical Time-Frequency Model', in *Proceedings of the 103rd Convention of the Audio Engineering Society (AES)*, New York, NY, September 1997. www.aes.org/e-lib/browse.cfm?elib=7150

[Jot and Cohen, 2026] Jean-Marc Jot and Michael Cohen, 'Spatial Sound Rendering for Audio Augmented Reality', in Michael Cohen (editor), *Augmented Reality – Situated Spatial Synergy*, IntechOpen, 2026. https://doi.org/10.5772/intechopen.1014218

[Jot and Fejzo, 2011] Jean-Marc Jot and Zoran Fejzo, 'Beyond Surround Sound – Creation, Coding and Reproduction of 3-D Audio Soundtracks', in *Proceedings of the 131st Convention of the Audio Engineering Society (AES)*, New York, October 2011. www.aes.org/e-lib/browse.cfm?elib=15989

[Jot et al., 1996] Jean-Marc Jot, Jean-Pascal Jullien, and Olivier Warusfel, 'Patent US5812674 – Method to Simulate the Acoustical Quality of a Room and Associated Audio-Digital Processor', *France Télécom*, August 1996.

[Jot et al., 1995] Jean-Marc Jot, Véronique Larcher, and Olivier Warusfel, 'Digital Signal Processing Issues in the Context of Binaural and Transaural Stereophony', in *Proceedings of the 98th Convention of the Audio Engineering Society (AES)*, Paris, France, February 1995. www.aes.org/e-lib/browse.cfm?elib=7786

[Jot and Lee, 2016] Jean-Marc Jot and Keun-Sup Lee, 'Augmented Reality Headphone Environment Rendering', in *Proceedings of the AES Conference on Audio for Virtual and Augmented Reality*, Los Angeles, CA, September 2016. www.aes.org/e-lib/browse.cfm?elib=18506

[Jot et al., 2022] Jean-Marc Jot, Alexey Lukin, and Christopher Landschoot, 'Binaural Externalization Processing – from Stereo to Object-Based Audio', in *Proceedings of the 153rd Convention of the Audio Engineering Society (AES)*, New York, NY, October 2022. http://www.aes.org/e-lib/browse.cfm?elib=21939

[Jot et al., 2006] Jean-Marc Jot, Adam Philp, and Martin Walsh, 'Binaural Simulation of Complex Acoustic Scenes for Interactive Audio', in *Proceedings of the 121st Convention of the Audio Engineering Society (AES)*, San Francisco, CA, October 2006. www.aes.org/e-lib/browse.cfm?elib=13784

[Jullien, 1995] Jean-Pascal Jullien, 'Structured Model for the Representation and the Control of Room Acoustical Quality', in *Proceedings of the 15th International Congress on Acoustics (ICA)*, pp. 517–520, Trondheim, Norway, June 1995.

[Jullien et al., 1992] Jean-Pascal Jullien, Eckhard Kahle, Suzanne Winsberg, and Olivier Warusfel, 'Some Results on the Objective Characterisation of Room Acoustical Quality in Both Laboratory and Real Environments', *Proceedings of the Institute of Acoustics*, volume 14(Part 2), 1992.

[Kahle, 1995] Eckhard Kahle, *Validation d'un modèle objectif de la perception de la qualité acoustique dans un ensemble de salles de concerts et d'opéras*, Ph.D. thesis, Laboratoire d'Acoustique de l'Université du Maine, Le Mans, France, June 1995.

[Kahle, 2013] Eckhard Kahle, 'Room Acoustical Quality of Concert Halls: Perceptual Factors and Acoustic Criteria – Return from Experience', *Building Acoustics*, volume 20(4) pp. 265–282, December 2013, ISSN 1351-010X. https://doi.org/10.1260/1351-010x.20.4.265

[Kirsch et al., 2021] Christoph Kirsch, Josef Poppitz, Torben Wendt, Steven van de Par, and Stephan D. Ewert, 'Spatial Resolution of Late Reverberation in Virtual Acoustic Environments', *Trends in Hearing*, volume 25 pp. 1–17, 2021. https://doi.org/10.1177/23312165211054924

[Kuttruff, 2016] Heinrich Kuttruff, *Room Acoustics*, CRC Press, Boca Raton, FL, 6th edition, 2016. https://doi.org/10.1201/9781315372150

[Larcher et al., 2000] Véronique Larcher, Jean-Marc Jot, Jérôme Guyard, and Olivier Warusfel, 'Study and Comparison of Efficient Methods for 3-D Audio Spatialization Based on Linear Decomposition of HRTF Data', in *Proceedings of the 108th Convention of the Audio Engineering Society (AES)*, Paris, France, February 2000. https://aes2.org/publications/elibrary-page/?id=9241

[Lavandier, 1989] Catherine Lavandier, *Validation perceptive d'un modèle objectif de caractérisation de la qualité acoustique des salles*, Ph.D. thesis, Université du Maine, Le Mans, France, June 1989.

[Lee, 2025] Hyunkook Lee, 'A Unified Conceptual Model of Immersive Experience in Extended Reality', *Computers in Human Behavior Reports*, volume 18, May 2025, https://doi.org/10.1016/j.chbr.2025.100663

[Liu and Manocha, 2022] Shiguang Liu and Dinesh Manocha, *Sound Synthesis, Propagation, and Rendering*, Springer, 2022. https://doi.org/10.1007/978-3-031-79214-4

[Llewellyn and Paterson, 2021] Gareth Llewellyn and Justin Paterson, 'Towards 6DOF: 3D Audio for Virtual, Augmented, and Mixed Realities', in Justin Paterson and Hyunkook Lee (editors), *3D Audio*, Chapter 3, pp. 43–63, Routledge, New York, 1st edition, June 2021. https://doi.org/10.4324/9780429491214-3

[Lochner and Burger, 1958] J. P. A. Lochner and J. F. Burger, 'The Subjective Masking of Short Time Delayed Echos by Their Primary Sound and Their Contribution to the Intelligibility of Speech', *Acustica*, volume 8(1) pp. 1–10, 1958.

[Lokki, 2014] Tapio Lokki, 'Tasting Music Like Wine: Sensory Evaluation of Concert Halls', *Physics Today*, volume 67(1) pp. 27–32, January 2014, ISSN 0031–9228. https://doi.org/10.1063/pt.3.2242

[Lokki et al., 2012] Tapio Lokki, Jukka Pätynen, Antti Kuusinen, and Sakari Tervo, 'Disentangling Preference Ratings of Concert Hall Acoustics Using Subjective Sensory Profiles', *Journal of the Acoustical Society of America*, volume 132(5) pp. 3148–3161, November 2012. https://doi.org/10.1121/1.4756826

[Massé et al., 2020] Pierre Massé, Thibaut Carpentier, Olivier Warusfel, and Markus Noisternig, 'Denoising Directional Room Impulse Responses with Spatially Anisotropic Late Reverberation Tails', *Applied Sciences*, volume 10(3) pp. 1–17, February 2020. https://doi.org/10.3390/app10031033

[Max/MSP] Cycling'74, 'Max/MSP by Cycling'74'. https://cycling74.com

[Meyer-Kahlen et al., 2024] Nils Meyer-Kahlen, Sebastian J. Schlecht, Sebastià Amengual Garí, and Tapio Lokki, 'Testing Auditory Illusions in Augmented Reality: Plausibility, Transfer-Plausibility, and Authenticity', *Journal of the Audio Engineering Society*, volume 72(11) pp. 797–812, November 2024. https://doi.org/10.17743/jaes.2022.0178

[Nicol, 2018] Rozenn Nicol, 'Sound Field', in Agnieszka Roginska and Paul Geluso (editors), *Immersive Sound – The Art and Science of Binaural and Multi-Channel Audio*, Chapter 9, pp. 276–310, Routledge, New York, 1st edition, October 2018. https://doi.org/10.4324/9781315707525

[Noisternig et al., 2003] M. Noisternig, A. Sontacchi, T. Musil, and R. Holdrich, 'A 3D Ambisonic Based Binaural Sound Reproduction System', in *Proceedings of the AES 24th International Conference on Multichannel Audio*, Jun 2003.

[Nolan et al., 2018] Melanie Nolan, Efren Fernandez Grande, Jonas Brunskog, and Cheol-Ho Jeong, 'A Wavenumber Approach to Quantifying the Isotropy of the Sound Field in Reverberant Spaces', *Journal of the Acoustical Society of America*, volume 143(4) pp. 2514–2526, 2018, ISSN 0001–4966. https://doi.org/10.1121/1.5032194

[Peacock et al., 2006] D. Peacock, P. Harrison, A. D'Orta, V. Carpentier, and E. Cooper, *OpenAL Effects Extension Guide*, Creative Labs, July 2006.

[Pind et al., 2022] Finnur Pind, Johannes Einarsson, Steinar Guojonsson, Matthias Cosnefroy, Jesper Pedersen, Jon Stefansson, and Alessia Milo, 'A Novel Wave-Based Virtual Acoustics and Spatial Audio Framework', in *Proceedings of the AES Conference Audio for Virtual and Augmented Reality*, Seattle, CA, August 2022. www.aes.org/e-lib/browse.cfm?elib=21871

[Polack, 1993] Jean-Dominique Polack, 'Playing Billiards in the Concert Hall: The Mathematical Foundations of Geometrical Room Acoustics', *Applied Acoustics*, volume 38(2–4), 1993. https://doi.org/10.1016/0003-682x(93)90054-a

[Potard and Burnett, 2004] Guillaume Potard and Ian Burnett, 'Decorrelation Techniques for the Rendering of Apparent Sound Source Width in 3D Audio Displays', in *Proceedings of the 7th International Conference on Digital Audio Effects (DAFx)*, Naples, Italy, October 2004.

[Pulkki, 1997] Ville Pulkki, 'Virtual Sound Source Positioning Using Vector Base Amplitude Panning', *Journal of the Audio Engineering Society*, volume 45(6) pp. 456–466, June 1997. http://www.aes.org/e-lib/browse.cfm?elib=7853

[Pulkki and Karjalainen, 2008] V. Pulkki and M. Karjalainen, 'Multichannel Audio Rendering Using Amplitude Panning [DSP Applications]', *IEEE Signal Processing Magazine*, volume 25(3) pp. 118–122, May 2008, ISSN 1053–5888. https://doi.org/10.1109/msp.2008.918025

[Rafaely et al., 2022] Boaz Rafaely, Vladimir Tourbabin, Emanuel Habets, Zamir Ben-Hur, Hyunkook Lee, Hannes Gamper, Lior Arbel, Lachlan Birnie, Thushara Abhayapala, and Prasanga Samarasinghe, 'Spatial Audio Signal Processing for Binaural Reproduction of Recorded Acoustic Scenes – Review and Challenges', *Acta Acustica united with Acustica*, volume 6 p. 47, 2022. https://doi.org/10.1051/aacus/2022040

[Raghuvanshi and Snyder, 2018] Nikunj Raghuvanshi and John Snyder, 'Parametric Directional Coding for Precomputed Sound Propagation', *ACM Transactions onGraphics*, volume 37(4), September 2018. https://doi.org/10.1145/3197517.3201339

[Riedmiller et al., 2015] Jeffrey Riedmiller, Sripal Mehta, Nicolas Tsingos, and Prinyar Boon, 'Immersive and Personalized Audio: A Practical System for Enabling Interchange, Distribution, and Delivery of Next-Generation Audio Experiences', *SMPTE Motion Imaging Journal*, volume 124(5) pp. 1–23, July 2015, ISSN 1545–0279. https://doi.org/10.5594/j18578

[Robinson et al., 2012] Charles Q. Robinson, Sripal Mehta, and Nicolas Tsingos, 'Scalable Format and Tools to Extend the Possibilities of Cinema Audio', *SMPTE Motion Imaging Journal*, volume 121(8) pp. 63–69, 2012. https://doi.org/10.5594/j18248XY

[Roginska, 2018] Agnieszka Roginska, 'Binaural Audio Through Headphones', in Agnieszka Roginska and Paul Geluso (editors), *Immersive Sound – The Art and Science of Binaural and Multi-Channel Audio*, Chapter 4, pp. 88–123, Routledge, New York, 1st edition, October 2018. https://doi.org/10.4324/9781315707525

[Roginska and Geluso, 2018] Agnieszka Roginska and Paul Geluso (editors), *Immersive Sound: The Art and Science of Binaural and Multi-Channel Audio*, Routledge, New York, 1st edition, October 2018. https://doi.org/10.4324/9781315707525

[Romblom et al., 2016] David Romblom, Catherine Guastavino, and Philippe Depalle, 'Perceptual Thresholds for Non-Ideal Diffuse Field Reverberation', *Journal of the Acoustical Society of America*, volume 140(5) pp. 3908–3916, November 2016. https://doi.org/10.1121/1.4967523

[Savioja et al., 1999] Lauri Savioja, Jyri Huopaniemi, Tapio Lokki, and Riitta Väänänen, 'Creating Interactive Virtual Acoustic Environments', *Journal of the Audio Engineering Society*, volume 47(9) pp. 675–705, September 1999. https://www.aes.org/e-lib/browse.cfm?elib=12095

[Savioja and Svensson, 2015] Lauri Savioja and U. Peter Svensson, 'Overview of Geometrical Room Acoustic Modeling Techniques', *Journal of the Acoustical Society of America*, volume 138(2) pp. 708–730, August 2015. https://doi.org/10.1121/1.4926438

[Schroeder et al., 1974] M. R. Schroeder, D. Gottlob, and K. F. Siebrasse, 'Comparative Study of European Concert Halls: Correlation of Subjective Preference with Geometric and Acoustic Parameters', *Journal of the Acoustical Society of America*, volume 56(4), October 1974. https://doi.org/10.1121/1.1903408

[Schubert et al., 2001] Thomas Schubert, Frank Friedmann, and Holger Regenbrecht, 'The Experience of Presence: Factor Analytic Insights', *Presence: Teleoperators and Virtual Environments*, volume 10(3) pp. 266–281, June 2001, ISSN 1054-7460. https://doi.org/10.1162/105474601300343603

[Spat for Live] Ableton, 'Spat Bundle by Music Unit'. www.ableton.com/packs/spat-bundle

[Spat Revolution] Spat Revolution, 'Spat Revolution by Flux::'. www.flux.audio/project/spat-revolution

[Spors et al., 2013] Sascha Spors, Hagen Wierstorf, Alexander Raake, Frank Melchior, Matthias Frank, and Franz Zotter, 'Spatial Sound with Loudspeakers and Its Perception: A Review of the Current State', *Proceedings of the IEEE*, volume 101(9) pp. 1920–1938, September 2013. https://doi.org/10.1109/jproc.2013.2264784

[Tsingos, 2018] Nicolas Tsingos, 'Object-Based Audio', in Agnieszka Roginska and Paul Geluso (editors), *Immersive Sound – The Art and Science of Binaural and Multi-Channel Audio*, Chapter 8, pp. 244–275, Routledge, 1st edition, October 2018. https://doi.org/10.4324/9781315707525

[Väänänen and Huopaniemi, 2004] R. Väänänen and J. Huopaniemi, 'Advanced AudioBIFS: Virtual Acoustics Modeling in MPEG-4 Scene Description', *IEEE Transactions on Multimedia*, volume 6(5) pp. 661–675, September 2004. https://doi.org/10.1109/TMM.2004.834864

[Vorländer, 2008] Michael Vorländer, *Auralization: Fundamentals of Acoustics, Modelling, Simulation, Algorithms and Acoustic Virtual Reality*, Springer, Berlin, 2008. https://doi.org/10.1007/978-3-030-51202-6

[Vorländer and Mechel, 2008] Michael Vorländer and Fridolin P. Mechel, 'Room Acoustics', in Fridolin P. Mechel (editor), *Formulas of Acoustics*, Chapter M, pp. 873–943, Springer Verlag, Berlin, 2nd edition, 2008. https://doi.org/10.1007/978-3-540-76833-3

[Walther and Faller, 2011] Andreas Walther and Christof Faller, 'Assessing Diffuse Sound Field Reproduction Capabilities of Multichannel Playback Systems', in *Proceedings of the 130th Convention of the Audio Engineering Society (AES)*, London, UK, May 2011. www.aes.org/e-lib/browse.cfm?elib=15895

[Xie, 2022] Bosun Xie, *Spatial Sound Principles and Applications*, CRC Press, Boca Raton, FL, 1st edition, 2022. https://doi.org/10.1201/9781003081500

[XP4L] XP4L, 'XP4L'. www.xp4l.com

[Zhang et al., 2017] Wen Zhang, Parasanga N. Samarasinghe, Hanchi Chen, and Thushara D. Abhayapala, 'Surround by Sound: A Review of Spatial Audio Recording and Reproduction', *Applied Sciences*, volume 7(532) pp. 1–19, May 2017. https://doi.org/10.3390/app7050532

[Zotkin et al., 2004] Dmitry N. Zotkin, Ramani Duraiswami, and Larry S. Davis, 'Rendering Localized Spatial Audio in a Virtual Auditory Space', *IEEE Transactions on Multimedia*, volume 6 pp. 553–564, August 2004. https://doi.org/10.1109/TMM.2004.827516

[Zotter and Frank, 2019] Franz Zotter and Matthias Frank, *Ambisonics: A Practical 3D Audio Theory for Recording, Studio Production, Sound Reinforcement, and Virtual Reality*, Springer, Cham, Switzerland, 1st edition, 2019. https://doi.org/10.1007/978-3-030-17207-7

Chapter 9

# From measurement to immersion

## Historical acoustics and modern applications

*Brian F.G. Katz*

When designing compelling spaces, the acoustic characteristics are essential to creating a rich auditory experience. My contribution on these types of projects today lies in room acoustic characterization, with a focus on capturing the unique acoustic properties of a space. Unlike live performance recording, which involves different principles, my work is centered on systematically understanding and documenting how sound behaves in a given environment.

I approach room acoustic surveys with a meticulous and methodical process. The core of my work in this area is impulse response surveying, which emphasizes efficiency and precision rather than design. Time management plays a pivotal role in ensuring that thorough measurements are captured without errors. Planning, note-taking, and detailed documentation are critical, as even minor lapses in organization can result in the need to discard valuable data. Even seemingly small tasks, such as managing cabling, require careful planning to avoid complications.

### Concert hall acoustics

In concert hall acoustics, a truly accurate understanding requires measurements from across the entire venue – not just the premium seats. I prioritize maintaining a consistent density of measurement points, ensuring comprehensive coverage. Using multiple types of microphones allows for a finer grid of measurement, although when limited by equipment, the grid becomes coarser. Regardless, I aim to maintain a well-distributed set of measurement points.

Homogeneity is key in concert hall design. The acoustic experience should be consistent throughout the venue, from the cheap seats to the expensive ones. If a particular area shows potential issues, we may focus our measurements on that zone, but, overall, I strive for broad coverage, as time constraints often limit in-depth exploration of specific areas.

In this field, we frequently work under tight deadlines. Whether we have two days or just two hours to capture measurements, efficiency is paramount. If I have a limited window, I focus on a straight-line measurement approach, moving from the stage to the exit and documenting regular data points. In my experience, the progression of how parameters change throughout the space is often more valuable than their absolute values at any given point.

When I was a consultant, my work focused on room acoustic design with an emphasis on creating comfortable, optimal spaces for clients. This involved auralization, where we

DOI: 10.4324/9780429297526-11

demonstrated the potential impact of design decisions through simulations and sound modelling. I continue to work with smaller consulting firms, helping them with technical measurements and presentations, guiding them to make scientifically informed decisions while remaining neutral.

## Stavanger Concert Hall project

One of the most complex concert hall projects I've worked on was measuring the Stavanger Concert Hall, a stunning venue featuring three balconies. While initial measurements were straightforward from the first row, logistical challenges such as cable management became more complex as we moved to the rear rows and upper levels.

Initially, we used a single omnidirectional speaker in the center of the stage, but this didn't capture the full depth of an orchestra's acoustics. We shifted to using distributed speakers to create a more accurate representation of sound in the hall. This transition significantly lengthened the measurement process, as we had to determine the optimal speaker placement and develop a routine for moving and setting them up efficiently.

The Stavanger project posed numerous challenges, requiring the use of eight omnidirectional microphones, figure-of-eight patterns, a dummy head, and an Ambisonic microphone to achieve comprehensive coverage. Unfortunately, one of the speakers fell and broke mid-measurement due to a cable mishap, requiring us to reconfigure our setup. Despite the difficulties, we worked with a team of six people over 48 hours, developing a rhythm that allowed us to successfully capture nearly 4,000 impulse responses, which I am still analyzing to this day.

Thoroughness is the backbone of my approach, along with pushing technical boundaries and using appropriate technologies. Having said that, I avoid overcomplicating things with multi-channel, MIMO systems that focus on the directivity of sources but sacrifice spatial distribution. Given the choice, I would opt for 12 speakers on stage to better capture the full soundscape, rather than a 12-channel speaker sphere driven independently.

## Chamber music space design

One of the more intriguing projects involved a small chamber music space in a large office block, where the client debated whether to implement a box-in-a-box isolation design. We created an auralization where the client could walk into a virtual environment and listen to a violin excerpt with and without the proposed isolation. The concept was simple yet effective in conveying the benefits of the design.

For this project, the location's proximity to a food court introduced significant background noise, which needed to be addressed. We selected a piece of music with quiet sections and long pauses to highlight the outside noise. The experience was enlightening. The background noise, which was already on upon arrival in the room, was almost imperceptible to the client. But once it was turned off, the client noticed how much it affected their perception of the music. By the end of the session, the client understood the value of isolating the space.

We set up an eight-speaker Ambisonic cube, focusing on creating an immersive listening experience rather than striving for perfect directional accuracy of the intrusive noise. The client listened to a three-minute violin excerpt, first unknowingly with the ambient noise present and then again with the noise removed. Upon hearing the difference, he immediately recognized the subtle improvements in clarity that the design would provide. Even though he didn't proceed with the investment, he left with a deeper appreciation for the value of high-quality acoustic design.

## Binaural sound-based storytelling

Over the years, we've worked on various projects like this, some more creative and communicative than others. One standout was the creation of a binaural plug-in that we were particularly proud of. With a modest budget for outreach, we decided to take a different approach than the typical how-to video. Instead, we opted to craft something engaging and entertaining that would showcase the full potential of the tool. The idea was to create a fun, immersive experience, introduce binaural sound, and spark curiosity. We wanted people to enjoy the experience and, if interested, inquire about the tool we used and how we achieved the results.

This project marked one of my first ventures into artistic direction, which was a departure from my usual research-focused work. Initially, I saw myself as the scientist, the researcher. But in this case, I took on the role of executive producer. We enlisted an audio engineer, Lucie Hardoin, with extensive experience in binaural sound. She had never worked with synthesis before but had a strong background in binaural head recordings. This project offered a unique opportunity for her to explore that creative side.

She designed the project with a clear vision, incorporating the features of our tool in ways that were both intuitive and effective. Her knowledge of binaural sound meant there was no learning curve – she understood the nuances, like avoiding placing sound sources directly in the center, which can sound unnatural. She kept the scene dynamic, rich, and engaging. The outcome was a success, and the project, *Binaural steps episode 1: Emily's world*, received positive feedback.

Earlier, we had collaborated with a sound engineering school, where one of my former students now teaches. For their end-of-year project, we tasked a group of students with using our tool to create a sound-based story. The result, *Station 508: An Adventure in Space*, was an alien adventure that turned out to be surprisingly well-executed.

## Notre-Dame Cathedral projects

Before the Notre-Dame Cathedral fire, we were involved in the *BiLi Project* (around 2012–2013), an industrial research initiative aimed at bringing binaural technology out of the lab and into practical, real-world applications. This collaboration involved research labs, French television, Radio France, the National Conservatory, and several startups in the audio equipment space. The goal was twofold: conduct in-depth case studies on binaural audio and explore its creative applications.

During this project, the Conservatory was recording a live concert at Notre-Dame, and we saw a unique opportunity to highlight binaural technology. The sound engineering school, which uses such events for student training, handled the music recording. On a separate evening, we set up to perform acoustic measurements throughout the Cathedral. The idea was to create a highly accurate 3D model of the acoustics, which we could later use to reinsert the concert's audio with precise spatial characteristics, allowing for a virtual 'fly-around' of the performance.

This production began with a simple but compelling idea: we wanted to convey the unique acoustic challenges of the cathedral to a wider audience. After attending a concert at the Notre-Dame Cathedral in Paris, we realized that its eight-second reverb time makes it pretty unsuitable for classical music. In the first few rows, the sound was stunning, but as you moved farther back, the clarity rapidly diminished – the notes blurred into a mush, leaving only indistinct sound.

Our goal was to communicate this auditory experience. Instead of using typical mixes for a radio or disc-style production – where everything sounds 'perfect' and overly polished – we wanted to show what it *actually* feels like to be in the cathedral. Our approach was to present a real concert recording, but with the perspective of someone experiencing the acoustics in situ, with all its imperfections, yet still maintaining an artistic appeal. The idea evolved into a *fly-around* experience, where listeners are virtually placed on a magic carpet and guided through the concert while listening to the performance, resulting in the *Ghost Orchestra Project – 360° Virtual Ambisonic concert reconstruction in Notre-Dame*. What seemed like a small concept at first grew into a six-month production – one that wasn't initially part of the original project but turned into a key part of the narrative.

## Acoustic measurements

We spent several hours measuring the acoustics of Notre-Dame, but, in retrospect, we didn't capture the full scope of the space. There were areas we didn't reach due to time constraints, and some of the measurements were cut short when an organist unexpectedly began a tour. We had only a small window – 10 p.m. to midnight – and were told by the security guard that we had to leave promptly at midnight. Despite these limitations, we were able to gather valuable data, which would later prove crucial.

The true value of these measurements became evident in the aftermath of the 2019 fire at Notre-Dame. I distinctly remember receiving a phone message while recalling those measurements: 'you have those measurements, right?' That exchange emphasized the importance of the work we'd done – something I hadn't fully realized at the time.

When I investigated further, I found that there were only a few historical references to the acoustics of the cathedral, and they lacked the necessary detail or methodology. I even discovered that, back in the 1980s, some measurements had been taken in the lab where I worked, but the reports were mostly unusable due to a rather nontraditional measurement and analysis approach. The most significant element we had was the original old reel-to-reel recording from those measurements. We had it digitized, and in the process, we found three *balloon pops* on the tape, which, surprisingly, turned out to be valuable audio markers for our analysis.

At the time, I hadn't realized just how rare and vital this data would become. In hindsight, it was clear how much our measurements would contribute to the restoration efforts when the fire devastated the Cathedral in 2019.

## Science advisory group and restoration efforts

As the aftermath of the fire at Notre-Dame began to unfold, we quickly recognized that our work had positioned us uniquely to offer crucial assistance. The restoration of the cathedral would require a scientifically accurate acoustic model, one that could guide the rebuilding process. In that earlier work, my PhD student Bart Postma had probably spent six months meticulously calibrating that model, putting more effort into it than most consulting firms would for such a project. Our work was more than just data collection; it had the potential to contribute to the entire restoration.

At that moment, we were ready to step in and offer our expertise to the French authorities. Our role in the scientific advisory group of Notre-Dame began to take shape, and with the collaboration of Mylène Pardoen, a soundscape archaeologist, we offered our insights and assistance to ensure that the cathedral's acoustics would be restored as faithfully as possible.

In the early period after the fire, the general sentiment was uncertain. There was no clear path for the building's restoration, and architectural proposals were being presented with a sense of urgency. Some of the ideas were radical – concepts that no one had seriously considered before, like adding a glass roof or a forest atop the building. While these ideas were creative, they lacked a clear sense of feasibility.

Ultimately, after several months, President Macron made the decision to rebuild Notre-Dame 'as it was before', to its former state. Though the exact specifications weren't finalized, it was clear that the structural shape and materials would remain largely unaltered. Our role then shifted. We no longer needed to focus on refining a predictive model, with the acoustic issues being more concerned with the finer details – like the impact of cleaning the cathedral's surfaces or minor structural changes. At that point, the restoration effort was focused on nuance, detail, and preservation. We were then able to document the acoustic progress, through repeated site visits and measurements. We were also asked to evaluate the acoustic impact of several proposals for the new choir organ using our model (the original organ was damaged in the fire).

### *Research project: a journey through time and sound*

We proposed a significant research initiative that aimed to integrate both historical and forward-thinking perspectives, *The Past Has Ears at Notre-Dame* project. However, since the forward-thinking aspect wasn't as central, we were ultimately able to focus heavily on the history and evolution of sound within the context of the cathedral. This led us to explore an intriguing question; what were the acoustics 'before' the fire, and before that, and before that? In other words, how has the acoustics of Notre-Dame changed over its 850-year existence? In fact, before Notre-Dame, there was another church on the site. The key point was to examine how the acoustics we now experience might be quite different from what was experienced in the past.

This exploration wasn't just about architecture; it also involved understanding how singing styles, preaching, and decoration evolved alongside the acoustics of the space. With a team that included my PhD student Sarabeth S. Mullins and post-doc Elliot K. Canfield-Dafilou, we found ourselves diving into an interdisciplinary study that spanned history, architecture, and acoustics, with a strong focus on understanding how these elements influenced each other over centuries.

Given the scale and significance of this research, we felt from the outset that it was crucial to make our findings accessible to the public, who were in a sense, financing the work. We decided to create an outreach production – a radio-fiction podcast – to share our research in a format that was engaging yet informative. This was an opportunity to allocate part of the research budget toward public-facing media.

We teamed up with *narrative*, a company experienced in creating binaural museum installations, a type of immersive audio experience where listeners are guided through historical stories. They had worked on notable projects like an installation at a Parisian chateau, where binaural audio created an immersive historical narrative. Our goal was slightly different: rather than focusing purely on the immersive qualities of binaural audio, we wanted to highlight the research results themselves.

### *Tools for storytelling: mixing research with creativity*

We didn't just hand over our research data; we also developed custom tools to make it usable for the creative team. Specifically, we built a VST plug-in, *RoomZ*, that enabled sound engineers to pan through impulse responses and simulate the acoustics of different eras in Notre-Dame's history. The plug-in was designed to be easy to use in a DAW, giving the sound engineers the ability to simulate the acoustics of specific locations and time periods as they walked through the cathedral in the story.

At the same time, we were capturing period-specific music in our anechoic room, recording choirs singing early music. One memorable day, we brought in a children's choir, and it was both a fun and challenging experience to capture just a few minutes of high-quality audio with a group of energetic kids.

Coincidentally, one of our team members, Valérie Le Page, a doctoral student in musicology who also teaches voice, stepped in as the choir director when the original one was unavailable. She had remarkable control over the children, getting them to focus with just a glance. With this wealth of musical content, we were able to integrate period-specific vocal recordings into the project, which were essential for creating a rich auditory experience.

As the executive producer, my role was to step back and trust the professionals, but the creative process wasn't without its challenges. The push and pull between the artistic direction of the story and the need to convey research-driven information was constant. The goal was not just to create a compelling narrative but also to ensure that the information we wanted to share – particularly about the acoustics and history of Notre-Dame – was effectively integrated into the experience.

While we wanted the story to be engaging and emotional, we didn't want to stray too far from the science. The mix had to feel *plausible* rather than perfectly realistic. We were striving for an experience that could *sound* like the real thing but with a bit of artistic liberty to enhance the emotional impact. In the world of binaural audio, this means accentuating certain elements – like how reverb behaves in a space – without distorting the physical reality too much.

For instance, in Notre-Dame, reverb is a key aspect of the acoustics. If you were to hold a conversation in the cathedral, you wouldn't hear a lot of reverb; it's only noticeable when you're far from the speaker or speak very loudly. However, in our binaural mix, we had to balance how much reverb to introduce so that it sounded authentic, even though listeners were only hearing the audio through headphones.

Since the radio fiction, we've also made a binaural immersive audio guide on the sonic history of the cathedral, *Notre-Dame Whispers*, with my doctoral student Julien De Muynke and the team at *Talkartive*. Finally, along with my colleague David Poirier-Quinot directing the production with the Conservatoire de Paris and the team at *Dada! Animation*, we've produced a one-hour film-concert, *Vaulted Harmonies*, a kind of 'archeoconcert' taking the public through eight centuries of musical and architectural evolution with 11 musical pieces recorded in studio and rendered in the appropriate period's acoustics, all with animated graphics highlighting the cathedral across time.

### *Crafting the experience: the challenge of 'concert realism'*

One of the most complex aspects of the *Ghost Orchestra* project was achieving *concert realism*. The goal wasn't to replicate what one would hear sitting in a specific seat in a concert hall but to convey a more holistic sense of the performance. We wanted to capture the feeling of being at the concert, hearing the orchestra, the choir, and the soloist in their correct positions but without the constraints of a single perspective.

There were many technical hurdles to overcome, especially in the editing process. For instance, we realized that our mix included over 45 tracks in third-order Ambisonics, which required a massive amount of data. As the timeline for the production was tight, we had to predefine certain things like the listener's trajectory through the cathedral, and we sometimes found ourselves in the wrong place sonically at critical musical moments – such as missing the soloist when they were performing from a distant location in the cathedral. These missteps were part of the learning curve, but they ultimately helped us refine our approach.

Creating a compelling binaural experience requires a deep understanding of how the brain processes sound. We've been building our own rendering engine, *Anaglyph*, which incorporates all the lessons from our research into how sound behaves in space. One critical aspect we've worked on is how to simulate distance cues more accurately. Simply adjusting volume levels doesn't truly convey the sense of distance; instead, we've developed methods that account for how sounds moving around the head change with distance, while preserving their perceived loudness.

Another challenge is ensuring that the reverberation of sounds is accurate. For instance, reverb behaves differently depending on the direction of the sound source. If a sound is coming from the left, the reverberation in the right ear will differ from the reverberation in the left ear. Our rendering engine accounts for this, creating more natural and immersive soundscapes.

One of the most difficult challenges in binaural sound is externalization – making sounds over headphones appear to come from outside the listener's head without visual cues. Studies suggest that experts may have a better ability to externalize sound, but for most people, it's a learned skill. To help listeners, we've focused on creating binaural experiences that guide them through the soundscape, encouraging a sense of immersion and helping them accept the auditory illusion.

### *Urban soundscapes and psychological research*

We've also conducted research on how binaural audio can be used to recreate urban soundscapes. In a controlled lab setting with then doctoral student Catherine Guastavino, we compared the psychological perception of urban noise using different types of audio arrays: stereo, a 2D circular array, and a full Ambisonic spherical array. Our goal was to test whether it's possible to replicate the experience of an outdoor environment in a lab using advanced sound technologies. The results were promising but also highlighted the importance of creating an experience where listeners can fully immerse themselves in the soundscape without any preconceived expectations of what the space should look like. So we built a kind of visual capsule, with no vertical or horizontal right angles, while also masking all the speakers from the listener.

## Conclusion: striving for plausibility, not perfection

In the end, the goal of our research and its public-facing productions is to create a *plausible* experience, one that could very well represent what you would hear in a particular space at a particular time. Realism in binaural audio is elusive because the listener's experience will always be affected by their environment, their cognitive processing, and even their expectations. However, by focusing on plausible, immersive soundscapes, we aim to give listeners a richer understanding of both the historical and artistic aspects of Notre-Dame, as well as the complexities of acoustic research.

By constantly refining our tools and techniques, we continue to push the boundaries of binaural audio, ensuring that each new project – whether it's an academic study or an immersive audio production – offers a compelling and truthful sonic experience.

Chapter 10

# Volumetric sound capture

*Tomasz Żernicki*

## Sound for virtual reality and metaverse

Imagine stepping into a virtual concert where you can not only hear the music but also feel the crowd's energy, move freely around the stage, and experience the performance from any angle you choose – all without leaving your living room. What if the sound changes dynamically as you explore this digital space, just as it would in the real world? Welcome to the metaverse, where the boundaries between physical and virtual realities dissolve, and sound plays a pivotal role in shaping immersive experiences.

In this contribution, we will embark on a journey to explore the transformative power of sound in the metaverse and virtual reality (VR). We will begin by gaining insights into the concept of the metaverse and how it is evolving as a convergence of physical and virtual worlds. Next, we will trace the history of virtual reality, examining the development of VR technologies and their impact on how we experience sound. A key focus will be the importance of 3D audio and its role in creating realistic and engaging virtual environments. We will also delve into the main challenges of audio in VR, including issues related to auditory display, synchronization, and distance perception.

Further, we will explore advanced concepts such as six degrees of freedom (6DoF) and volumetric sound, understanding how these technologies enable more immersive and interactive audio experiences. Technical implementations will be discussed, including the use of Ambisonics and the MPEG-I Immersive audio coding standard in practical scenarios. To ground these concepts in reality, we will review case studies such as 6DoF audio productions with the Poznań Philharmonic Orchestra and collaborations with artists like Jordan Rudess. Finally, we will provide practical recommendations for creating 6DoF audio content, covering aspects of recording, mixing, and production workflows.

### *Why this matters?*

As the metaverse continues to grow, sound is becoming an essential element that enhances realism and user engagement. Whether you're a sound engineer, designer, or a professional interested in immersive media, understanding the role of sound in virtual environments is critical. This contribution will equip you with the knowledge and tools to navigate the complexities of immersive audio, anticipate future trends, and contribute to shaping the soundscapes of the metaverse.

The advent of sound processing tools, particularly those designed for object-based sound, initially catered to the gaming sector's needs. These tools are now instrumental

DOI: 10.4324/9780429297526-12

across diverse industries, including film, music, radio, and television, showcasing their adaptability and significant impact. The gaming industry has a central role in shaping the metaverse. Within this context, the gaming industry's contribution to the development of the metaverse becomes most apparent, signifying the far-reaching influence of its technologies.

To embark on this journey, let's first delve deeper into understanding what the metaverse truly is and how it serves as the foundation for these immersive audio experiences

## Metaverse

The metaverse is an expansive digital realm that transcends traditional online interactions, offering a multimodal experience where users engage with various senses – particularly vision, sound, and touch. It goes beyond simple digital experiences to create immersive environments that blend the physical and virtual worlds. In this contribution, we will explore the concept of the metaverse, its origins, and its potential impact on society and culture, with a focus on how these developments relate to sound and immersive audio experiences.

## Definition and scope of the metaverse

### Broad definition

The metaverse can be defined as a collective virtual shared space that results from the convergence of virtually enhanced physical reality and physically persistent virtual spaces. This includes not just virtual reality (VR) and augmented reality (AR) environments but also the broader internet and all digital interactions within it. In simple terms, the metaverse is where the digital and physical worlds meet, allowing people to interact in a virtual space that feels just as real as the physical world.

### Key characteristics

Understanding the metaverse requires recognizing a few key characteristics that distinguish it from other digital platforms:

- **Persistence:** The metaverse continues to exist even when you're not logged in. Unlike traditional online games or platforms, where the world only exists when you're actively engaged, the metaverse is always on, always evolving.
- **Synchronicity:** Events in the metaverse happen in real time, meaning that, when you interact with something or someone in the metaverse, it's happening at that moment, just as it would in the physical world.
- **Interoperability:** One of the most exciting aspects of the metaverse is that it allows seamless transitions between different virtual spaces. For instance, you could move your avatar (a digital representation of yourself) from one game environment to another without losing your identity or any virtual possessions. This feature supports a unified digital existence across multiple platforms.

## Historical context and evolution

### Origins in science fiction

The concept of the metaverse has its roots in science fiction. One of the earliest mentions of a fully immersive virtual world is in Neal Stephenson's 1992 novel *Snow Crash*, where the term 'metaverse' was first coined. In Stephenson's vision, the metaverse was a massive virtual world where people interacted through avatars in a 3D space, much like today's virtual reality environments.

Similarly, films like *The Matrix* (1999) envisioned a world where people lived in a simulated reality so convincing that they couldn't distinguish it from the physical world. These early imaginings helped shape the way we think about virtual worlds today and set the stage for the development of the metaverse.

### Evolution of virtual worlds

The development of virtual worlds has come a long way since the days of text-based environments like MUDs (multi-user dungeons), which were popular in the 1980s and early 1990s. These early online spaces allowed users to interact with one another in real time, but the experience was entirely text-based, with no visual or auditory components.

The late 1990s and early 2000s saw the rise of MMORPGs (Massively Multiplayer Online Role-Playing Games) like *World of Warcraft*, which introduced richly detailed visual worlds where thousands of players could interact simultaneously. Platforms like *Second Life*, launched in 2003, further expanded the idea of virtual worlds by allowing users to create their own content, from buildings to entire landscapes, and even to conduct virtual business.

Later, platforms like *Fortnite* and *Roblox* represented the next evolution of virtual worlds, where users not only play games but also create their own experiences, socialize, and even attend virtual events like concerts. These platforms became increasingly immersive, integrating elements of VR and AR to create more realistic and engaging environments.

## History and development of virtual reality

As discussed by Warren in [1, 2], The origins of virtual reality trace back to the 1950s with the Sensorama built by Morton Leonard Heilig [3], an innovative experience that combined visual, auditory, and tactile stimuli to immerse the user. This device, featuring speakers and vibrating seats, was designed to create a fully immersive experience by eliminating external distractions. Despite its initial popularity, the technology did not achieve widespread adoption then. Additionally, in 1968, Ivan Sutherland introduced *The Sword of Damocles* [4], a virtual reality prototype, a system that displayed output from a computer program in a stereoscopic display. The perspective the software showed the user would depend on the position of the user's gaze – which is why head tracking was necessary. The HMD had to be attached to a mechanical arm suspended from the ceiling of the lab partially due to its weight and primarily to track head movements via linkages.

The 1990s were marked by the emergence of virtual reality helmets. These early models were cumbersome and often induced discomfort, including motion sickness, limiting their practicality and leading to a temporary decline in interest. A notable example is Mega Visor Display (MVD) introduced by Sega and the UK company Virtuality [5] or Forte Technology's VFX1 headset (1995) [6].

The resurgence of interest in augmented reality (AR) and virtual reality (VR) technology became particularly notable in the mid-2010s. Major developments during this period included the launch of Oculus's first VR headset in 2016, which marked a significant step in consumer-ready VR technology. Additionally, Magic Leap, founded in 2010, focused on pioneering mixed reality experiences. Microsoft introduced its Hololens in 2016, offering a new approach to AR through holographic computing. Similarly, the HTC Vive, released in the same year, further expanded the VR market with its innovative tracking and room-scale VR capabilities. These advancements collectively signified a renewed and growing interest in AR and VR technologies, highlighting their potential across various applications.

The pandemic of 2020 catalyzed this renewed interest in virtual reality, driven by the necessity for digital cooperation and entertainment during periods of isolation. This era witnessed a significant increase in virtual concerts, which gained popularity as traditional live events were suspended. In 2020, The platform Fortnite hosted a landmark virtual concert, attracting 12 million participants globally and showcasing the vast potential of virtual events. This period also saw a rise in virtual reality equipment adoption, indicating a renewed and growing consumer interest despite past challenges in the 1990s.

The release of Apple's Vision Pro in 2024 marked a significant milestone in AR technology, introducing what Apple described as a 'spatial computer'. This device integrated hardware and software, which further improved on aspects like resolution and delay that are both generally considered crucial for seamless experiences.

### *Blurring boundaries between physical and virtual spaces*

The metaverse represents the merging of physical and virtual worlds, where the lines between the two are increasingly blurred. Imagine attending a concert where half the audience is physically present in a stadium, and the other half is participating via VR from around the world. Both groups experience the same event, hear the same sounds, and interact with one another in real time, despite being in different locations. This convergence is a key aspect of the metaverse, where experiences are not confined by physical boundaries.

For audio engineers, this blurring of boundaries presents both challenges and opportunities. Sound must be designed to integrate seamlessly between these spaces, ensuring that virtual sounds are believable within physical contexts and vice versa. The acoustics of a virtual concert hall might need to match or complement the acoustics of a real-world venue, creating a cohesive experience for all participants, whether they are present physically or virtually.

### *Parallel virtual environments*

In the metaverse, multiple virtual environments can exist simultaneously, each with its own unique characteristics and rules. These are sometimes referred to as parallel worlds. For example, one virtual environment might be a bustling city where users can work and socialize, while another might be a serene forest where users go to relax and meditate.

These worlds coexist within the metaverse, and users can move between them as easily as walking from one room to another in the physical world.

From an audio perspective, each of these parallel environments requires its own soundscape – a unique auditory signature that reflects its particular environment. Designing sound for these different worlds involves considering the physical properties of the environment (e.g., the echo in a virtual canyon versus the muffled sounds of a dense forest) and ensuring that transitions between these soundscapes are smooth and immersive for the user.

## Social and cultural impact of the metaverse

### Changing social interactions

Only time will tell whether the metaverse has revolutionized the way people interact socially, as it is expected to do. Virtual spaces within the metaverse may become central to how people meet, work, and play. For example, instead of commuting to a physical office, people might meet in a virtual workspace, where they can collaborate in real time, regardless of their physical location.

For audio engineers, this shift means that the design of soundscapes plays a crucial role in shaping these virtual social spaces. The sound design must consider how different environments influence social interactions – whether it's the background hum of a virtual café where people gather to chat or the more formal, acoustically neutral setting of a virtual boardroom.

### Cultural implications

The metaverse also has significant cultural implications. It is giving rise to new forms of digital art, virtual economies, and entertainment that are unique to these environments. For instance, digital artists are creating virtual sculptures and environments that can only exist in the metaverse. Virtual economies, where users can buy, sell, and trade digital goods and services, are becoming a significant part of this new culture. Virtual concerts and events, where users can attend from anywhere in the world, are also becoming increasingly popular.

As an audio engineer, understanding these cultural shifts is important because they directly impact how sound is used in the metaverse. For example, the sound design for a virtual concert must consider the acoustics of the virtual venue, the type of music being played, and the cultural context in which the event is happening. The ability to design immersive and culturally relevant soundscapes will be crucial for creating engaging and meaningful experiences in the metaverse.

Building upon these cultural shifts, it's essential to explore the technological advancements that have made the metaverse possible. The history and development of virtual reality offer valuable insights into how we've arrived at this point.

## The metaverse standard forum

Establishing the Metaverse Standard Forum in June 2022 can be considered a milestone and a testament to the renewed belief in the opportunities in the burgeoning field of VR and AR, as the forum aims to develop virtual reality standards and has attracted over a thousand companies within its first month. This collaborative effort underscores the collective ambition to advance virtual reality technology, ensuring its compatibility and effectiveness across various applications. The forum represents a vital platform for innovation, sharing insights, improving interoperability, and addressing the technical challenges inherent in virtual reality development.

In June 2022, around 1,000 companies joined the Metaverse Standard Forum, which indicates the importance and momentum in technology development for Maetaverse.

### *Outlook on the metaverse*

The ultimate goal for the metaverse is to create a seamless digital alternative to the physical world and to become a significant evolutionary step in how we interact with digital technology, aiming to create a seamless integration of digital and physical realities. This entails enabling individuals to participate in events remotely yet with the sensation of physical presence. The previously unattainable immersive experience is now within reach, thanks to technological advancements. To give a concrete example of a standard that directly enables these immersive experiences, developing the MPEG standard for immersive audio and visual experiences empowers users to choose their vantage points within a virtual event. Unlike traditional settings where the director fixes the perspective, users in the metaverse can navigate freely, deciding how and from where they experience the event.

This constitutes a significant shift toward a user-defined experience and represents a departure from conventional media consumption, paving the way for interactive and personalized engagement. In traditional media, the audience is typically passive, consuming content in a linear, predetermined way – whether it's watching a movie, reading a book, or listening to music. The narrative, perspective, and experience are controlled by the creators, with little to no input or interaction from the user.

In contrast, the metaverse allows users to become active participants in the experience. Instead of being confined to a single, creator-defined viewpoint, users can now choose how they engage with content. For instance, in a virtual concert within the metaverse, users might start by viewing the performance from a front-row seat, but, with a simple command, they could switch to a backstage perspective or even position themselves on stage next to the performers. This level of interactivity extends beyond just the visual; the sound experience also adapts dynamically, offering a different acoustic perspective based on the user's location within the virtual space.

A real-world example of this shift is Travis Scott's virtual concert in *Fortnite*, which took place in April 2020. Unlike traditional concerts where attendees have a fixed viewpoint, this event allowed millions of players to experience the performance from different perspectives within the game. Some users chose to hover above the stage, while others moved through the virtual environment, exploring different angles and experiencing the concert in a personalized way. The sound design played a crucial role, with audio changing depending on the user's position, creating a truly immersive experience.

This level of user control and interaction marks a significant departure from how we've traditionally consumed media. It transforms content consumption into an active, personalized experience, where users not only choose how they engage with content but also influence how it evolves. This user-defined experience is set to become a cornerstone of the metaverse, where the boundaries between creator and consumer blur, leading to new forms of storytelling, social interaction, and economic activity.

As the metaverse grows, content creators will need to adapt, offering experiences that are not just immersive but also flexible and interactive, allowing users to shape their own journeys through digital worlds. This evolution in media consumption opens up vast opportunities for innovation, particularly in areas like immersive 3D audio, which will be crucial in making these experiences as lifelike and engaging as possible.

#### *Persistent audio across worlds*

Persistent audio in the metaverse refers to sound elements that transition smoothly as users move between different virtual environments, creating a cohesive and immersive experience. For instance, as a user shifts from a bustling virtual city to a serene forest, the ambient sound should seamlessly adapt – footsteps soften, and background music shifts in tone. This continuity is crucial for maintaining immersion.

The main challenge in achieving persistent audio lies in managing different acoustic properties between environments, such as the echoes of a city versus the subtle sounds of nature. Advanced AI-driven audio engines enable real-time adjustments with latency as low as 20 ms, ensuring that audio parameters like reverb and frequency response dynamically adapt to the changing environment.

As the metaverse expands, persistent audio will be vital for fluid, engaging experiences. The future will see more personalized, AI-driven audio that enhances user immersion across multiple worlds, with real-time processing capabilities supporting vast data ranges, from frequency responses to reverb times.

In summary, persistent audio is key to the future of the metaverse, providing seamless and immersive experiences as users navigate complex virtual worlds. While the potential of sound in the metaverse is immense, realizing this potential comes with its own set of challenges. In the next section, we'll explore the main obstacles faced in audio for virtual reality and discuss strategies to overcome them.

## Main challenges in audio for virtual reality

Creating immersive audio experiences in virtual reality (VR) presents several challenges that are crucial for delivering a believable and engaging environment. These challenges can be broadly categorized into three key areas: auditory display, synchronization, and distance perception. Each of these elements requires careful attention and advanced solutions to ensure that the sound in VR environments enhances the overall experience.

### *Auditory display*

A primary expectation in VR is that visual stimuli will be matched with corresponding auditory feedback. In many gaming environments, the focus is often on essential sounds, such as gunfire, which are critical for gameplay. However, to fully replicate an environment and foster a sense of presence, it is inadequate to rely solely on these singular sound sources. The complexity of real-world audio, with its layers of ambient sounds, echoes, and subtle auditory cues, must be captured to create a truly immersive experience.

Achieving this level of immersion requires that audio not only mirrors the visual environment but also enhances it. This involves creating dynamic soundscapes that react to user interactions and changes in the environment. For instance, in a VR setting, the sound of footsteps should change depending on the surface being walked on – wooden floors, gravel paths, or wet concrete. Additionally, the sound should interact with the environment, such as echoing off walls or being absorbed by soft materials. This approach ensures that the auditory display is as rich and varied as the visual experience, contributing significantly to the user's sense of presence.

## Synchronization

Synchronization between sound and visuals is another significant challenge in VR. When interacting with objects in a virtual space, such as opening a door, the sound must align perfectly with the visual action. This includes matching the sound's timbre, duration, and reverberation to the physical laws and expectations of the virtual environment. Any mismatch between what the user sees and hears can break the immersion and make the experience feel unnatural.

One of the critical challenges in synchronization is minimizing latency – the delay between the visual and auditory stimuli. Even a small delay can disrupt the user's experience, making interactions feel sluggish or out of sync. To overcome this, VR systems must optimize the audio processing pipeline, reducing latency to imperceptible levels. Balancing the complexity of the soundscape with system performance is also crucial. While more detailed and realistic sounds add to the immersion, they can also increase the computational load, potentially introducing latency. Techniques such as predictive audio rendering, where the system anticipates user actions and preprocesses the corresponding sounds, can help maintain synchronization without compromising performance.

## Distance perception

Distance perception in VR audio is vital for creating a convincing sense of space. In the real world, we intuitively gauge how far away a sound source is based on its volume, clarity, and the way it interacts with the environment. Replicating this in a virtual environment is essential for maintaining realism and helping users navigate the space effectively. For example, the sound of a waterfall from afar might be a soft, indistinct noise, but as the user approaches, the sound should gradually become louder and more detailed, reflecting the changing proximity.

### *Techniques for accurate distance perception*

To accurately represent distance in VR, sound designers must carefully manage how sound attenuates (fades) with distance, ensuring it behaves realistically within the virtual environment's scale. Reverb and echo effects are also critical, as they provide important cues about the size and nature of the space. For example, a large, open area might have a long reverb tail, while a small, enclosed room would produce a short, sharp echo. Advanced spatial audio techniques, such as binaural recording or real-time 3D audio rendering, can enhance distance perception by simulating how sound waves interact with the user's head and ears, further contributing to a sense of depth and realism.

To address these challenges comprehensively, technologies like six degrees of freedom (6DoF) and volumetric sound have emerged as promising solutions. Let's delve into how these innovations are shaping the future of immersive audio.

The solution for audio for virtual reality is a combination of:

- six degrees of freedom (6DoF) sound and
- volumetric sound.

These components are crucial for creating immersive audio experiences that accurately replicate real-world environments. However, despite the advancements in virtual reality productions, games, or films, achieving a fully volumetric sound or implementing 6DoF remains a significant endeavor.

These technologies will be described in the upcoming sections.

## Volumetric sound

### *Overview*

Volumetric sound represents a sound source with volume, characteristics, and structure that vary based on the listener's distance.

There is no widely recognized formal definition of 'volumetric sound' in the literature or standard dictionaries. The term 'volumetric' generally relates to the measurement of volume, but when applied to sound, it can imply a more complex auditory experience that involves multiple sound sources and spatial characteristics.

In essence, volumetric sound can be understood as an approach to sound reproduction that seeks to maintain the spatial characteristics and complexity of the original sound environment, particularly in applications like virtual reality and immersive audio experiences. This approach contrasts with traditional methods that may simplify sound sources to single points, thereby losing the depth and richness of the original soundscape.

Unlike traditional sound sources that emit from a single point, volumetric sounds can emit from every point on the object's surface in an environment. This means that the sound dynamically changes based on the dimensions of the part and the listener's position relative to it.

The sound plays all around the part when a listener is inside the part. As the listener moves away, the sound becomes more directional, and the perception of the sound's width changes depending on the part's size and the listener's proximity.

In other words, volumetric sound represents a complex sound structure with variable directivity. This means that the sound dynamically changes based on the dimensions of the part and the listener's position relative to it. However, volumetric sounds combine a mixture of complex sounds that can further form another pattern in 3D space.

This concept is illustrated by the analogy of a piano versus a loudspeaker. A complex piano with numerous sound sources cannot be accurately represented by a single loudspeaker proclaiming, 'This is how the piano sounds'. Each piano component emits sound differently, creating a rich auditory experience that changes based on the listener's position relative to the instrument. This complexity makes piano recording challenging, requiring various techniques to capture the instrument's nuanced sounds. Thus a piano is an example of a volumetric object, embodying a sound source composed of multiple interacting elements rather than a singular point source [7, 8].

Another instance of volumetric sound can be observed in urban environments. From a distance, the city's soundscape might present as a unified noise. However, as one approaches, the soundscape's complexity unfolds, revealing individual elements like distinct building sounds. This transition from a monolithic to a detailed auditory perception underlines the essence of volumetric sound.

Roblox's implementation of volumetric sound [9–11], which adjusts volume based on distance without altering the sound's directional qualities, highlights the limitations of this technology. Despite its pioneering efforts, authentic volumetric sound encompasses more than mere volume adjustments, aiming for a holistic representation of sound that mirrors real-world experiences.

Volumetric audio can also be implemented using procedural audio, which refers to generating sound in real-time through algorithms or scripts rather than relying on pre-recorded audio files. This approach allows for highly dynamic and interactive soundscapes that can adapt to a user's actions and the environment. Volumetric sound represents a complex sound structure with variable directivity, meaning that the sound dynamically changes based on the dimensions of the sound source and the listener's relative position. This type of sound combines a mixture of complex audio elements that can form intricate patterns in 3D space.

A prime example of procedural audio can be seen in *The Witcher 3*, where the wind's sound dynamically alters based on the player's location within the game world. These volumetric sounds can be generated and adjusted in real time by employing procedural audio, offering a more immersive experience that reacts to the listener's movements. The challenge lies in developing a universal model that can be applied beyond a single game, extending to varied scenarios for authentic sound representation in virtual environments.

## Six degrees of freedom (6DoF) sound

### *Overview*

The creation of live sound objects poses a notable challenge. Three degrees of freedom (3DoF) allow users to rotate their heads along three axes – yaw, pitch, and roll. However, this limitation restricts movement to these axes alone, without the ability to move laterally, vertically, or forward and backward. 6DoF sound introduces the capability to move within three additional dimensions – forward, backward, sideways, up, and down, or $x$, $y$, and $z$(see Figure 10.1) – enriching the user's experience in virtual events or computer games. Achieving this level of auditory realism in a live recording presents a considerable challenge. The goal is to create an immersive sound environment that mirrors the freedom of movement in the physical world. However, doing so requires advanced sound modeling techniques beyond traditional approaches.

Consider the example of synthetic sound in a vintage game in illustrating this concept. This game features sound with 6DoF, allowing users to navigate freely and experience sound from any direction. However, this simulation lacks the consideration of room acoustics, focusing solely on direct sound sources, such as the noise from a gun, without replicating the nuances of a natural environment. More sophisticated models are needed to transcend this limitation, ones that incorporate room acoustics to simulate how sound

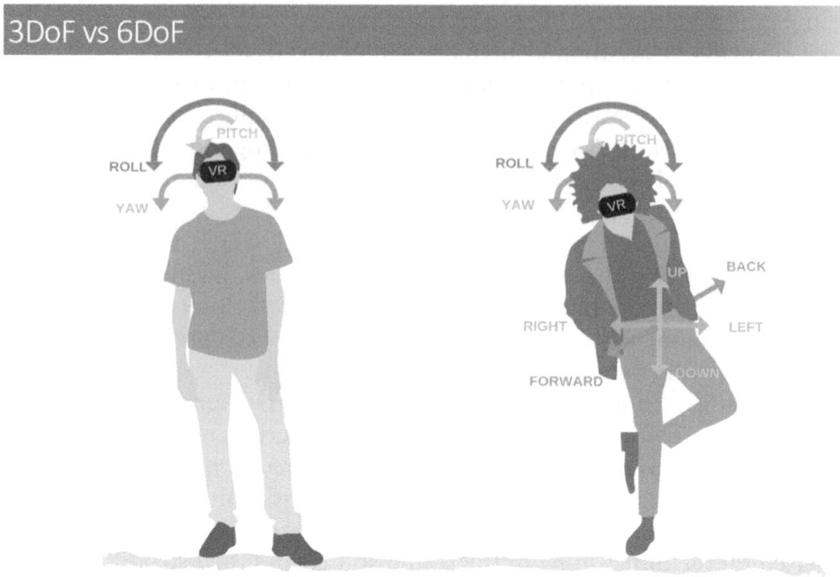


*Figure 10.1* 3DoF (yaw, pitch, roll) versus 6DoF (yaw, pitch, roll, *x*, *y*, *z*)

behaves and propagates in various spaces. These models aim to capture direct sounds and the reflections and diffractions that occur in an actual environment, thereby enhancing the realism of the virtual soundscape.

## ***Navigable 6DoF audio using Ambisonics interpolation***

### *Ambisonics*

Ambisonics [12] provides a robust solution for capturing 3D audio. Unlike conventional microphones that record sound in a single channel from a fixed point, offering limited or no information on sound directionality, Ambisonics microphones employ a multi-channel recording approach. For instance, a third-order Ambisonics microphone captures audio across 16 channels. This comprehensive capture method allows for a detailed representation of sound directionality. To experience the depth and nuance of an Ambisonic recording, the audio must be decoded through specific equipment, such as headphones or a multi-channel audio system, enabling listeners to accurately perceive the spatial orientation of sounds.

### *Ambisonics interpolation*

Ambisonics microphones capture the sound environment from a specific location, allowing for rotational movement but not positional changes. Deploying multiple Ambisonics microphones across a space makes navigating between different audio perspectives possible. This technique, known as Ambisonics interpolation [13, 14], facilitates calculating sound position and characteristics at various points within the space. Such operations are cost-effective and can be executed in real time.

An illustrative application of this method involved placing numerous microphones inside and outside a car, creating an expansive sound environment navigable in post-production. This setup allows sound engineers to specify desired audio perspectives, manipulating the sound field as if moving through it physically.

The main benefit of using Ambisonics interpolation is a near-perfect reconstruction of room acoustics. It is not required to simulate acoustic through computational expensive auralization process. The reverberation information is already included in the Ambisonics recordings.

Despite the immersive realism achieved through Ambisonics interpolation, a significant challenge arises in the modeling and altering Ambisonics sound. While object-based sound representation offers greater flexibility in sound manipulation, it lacks Ambisonics sound's authenticity and immersive quality. Therefore, while Ambisonics interpolation excels in realism, it presents limitations in the ease of modifying recordings, illustrating a trade-off between realism and adaptability in sound design.

The proposed Ambisonics interpolation algorithm aims to facilitate six degrees of freedom (6DoF) audio experiences by employing multiple simultaneous recordings from Higher-Order Ambisonics (HOA) microphones. Here's a detailed explanation.

### GENERAL OVERVIEW OF THE ALGORITHM

The Ambisonics interpolation algorithm is based on simultaneous recordings from multiple HOA microphones distributed across the sound scene in a grid formation. This method aims to provide a listener with a seamless, realistic auditory experience when navigating within a 6DoF audio scene. Unlike traditional single-point Ambisonics that only capture rotational movements, this algorithm allows listeners to move freely through space, experiencing both translational and rotational audio cues, which is essential for fully immersive audiovisual applications like virtual reality.

The core of the approach is to use multiple spatially distributed microphones and perform interpolation between their signals based on the listener's position relative to the microphones. The interpolation process involves combining audio signals from the closest microphones to the listener's position, applying distance-dependent filtering, scaling, and rebalancing operations to achieve a coherent sound field throughout the scene (see Figure 10.2).

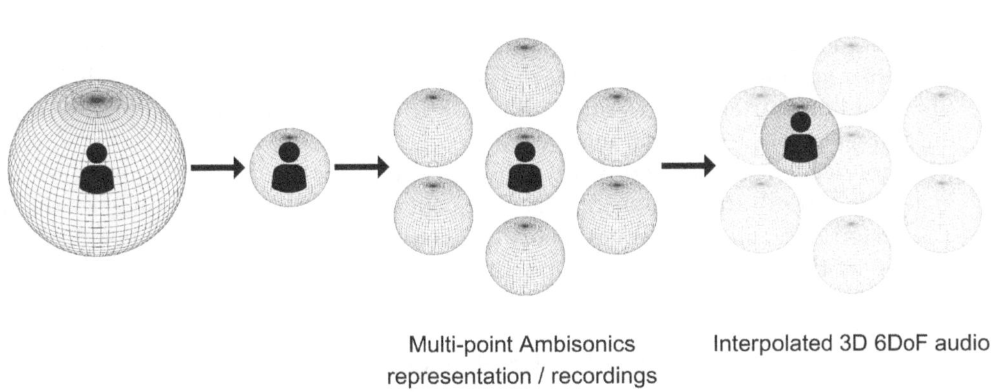


*Figure 10.2* 6-DoF Ambisonics interpolation algorithm – general idea

### MICROPHONE GRID AND IMPORTANCE OF DISTANCE

The recording setup uses a grid of HOA microphones placed uniformly across the audio scene. Typically, the microphones are arranged either in an equilateral triangle or square grid, ensuring that the entire volume of interest is covered evenly. If full 6DoF movement is required (including height), the grid can be stacked in layers to provide coverage along the vertical axis.

The distance between microphones and the listener's position is crucial for this algorithm. The microphone grid's arrangement, combined with interpolation based on distance, ensures that the listener experiences a natural attenuation of sound sources and coherent transitions between different listening points. Essentially, the microphones act as reference points, and the interpolation weights the contribution of each microphone based on how far it is from the listener's position. This distance-dependent approach allows the algorithm to effectively simulate the changing acoustic properties of the environment as the listener moves, creating a realistic sense of depth and spatial presence.

### DETAILED EXPLANATION OF AMBISONICS INTERPOLATION

The playback system performs Ambisonics signal interpolation in the time domain, using components from all the HOA microphones in the grid to create an interpolated Ambisonics signal. Here's how the key parameters influence the quality of the output:

1. **Distance-dependent scaling function:** The scaling function is used to control the contributions of the individual microphones to the final interpolated sound. The function attenuates contributions from microphones that are farther away from the listener, ensuring that the closest microphones have the most influence. This method prevents ambiguity in source localization and maintains a natural progression of the audio scene as the listener moves. Parameters such as the attenuation threshold and slope significantly impact the smoothness of transitions and the overall auditory coherence.
2. **Component rebalancing:** Rebalancing involves adjusting the ratios between the 0th-order omnidirectional component and the directional components of higher orders. As the distance between the listener and a given microphone increases, the influence of directional components is reduced, and the omnidirectional component becomes more dominant. This helps to avoid localization ambiguities when transitioning between two adjacent microphones. The parameters involved in this rebalancing (thresholds and attenuation slopes) influence the perceived localization precision and the consistency of the spatial image.
3. **Distance-dependent filtering:** A distance-dependent low-pass filter is applied to simulate the natural attenuation of high frequencies over distance, similar to the air absorption effect in real environments. The cutoff frequency of this filter decreases linearly as the distance from the microphone increases, which helps simulate how high-frequency content diminishes with increased distance in real-world scenarios. This contributes to the realism of the reproduced scene, as the listener perceives sources farther away with reduced high-frequency energy. The parameters here include the cutoff frequency threshold and the slope, which directly affect the perceptual realism of distance cues.

4. **Threshold parameters:** These thresholds define when specific attenuation behaviors (such as distance-dependent attenuation and rebalancing) start to take effect. Fine-tuning these values is crucial for ensuring smooth transitions as the listener moves through the recorded space. If the thresholds are set too aggressively, the transitions between sound fields may become abrupt or unnatural, while too lenient values may lead to insufficient differentiation between closer and farther microphones, resulting in a lack of spatial depth.
5. **Interpolation technique:** The interpolation itself is a linear combination of Ambisonics signals from different microphones, with the weights being determined based on distance functions. By prioritizing contributions from the nearest microphones, the listener's position can be continuously updated, ensuring a consistent auditory image. The choice of linear versus more complex (e.g., spline-based) interpolation methods impacts both computational efficiency and the quality of smoothness when the listener moves.

### *How to create six degrees of freedom audio content*

Creating content with six degrees of freedom (6DoF) audio involves utilizing multiple Ambisonics microphones in conjunction with 360° cameras. This process is streamlined by leveraging tools and techniques already established for producing 360° videos. By strategically placing cameras and microphones at designated points within the recording environment, both the visual and auditory elements necessary for creating immersive experiences are captured. The result is a collection of 360° videos, each embedded with Ambisonics sound, providing a comprehensive sensory experience.

To enhance the versatility of these recordings, specific applications are employed, enabling the seamless integration and transition between individual video clips. This functionality facilitates the creation of a navigable audiovisual space where users can move and look around freely, experiencing the content from any chosen perspective. This approach simplifies the production of 6DoF audio content, making it accessible for creators aiming to deliver high-quality immersive experiences.

Having understood the technical underpinnings of 6DoF and volumetric sound, it's beneficial to examine real-world applications. The following section showcases use cases that demonstrate these concepts in action.

## Implementation of full VR experience with 6DoF audio

### *Use case: Philharmonic Orchestra*

#### *Overview*

During the COVID-19 pandemic, a recording session was conducted with the Poznań Philharmonic Orchestra using an array of 30 third-order Ambisonics microphone arrays. This complex setup resulted in a total of 600 audio channels. Each microphone array, equipped with USB connectivity, was connected directly to USB hubs, which simplified the setup and reduced the need for additional adapters. The entire system was managed by a single, efficient laptop, capable of handling the significant data processing required for the session.

The microphones were arranged in equilateral triangle formations, with each side measuring 2 m. This configuration was carefully chosen to optimize the input parameters for the Ambisonics interpolation algorithm, ensuring even coverage of the entire recording space. Spacing the microphones in this way minimized gaps in the sound field and captured the rich spatial characteristics of the performance.

The visualization of the orchestra and the recording environment was rendered in a high-fidelity 3D software platform. This platform enabled the simulation of a comprehensive virtual environment, integrating modeled microphones and characters to enhance the realism of the audio experience. (Figure 10.3). This setup enabled dynamic rendering of sound, making the virtual experience more immersive for the user, much like navigating around a live event.

This approach demonstrates [15] the potential for creating immersive soundscapes that the listener can navigate. For instance, adopting a virtual reality headset enables users to move within the recorded space, altering their auditory perspective based on their location. By employing a virtual microphone that tracks the virtual camera's movement, it can capture sound from any point in the space, offering a flexible and immersive audio experience. This technique allows for post-recording manipulation of the sound field, enabling sound engineers to extract audio from specific points and tailoring the final output to the desired immersive experience.

### *Hardware used*

The recording of the Poznań Philharmonic Orchestra employed six degrees of freedom (6DoF) technology to create an immersive and dynamic audio experience. This technology was crucial for allowing listeners to move freely within the virtual space, experiencing the orchestra from multiple angles, as if they were physically present at the venue. This technology captures not just the audio elements but also the spatial relationships within the concert hall, providing users with a realistic and interactive experience.

We utilized 30 third-order Ambisonics microphones to capture the orchestra's performance. These microphones were strategically positioned to provide full coverage of both the stage and the surrounding audience area. Third-order Ambisonics microphones can capture sound from all directions, which is essential for recording a live orchestra in a way that preserves spatial detail. The microphones were arranged in a carefully designed grid to optimize the spatial resolution of the recording, allowing for an accurate representation of the live sound environment.

Specification of the microphones used:

- 19 omnidirectional capsules
- Sampling frequency 48 kHz
- 24 bits
- USB connectivity
- Matched sensitivity parameters for all microphone capsules

We used third order Ambisonics microphones because they offer the best trade-off between spatial resolution and complexity of recording. Third-order Ambisonics produce 16 audio channels.

After the recording, author was involved in developed a virtual reality (VR) application [16, 17] that allowed users to navigate through the volumetric audio space of the orchestra, effectively creating a 'walkthrough movie'. This VR application enabled users to move between different positions in the virtual concert hall, adjusting their auditory perspective based on their movement. By using the 6DoF technology, users were able to virtually step inside the orchestra, listen to individual sections, and gain a unique perspective that traditional recordings cannot offer.

### *General overview of the 6DoF audio production process*

To facilitate the VR experience, the physical venue was simulated within the game engine (Unreal Engine 4) (Figure 10.3) and integrated with Wwise using the Higher-Order Ambisonics (HOA) interpolation and renderer plug-in. Unreal Engine 4 recreated the venue with high fidelity, including all structural and spatial elements, while Wwise provided the audio processing framework that allowed for accurate and dynamic sound reproduction. The integration of Unreal Engine and Wwise was crucial for creating a navigable VR demo, enabling users to explore both the audio and visual aspects of the concert interactively.

The initial steps of the project involved gathering comprehensive information about the recording environment. This included creating a scaled map of the venue to plan the placement of both the microphones and the musicians. The scaled map was essential for determining optimal microphone placement to ensure even coverage of the orchestra and the audience. It also helped in avoiding acoustical dead spots and ensuring that the recording accurately captured the depth and ambience of the concert hall.

The primary goal of the project was to capture the acoustics of the venue in a way that allows listeners to walk among the musicians virtually. This required a strategic

*Figure 10.3* Design of full VR experience in 3D modeling tool

placement of the microphones, ensuring that each section of the orchestra was covered and that the spatial characteristics of the venue were preserved. To achieve this, the microphones were arranged in a triangular grid pattern across the stage, providing consistent coverage and minimizing the possibility of gaps in the sound field. The use of 30 third-order Ambisonics microphones enabled the capture of high-resolution spatial audio that could later be rendered for an immersive VR experience. Additional microphones were positioned among the audience.

### *Planning and preparation for the 6DoF audio recording session*

Detailed maps (Figure 10.4) were created to indicate the precise locations of the microphones and the positions of the musicians. The spacing between microphones was limited to a maximum of 2 m to ensure a high level of spatial accuracy in the recordings. This configuration provided the necessary overlap between microphone coverage areas, which was crucial for accurate Ambisonics interpolation during post-production.

The microphones were positioned at a height of 1.5 m, which was chosen to approximate the ear level of a seated audience member. This height ensured that the recording captured sound in a natural way for listeners, enhancing the immersive quality of the VR experience. Given the large number of microphones and their strategic placement, careful cable routing was essential. USB and Jack extenders were used to connect the microphones to the recording system, and meticulous planning was required to prevent cables from interfering with the musicians or being visible in the 360° video.

In preparation for the recording, a primary recording computer was set up along with backup systems to ensure that no data would be lost in the event of equipment failure. The recording computer needed to handle a significant data load, as each microphone produced 20 channels of audio at a 48 kHz sampling rate with 24-bit depth. This resulted in a substantial amount of data, requiring high-capacity storage solutions and reliable

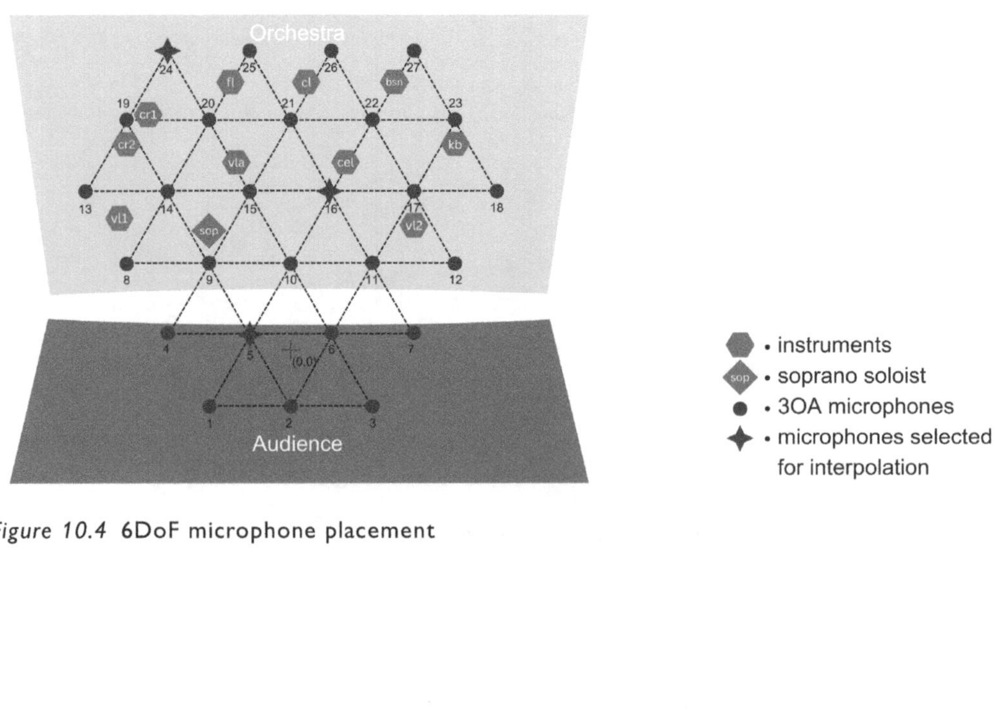


*Figure 10.4* 6DoF microphone placement

backup procedures. A test run was conducted before the actual recording to verify that all equipment was functioning correctly and to address any potential technical issues.

### *Audio rendering explanation in the audio authoring tool and game engine*

The audio rendering process began by importing the third-order Ambisonics files into the audio authoring tool (Wwise), where a new audio bus was created and configured for the appropriate channel setup. This setup allowed for accurate routing of the Ambisonics signals through the audio authoring tool. The 6DoF Ambisonics interpolation renderer plug-in was then configured, specifying the number of arrays used (in this case, 30) and defining game parameters that enabled dynamic manipulation of the audio in response to user movement.

The game engine (Unreal Engine 4) played a significant role in the visual representation of the concert hall, with an emphasis on accurate positioning of microphones and musicians. This detailed 3D-rendered scene was essential for creating a visually immersive experience that matched the audio environment. The accurate placement of visual elements ensured that the user's visual perspective aligned with the spatial audio, maintaining immersion.

The connection between the game engine and the audio authoring tool was established using Real-Time Parameter Controls (RTPC). RTPC allowed the audio to dynamically adapt based on the user's movement within the virtual environment. As users navigated the VR space, the audio rendered in real time, adjusting to changes in the user's position and providing a realistic simulation of moving around a live orchestra. This setup allowed for an interactive and immersive experience where users could observe how sound localization and balance shifted as they explored different areas of the concert hall.

Overall, the combination of careful planning, advanced audio technology, and detailed visual rendering made it possible to capture and reproduce the performance of the Poznań Philharmonic Orchestra in an immersive VR format. This approach not only provided an engaging user experience but also offered a valuable demonstration of the potential of 6DoF audio technology for future audiovisual productions.

## ***Technical recommendations for full VR 6DoF audio***

This section outlines the main conclusions and lessons learned from the 6DoF audio recording and production session.

### *Capturing the environment using multiple microphones: considerations for placement, number of microphones, and setup*

In the context of the 6DoF recording of the Poznań Philharmonic Orchestra, several critical factors were considered to capture the environment accurately using multiple Ambisonic microphones.

**Microphone placement:** In this recording, 30 third-order Ambisonics microphones were placed in a regular triangular grid within the Aula of Adam Mickiewicz University. The maximum distance between adjacent microphones was 2 m, except for three microphones placed in the audience area, which were spaced about 4 m apart. Each microphone was positioned at a height of 1.5 m.

This grid setup ensured even coverage of the recording space, minimizing the chances of gaps in the sound field. Placing microphones at 1.5 m, approximately at ear height for seated listeners, likely helped in capturing sound reflections and ambient noise at a level similar to what an audience member would experience. Uniform placement is crucial for achieving smooth transitions between sound fields captured by different microphones, which is vital for maintaining the realism of the 6DoF audio experience. Any deviations could cause inconsistencies in sound localization, thereby reducing the immersion.

**Number of microphones:** The use of 30 microphones was necessary to cover the large and acoustically complex space of the concert hall. This generated more than 600 GB of audio data due to the 20-channel output from each microphone with a 48 kHz sampling rate and 24-bit depth.

The number of microphones directly impacts the spatial resolution of the recorded audio. A higher number of microphones allows for a more detailed and accurate sound field, which is particularly important in large environments like concert halls. This ensures that the final audio experience is rich in detail, accurately representing the spatial characteristics of the performance.

**Microphone grid:** Some microphones had to be placed closer together than the standard 2-m spacing due to physical constraints within the venue, such as the placement of musicians and instruments.

Adapting the microphone grid to the physical layout of the environment is essential for capturing all critical areas of the sound field without leaving gaps. This flexibility in placement allows for a more realistic and complete recording while accommodating real-world limitations.

**Computational redundancy:** Three computers were used during the recording session – one main and two backups – to handle the large data volume and ensure the session could proceed without interruptions.

Ensuring that there is sufficient redundancy in the recording setup is crucial to prevent data loss, which could compromise the entire recording session. Given the large data volume involved in 6DoF recordings, having reliable backup systems in place is necessary to secure the integrity of the project.

### *Ambisonics interpolation in a 6DoF experience*

Ambisonic interpolation is crucial in creating a seamless 3D audio experience by blending the audio signals captured from multiple Ambisonic microphones. This technique was central to the 6DoF Philharmonic recording.

The interpolation technique used in the 6DoF Philharmonic recording applied a distance attenuation method. The interpolated signal was computed as a sum of contributions from all Higher-Order Ambisonic (HOA) signals captured by the microphone grid. The closest microphones had the most significant influence on the sound field, with adjustments made based on the listener's position within the virtual environment.

This approach ensures that, as the listener moves within the virtual space, the sound they hear reflects their position relative to the sound sources, creating a realistic and immersive experience. This method prevents abrupt changes in the audio, maintaining the continuity of the sound field and enhancing the realism of the 6DoF experience.

Four key parameters were adjustable within the 6DoF rendering algorithm: volume threshold, volume range, HOA threshold, and HOA range. These parameters allowed fine-tuning of the rendered sound field to maintain spatial accuracy as the listener navigates the environment.

Adjusting these parameters is essential for optimizing the listener's experience. Fine-tuning the volume and HOA thresholds ensures that the closest microphones dominate the sound field, which enhances the clarity and accuracy of the audio experience as the listener moves through the virtual space.

The interpolation algorithm was designed to ensure smooth transitions between different Ambisonic signals as the listener moves, preventing perceptible jumps or changes in the audio experience.

Smooth transitions are critical to maintaining immersion in a 6DoF audio experience. Abrupt changes in the sound field can disrupt the illusion of a continuous, navigable space, reducing the overall quality and realism of the virtual environment.

The presence of 32 musicians in a large, reverberant concert hall made sound localization more challenging, which impacted the subjective scores during evaluation.

Ensuring precise sound localization is important for maintaining the spatial integrity of the audio experience. In complex acoustic environments, like a concert hall with many reflective surfaces and multiple sound sources, maintaining localization accuracy is challenging but essential for creating a realistic and immersive experience.

### *Challenges in creating 6DoF audio experiences*

Several challenges were encountered during the 6DoF Philharmonic recording, which are common in creating 6DoF audio experiences:

Processing and rendering the interpolated sound field in real-time, especially with high-order Ambisonics, requires significant computational power. High computational demand can limit the ability to deliver real-time audio experiences, especially on less powerful hardware. Ensuring that the processing is optimized for performance is crucial for maintaining a smooth and immersive audio experience.

The recording session generated over 600 GB of audio data, requiring efficient storage and backup solutions. Managing large datasets is essential to ensure that all recorded material is securely stored and easily accessible for post-production. Poor data management could lead to data loss or corruption, jeopardizing the entire project.

The large and acoustically complex environment of the concert hall, combined with the presence of 32 musicians, made sound localization and overall audio quality harder to manage. The acoustic complexity of the environment can affect the quality and accuracy of the final audio experience. Properly addressing these challenges ensures that the recorded audio accurately reflects the environment, providing a more authentic and immersive experience.

In the 6DoF Philharmonic recording, integrating the audio with the visual component in Unreal Engine 4 was necessary for a cohesive virtual reality experience. Seamless audiovisual integration is critical for immersion in VR experiences. Any lag or mismatch between the audio and visuals can break the illusion of reality, diminishing the overall experience.

### *Considerations and pitfalls to watch out for*

When working on 6DoF audio experiences like the Philharmonic recording, several factors need careful attention:

Ensuring that the microphone grid is precisely laid out according to the planned design is vital, as small deviations can lead to inconsistencies in the recorded sound field. Precision in microphone placement is crucial for maintaining the spatial integrity of the recorded sound. Inaccuracies can cause gaps or overlaps in the sound field, leading to a less realistic and immersive experience.

Be aware of reflective surfaces in the recording environment, as they can significantly impact the captured audio quality. Acoustic reflections can introduce unwanted artifacts or distortions in the recorded sound, which can be challenging to correct in post-production. Managing these reflections through careful placement and acoustic treatment is essential for a clean and accurate recording.

Properly managing and backing up the large amounts of data generated during recording is crucial. Proper data handling is critical to avoid the loss of valuable recorded material. Given the complexity and scale of 6DoF projects, losing even a small portion of the data can significantly impact the final product.

### *Tips and tricks for effective 6DoF audio recording*

Based on the experience from the 6DoF Philharmonic recording, here are some practical tips:

- **Pre-production planning:** Spend significant time planning the microphone placement and setup, taking into account the specific acoustics and layout of the recording venue. Thorough pre-production planning ensures that the recording session proceeds smoothly and that all potential issues are addressed before they become problematic. This planning phase is crucial for achieving high-quality results.
- **Use of virtual models:** Create a virtual model of the recording environment to aid in microphone placement and setup. Using virtual models allows for precise planning and visualization of the recording setup, reducing the chances of errors during the actual recording. This approach can also help in experimenting with different setups before committing to a final arrangement.
- **Dynamic adjustment:** Be prepared to make on-the-fly adjustments during the recording session to accommodate unforeseen issues or changes in the environment. Flexibility during the recording process is important for adapting to real-world challenges that may not have been anticipated during the planning phase. Being able to make quick adjustments ensures that the recording session can continue smoothly without compromising the quality of the final product.

### *Experiences that work well with 6DoF audio*

The 6DoF recording technique is particularly well-suited for:

- **Live performances:** The 6DoF Philharmonic recording demonstrated the effectiveness of this technique for capturing live performances, such as those by the Poznań Philharmonic Orchestra. The ability to capture and reproduce the spatial characteristics of a live performance in a concert hall provides a highly immersive experience, allowing listeners to feel as if they are part of the audience, with the freedom to explore different perspectives within the venue.

- **Interactive VR experiences:** This technique is ideal for creating interactive VR experiences where users can move freely within a virtual space, and the audio needs to adapt seamlessly to their movements. Providing a dynamic audio environment that changes based on the user's position enhances the immersion and realism of VR experiences, making them more engaging and lifelike.
- **Immersive storytelling:** 6DoF audio is effective in immersive storytelling, where spatial audio can guide the narrative and enhance the emotional impact. In storytelling, sound plays a crucial role in setting the mood and guiding the audience's attention. Using 6DoF audio allows for a more interactive and emotionally engaging narrative, as the sound environment can change dynamically with the story.

By integrating these practices, creators can enhance the overall quality of immersive audio experiences. In conclusion, let's reflect on the key takeaways and future directions for 6DoF audio production.

#### *Summary of key recommendations*

The 6DoF Philharmonic recording of the Poznań Philharmonic Orchestra showcases the potential of using multiple Ambisonic microphones to create immersive, navigable audio experiences. By carefully planning microphone placement, using advanced interpolation techniques, and ensuring seamless integration with visual elements, it is possible to deliver high-quality 6DoF audio that enhances the listener's experience in virtual environments. However, attention must be paid to the challenges posed by large acoustic spaces, data management, and the computational demands of real-time audio processing. With the right approach, 6DoF audio can be effectively used in live performances, VR experiences, and immersive storytelling, offering a rich and engaging experience for the audience.

## Implementation of navigable 6DoF sound with teleportation (multipoint 360 audio and video)

### *Use case: Jordan Rudess and Friends project*

For this project, 360° cameras were utilized, acknowledging the inherent limitation that movement could only occur between fixed points, not freely, due to the constraints of 360 video technology.

In July 2021, Jordan Rudess of Dream Theater, along with Steve Horelick and Jerry Marotta, both associated with Genesis, gathered at the Dreamland Recording Studios, in Woodstock, New York. It was a spectacular concert event, spanning classical to impressionistic piano genres to progressive, experimental rock with spacious synth soundscapes. The goal was to capture sound at precise locations, ensuring that the acoustics of the Old Church, the chosen venue, were faithfully represented.

During playback, listeners could alter their point of listening, simulating the experience of being physically present among the performers. This recording was notably conducted during the COVID-19 pandemic when travel to the United States was restricted. Consequently, a local company was engaged to set up the microphones while the production was managed remotely from a studio in Poznań. This approach overcame logistical challenges, ensuring the project's completion despite the prevailing global circumstances.

## Setup and hardware used

- One standard laptop with a typical office configuration was used for the recording. The key aspect to remember when choosing a computer is the hard drive speed that handles the number of audio channels you record. For example, in this case, 0.0805 gigabytes per second, or 289.8 GB per hour, were recorded. That is 30 microphones, 20 channels per microphone, 48,000 samples per second, and 24 bits per sample.
- 6 × third-order Ambisonics microphones with synchronization
- 5 × 360 8 K cameras, without fan
- Six loudspeakers connected to the surround system that generated additional ambient sound

Figure 10.5 demonstrates how the microphone and cameras were placed among musicians.

### *Technical recommendations for multipoint 360 audio and video*

This section describes the main recommendations on how to prepare the multipoint 360 audio and video using lessons learned from use cases similar to Jordan Rudess and Friends project.

Additionally, there is a set of guidelines for sound engineers using Ambisonics recording techniques for music production. By using spherical microphone arrays, beamforming, and virtual microphones, a flexible, high-quality mixing environment can be created. The following sections provide practical advice for leveraging Ambisonics to isolate sound sources, manage reverberation, and achieve a clean and immersive audio experience.

#### *Production workflow for multipoint 360 audio and video*

The workflow of creating multipoint 360 audio and video content leverages well-known tools and techniques in sound and video production, but what sets it apart is

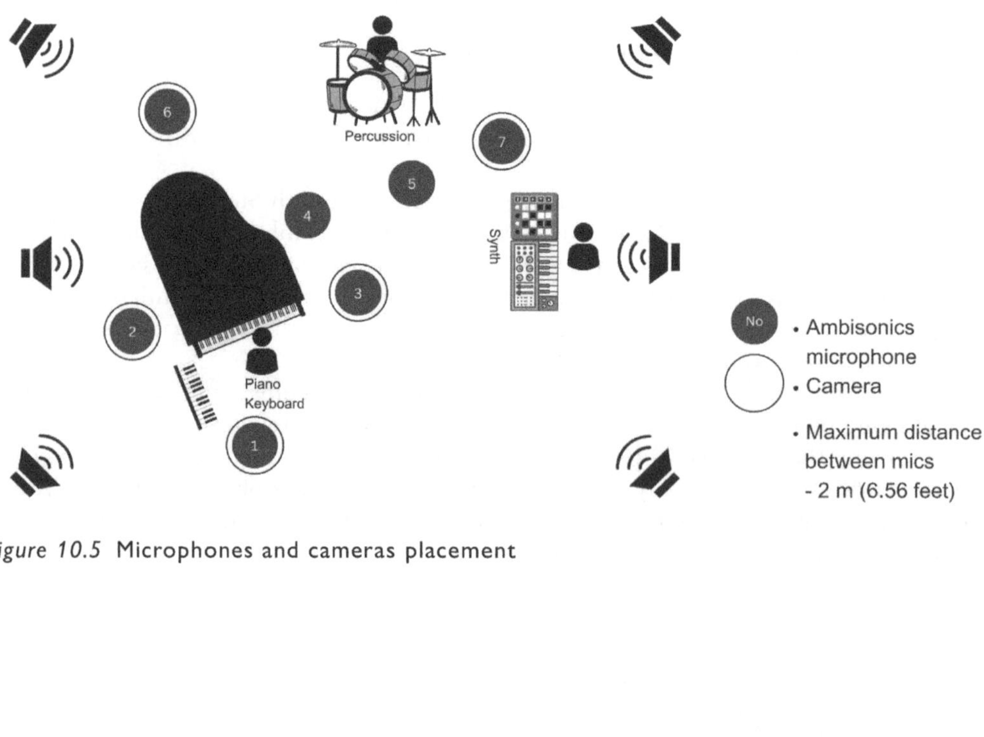


*Figure 10.5* Microphones and cameras placement

the integration of real-time 6DoF interpolation and spatial audio recording methods. By carefully managing these stages, sound engineers can create fully immersive and navigable audio experiences that elevate the realism of VR environments. Here's a breakdown of the workflow, from the initial recording to the final VR application upload:

### RECORDING SETUP: MICROPHONES, CAMERAS, CABLING, AND SYNCHRONIZATION

The setup begins with strategically positioning 360° cameras and Ambisonics microphones to capture the full soundscape and visual environment. Each camera should be paired with a corresponding Ambisonics microphone, forming a coherent audio-visual capture unit. Placement should consider spatial relationships to achieve an immersive experience from an artistic point of view, with microphones capturing sound from all directions and cameras providing a complete spherical video view. However, in the use case with teleportation, we do not have to worry about 6DoF interpolation.

Since we use 360° cameras, nothing can be hidden behind the camera as in traditional 2D production – everything will be visible. To manage equipment discreetly, it is essential to have a dedicated control room that is not visible from the recording room. High-quality cabling is still necessary to ensure clear audio capture and proper synchronization, but cables must be meticulously organized to avoid visibility in the shot. Careful cable management also prevents tripping hazards and interference.

Since we are recording 6DoF audio, the acoustics of the room matter significantly, and audio should be perfectly synchronized sample by sample from each microphone. In this case, we used timecode systems, and then software-based solutions to ensure perfect alignment, reducing the risk of drift over long recordings. In that use case, it is recommended to have a small buzzer next to each microphone to provide a synchronization signal. Synchronization tools may include clapboards for manual sync points or more advanced setups with Genlock for video and word clock for audio.

For future recording setup, it is recommended that all Ambisonics microphones share the same clock for their analog-to-digital converters.

### DAW SETUP WITH 6DOF INTERPOLATION PLUG-IN FOR CRITICAL LISTENING DURING RECORDING

A dedicated DAW is set up to capture, monitor, and manage Ambisonics audio streams. This time, we do not need 6DoF interpolation. However, we should be able to listen to each microphone signal using an Ambisonics converter for a particular plug-in and an Ambisonics-to-binaural decoder.

It is important to use a DAW that supports multi-channel tracks that are suitable for your microphone. In the use case described in this section, we needed a DAW that records at least 20 channels per track.

Engineers monitor the recording through high-quality headphones or multi-channel speaker setups, using the 6DoF interpolation plug-in to simulate the final immersive audio experience. This setup helps make immediate adjustments if necessary, such as repositioning microphones to reduce unwanted reflections or adjusting camera angles for better visual and auditory coherence.

POST-PRODUCTION: EDITING AMBISONICS AUDIO AND 360 VIDEO

Each camera's Ambisonics audio recording is processed separately in the DAW. Engineers use spatial audio processing techniques, including equalization, compression, and noise reduction, to optimize the audio quality. The 6DoF interpolation algorithm is then applied to blend multiple Ambisonics sound fields, creating a navigable soundscape.

The video editing software stitches and synchronizes the 360° footage from all cameras, ensuring a smooth visual experience. Color grading and visual effects are added as needed to enhance the visual appeal.

The edited Ambisonics audio is synchronized with the 360° video, ensuring that the spatial sound matches the visual cues. Tools like Reaper or Pro Tools may be used for audio editing, while Adobe Premiere Pro or DaVinci Resolve handles video editing. A specialized VR plug-in or software like Reaper's Ambisonic Toolkit facilitates seamless audio integration.

The final step involves preparing the content for distribution on VR platforms like Oculus TV. The audio and video files are encoded into formats compatible with the target platform, often using H.265 for video and Ambisonics B-Format for audio.

Before uploading, a thorough quality check ensures that the audio and video sync perfectly, with no noticeable artifacts or inconsistencies. VR headsets are used to test the experience, ensuring that the 6DoF sound transitions smoothly as users move between different points.

The content is then uploaded to platforms like Oculus TV, where users can experience it in full 360° immersion, with spatial audio that dynamically adjusts based on their movements and position within the virtual environment. The playback application has to support camera switching from the user's point of view, allowing them to choose different perspectives seamlessly.

### *Mixing Ambisonics sound*

Multipoint Ambisonics recording offers substantial flexibility when it comes to adjusting individual Ambisonics recordings. Unlike a fully navigable 6DoF audio setup, where smooth transitions between different zones are critical to maintaining a seamless experience, multipoint Ambisonics allows for more aggressive sound manipulation techniques. This flexibility is largely because we don't have to ensure flawless continuity between zones, allowing for a more tailored approach to sound adjustments.

In this context, the main priority remains the spatial integrity of sound sources in relation to the visual elements. Ensuring that the perceived position of each sound source matches its visual counterpart is key to maintaining immersion. Compared to navigable 6DoF audio, where continuity is paramount, multipoint Ambisonics allows for more freedom in using sound source separation techniques in Ambisonics space. This means we can adjust the gain of particular directions, extract specific sound sources, and further process them in the digital audio workstation (DAW) using linear processing filters and equalizers (EQ). This increased level of manipulation empowers sound engineers to refine the audio experience without the same level of restriction typically associated with interactive and navigable VR environments.

The use of spherical microphone arrays can offer simplified and compact recording setups and, at the same time, lead to powerful mixing options. The flexible

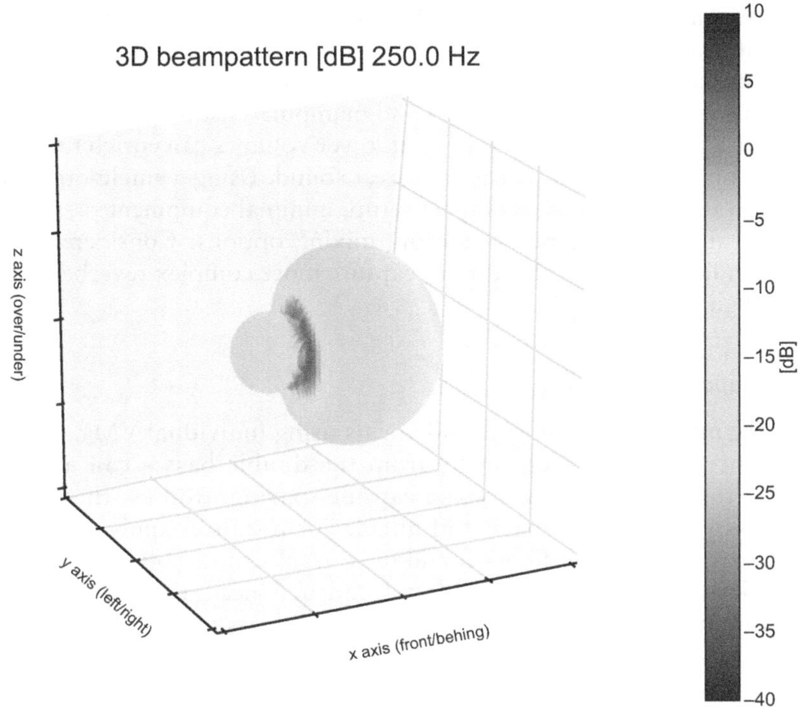


*Figure 10.6* Virtual microphone

combination of source-separated signals with varied degrees of beamforming focus enables independent levels of control while maintaining the spatial coherence and reverberation qualities of the recorded spaces. The signal separation is done by a dedicated piece of software that creates beamformings (here, referred to as virtual microphones, or VMs). The virtual microphones' directivity can be adjusted by 1° and steered in any direction of the recorded sound field, separating individual sound sources [9–11].

Figure 10.6 presents an example of a virtual microphone with a 3D pattern for 250 Hz.

### *Leveraging beamforming for sound source isolation and reverberation control*

Before the mixing process, it is advised to set up multiple virtual microphones (VMs) in the Ambisonics space. That setup allows for gain or EQ adjustment in different directions. The higher the order of Ambisonics recording, the more VMs can be set up. For instance, 6 VMs can focus on specific instruments (violins, viola, double bass, cello) and the narrator, while four can be used for capturing reverberant sound from broader directions. Additionally, 5 VMs from the secondary array can be configured solely for ambient reverberant sound. Use broader beamforming with arbitrary symmetrical arrangements to create stereo pairs that introduce a balanced reverberant signal into the sound mix. Narrow beamforming can employ custom algorithms to determine the direction of arrival (DOA) for each sound source, defining precise azimuth and elevation angles for focused directivity. This setup efficiently separates and manipulates sound sources. Adding ambiance stereo pairs from the secondary array at lower volumes can enrich timbre and spatiality without compromising the clarity of direct sound. Using a single-order Ambisonics array offers advantages such as a compact setup, minimal equipment presence in performance spaces, and extensive post-recording mixing options. Consider future research into using nonadjacent spherical arrays to capture more complex reverberant sounds for surround configurations.

#### *Audio source separation techniques*

When isolating multiple instruments and vocals using individual VMs, challenges such as low-frequency crosstalk – especially from the double bass – can arise due to the dimensions of the microphone array and capsule spacing. Address these challenges by using high-pass (HP) and low-pass (LP) filters. For instance, apply 150 Hz HP filters for vocal tracks, 200 Hz HP filters for ukulele tracks, and 100 Hz LP filters for the double bass. These filters help mitigate low-frequency bleed. Equalization (EQ) further refines separation – apply EQ cautiously to avoid cross-impact between tracks. For ukulele tracks, use a 1 kHz band attenuation to reduce vocal content and boost the 4 kHz band to emphasize transients. For vocal tracks, attenuate at 400 Hz and boost at 1 kHz to enhance clarity. These techniques help maintain the distinction between different instrument groups.

#### *EQ strategies for improved separation*

Use careful EQ adjustments to ensure that enhancing one instrument does not degrade the quality of others. Separate tracks into two main groups: vocal and ukulele. In the ukulele group, attenuate at 1 kHz to minimize vocal overlap and boost the 4 kHz band to enhance transient sounds, resulting in better instrument clarity. For the vocal group, attenuate at 400 Hz to clear muddiness and amplify the 1 kHz band to highlight vocals. Keep EQ adjustments within a range of +/–5 dB to preserve overall sound balance. Precision in EQ processing is crucial, as significant alterations can compromise the entire mix.

#### *Dynamic range compression and gate effects*

Apply dynamic range compression (DRC) to control the dynamic range of tracks. For example, use DRC primarily on the double bass track, which, after low-pass filtering and EQ, should be well-separated from other sources. This ensures consistent levels throughout the mix. Additionally, DRC should be applied to the vocal group to make it more prominent without overwhelming the mix. Leave other tracks unaffected to preserve their natural dynamics. Use the gate effect on the double bass track to accentuate transient signals and suppress inactive segments, reducing background noise and cleaning up the track. These techniques contribute to a polished and cohesive final mix.

### *Practical applications and case studies*

Using spherical microphone arrays and beamforming techniques provides significant benefits for classical music recordings. These methods allow for creative control in post-production, leading to mixes that are spatially rich and clear. Traditional microphone setups often fail to achieve the desired sound quality or require extensive equipment and setup time, whereas Ambisonics arrays offer superior results. This practical approach helps sound engineers understand how to apply these methods to achieve optimal results in their audio projects.

## Summary

This contribution has provided a comprehensive exploration of the integral role sound plays in shaping immersive experiences within virtual reality (VR) and the metaverse. From understanding the core concepts of the metaverse and tracing the historical evolution of virtual reality to the technological advancements in spatial audio, we have examined the complex yet rewarding challenge of crafting realistic auditory landscapes.

Key insights include the necessity of persistent audio and seamless transitions across virtual environments to enhance user immersion. We delved into the challenges inherent in VR audio, such as auditory display, synchronization, and accurate distance perception, which require sophisticated solutions like six degrees of freedom (6DoF) and volumetric sound. The contribution also highlighted practical implementations, illustrated by case studies such as the Poznań Philharmonic Orchestra and the Jordan Rudess project, demonstrating how advanced audio techniques are applied in real-world settings.

We explored volumetric sound and its nuanced applications, the complexity of capturing spatial characteristics using Ambisonics, and how Ambisonics interpolation allows for dynamic auditory experiences. By understanding the principles and technical requirements of 6DoF audio, professionals can better anticipate the demands of VR soundscapes and make informed decisions about recording and rendering techniques.

The journey through sound for the metaverse has underscored the importance of merging audio with visual cues for an integrated and immersive experience. As VR technology and the metaverse evolve, sound will continue to be a pivotal component, offering endless possibilities for innovation in creating deeply engaging virtual environments.

## References


1. Warren Robinett, 'Interactivity and Individual Viewpoint in Shared Virtual Worlds: The Big Screen vs. Networked Personal Displays,' *SIGGRAPH Computer Graphics*, 28, 2 (May 1994), 127–130. https://doi.org/10.1145/178951.178969
2. Brenda Laurel, 'Computers as Theaters,' https://www.cs.cmu.edu/~social/reading/Laurel-ComputersAsTheatre.pdf
3. Sensorama, https://en.wikipedia.org/wiki/Sensorama
4. Ivan Sutherland, 'The Sword of Damocles (Virtual Reality),' https://en.wikipedia.org/wiki/The_Sword_of_Damocles_(virtual_reality)
5. Sega VR, https://en.wikipedia.org/wiki/Sega_VR
6. VFX1 Headgear, https://en.wikipedia.org/wiki/VFX1_Headgear
7. Manuel Brandner, Nils Meyer-Kahlen, and Matthias Frank, 'Directivity Pattern Measurement of a Grand Piano for Augmented Acoustic Reality," in *Fortschritte der Akustik (DAGA) [Advances in Acoustics (DAGA)]*, Hanover, Germany, 2020, pp. 138–141.

8. Agnieszka Rogińska, Justin Matthew, Jim Anderson, and Alex U. Case, 'High-Resolution Radiation Pattern Measurements of a Grand Piano – The Effect of Attack Velocity,' *Journal of the Acoustical Society of America*, 133, 5 (May 2013), p. 3234.
9. Michael Gerzon, 'Ambisonics. Part Two: Studio Techniques,' *Studio Sound*, 17, 8 (August 1975), pp. 24–26, 28, 30; *Correction: Studio Sound*, 17, 10 (October 1975), p. 60.
10. Eduardo Patricio, Andrzej Ruminski, Adam Kuklasinski, Lukasz Januszkiewicz, and Tomasz Zernicki, 'Toward Six Degrees of Freedom Audio Recording and Playback Using Multiple Ambisonics Sound Fields,' *Audio Engineering Society Convention*, 146 (2019).
11. Kaspar Müller and Franz Zotter, 'Auralization Based on Multi-Perspective Ambisonic Room Impulse Responses,' *Acta Acustica*, 4, 6 (2020), p. 25. doi: 10.1051/aacus/2020024.
12. B. Mróz, M. Kabaciński, T. Ciotucha, A. Rumiński, and T. Żernicki, 'Production of Six-Degrees-of-Freedom (6DoF) Navigable Audio using 30 Ambisonic Microphones,' 2021 Immersive and 3D Audio: From Architecture to Automotive (I3DA), Bologna, Italy, 2021, pp. 1–5. doi: 10.1109/I3DA48870.2021.9610970.
13. Jakub Zamojski, Piotr Makaruk, Lukasz Januszkiewicz, and Tomasz Zernicki, 'Recording, Mixing, and Mastering of Audio Using a Single Microphone Array and Audio Source Separation Algorithms,' in *Audio Engineering Society Convention 143*. Audio Engineering Society, October 2017.
14. Tomasz Ciotucha, Andrzej Ruminski, Tomasz Zernicki, and Bartłomiej Mróz,'Evaluation of Six Degrees of Freedom 3D Audio Orchestra Recording and Playback Using Multi-Point Ambisonics Interpolation,'*AES Convention*, 150 (May 2021).
15. 6DoF Microphone Placement Documentation – Sound Objects, Volumetric Audio Definition, https://create.roblox.com/docs/sound/objects
16. The Volumetric Sounds Beta Is Here, https://devforum.roblox.com/t/the-volumetric-sounds-beta-is-here/1695240
17. Volumetric Audio on Roblox?! | Roblox Studio Tutorial, https://www.youtube.com/watch?v=yhJDJl1N5Qw

Chapter 11

# From live sound to virtual and distributed experiences

*Gavin Kearney*

You'll find that most people working in immersive audio these days come from music, film and television, or perhaps even gaming. I was one of the music people. Although I loved studying electronic engineering at University, I came to realize that I was really craving the live performance industry.

So I took a break from my electronic engineering degree and started working as a live sound engineer, first as a backline technician, then on to monitor engineering, and eventually to Front of House. It was through this journey that I developed a deep interest in acoustics.

I'm an avid cinemagoer and love the immersive experience it offers. But I found that the immersive experience you get in live music is a different beast and can sound great if you have the right acoustic environment and the right material to work with. But it can also be really variable. For instance, engineering an ensemble in a church can sound fantastic, but engineering a great band in a poorly designed venue can be a struggle to make it sound good. This led me to questioning why aren't we exploiting more immersive technologies in the live performance sector, and how can we use them to improve live experiences?

So during my master's studies, I began extensive research on systems such as Vector Base Amplitude Panning and Delta stereophony. The latter was a sound reinforcement system from the 1970s and 1980s, largely based on manipulating the precedence effect. It involved creating 'source or simulation radiators' at the stage area with delayed reinforcement throughout the venue. Another was Ambisonics. This was an entirely different paradigm for recording and reproducing immersive sound, and I hadn't heard of anyone using it for live events.

So my colleague at Trinity College in Dublin, Enda Bates, and I started investigating whether these systems could be used for live sound reinforcement, particularly for contemporary music concerts. We questioned whether these systems could create a good sense of immersion for a distributed audience, not just those in the sweet spot. Could we pan sound sources around the audiences and break away from the traditional approach where you hear the sound where you see the musicians?

We conducted a series of psychoacoustic tests and discovered some really interesting things. Delta stereophony was great, but it colored the sound quite badly. Ambisonics was excellent, but only if you go to the higher orders. Once you start to get to about third order, it becomes usable across an audience. But for localization, we found that VBAP performed best for most scenarios. This early research significantly shaped my understanding and approach to immersive sound.

DOI: 10.4324/9780429297526-13

After our successful research, I returned to industry for a while and then was offered the opportunity to pursue a PhD at Trinity, which I accepted. I thought it would be interesting if one could record a live experience, but with the option for consumers of the media to choose their own acoustic sweet spot. Instead of a sound engineer dictating where the experience should be, how it should be situated, and where the guitar and vocals should be, you could have an interactive framework that allows you to move around musicians and choose your own acoustic sweet spot.

The main research question was, could you replicate a real acoustic environment with a virtual acoustic environment? And if so, how well could you do it? My entire thesis revolved around taking measurements of acoustic spaces and finding the most efficient framework to interpolate across measurements for interactive experiences. It was all based on Higher-Order Ambisonics, which was incredibly exciting.

## Research

At the end of my PhD, I started working as a postdoctoral research fellow. I worked on a project called 'Metropolis: at Trinity College, which involved the digital twinning of real spaces in a completely virtual world. This was around 2010, so it was quite cutting-edge at the time.

Shortly after that, I accepted a position at the University of York. The first big project I worked on was with Mercedes-Benz at the Grand Prix. They contacted us out of the blue with an issue: Lewis Hamilton had just transferred teams and was not happy with the sound of their simulators. They asked if we could make them more realistic and immersive, so we agreed to take a look. We had the privilege of putting binaural microphones in Hamilton's ears and recording his experience as he went around the race-tracks at Monza and Silverstone. We were also able to go to where they tested the engines in Brackley and record the raw engine sounds. We obtained impulse response measurements of the car chassis, and we were able to put together a real-time virtual engine using FMOD. It was a really fascinating and rewarding experience to see it all come together.

Our work with the sound of the V6 engines was a resounding success. In fact, it was during the season when they switched from the V8 to the V6 engines that we predicted how the V6 would sound on the racetrack, and they sounded terrible. When we presented our findings, they initially didn't believe us. But when they took the real cars onto the tracks and our predictions were spot on, they ended up thanking us. Following that, Mercedes started winning Grand Prix after Grand Prix! I'm not saying it was totally because of us, but we were part of the equation, which was great.

This success lent credibility to our work. And it was around that time, we realized that the existing binaural frameworks for virtual Ambisonic rendering, particularly the Head-Related Transfer Functions (HRTFs), were not up to par. The databases that were available were good, but many of them had gaps in the sphere. With Ambisonic sound fields, you want to have a nice spherical HRTF set to work with, essentially a virtual loudspeaker array. But most of the available datasets only went down to around –45° elevation.

## Spatial Audio for Domestic Interactive Entertainment (SADIE) project

I embarked on a project called SADIE: Spatial Audio for Domestic Interactive Entertainment. Initially, this project was supposed to be about crosstalk cancellation for the home. As part of this, I recorded a whole load of HRTFs (see Figure 11.1) that were Ambisonic friendly and open-sourced them.

Then Google approached. They had just built the VR pipeline, and while the THRIVE HRTFs they were using were good, they compared these with the SADIE ones via an independent study with DELTA SenseLab in Denmark. The results showed that the quality of experience was much greater with the SADIE HRTFs and was a significant validation of our work and its impact on the industry.

It was a fantastic development. They switched over the HRTF filters overnight, and suddenly everyone engaging with YouTube VR or using Google resonance tools were essentially using the 'ears' of the University of York. This was a significant achievement for us as it opened up opportunities to collaborate with a host of different companies who were aware of our work, including Dolby, Meridian, Sony, Abbey Road, and many others.

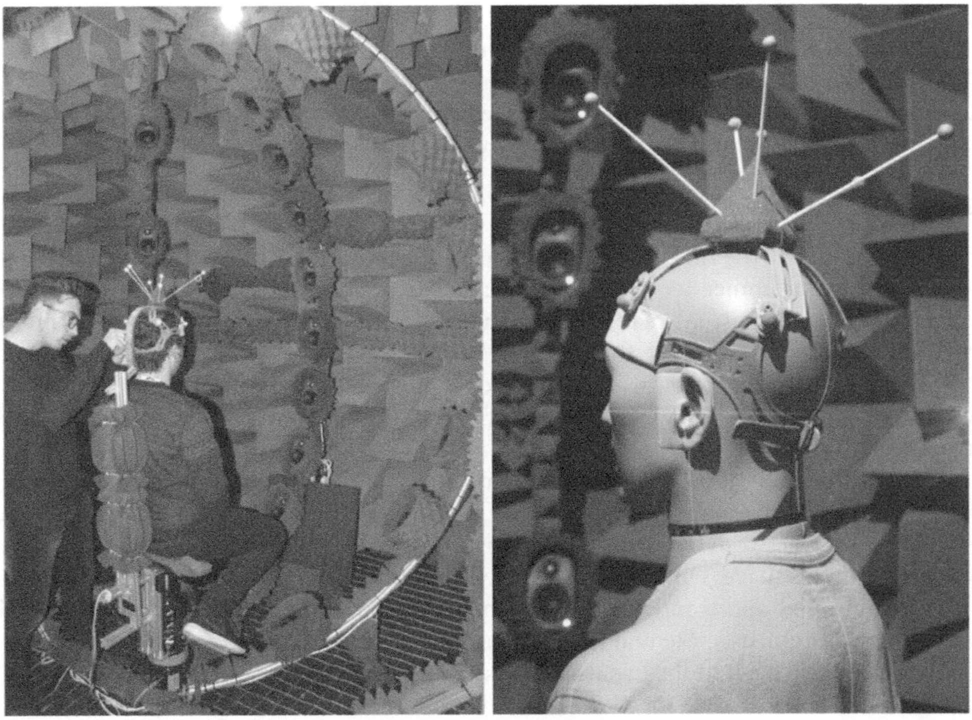

*Figure 11.1* Measurement of the SADIE II database: Left: Subject being prepared for a measurement session in the University of York Anechoic Chamber. Right: KEMAR binaural mannequin with headtracking markers mounted.

### Audio description

Another major project that I've worked on 'was Enhancing Audio Description'. This project, in collaboration with Professor Mariana Lopez at the School of Arts and Creative Technologies at York, aims to use 3D audio to create more accessible versions of film and television soundtracks for visually impaired audiences.

When visually impaired individuals go to the cinema, they typically hear a down-mixed version of the soundtrack on headphones, along with a third-person narration telling them what's happening in the film. We thought about how we could make this a much more accessible experience and decided to take the audio elements of the soundtrack and spatialize them in a way that is understandable for people who are visually impaired.

In our lab, we ran some simple tests with visually impaired people to determine their perception of how many people are talking in an auditory scene. For the tests, we panned all the dialogue centrally, without any overlap. Initially, with a scene of four people talking, the error rate was a high 44%. However, once we started to spatialize the dialogue, the error rate dropped by half to 22%[1].

This made it clear to us that if you spatialize elements in a mix, the intelligibility of the narrative becomes much easier to follow, and this insight has been instrumental in our ongoing work in immersive audio.

When we played demonstrator EAD immersive soundtracks over headphones with accompanying visuals, the visually impaired individuals found it to be a great alternative to standard audio description. But interestingly the people who had no sight loss absolutely loved it, which was quite a surprise.

Expanding these methods to loudspeaker systems can create a more inclusive experience. We're exploring the limits of spatializing elements of the soundtrack for both visually impaired people and those who can see both the visual elements and hear the aural elements. Our goal is to make it a fully inclusive experience.

### Networks for music performance

At York AudioLab, the main focus of our research since 2017 has been immersive audio over networks for immersive music performance experiences. This involves bringing performers together in virtual reality, evaluating user experiences, and understanding the workflows needed to facilitate this collaboration.

We've been investigating various conditions, such as latency, immersive audio, video, and avatar representation. We're also interested in how these experiences differ between professional and amateur musicians.

I have the pleasure of collaborating with my colleague, Professor Helena Daffern, who is also at the AudioLab. She's a voice scientist, and I, of course, am an immersive audio scientist. The synergy between our areas of expertise has been instrumental in creating immersive experiences in this space. Our combined knowledge allows us to explore new frontiers in the realm of immersive audio and music performance. It's an exciting and continually evolving field that holds immense potential for the future of music and audio technology. Indeed, it's all come full circle back to the live performance experiences I was discussing earlier.

### Immersive sound attributes

When we think about immersive sound experience, there are a couple of aspects to consider. Immersion, in my understanding, is when you're deeply engaged or absorbed in an experience. However, it's important to note that this deep engagement doesn't necessarily require immersive audio. One can be fully engaged when listening to a compelling piece of music or an intriguing podcast without the need for it to be spatial audio.

On the other hand, spatial audio, by its very nature, is immersive. All good spatial audio creates an immersive experience and in my opinion is the key distinction here. When we design spatial audio technologies, techniques, and workflows for immersive audio, the goal should be to create a sense of presence and envelopment. The listener should feel surrounded and feel like they are a part of the spatial audio environment.

So while there are two separate elements at play here – immersion in an experience and the immersive quality of spatial audio – it's clear that, for spatial audio to be effective, it needs to be immersive.

The attributes that I believe are important include a sense of presence and envelopment. This means that if you're trying to superimpose a virtual acoustic environment onto someone, they should no longer feel like they're in their current environment. If they close their eyes, they should truly feel as if they're at Carnegie Hall or York Minster. The experience should be completely enveloping, engaging, and realistic. However, this doesn't mean that a truly immersive spatial audio experience can't be completely hyperreal or something that's not based in reality. For instance, when we go to the cinema to see a sci-fi movie like *Star Wars*, the sound effects are literally out of this world. It's an engaging, immersive experience that's scintillating.

Measuring this is a real challenge. Obviously, there's looking at things like qualitative data, asking people about the quality of their experience and how they feel. But oftentimes, working in virtual and augmented reality doesn't give you the big picture. Sometimes when we look at biometric markers, when we're analyzing the quality of experience of immersive audio in a VR scenario, and we're measuring heart rate, skin conductance, and maybe EEG data, that data might tell us something very different from what people are reporting verbally.

When some people put on a VR headset for the first time, the visual component might overwhelm them, which can be a problem when evaluating immersive audio. I remember a similar situation, the first time I actually experienced Ultra-High Definition video at the AES Spatial Audio conference in Japan, around about 2010. I remember sitting in the audience, and the visuals were just overwhelmingly stunning. This was accompanied with 22.2 surround sound, and I had to catch myself halfway through the presentation to really focus on what I was hearing, because up to that point it could have just been 5.1.

This highlights part of the problem with cross-modal perception scenarios. When we experience immersive audio, we have to be very cognizant of what the other modalities are doing, especially the visual. For example, if I play somebody an immersive audio experience seated in our 50-channel sphere, and I'm using HRTFs that we've measured in that sphere, I can absolutely convince them that they are listening to the loudspeakers, not to headphones. But if I try to recreate that same experience in my office, the illusion just breaks down, demonstrating just how important the visual modality is. This is a big part of the problem with measuring the quality of experience in immersive audio systems. It's understanding how the rest of the modalities integrate to create that experience.

There's indeed a distinct difference in criteria or design aesthetics between a live show and a more controlled virtual experience. The sensation of bass, for instance, is a key element in a real live show, where you actually feel the low-frequency vibration in your

*Figure 11.2* VR user engages in immersive and interactive soundscapes in York AudioLab's 50-channel spherical loudspeaker array.

chest. This sensation cannot be replicated in a virtual environment, unless you're using devices like SUBPAC suits.

The first challenge is the acoustic environment itself. Ensuring that you're rendering a close approximation of the real-world environment is not an easy task. For a 3DoF experience, we can use Ambisonic spatial impulse responses measured from a set listening position in the real room. But if we want to replicate a 6DoF experience, then we might measure grids of impulse responses and have a renderer interpolate across them or use appropriate parametric methods so that we can use efficient reverbs that sound like the real thing. And, of course, we need to consider how the acoustics of a venue will change with an actual audience in situ and how this effects our reference measurements when modeling.

Then there's the social aspect of it, which hasn't been fully cracked in the virtual world yet. Being in a room and having a musical experience with 10,000 other people hasn't been successfully replicated in virtual reality enough to create a truly authentic experience. The feeling of being part of the crowd, engaged in some emotional feedback with the artist, is something that is really hard to replicate. This is immersive and engaging, regardless of the spatial audio aspect of it. This isn't only about efficient rendering algorithms but also about appropriate network architectures and codecs that facilitate low-latency, low-bandwidth interactions.

We face a fundamental problem in the world of virtual reality, in that all of the traditional paradigms for evaluating immersive audio fall away. In the typical paradigm where we have a reference, some beautiful piece of audio, our different conditions, which we degrade relative to that piece of audio, don't apply here. If you want to evaluate your rendering system in a virtual reality environment, what's your reference? The real world?

So I think a whole series of research problems need to be solved on evaluating immersive experiences in virtual environments that have a basis in the real world. The closest that we've got is when we've had performers interact together in both the real and virtual worlds, and we've measured biometric data looking at heart rate and skin conductance of both conditions. We had the musicians first perform together in a church in York, then took them into a virtual version with their positions completely replicated by 360 camera footage and spatial impulse response measurements taken from their real world positions in the ensemble. They listened back to audio and video of their colleagues performing as if they were right beside them with a virtual acoustic foldback of their own singing.

## Environment

Typically, we see that when performers go into a virtual environment they're more stressed than they are in the real world. The technology is getting in the way. Until we get to a point where the visuals are incredibly realistic and the audio sounds completely natural, we're going to have that barrier, especially to create that experience of having hundreds of people around you engaged in a live performance.

## Context

What constitutes good spatial audio can indeed depend on the context. If it's in the context of a narrative, such as when you're engaged in a cinematic experience, good spatial audio, first and foremost, doesn't detract from that experience. The technology shouldn't be showing itself off just for the sake of it.

Outside of that, in an experience where you want the sound to be as realistic (or hyper-realistic, depending on your expectation), good localization is key. The audio and the visual need to match up well, and there needs to be good congruence. This might mean that you need individual personalized Head-Related Transfer Functions (HRTFs) in the binaural rendering. Or where you need enough spatial resolution within a loudspeaker rendering to deliver good audiovisual congruence.

But there are things that should also be considered from a sound design experience to make them sound as natural as possible. Aspects like reverberation. for example, should sound realistic, not metallic or harsh.

But a lot of researchers will go ahead and do plots of how sound fields should behave and how they should be giving the correct Sound Pressure Level (SPL) and the correct binaural parameters and so forth. All of this is important, but when you listen to the output, the spatial audio absolutely has to sound good. For me, it all comes down to the final user experience. No matter how accurate your localization is, no matter how beautifully efficient the rendering of your reverb algorithm is, if it doesn't sound good, it's just not going to fly.

In nearly every single spatial audio test we've done that ranks the importance of different attributes, 'naturalness' and 'timbre' are usually streets ahead of attributes like 'envelopment' or 'localization', so the tonal quality has to be right. Ultimately, it's about creating a sound experience that is pleasing and engaging to the listener.

It's quite telling in binaural audio studies where a non-binauralized stereo reference is included that the stereo reference is generally ranked the highest. So with no binaural spatialization, just based purely on timbre, users prefer it.

Right now, we're looking at how we can get to the least amount of filtering possible to preserve the original timbre of the source audio. Again, the spatial audio shouldn't get in the way. It should be there to enhance the experience and provide good localization. But at the end of the day, it's a tool, and so it should serve whatever the creative expression is.

## Multimodal experience

When considering a multimodal experience, it's crucial to think about the design aspect in a different way. Certainly, when the visual component is present, you can get away with quite a lot. However, it depends on the context. For instance, if it's for a training application or a gaming application where you really need the visuals and the audio to align perfectly, then, yes, you've got to have that localization nailed. But if there's a lesser need, such as for a screen presentation or cinema, well we've been using the ventriloquist effect to great effect for decades. We've no problem with the dialogue coming from the center channel when the actors are on separate sides of the screen. So clearly, there are contexts where audiovisual congruence doesn't matter as much.

Contemporary designs for immersive audio experiences focus more on the congruent aspect. We have been looking at what happens if, in a cinema scenario, we pan the dialogue to the actors. Will it diminish the experience for a distributed audience? It makes for a more inclusive experience for people who are visually impaired because they've got spatialized dialogue. Now, they can really tell things are spatial and things are moving around. But do people who don't have visual impairments mind?

So, for film, broadcast and games, it seems we're progressing toward integrating object-based audio into production workflows from the get-go. One idea is to tag audio objects to correspond with visual objects, making them a seamless part of the production process. So why not incorporate immersive audio right from the start of production? This approach marks a new way to craft experiences on expansive soundstages.

Consider the advancements in virtual production with large LED volumes adjusting CGI to match the camera's perspective, enhancing the actor's environment in real time. Why not extend this concept to include immersive sound design during filming? Imagine leveraging AI technology during shooting to develop assets for an immersive soundtrack. In post-production, the assets created during filming would be refined further. Mixers and sound designers would have the flexibility to mold these elements into their desired compositions. This direction represents a transformative approach to our design philosophy.

## Design philosophy

There are a few elements to my overall design philosophy in terms of immersive experience. First and foremost, it depends on the context. The technical prowess of the immersive audio system should not hinder creating a sense of connection and engagement with the end-user. It shouldn't detract from a narrative or any artistic expression you're trying to achieve. In whatever way we design things, that should not be compromised.

### *User experience*

The user experience is at the heart of everything, including what their expectations are within the context of what they're listening to. If they go to a cinema, they expect hyper-real sound effects and spatial audio, regardless of the cinematic content. Even the simplest films have hyper-real Foley, so we need to consider the context of that experience.

### *Universal design*

We also need to consider accessibility from the user's perspective. We try to embody the principles of universal design, so nobody gets left behind. We aim to create accessibility features that ensure immersive audio experiences are inclusive and can be enjoyed by people from all different backgrounds and abilities. Whether you're visually impaired, aurally impaired, or neurodiverse, you should be able to engage and understand the narrative through the technology. It shouldn't be a barrier to you.

Technical excellence is certainly important. Ensuring accurate localization or optimal performance of whatever rendering algorithm you're using in the immersive experience is crucial. It should have minimal latencies and provide optimal audio quality, perhaps depending on the bandwidth restrictions if you're streaming or connecting people across the network. This is particularly important for live performance experiences where you're potentially running the risk of having glitches in the system that would break the immersive experience. The last thing you want in a network music performance experience is low latency, real-time clicks because someone else on the network has decided to upload a lot of content to YouTube. So, due to potential bandwidth constraints, you need to ensure technical excellence in whatever delivery mechanism you're providing for the immersive audio.

### *Scalability*

Then, you have to think about scalability. If I'm designing an immersive audio experience, can it be consumed on a mobile device with the same level of engagement as it could be consumed in a full 6DoF virtual reality experience? Can we design for the whole continuum of potential interactions with different devices? So we need to ensure good quality of experience across the whole gamut of potential reproduction scenarios within that scalability.

## Artistic content

The relationship between the technology, content (or the 'repertoire)' and the user was quite important for the distributed concert at Huddersfield University during the 2023 AES conference on Spatial Audio.

In this experiment, we remotely connected four performers in a studio booth at the AudioLab in York with four performers at the Phipps hall in Huddersfield over the internet using JackTrip (see Figure 11.3). An audience was in situ at Phipps hall, who heard all the singers rendered together with the virtual acoustics of BBC Maida Vale studio 2 over the venue's immersive loudspeaker system. The four York singers were represented by four Genelec loudspeakers on stage. At York, an audience was located in an adjacent classroom to the studio. Both the singers and audience in York experienced a virtual stage over headphones, which consisted of an arc of performers in front of the listener with the York singers on the left and the Huddersfield singers on the right. Initially, we erred on the side of caution when we started the concert by choosing a slower-paced repertoire. This included pieces such as 'Crucifixus' by Antonio Lotti and 'If Ye Love Me' by Thomas Tallis. This was to ensure that if there were any latency issues, the performers could potentially compensate. However, we became adventurous toward the end and

*Figure 11.3* 'Ex Corde Vocal Ensemble performing live as part of the Network Music Performance between York AudioLab and Phipps hall in Huddersfield: Four loudspeakers on the stage are fed remote audio from four singers located live at York, and the entire concert is rendered for the audience with the acoustics of BBC Maida Vale Studio 2.

introduced some fast-paced material such as 'Tutto Lo Di' by Orlando Di Lasso, which worked absolutely fine.

Tempo of repertoire can be a design consideration. However, it goes back to the point that we're erring on the side of caution due to the technical limitations involved. We don't want to be in that position. We want to create experiences that are not bound by the problems of network latency. While we can't overcome the speed of light yet, other technical limitations introduce latencies and delays that we can overcome.

We need to be cognizant of putting our best foot forward when we create these experiences, but not let them get in the way of presenting the material we want to put on. We aim to convey the artistic expression that we want the audience to receive.

## Evaluation

In terms of evaluating an immersive experience, we must clearly define what are the actual effects and attributes you want to evaluate? Often, in our tests, when we have an inkling of what we're looking for, we tend to do what's known as an *attribute elicitation* phase. We'll get a few people into a room, play some examples of the immersive experience with different conditions, and ask them to identify the differences between these examples. This guided discussion on the attributes we should be extracting helps us evaluate it with a larger number of people.

Choosing the right participants is also crucial. As a university, we have many Music Technology students who represent one demographic, but, ideally, you want a diverse group to represent your target audience. Capturing a range of perspectives in the evaluations is key.

We tend to conduct evaluations in laboratory conditions, controlling the environment for the immersive audio experience, the visual aspects, and other modalities. Ensuring replicable results is scientifically important for us. However, it's also exciting to see how the technology performs in real-world conditions that we hadn't anticipated.

The evaluation also requires the right tools. You need the right equipment, high-end audio systems, and the appropriate type of headphones and loudspeakers that reproduce what you're trying to evaluate. If you're going to do Higher-Order Ambisonics, for example, you need to have a spherical loudspeaker array. Interestingly, the spherical rig in our lab is primarily for headphone research, validating that what we're getting on headphones is perceptually equivalent to what is reproduced over the loudspeaker system. So having the right equipment and calibrating that equipment is crucial.

Then there are the test scenarios themselves. What are you looking to evaluate, and what is the context? If you're evaluating an immersive audio codec, for example, you should test it in a range of different scenarios, such as teleconferencing, gaming, cinematic representation, or virtual augmented reality. Testing all of these will give you a full perspective of the immersive audio experience and how the codec would behave in those different scenarios.

Psychophysical measurements are very informative when it comes to evaluating experiences. Factors like people's heart rate, pupillometry, skin conductance, and EEG data can tell us something about the emotional state of people in the immersive experience. I believe a lot can be correlated with qualitative data, such as questionnaires and surveys. This could potentially provide a basis for evaluating an experience in non-invasive ways and enable us to say 'this experience is immersive because of these psychophysical parameters'.

## Iterative process

I would say that a large percentage of the time, the question we end up asking is not the original research question we had in mind. This is because as we start to develop something, it often becomes apparent that there's a series of attributes that need to be evaluated before we can address the actual research question of interest.

For example, if you're considering an immersive performance system that you want to test under real network conditions, you might initially want to test how well the reverberation algorithm is working relative to measured spatial impulse responses. However, you might then realize that the latency of the system is getting in the way such that you can't really evaluate the performer's experience of the reverberation until you've evaluated the latency. Often, these technical barriers can obstruct the actual question that you want to answer. Frequently, we start by playing around with the system we have created and identify the barriers thus determining the next steps before we get to the actual research question itself.

Along the way, we run attribute elicitation phases, trying to figure out the most important perceptual attributes that we should be looking for. We may also run objective metrics. For instance, we've looked at things like binaural parameters, Interaural Time Difference (ITD), Interaural Level Difference (ILD), Inter-Aural Cross Correlation (IACC). We're seeing differences between a rendering system and reality. We want to see if these are apparent, so we will do localization tests or general perceptual comparisons of the degradation of the audio or changes to any aspects of the reverberation algorithm, for example.

From there, we refine the test. It's always an iterative process. We'll likely pilot things, run a test phase, and then go back and change things about the test before we do a final test phase. Often, we may iterate three times before we come to a satisfactory conclusion.

### Different audiences

Our testing methods differ when we have expert listeners versus when we release to the general public. It largely depends on the application area. For instance, we've done significant work with children who have autism. In these scenarios, we design experiences where we can obtain quantifiable data without it being a challenge for them to understand what we mean by immersion or good spatial audio quality. We're looking at differences between extremes, such as just stereo versus spatial audio. In this instance, having a non-expert listener is absolutely fine. However, when we're looking to assess a system where we're aiming for minuscule improvements that will accumulate over time and result in an improved system, we'll really need expert listeners who can scrutinize those systems.

By the time it goes into the consumer world, we have already undertaken a series of tests with expert listeners. Often, this is out of necessity because of practical restraints. Expert listeners are supposed to get us to the same point as a larger population of non-expert listeners. Although this is often true, it's not always the case. By the time we get to test with a large population, we've already gone through several iterations of testing with expert listeners.

### Maida Vale VR experiment[2]

We started working on a virtual recording studio project just before the beginning of the COVID-19 pandemic. We had conversations with Abbey Road studios, pondering, 'What would it be like to record a band completely in virtual reality?' We started to develop some work with Abbey Road, but, due to the pandemic, we weren't able to pursue the research with them. With Abbey Road's blessing, we decided to pursue this work with the BBC at Maida Vale Studios. We wanted to see if we could do a recording at Maida Vale of a real band and then completely replicate that recording in virtual reality to see whether we could get the same results.

Maida Vale is a historic recording facility, and it's been home to the BBC singers for the last 100 years. Everyone from Zeppelin to the Beatles to Nirvana to Bing Crosby has played there. We had the opportunity to measure all of the studios as they were and to do some digital twinning research on them. This was very exciting from an acoustical heritage point of view, but it also fulfilled our objective of trying to create an immersive audio experience where performers can dial into a virtual version. We decided to focus on one of the studios, Studio Four, which is well known for rock and roll recording and the BBC Live Lounge sessions.

We went to Maida Vale and recorded a whole grid of impulse response measurements, around 1,200 individual measurements (see Figure 11.4). We did Ambisonic measurements and binaural measurements, including seven source points, four performer positions, and some additional measurement locations for representing a drum kit.

*Figure 11.4* Members of the York AudioLab team setting up for measuring grids of spatial impulse responses at BBC Maida Vale Studios

We also did visual capture, including photogrammetry, LiDAR, and geometric measurements with texture photos. It turned out that the geometric measurements with textures were the best for creating the experience within a game engine. We had relatively crude avatar representations, but they were sufficient for the performers to see the movement of each other within the game engine.

The band Wolph was involved in this project. They are a four-piece band consisting of drums, bass/synthesizer, guitar, and keyboards, with each musician on vocals too. We first recorded them together in Studio Four and then separated them out into their own individual booths and other studios. They were all connected together over a local network. Then they put on Quest headsets to run the same recording setup in the virtual world. This time, they were listening through the virtual acoustics of the space and seeing the digital twin in VR.

We evaluated different aspects of the virtual experience. This included three different latency conditions: 19, 24, and 29 ms. We also examined mono and Higher-Order Ambisonic representation within the virtual world.

We also conducted a qualitative questionnaire asking the question, 'Were you aware of the delay between performers?' The results were interesting. The feedback was dependent on the instrument. Especially with the drums, the effect of latency was really apparent for the drummer. But what was really interesting was that. even though we changed the latency conditions, they were all able to perform in time because the drummer was so good. The song was at 110 BPM, and he consistently maintained that tempo. As a fun experiment, we took the drummer away and asked the other musicians to play in the long latency (90 ms) condition. They started playing in time, but then they just slowed down. It was fascinating to see.

We also assessed the presence of the avatars. Despite the limited range of motion of the avatars, having that little bit of expression and visual interaction positively influenced the performance.

One of the key elements for the performers was to set a pass-through mode in the VR world, so they could see their real-world instruments. There were all sorts of problems with that because we were using the infrared cameras on the head-mounted displays. The depth perception was off, and it was all fuzzy and grainy. The pass-through portals would move away from the instruments over time, so there were also issues with that. This was a good example of technical barriers getting in the way of creating the experience the way we wanted to.

## Prioritize content and audio quality

The most important lesson I've learned over the years is that technology shouldn't get in the way; it should enhance the experience, not overshadow it. If you develop audio technologies without considering the end-user experience, you're just leading to a disconnect between the technology and the content, and that's never good.

You've got to prioritize the content quality, the storytelling aspect, and the user engagement. If your spatial audio implementation is poor and the technical quality isn't good, then you may have incongruent spatial audio cues that will detract from the experience. Sometimes, in some contexts, that's okay; in other applications, maybe not. You just have to be careful about the context. Perhaps you need to have personalized HRTFs for a particular training scenario. Or maybe you just need a simple quad setup for creating an immersive experience you know will work well for an audience at the time.

But it's also really important to make sure that the technology is accessible and available to a wide range of users. When we develop immersive technologies or create immersive experiences, we need to ensure that we're not excluding particular demographics such as people with hearing or visual impairments or those who are neurodiverse.

Additionally, if you're going to release an immersive audio experience into the consumer space, then test across a range of devices and platforms, and try to get all those bugs out before you release it. If you've got inadequate testing in the immersive experience, you'll have unforeseen issues and ultimately a subpar user experience. Audio quality is particularly important and can really vary across different platforms and devices. It will diminish the user experience if it's not done right. Even the translation between loudspeakers and headphones needs to sound good. Use HRTFs if that's what you think is needed, but ask yourself the question whether it is detracting from the timbre of the experience? A non-personalized HRTF can sound strange to some people. In some cases, the immersive experience may be better being stereo. This goes back to the principle of not letting the technology get in the way. Just because you're using spatial audio doesn't mean you need to use bad spatial audio.

## References

[1] López, M., Kearney, G. and Hofstädter, K. (2020) 'Seeing Films through Sound: Sound Design, Spatial Audio, and Accessibility for Visually Impaired Audiences' *British Journal of Visual Impairment*. https://doi.org/10.1177/0264619620935935

[2] Cairns, P., Hunt, A., Johnston, D., Cooper, J., Lee, B., Daffern, H. and Kearney, G. (2023) *Evaluation of Metaverse Music Performance with BBC Maida Vale Recording Studios [PDF]*. AudioLab, School of Physics Engineering and Technology, University of York, York, UK; Paper, 2023. Available: https://aes2.org/publications/elibrary-page/?id=22139

Chapter 12

# Acoustic and perceptual factors in sound design for XR

*Marta Gospodarek, Olivier Warusfel, and Agnieszka Roginska*

## Introduction

The evolution of technology coincides with the emergence of innovative artistic expressions in extended reality (XR) realms. Within this dynamic landscape, sound design plays a pivotal role, serving as the bridge between technological advancements and artistic expression. Unlike sonic art, which stands as an independent form of creative expression, sound design is focused on crafting audio experiences with a purpose beyond themselves (Gibbs, 2007). While sound design is traditionally associated with the realms of film and television, its scope extends far beyond, encompassing diverse fields and applications. In virtual reality (VR), sound design is instrumental in constructing immersive auditory layers that transport users into virtual realms seamlessly. Conversely, in augmented reality (AR), sound design faces a unique challenge: smoothly integrating virtual elements into the real environment while maintaining the integrity of the user's auditory experience.

The main goal of the sound design for XR experiences is to enhance immersion and presence in the experience (Tatlow, 2024). To achieve this, the audio layer needs to correspond with visual cues and engage the user on the sensory level. Consequently, the use of spatial audio is critical to provide a three-dimensional auditory space around the user, which is essential for immersion. The audio objects need to be properly spatialized to ensure that the localization of sound sources is accurately aligned with the position of visual cues. Acoustic properties of the space are required to be consistent with the virtual world (in the case of VR) or with the real environment around the user (in the case of AR) (Serafin et al., 2018). Furthermore, sound design plays a crucial role in evoking emotions and enhancing storytelling (Felnhofer et al., 2015). It provides additional context about actions and environments, supplementing the visual layer. Moreover, sound design for XR supports the interactivity of the experience by giving auditory feedback to the user's actions.

However, practical implementation of sound design for XR experiences may encounter a number of challenges that need to be considered when creating the audio layer. Technical limitations on VR/AR platforms, including constraints on audio processing capabilities, spatialization techniques, and hardware compatibility, may restrict sound design options. It is essential to balance audio quality with performance optimization to ensure smooth playback on VR devices, which often have limited processing power and memory resources. Seamless integration of audio with other elements of the VR experience, such as visuals and interactive mechanics, requires meticulous coordination and collaboration among designers, developers, and audio engineers. Additionally, the

DOI: 10.4324/9780429297526-14

immersive nature of VR requires extensive testing and iteration of sound design to ensure that audio cues are plausible, coherent, and able to enhance the overall user experience. Understanding which aspects of audio rendering are most crucial for the immersive experience is essential to effectively manage the trade-offs that are often unavoidable in audio implementation for XR productions.

Acoustic simulation within the XR environment is one of the most fundamental aspects determining the quality of the experience. In both VR and AR, auditory cues must align with the user's perceived space to maintain immersion. However, the requirements for cohesion differ significantly between VR and AR environments.

In VR, users are completely immersed in a synthetic environment, without access to any auditory or visual cues from the real world. This isolation provides sound designers with greater flexibility: the acoustic characteristics of the virtual scene need only to be internally coherent with the virtual visuals, not with the real world. A broader range of simulated acoustic parameters can thus be perceived as plausible, as long as they are consistent within the virtual context.

In contrast, AR overlays virtual elements onto the user's actual physical surroundings, which are both seen and heard in real time. As a result, virtual sounds are directly compared with the real-world acoustic environment – such as room reverberation or direct sound propagation. This establishes a far more demanding baseline for perceptual congruency. In AR, the Virtual Acoustic Environment (VAE) – a computational model of the sound field designed to replicate real acoustic properties (Savioja et al., 1999) – must closely match the physical characteristics of the user's surroundings. Even slight mismatches between virtual and real sounds can lead to perceptual dissonance, undermining realism and immersion. Therefore, while VR permits a more flexible design space due to its sensory isolation, AR imposes stricter requirements for accuracy and coherence of acoustic cues.

This chapter explores the perceptual and acoustic factors influencing sound creation in XR environments. The first section examines methods for modeling VAEs and their impact on sound perception. The second section discusses various measures of perceived realism in XR auditory layers. The third section provides a summary of a case study on acoustic and perceptual factors affecting sound plausibility in AR. Finally, the chapter concludes with a case study focused on design principles and challenges in crafting sound for collocated VR experiences.

## Virtual acoustic environment

Every virtual acoustic system comprises three fundamental modules: source modeling, environment modeling, and listener modeling. In the context of AR audio, where the aim is to seamlessly integrate real and virtual elements within the environment, these modules strive to emulate the characteristics of reality, including the real source, the physical space the user occupies, and the individual listener. The closer each module aligns with reality, the more immersive and convincing the user experience becomes. Furthermore, these modeling components should exhibit dynamic responsiveness, meaning they must update in real time to reflect changes in the sound scene, particularly when sources or the listener are in motion.

The upcoming sections discuss the methods for modeling each of the modules and their influence on the perception of sound.

### *Sound source modeling*

Source modeling techniques usually entail the generation of a sound and the modeling of its radiation pattern. Sound can be reproduced from pre-recorded material or generated using different sound synthesis techniques. The audio material then needs to be processed to recreate the directivity characteristics of the real source. Dynamic directivity allows obtaining a different spectrum of sound depending on the rotation of the source in relation to the listener. In most implementations, sound sources are modeled as omnidirectional point sources. It is a sufficient method for some applications, but to achieve a higher level of realism, more detailed rendering is required. Radiation patterns vary in complexity depending on the type of source. For common sources such as voices or loudspeakers, the energy is distributed mostly in the frontal hemisphere, and high frequencies are attenuated with the increase of angular distance from the front (Monson et al., 2012). Modeling directivity is particularly challenging for musical instruments, where the radiation pattern is not only frequency dependent but also varies with articulation, and performance parameters (e.g., fingering for woodwind instruments or the presence of a mute in brass instruments) (Karjalainen et al., 1995). For example, the directivity of a violin changes drastically even within a very small frequency shifts, resulting in a phenomenon where the instrument's vibrato – originally a frequency modulation – is accompanied by a corresponding spatial modulation, sometimes referred to as 'spatial vibrato' (Weinreich, 1997).

There are several ways of implementing source directivity (Rindel et al., 2004). Previous studies have suggested the use of multi-channel source directivity auralization (Otondo and Rindel, 2005). In this method, the first step is to characterize the source with multi-channel anechoic recordings. After that, the source's radiation sphere is divided into angular segments. Each segment is assigned to one microphone position. The resulting Room Impulse Response (RIR) is then convolved with the corresponding channel of the multi-channel recording. The disadvantage of this method is that it leads to abrupt changes in level and spectrum when the source orientation is modified. To alleviate this problem, another solution was proposed (Postma and Katz, 2016). Instead of using multi-channel recording, an overlapping beamforming approach to multi-channel source decomposition is suggested. In this method, the source's radiation sphere is divided into 12 beam patterns. The beams have minimal overlap while remaining equal in gain sum in all directions. The directivity is applied by controlling gains for each of the 12 components.

Many 6DoF audio renderers implement directivity using a simple parametric function. Zero- and first-order directivity patterns (omnidirectional, cardioid, bidirectional, etc.) can be computed by calculation of the weighted sum of the omnidirectional and dipole pattern (Southern and Murphy, 2009). Yet, in this approach, the directivity is applied to all frequencies with the same weights.

More advanced approaches include creating the acoustic radiation pattern in the spherical harmonics domain for each harmonic partial of every played tone. An assessment of the complexity of the acoustic radiation pattern can be conducted based on the number of excitation points utilizing the centering algorithm (Shabtai et al., 2017). However, for simpler sources, it may be sufficient to implement smooth, frequency-dependent weighting of the spherical harmonics – without explicitly modeling each individual harmonic partial.

### *Influence of directivity of the sound source on perception*

The study by Postma et al. (2017) included perceptual listening tests where participants evaluated omnidirectional, static, and dynamic directivity applied to the anechoic recording of an actor's performance in theater space. The ratings included plausibility, distance, apparent source width, and listener envelopment. The results showed that dynamic directivity was rated significantly higher for plausibility, distance, and apparent source width in comparison with static voice directivity and omnidirectional sources. While it seems intuitive that implementation of dynamic directivity should improve plausibility, the question arises if including the listener's movement in 6DoF will affect sensitivity to source directivity changes. The study by Robotham et al. (2019) focused on answering this question by finding the just noticeable difference (JND) for static and dynamic (with listener's movement) evaluation of source directivity. In the study, directivity was independent of frequency; thus the change of directivity caused modification of loudness rather than timbre. The results showed that, for the static presentation of steady signal (noise), the JND was –0.6dB while in the 6DoF condition, this difference was not audible, which suggests that the threshold is higher. For non-steady signals, the JND difference remained at –2.6 dB and seemed not to change between static and dynamic conditions.

### *Room modeling*

The goal of room modeling in virtual acoustics systems is to simulate the sound propagation behavior in an acoustic space. Sound wave initialized by the source interacts with the space around. When it arrives at the boundary, it reflects and diffracts. As a result, the sound wave does not arrive at the listener at one time but instead bounces off the obstacles, creating reflections. At first, the reflections come to the receiver at distinct time intervals but slowly their energy decreases while the echo density increases, ending in a diffuse reverberation. That is why the Room Impulse Response (RIR) is usually divided into three sections: direct sound, early reflections, and late reverberation. The amount and spatio-temporal distribution of reflections depend on the room's shape, as well as the diffusion and absorption coefficients of the surface materials. The direct sound segment of the RIR is usually modeled separately to account for the changing positions of the source and receiver. The late reverberation is generally considered diffuse, meaning that individual reflections are no longer perceived separately and are assumed to be uniformly distributed around the listener. Because these reflections no longer require individual specification, their collective behavior can be modeled statistically, as an ergodic process. That is why the late reverberation time–frequency envelope is considered independent of the position in the room (Barron and Lee, 1988). The early reflections part of the RIR is generally the most challenging to simulate. As the spatio-temporal pattern of the early reflections is perceptually salient, the timing and the direction of arrival for each reflection need to be properly rendered. Besides that, the pattern of the reflections changes depending on the position in the room of the source and receiver. Very accurate rendering of early reflections requires very detailed measurements or high-cost calculations and usually is not practically feasible. That is why different ways of approximation are investigated in order to simplify the rendering without losing perceived spatialization quality.

There are three basic approaches to modeling the acoustics of the room for virtual environments:

- Based on physical models
- Based on the convolution with the premeasured RIR
- Based on algorithms, e.g., delay networks

The ensuing sections provide a quick overview of each of the methods.

### *Physical-based approaches to room modeling*

The acoustical modeling of sound propagation can be approached through two main frameworks: wave-based and geometric-based methods. In wave-based techniques, sound propagation is physically modeled using spatio-temporal approximations. These methods are fundamentally based on solving the wave equation – a second-order linear partial differential equation – after applying time and/or spatial discretization. The wave-based modeling is able to provide more accurate results, but, at the same time, it is much more computationally expensive (Savioja and Svensson, 2015). Geometrical acoustics (GA) aims to compute the various propagation paths based on the initial model of the space, receiver, and source. In GA, sound is assumed to propagate as rays, and all of the properties of sound waves are neglected. This assumption is valid in high frequencies where the sound wavelength is much smaller than the dimensions of the elementary surfaces of the room. The problem arises in low frequencies where the approximation is not as relevant. Despite that, this technique is widely used in practical implementations of room acoustics as being faster and more efficient (Pompei et al., 2009).

A number of different approaches exist within the GA framework. One of them is the image method (IM) which replaces the physical boundaries of the environment with an equivalent infinite lattice of image sources (Allen and Berkley, 1979). It aims to find purely specular reflection paths between the source and receiver. The main assumption is that all of the boundaries are perfectly flat and rigid. Diffraction and interference of sound waves are neglected. The main advantage of the method is its relatively simple implementation and sufficient quality, especially for the early reflections segment of the IR. The ray-tracing method is an alternative algorithmic implementation. The method tracks the rays emitted by the sound source as they reflect off the surfaces of a modeled space. The ray-tracing method is more efficient than IM, although it creates geometrical path approximations that can cause delay errors. On the other hand, acoustic radiosity is a method that may be used to account for sound scattering on non-perfectly flat and rough surfaces. The method assumes perfectly diffuse reflection from all surfaces of the enclosure (Miles, 1984; Nosal et al., 2004).

### *Convolution-based approaches to room modeling*

The convolution-based methods aim to reconstruct the acoustics of the space from premeasured RIRs. The types of RIRs used previously for acoustic rendering in virtual systems include omnidirectional RIRs (Pörschmann and Wiefling, 2015), binaural room impulse responses (BRIRs) (Neidhardt and Knoop, 2017; Neidhardt et al., 2018; Werner et al., 2021), and spatial room impulse responses (SRIRs) (Nowak and Klockgether,

2017; Engel et al., 2019). With this technique, the reference IR is usually divided into two or three segments: direct sound and reverberation or direct sound, early reflections, and late reverberation. The direct sound is simulated separately to ensure the proper sound positioning by using convolution with Head Related Impulse Response [HRIR]. The reverberation is usually left unchanged from the reference IR. As the spatio-temporal structure of the early reflections segment is perceptually relevant and most difficult to simulate, different methods of simplification are used for this segment (e.g., keeping the temporal structure intact and using convolution with HRIRs to obtain one spatial pattern of the reflections independent from the position in the room [Pörschmann and Wiefling, 2015]).

### *Algorithmic approaches to room modeling*

In typical rooms, reflections build up over time until the *mixing time*, at which point a diffuse reverberant field is established. The late reverberation can be considered a diffuse sound field that is largely independent of both source and listener positions. Because the individual reflections are no longer perceptually distinguishable, the reverberation can be approximated using a system of delay lines. To realistically simulate the increasing reflection density over time, these delay lines must be looped, allowing for recursive energy buildup characteristic of a diffuse field.

Feedback Delay Networks (FDNs) are designed with parallel delay lines connected recursively through a feedback matrix. The state-of-the-art FDN implementation was proposed by Jot and Chaigne (1991). The authors suggested using a set of multiband absorptive filters connected with delay lines in order to control the frequency-dependent reverberation time. The advantages of FDN algorithmic reverb are the simple design, low computational complexity, and high quality of reverberation, which can be easily tuned to the real room reverberation characteristics. Although no new standards are established, new methods are proposed that aim to improve the FDNs (Lee et al., 2012; De Sena et al., 2015).

## *Acoustic modeling and perception*

One of the aims of acoustic research is to investigate the interaction between objective and subjective spatial characteristics of sound. Based on that, it is possible to draw conclusions about the influence of different objective parameters of rendering on the perception of virtual acoustics. Research in this area is very extensive. The next section gives a quick overview of the most important aspects.

Direct sound allows localizing the source in a space. Specular reflections that arrive 5–10 ms after the direct sound can create a perceived image shift and increase the apparent source width (Olive and Toole, 1989). These first reflections are not perceived as a separate event (Wallach et al., 1973), and consequently, they modify the timbre of the direct sound due to comb-filtering. The time delay between the direct sound and the first perceptually distinct reflection influences the perception of space dimensions and presence. Besides that, the temporal pattern affects the perceived size of the environment (Kaplanis et al., 2014). The spatial, temporal, and spectral pattern of early reflections modifies the envelopment of the space (Barron and Marshall, 1981). However, with the increase in the density of the reflections, perception is influenced more by the statistical parameters of the late reverberation. In large rooms, the perception of room size is governed more by the reverberation time but in small rooms by the pattern of early reflections (Yadav et al., 2013). All of these findings suggest that early reflections patterns and statistical properties of the reverberation tail have an influence on the perception of acoustic space.

The quality of the simulated acoustic environment might be especially crucial for an AR system in which the user can see and hear the real surroundings. The visual cues create a particular set of expectations. Lindau's definition of plausibility states that plausibility depends on how much a simulation meets the expectations of the listener. Previous research proved that congruency between visual and auditory stimuli (in this case acoustic space) significantly influences the plausibility (Werner et al., 2016). Results of the experiment by Neidhardt suggest that simulations without any reverberation were degrading the plausibility of sound during the subject's walk (Neidhardt et al., 2018). Different studies focused on investigating the accuracy of the spatial rendering, which is necessary to achieve a high level of realism and immersion. For example, Picinali et al. (2017) compared different approaches to reverberation, each offering a distinct level of spatial accuracy. The study was focused on the influence of reverberation accuracy on localization and realism. The methods used included multi-channel Ambisonics-based, stereo, and mono reverberation methods. The results showed that there was no perceived difference between four of five reverberation techniques which suggests that the high complexity of reverberation does not always correspond with the improvement of perceptual attributes. Similar results were obtained by Engel et al. (2019). The research aimed to investigate the trade-off between the complexity of reverberation and simulation quality. The reverberation was based on SRIR of zeroth to the fourth order of Higher-Order Ambisonics (HOA). Results suggested that the order of the HOA reverberation sound field did not have any influence on the perceived realism of the simulation.

### *Listener modeling with Head-Related Transfer Functions*

In the section 'Sound source modeling', the importance of the source directivity modeling has been emphasized, i.e., the way the source radiates the acoustic energy in space. Conversely, each receiver has its own directivity characteristics. An omnidirectional microphone will respond equally to a sound wave coming from any direction, whereas a cardioid microphone will favor its frontal direction. Similarly, our head presents a very specific directivity function. It originates from the scattering of sound waves on our shoulders, head, and ear pinnae and has to be precisely modeled in order to provide convincing spatial cues to our perception. The Head-Related Transfer Function (HRTF) characterizes the auditory spatial cues of a person for a defined sound source position. It includes Interaural Time Difference (ITD), Interaural Level Difference (ILD) – also known as Interaural Intensity Difference (IID) – and spectral shaping caused by frequency-dependent scattering and diffraction of sound waves by the listener's pinna, head, shoulders, and torso. Obtaining a complete set of HRTFs of a person requires a long measurement procedure. Microphones are inserted into ear canals to capture the measurement signal from different sound locations regularly distributed around the head. Various methods for quickly acquiring personalized HRTF measurements are currently being researched (Guezenoc and Séguier, 2018).

### *Influence of non-individualized HRTFs on perception*

The simplest method for implementing HRTFs is to use a generic dataset derived from measurements on a standardized manikin, such as KEMAR. Yet the differences between the actual HRTFs of a person and generic HRTFs can cause different kinds of distortions of spatial auditory perception. These include front-back confusions, problems with externalization, and localization on the vertical plane (Begault et al., 2001; Wightman and Kistler, 2018). The distortions not only affect the spatial attributes of sound but can also influence the sense of presence (Väljamäe et al., 2004). However, previous studies have shown that there are ways to improve the perception of sound with non-individualized HRTFs: sound localization with non-individualized HRTFs can be greatly improved by training with correct answer feedback (Andéol et al., 2014; Mendonça et al., 2012). Other studies have shown that natural head movements can significantly reduce front-back confusions (Wallach, 1940; Blauert and Butler, 1985) and enhance the externalization of sound sources (Wersényi, 2009). Incorporating a head-tracking system that dynamically updates the auditory scene in response to user movement can further mitigate front-back reversals and improve externalization in virtual auditory environments. Additionally, the presence of reverberation has been found to improve externalization, even when only early reflections are simulated (Begault et al., 2001).

Previous studies have indicated that generic HRTF may be sufficient when paired with temporally and spatially matched visual cues (Berger et al.,2018) for VR audio reproduction, as well as AR audio reproduction (Werner et al., 2021).

## ***Other factors in designing virtual acoustics system***

Several other factors should be taken into account when designing a virtual acoustics environment. The next sections describe their implementation and perceptual importance.

### *Latency and its perceptual salience*

Latency in virtual displays defines the amount of time that passes between the action of the user (i.e., body movements, head rotation) and the corresponding visual and auditory feedback. In the case of the audio layer, latency refers to the delay between a listener's physical movement – especially head rotations or positional shifts – and the corresponding adjustment of spatial auditory cues delivered to the ears. It is defined by the time required to update the position data, transmit it to the receiver (typically the computer handling the calculations), render the audio signal based on the updated position, and send the processed audio to the playback system. If the delay exceeds the detectable threshold, it can cause multisensory distortions coming from the disparity between the perceived sound image and the expectations from proprioception (the sense of our own movement). They can affect user response time (Brungart et al., 2004) and sound localization (Sandvad, 1996). Brungart et al. (2005) investigated the detection threshold for head-tracker latency and found that a delay of less than 80 ms in an auditory scene with a single virtual source is typically not detectable by an average listener. However, their results also indicated that this threshold decreased by approximately 25 ms when a low-latency reference signal was provided at the same location as the virtual source. This situation mirrors AR environments, where listeners are exposed to real sounds without delay. Based on these findings, we can infer that the threshold may differ for VR and AR systems. Notably, there was significant variability in subject ratings, with some participants detecting threshold values as low as 30 ms. The authors suggest that to ensure the latency is imperceptible for all listeners, a threshold below 30 ms may be necessary, depending on the listening conditions.

In a separate study, Yairi et al. (2007) conducted two experiments to assess the detection threshold and just noticeable difference (JND) of system latency. In these experiments, subjects were asked to rotate their heads and determine whether they could perceive the delay. The average detection threshold and JND across nine participants were found to be 45 ms and 59 ms, respectively. However, other studies, such as Rappin et al. (2023), reported higher thresholds, around 200 ms. This discrepancy was attributed to differences in participants' expertise and the nature of the specific task.

### *3DoF versus 6DoF environments*

The freedom of the user's movement in three-dimensional space is common for AR and VR systems. In 3 degrees of freedom (3DoF) environments, the rotation of the user's head is tracked, whereas in 6-degrees-of-freedom (6DoF) environments, the rotation and position of the user are tracked. Consequently, all of the components and parameters of the audio rendering system need to be implemented in real time and updated according to the change of rotation and location of the user and the source. The perception of the auditory layer is different from the ability to self-translate and requires highly responsive rendering systems (Neidhardt et al., 2018).

### *Audiovisual interaction*

Although no visual virtual objects are presented in an audio-only AR experience, participants can still see their actual surroundings, meaning their judgments are influenced by visual cues.

Perceptual phenomena like the *Colavita effect* where visual stimuli change the perceived location of a simultaneously presented sound indicate that the influence of visual cues on the perception of sound may be crucial (Colavita, 1974). The importance of visuo-auditory integration mechanisms can be assumed to be especially significant in AR environments where the virtual sources blend with the real environment (Begault, 1999). The influence of visual cues on audio perception in AR systems may include the following:

- The physical space around the listener creates a certain type of expectations for the acoustic rendering, i.e., the type of reflections expected, thus affecting the plausibility judgment (Larsson et al., 2001; Thery et al., 2017; Postma and Katz, 2017; Bailey and Fazenda, 2018; Martin et al., 2022).
- The presence of the visual object representing the sound source may affect the overall judgment of the plausibility of the scene (Bailey and Fazenda, 2018) and possibly improve localization, which might be distorted when using generic HRTFs (Colavita, 1974).
- The visual cues may affect the perception of other attributes of sound, i.e., acoustic distance or loudness (Brungart, 1998; Zahorik et al., 2005).

## Plausibility and presence

One of the essential measures of the XR experience quality commonly used in literature is *presence*, which relates to the feeling of 'being there' in a virtual environment (Wagner et al., 2009). The sense of presence involves several components: the sense that we are located in and act from within the virtual environment (VE), and the sense that we are concentrating on the VE and ignoring the real environment. On the other hand, immersion describes the extent to which the computer displays are capable of delivering an illusion of reality to the senses of a human participant (Slater and Wilbur, 1997). Important parameters of immersion include the extent of the field of view, the number of sensory modalities that the system simulates, the quality of rendering in each sensory modality, the extent of tracking, the realism of the displayed images, the frame rate, and the latency. The concept of 'presence' refers to the mental state in which a user behaves and feels as if physically present within the computer-mediated environment (Draper et al., 1998). The sense of presence is a subjective experience and only quantifiable by the user experiencing it. Immersion and spatial aspects contribute to the sense of presence, but it is the quality of interaction which plays a major role in eliciting presence.

Slater divides the sense of presence into two concepts: place illusion and plausibility. Place illusion relates to the feeling of actually being in a virtual place, whereas plausibility relates to the impression that what is happening is real. In the case of AR, place illusion is less important as the surroundings are already real. On the other hand, the virtual elements overlaid on this environment should seem to be really happening, which can be evaluated by plausibility (Slater, 2009). Lindau suggests another definition of plausibility as 'a simulation in agreement with the listener's expectation towards a corresponding real event' (Lindau and Weinzierl, 2012), which also applies to AR. It is contrasted with authenticity, which requires an immediate comparison with a real event and consequently needs a very high level of accuracy which may not be necessary for real-life scenarios. Authenticity is associated with an *external* reference to which the listener is immediately comparing the virtual event. On the other hand, when evaluating plausibility, the listener is relating to the *inner* reference, which is caused by the memory of previous experiences. Moreover, setting the listener's expectation as the measure corresponds well with the majority of AR system applications. Real surroundings, on which the virtual objects are imposed, create a shared set of expectations toward the auditory experience of the scene. Taking all of that into account, plausibility seems to be an appropriate measure of the overall user experience in audio AR.

## Case study I: acoustic and perceptual factors affecting plausibility in sound design for AR

Plausibility of sound is a complex percept influenced by both acoustic and perceptual factors (refer to Figure 12.1). Acoustic factors, which are tied to the design of a virtual acoustic environment, were discussed in the previous section. However, perceptual factors also play a crucial role in shaping plausibility judgments. These factors are inherently linked to the acoustic ones, as they represent how objective acoustic attributes are perceived. Some perceptual factors are directly related to specific acoustic properties, often exhibiting nonlinear relationships, while others arise from the combination of multiple acoustic elements. Since plausibility evaluates sound relative to an internal reference, it is reasonable to assume that this reference may be influenced by real-world sounds encountered during the augmented reality experience, thus influencing the evaluation.

In addition, user movement plays a crucial role in plausibility judgments. In most AR experiences, users move in 6DoF, which can significantly impact their perception of sound. Movement enables users to experience variations in acoustic cues as they change their position relative to the sound source. Specifically, as users move, they can perceive alterations in the spatial properties of the sound field, such as changes in intensity, timbre, and the directivity of the sound source. For example, moving around a sound source may reveal differences in intensity or timbre, providing insight into the source's frequency characteristics and its directivity. These spatial variations influence how natural or plausible the sound scene appears.

Another aspect of plausibility perception that remains mostly unexplored is its relationship to other sound attributes. Sound perception has been extensively studied, revealing a number of attributes crucial to overall sound quality judgment. However, the link between these attributes and the perception of plausibility remains an open question. It is unclear whether certain attributes hold more importance for plausibility or if all attributes contribute equally. Further research is needed to explore these relationships and better understand the factors influencing the plausibility of sound in AR experiences.

This chapter describes an experiment aimed at the investigation of plausibility perception in the context of the audio AR environment where real and virtual sources are presented together. This scenario demonstrates the most challenging case of audio implementation for XR environments as the user is constantly exposed to the real reference by the sounds of the environment, which reveal the properties of the space. By allowing free movement in the space, 6DoF environments provide more dynamic cues, possibly making the comparison between real and virtual sounds even easier than in a standing scenario. This section presents only the summary of the evaluation method and key outcomes. A full description of the testing protocol, with statistical analysis and discussion, is available at Gospodarek (2024).

### *Goals and research questions*

The main research question addressed by this experiment was the following:

What perceptual and acoustic factors are meaningful in the plausibility perception of sound design for augmented reality environments?

The study aimed to answer a set of subquestions, which can be divided into several areas of interest posed.

#### Perceptual evaluation

- How does a subject's freedom of movement affect the perceptual evaluation of an AR sound scene?
- What is the correlation between plausibility and other perceptual attributes of sound?
- Do the properties of real reference affect plausibility judgment?
- Does the listener's walking speed and the amplitude of yaw movement influence plausibility, or, conversely, does plausibility affect walking speed and yaw movement amplitude?

### Acoustics

- How do objective measures of acoustical parameters correspond to subjective evaluation of acoustic processing?
- How do the position and orientation of the source within the room influence the assessment of auralizations?

### *Summary of the methodology*

In order to address these research questions and explore the influence of acoustic and perceptual factors on sound plausibility, the study implemented a novel experimental design. Figure 12.2 illustrates the conceptual distinction between conventional experimental designs aimed at assessing *authenticity* and the proposed methodology centered on evaluating sound *plausibility*. In earlier studies focused on *authenticity* (Figure 12.2, left), the experimental paradigm typically involved direct comparisons in which a

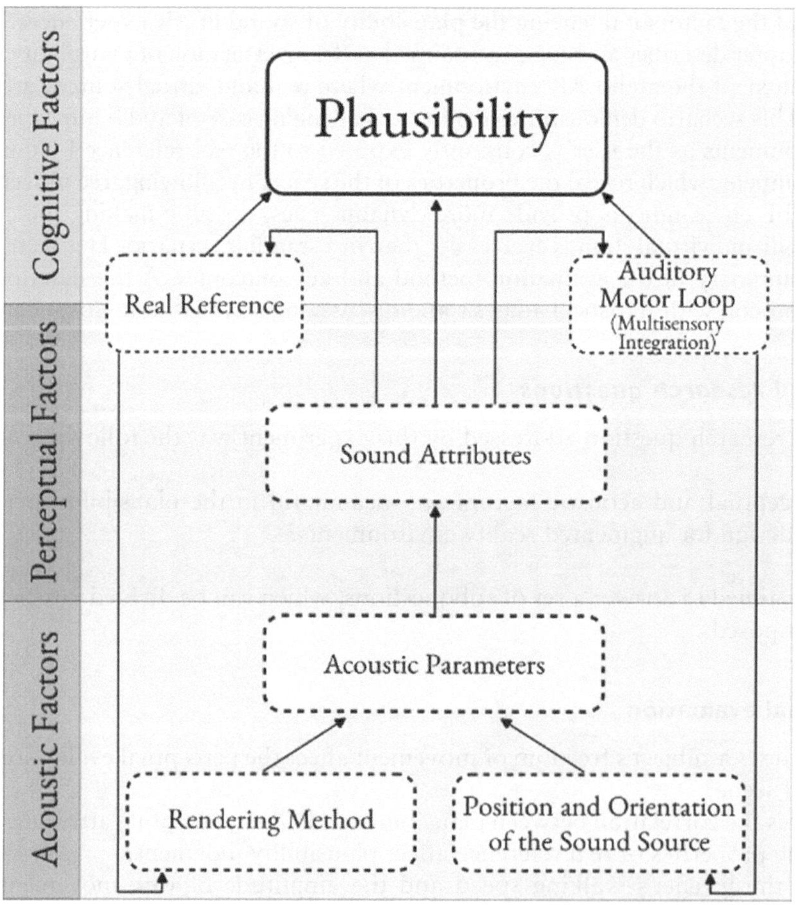


*Figure 12.1* Factors influencing plausibility perception

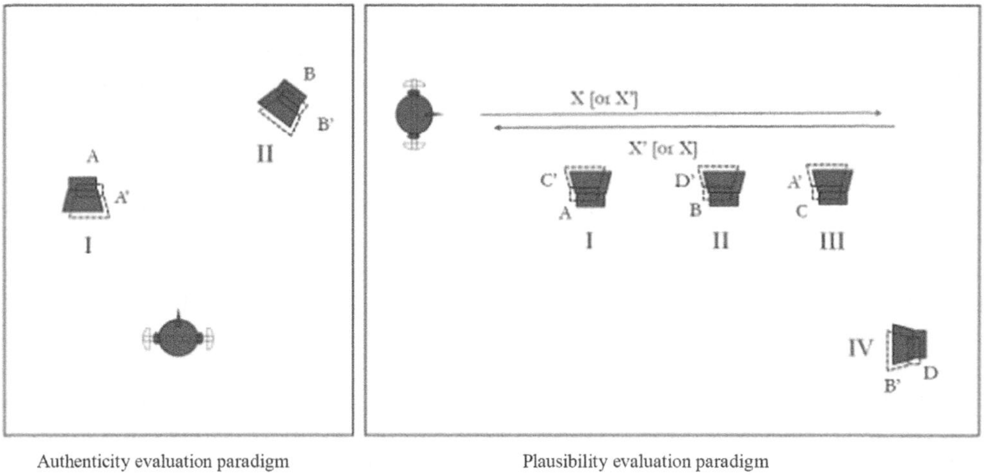


*Figure 12.2* Comparison of two experimental paradigms: Left: versus authenticity evaluation;. Right: versus plausibility evaluation

stimulus presented through a real sound source (e.g., A or B) was immediately followed or preceded by its virtual counterpart (A′ or B′) reproduced at the same location. These comparisons were made under controlled conditions where location, orientation, and the associated room acoustics were matched as closely as possible, allowing the listener to make an immediate and precise judgment about the similarity between real and virtual sources.

By contrast, the experimental design proposed for assessing *plausibility* (Figure 12.2, right) departs from this one-to-one matching. Here, a real source X is compared to its virtual counterpart X′, which is deliberately not placed at the same spatial position. During the test, listeners encounter the real source (X) while walking in one direction and the virtual source (X′) on their return path, or vice versa. For source pairs like A and A′ or C and C′, the walking trajectory passes near the sources, offering similar auditory perspectives. However, for pairs like B and B′ or D and D′, the listener hears them from substantially different distances and angles, making direct comparison more difficult.

During the study, participants listened to both real loudspeakers and their virtual counterparts presented over headphones. Pairs of stimuli were played consecutively from different positions in the room, as illustrated in Figure 12.2 (right). Although, within a single trial, participants did not hear the exact virtual counterpart of a real loudspeaker from the same position, the task was systematically repeated across all loudspeaker pairs. As a result, by the end of the experiment, all positions had been equally tested in both real and virtual modalities.

This methodological shift better reflects real-world listening conditions, such as those encountered in AAR scenarios. In such contexts, users rarely hear real and virtual sources from exactly the same perspective. While immediate comparison to a real reference was available in the experiment, it was intentionally non-identical to the virtual sound, thus more accurately simulating the conditions under which plausibility judgments are made in everyday environments.

The experiment was carried out in two phases (walking and standing) in order to evaluate the influence of participants' movement on perceptual judgment. During the walking phase, participants walked back and forth following a predefined path which allowed for the evaluation of the dynamic rendering of the stimuli. While they were walking, an audio clip was played once during the way forward from a real or virtualized loudspeaker placed at a given location and repeated during the way back also from a real or virtualized loudspeaker, but from a different location. After each trial, participants answered a short questionnaire to rate the two audio clips in terms of their respective plausibility and other audio quality attributes. During the standing phase, participants stood at a fixed point along the path, near the location of loudspeaker I (see Figure 12.2), while listening to the stimuli. Everything else followed the same procedure as in the walking phase.

The four loudspeakers positioned in the room were grouped into two pairs. For each trial, the audio clips were played consecutively on the two loudspeakers of a given pair (I–III or II–IV, see Figure 12.2). Pair I–III represents loudspeakers which provide a similar listening perspective in relation to the room and to the walking path. Thus the influence of the loudspeaker position on the plausibility and other attributes rating of the two audio clips is expected to be minimal. Pair II–IV represents a very different loudspeaker perspective (in terms of distance as well as orientation). Thus the influence of the loudspeaker position on the plausibility and other sound attribute ratings of the two audio clips may be higher. Participants were instructed to maintain a straight head position. The use of near-field open-ear headphones (AKG K1000), which enabled transparent hearing with compensation filters, facilitated a direct comparison between the simulation played through the headphones and real sound sources emitted by loudspeakers positioned in the room.

After each trial, participants filled out a questionnaire. The questionnaire not only investigated the plausibility perception but also included several perceptual attributes taken from the Spatial Audio Quality Inventory (SAQI) (Lindau et al., 2014), including blur, localization accuracy, externalization, reverberation difference, and timbre difference. The decision to include these attributes was aimed at exploring potential correlations between plausibility and other sound attributes. To facilitate such an investigation, plausibility was assessed on a continuous scale, departing from the binary *yes/no* approach often used in the literature. Additionally, recognizing plausibility as a continuum allowed for a more nuanced evaluation, preventing data from being unnecessarily constrained. This approach accounts for scenarios where virtual sound sources may be distinguishable from real references without necessarily being perceived as implausible. Moreover, real sound sources might themselves appear less plausible in certain contexts, such as when the direct path is obstructed. Thus the test protocol included not only comparisons between virtual and real sources but also evaluations of pairs involving two real sources or two virtual sources. By analyzing these results, the influence of the presence of a real source on plausibility judgments was investigated.

Another goal of the study was to validate the proposed experimental design by comparing two different approaches to the auralization of early reflections and late reverberation. The implementation of two different auralization methods allowed also to search for a correlation between objective acoustic parameters of auralizations and measurements and subjective ratings of the listeners.

The first method exploited a simple 3D numerical model of the room and ran a real-time beam-tracing method to calculate the time and spatial distribution of the reflections. The late reverberation was modeled with a Feedback Delay Network (FDN). The second auralization method characterized the experimental room with a single Spatial Room Impulse Response (SRIR), which was further manipulated in order to account for the relative listener-source distance. Both methods employed the same simulation of the direct sound propagation effect with a simple directivity model to simulate the loudspeaker radiation pattern. The two methods had different limitations and were chosen to investigate the possible influence of weaknesses of the simulations on plausibility judgment. With the first approach, the temporal structure of reflections was expected to be closer to reality as it was constantly updated according to the source and listener positions in the room. In contrast, the late reverberation was only an approximation of the measured IR as it was rendered through an FDN implementation. On the other hand, the SRIR auralization could accurately reproduce the time and frequency distribution of the late reverberation but could not account for the varying time and spatial distribution of early reflections along the walking path. The temporal structure of reflections was kept from the original SRIR and did not depend on the position of the listener and source as it would in reality. For the detailed description of the implementation and experimental design, refer to Gospodarek (2024).

### *Summary of the results*

A total of 4,798 ratings were collected from 25 expert listeners across two phases of the study. The data was analyzed using generalized linear mixed modeling (GLMM) and linear mixed modeling (LMM). The following section outlines the key conclusions drawn from the experimental results, organized according to the research questions.

#### How does a listener's freedom of movement affect the perceptual impression of an AR sound scene?

While walking, listeners could effectively utilize dynamic acoustic cues, such as changes in intensity, spectrum, and Direct-to-Reverberant Ratio (DRR), by altering their azimuth and distance relative to the sound source. Additionally, the parallax effect was pronounced for loudspeakers I, II, and III along the walking path (see Figure 12.2), enhancing the presence of sound sources. The primary factors influencing plausibility judgments during motion appeared to be the coherence between self-motion and auditory cues, as well as the alignment between visual and auditory cues.

In contrast, during the standing phase, listeners were confined to a single static perspective of the sources. Consequently, listeners judged mostly the perceptual aspects of the sound focusing on the intensity, spectrum, and DRR and relied more heavily on comparing the two sources presented together in each trial. This was proved by the results of analysis showing that the lower plausibility judgments for loudspeakers II and IV, even in real playback scenarios, were caused by the inaccuracy of intensity cues in the auralizations.

However, the ratings of loudspeakers I and III were very similar between the two phases. This was due to the unique experimental setup, where loudspeakers I, II, and III were aligned in a single line at varying distances. This setup facilitated reliable evaluation of different rendering methods for these loudspeakers as it mimicked a situation where one source could be heard from different distances. Consequently, the positioning of loudspeaker IV, which diverged from the others, led to more pronounced differences in evaluation due to the limited cues available during the standing phase.

**What is the correlation between plausibility and other perceptual attributes of sound?**

The analysis explored the correlation between plausibility difference and other attribute differences, revealing that each attribute independently contributes to explaining variance in plausibility. Blur ratings accounted for the largest portion of variance, followed by localization error, which was partially correlated with blur, and then externalization, timbre, and reverberation. Blur and localization error ratings were primarily influenced by the ambiguity between perceived sound source location and visual anchors, a crucial factor in plausibility evaluation. Externalization emerged as the next significant factor.

The most important conclusion is that plausibility is a complex subjective percept. As it relates to the comparison of sound to the inner reference, it encompasses different aspects of sound. All evaluated factors played a role in plausibility perception. The highest correlation with plausibility observed for blur and localization error, highlights the importance of congruence between visual anchors and sound. Externalization followed as a crucial aspect of binaural sound, prone to disruption and sensitive to room divergence. Surprisingly, reverberation emerged as the least significant attribute, consistent with previous findings indicating listeners' limited sensitivity to reverberation changes.

**Do the properties of real reference affect plausibility judgment?**

The properties of the real reference do affect plausibility judgment. Plausibility evaluation is based on the agreement between the perceived sound event and internal reference shaped by previous sound experiences. The experiment proved that in the context of AAR, the internal reference can be tuned by the real sound environment around the listener. The results of the experiment showed that manipulation of the real reference properties affected plausibility perception. By positioning two pairs of loudspeakers differently, the study enabled a comparison of plausibility perception when stimuli were presented with similar or different real counterparts. The results revealed that when real loudspeakers were positioned similarly, listeners found it easier to distinguish between auralizations compared to when the real counterparts were positioned differently. This suggests that the properties of the real reference, such as their positioning, can impact the perceived plausibility of virtual sounds. Therefore, adjusting the properties of the real reference, such as altering the loudspeaker's position within the pair, can be valuable in adapting the test to different real-life scenarios and objectives, where rendering accuracy may vary.

The results indicated also that the inclusion of real references significantly affected the perception of virtual sources. From all of the pairs of rendering methods presented in one trial, the pair of two auralizations received the highest scores of plausibility. This suggests that, in the absence of a real source for comparison (as in VR experiences), a virtual source may be perceived as more plausible than in conditions where such a comparison is available (as in AAR experiences).

**How do objective measures of acoustical parameters correspond to subjective evaluation of acoustic processing?**

The analysis showed that ratings of blur were mainly correlated with localizability. The imperfect localizability was caused by the errors of distance estimation for virtual and sometimes even real sources. The main acoustic cue for distance estimation was sound intensity. As the intensity values were compared between loudspeakers, the discrepancy between real and virtual loudspeakers compared in pairs led to localization errors and sometimes problems with loudspeaker recognition.

However, it has to be noted that the scope of the objective analysis was not able to encompass possible effects of the binaural rendering (in particular conversion between HOA and binaural for rendering room effects) as well as non-individual HRTFs which may also contribute to the blur and plausibility perception. There is a need for further research on other objective measures of acoustic simulation quality.

**How does the position of the source in the room and orientation influence the assessment of the auralizations?**

The positioning and orientation of the sound sources had varying impacts on the assessment. The position primarily influenced the distance between the loudspeaker and the listener, consequently affecting the DRR and sound intensity. Meanwhile, the orientation altered the angle between the loudspeaker and the listener, thus modifying the sound spectrum due to the directivity pattern.

Loudspeakers I, II, and III were placed close to the walking path and were positioned perpendicular to it. This setup provided participants with a strong parallax effect, enhancing the plausibility of virtual loudspeakers. Additionally, it minimized the significance of early reflections and reverberations, as they were overshadowed by direct sound.

In contrast, loudspeaker IV was situated farther from the path and oriented parallel to it. Here, early reflections and reverberations played a more substantial role in sound perception, influenced not only by distance but also by orientation, resulting in the decrease of high frequencies. Consequently, this led to a diminished parallax effect.

**What is an effective method for evaluating plausibility in 6DoF AR environments?**

The experimental design focused on replicating real-life scenarios of auditory augmented reality (AAR), where real sounds coexist with virtual ones, making direct comparisons challenging. By utilizing pairs of real and virtual loudspeakers with identical acoustical characteristics but varying positions, the study successfully imitated real-world AAR conditions.

The results highlight the subjectivity of sound plausibility, emphasizing its dependence on the specific context in which the sound is presented. The ability to tune the internal reference by including real counterparts in the evaluation task proved crucial in understanding plausibility perception. Moreover, the protocol enabled subtle differences in plausibility to be detected through the use of categorical scales, revealing a nuanced 'gray area' in plausibility perception that may be overlooked by binary evaluation paradigms.

Careful consideration of listener movement within the 6DoF space is crucial, and the testing protocol addressed this by implementing a predetermined walking path. This controlled movement ensured consistent evaluation conditions, allowing listeners to experience similar hearing conditions. Furthermore, the study effectively evaluated the chosen auralization methods. Both methods achieved the intended goals, eliciting subtle but significant differences in perception between real and virtual sounds, despite simplifications of the implementation.

Overall, the proposed protocol offered a comprehensive and adaptable approach to evaluating plausibility in 6DoF AR environments, providing valuable insights into the plausibility perception and its implications for AAR systems.

**How do the listeners' speed of walking and amplitude of yaw movement affect the evaluation?**

The analysis of tracking data across both phases uncovered a consistent pattern in listeners' behavior. During the walking phase, listeners exhibited faster movement when exposed to real playback, correlating with higher scores for sound attributes such as plausibility, loudspeaker recognition rate and lower for localization error and blur. Similarly, in the standing phase, reduced yaw movement amplitude during real playback corresponded to the similar change in ratings. Unconsciously, listeners walked faster and minimized head movements when encountering easier tasks with real playback, while they adopted slower speeds and increased head movements during auralization playback, indicative of a more challenging task.

These findings suggest a meaningful correlation between stimulus presentation and participant behavior, implying that analyzing listener behavior could provide insights into the perceived nature of the stimulus. However, further research in this area is needed to broaden the understanding of these relationships.

### *Implications for sound design in AR*

The findings of the study underscored the significance of aligning visual and acoustic cues for an immersive AR experience. Notably, the accurate portrayal of distance emerged as a critical factor in determining plausibility judgments. While distance perception primarily relies on intensity cues, the role of parameters such as DRR may vary depending on the stimulus type. Consequently, detailed attention to both direct sound and reverberant energy is essential for convincingly rendering virtual sound sources in AR environments.

Furthermore, the position and orientation of sound sources can influence the relative importance of direct and reverberated sound. Close-range sources need to prioritize the rendering of direct sound, whereas distant sources require careful consideration of early reflections and late reverberation. Additionally, when directional sound sources face away, the pattern of early reflections becomes particularly crucial, posing a challenge for sound design.

The context of the AR scene also needs consideration. If the scenario anticipates significant movement of the user, accurately simulating the parallax effect becomes pivotal for plausibility. This means prioritizing the 'near field' effects, including accurate reproduction of directivity patterns and faithful rendering of near field acoustic effects (e.g., near field binaural cues). Conversely, in more static scenarios, the sound perception is based more on a comparison of audio cues between real and virtual sources within a room. Consequently, the consistency between them gains importance.

Given the unpredictable nature of the user's environment in AR experiences, the sound design must accommodate scenarios with varying numbers of real sound sources. The likelihood of users hearing real sounds similar to virtual ones affects the detectability of discrepancies in the reproduced sounds. The closer the similarity between virtual and real sounds, the easier it becomes for users to perceive inconsistencies.

Lastly, the perception of plausibility depends on multiple attributes. Besides localization errors, blur, and reverberance, which were already discussed, aspects like externalization and accuracy of the directivity pattern also contribute significantly to sound scene plausibility. Ensuring fidelity across these attributes should enhance the overall immersive quality of the AR environment.

## Case study II: sound design and reproduction techniques for co-located narrative VR experiences

Virtual reality (VR) is expanding very fast in the fields of gaming and entertainment and numerous cinematic productions experiment with headsets to deliver new sorts of experiences. However, a lot of these works are designed to engage a single user at a time and do not usually invoke a sense of social gathering, a quintessential feature of cinema and theater. This section describes the audio design principles for a co-located VR experience which focus on delivering immersive audio experience. The full description is available at Gospodarek et al. (2019).

Technological developments observed in recent years now enable the creation of different types of VR production that allow large co-located audiences to experience a shared virtual environment presented in specially designed entertainment spaces (Layng et al., 2019; Gochfeld et al., 2018). These productions have the goal of bringing back the social aspects of cinema and theatre, which is achieved by designing an experience where the participants are able to see and hear one another as virtual avatars, spatially coherent with their actual physical location. Thus a cognitive impression of being in a shared experience is established, enabling a sense of presence and awareness of communication between fellow users.

The audio layer is especially important in VR experiences as it affects the subjective senses of *immersion*, *plausibility*, and *presence* (Kobayashi and Ueno, 2015), which are key to the success of co-located immersive VR. Spatial audio techniques, which allow users to match the position of sounds with their respective visual cues, can, in fact, improve these quality metrics (Brinkman et al., 2015), while poor audio production can negatively affect them (Zhao et al., 2017).

The following section reviews the factors and principles behind the implementation of audio systems for co-located narrative VR, whether cinematic or theatrical and in the second part, describes the implementation of the audio reproduction system for a short narrative VR art piece, *Cave*.

### *Background*

Immersive co-located VR is a type of production category which merges some elements of gaming interaction with the linear narrative elements of theatre and cinema. The defining element is the assumption that multiple participants are located in the same room and experience the same virtual content (through virtual or mixed reality devices) under their own individual perspectives and points of view, while also being able to see one another in the virtual space. Each participant is rendered in the shared virtual scene as a virtual avatar (usually humanoid), spatially matching their physical location and orientation in real time by means of motion-tracking technology.

Since simulating the social setting of a crowd inside a theatre is a goal for these systems, it is important that participants are treated as audience members and feel present as such in the space. To this goal, the audience is usually placed in 'seats' from which unique first-person views are dynamically rendered and the narrative content is placed onto a virtual 'stage'.

To technically achieve a multi-user visual reproduction, the headsets are connected using a network-synced infrastructure that allows for the simultaneous delivery of the content for all participants (Churchill and Snowdon, 1998). While the narrative content may or may not be linearly progressing (usually it is), the rendering of the audience members' avatars (e.g., their head rotation and off-axis shift) needs to be actively updated close to real time. Each client device reports its 3D location and orientation to a server and receives the location of every other device with their respective timestamps (Herscher et al., 2019). The rendering is finally facilitated at each device through time synchronization signals that make sure that there are no differences in perceived time between participants.

The implementation of such cinematic, or theatrical, experiences can exist under different variations. One particular experimental production, *Holojam in Wonderland*, shown at the 2017 New York's Future of Storytelling Festival, was portrayed as an 'Immersive Mixed-Reality Theatre' (Gochfeld et al., 2018). Two live-rendered actors and four audience members shared a virtual reality stage where a theatrical narration took place in a shared environment. While the actors represented the story characters, the audience was represented by avatars of butterflies, and all were allowed to move in six degrees of freedom (6DoF), explore the virtual world, and interact with a semilinear progression of events.

The sound was implemented through a quadraphonic loudspeaker system with an additional overhead speaker. The actors' dialogue was presented in dual form as live free-field speech alternated with pre-recorded dialogue lines played from the overhead speaker. This choice served the narrative purpose of simulating one actor's change in size both visually and aurally. No headphones were used as it was necessary for the free-field dialogue to be heard without the effects of occlusion and attenuation of the sound path to the ears, although it is possible to achieve transparent headphone reproduction using hear-through microphones (Rämö and Välimäki, 2014).

## Design factors and principles

### Sound principles for co-located VR theater

Each kind of VR experience requires a different approach to sound design and sound production due to the potentially different modalities of the medium used for storytelling. Most design implementations in VR are based on the game-audio framework (in the case of interactive experiences (Horowitz and Looney, 2014)) or on the cinematic framework (in 360° videos (Paterson and Kadel, 2019)). Co-located VR theater entails a set of design requirements for the audio layer which differs from the other types of VR productions:

#### TRANSPARENT HEARING

To enable communication within the audience, it is important that users can hear each other during the experience. The use of headphone playback is not appropriate in this context as it impairs free-field listening abilities of the participant. Even open-back headphones are shown to produce occlusion and attenuation effects in the ear canal (Gupta et al., 2018), making the blend between real and virtual sources more difficult to achieve. Although it is possible to equalize these effects with an attentive individualized calibration, a more flexible solution is to employ different kinds of non-obstructive sound reproduction devices such as hear-through earphones or headphones supplemented with microphones that enable transparent hearing (Rämö and Välimäki, 2014). A loudspeaker-only reproduction method would also provide transparent hearing but likely interfere with other requirements.

#### SPATIAL SOUND

The auditory localization of sound objects has to match the visual localization of the sound sources in order to achieve immersion and presence inside the experience (Brinkman et al., 2015). Spatial audio techniques need to be used to ensure proper perception of the sound localization and adequate proximity effects between the far and the near auditory fields.

#### CINEMATIC SOUND DESIGN

The sound layer has to support the storytelling and reflect a cinematic style of sound design, supporting the full spectrum of sounds which make a compelling experience. The implementation of sound for co-located cinematic VR is merging the approaches from traditional cinema and game audio. The VR narrative is linear, meaning it is played identically for every performance. This format creates an opportunity to design sounds which perfectly match the visual action, without the need for the sound randomization necessary in games (Horowitz and Looney, 2014). On the other hand, the experience is also interactive. Users have the ability to modify their orientation and position inside the scene, which means that their point of listening can change.

#### INDIVIDUAL AUDIO MIX

When each member of the audience's 'virtual seat' corresponds to their position in physical space, the sound mix delivered also must be matched to that position and orientation and thus differs for each member of the audience. As a result, each member of the audience receives an individual sound mix which represents their point of listening.

### *Proposed reproduction system*

The biggest challenge is to allow audience members to hear one another while delivering a high-quality spatial audio layer. This chapter suggests the employment of a hybrid reproduction system for immersive co-located VR experiences. The proposed system consists of a transparent hearing device and a loudspeaker array.

### SPATIAL AUDIO OVER HEADPHONES

Spatial audio content is more easily and flexibly deliverable through binaural audio techniques. Binaural audio technology allows the reproduction of spatial sound by encoding auditory cues into a stereo audio signal, thus changing the perceived localization of object sound sources (Begault et al., 2001).

The cues which depend on the anthropomorphic measurements of the person's head, pinna, and torso are unique for every individual. Head-Related Transfer Function (HRTF) characterizes the auditory spatial cues of a person for a defined sound source position. It includes interaural time difference (ITD), interaural intensity difference (IID), and spectral modulations. The limitation of binaural sound in most VR productions is the use of non-individualized HRTFs, which can cause distortions in perceived sound images such as front-back confusions, distortions in localization on the vertical plane, and weak externalization (Guezenoc and Séguier, 2018). Adequate reverberation (coherent to the visual environment) and head tracking can, to some extent, mitigate the drawbacks of using non-personalized HRTFs.

Headphone playback is the most common way of delivering spatial audio, mostly because it can ensure a perfect separation between the two binaural channels. The drawback of headphone reproduction is that even open-back headphones introduce significant attenuation of real-world sound sources (Gupta et al., 2018). The resulting coloration takes away from the plausibility of the experience, and in the case of the immersive co-located experience where the interaction between the audience members during the experience is crucial, a system which enables the delivery of audio signals without impairing the user's normal free-field hearing is necessary. One of the solutions to that problem is earphone drivers coupled with acoustically transparent earpieces (Martin et al., 2009). Another way of delivering the audio are nearfield open ear devices mounted in front of the ears, oriented towards the entrance of the ear canal. However, the small size of transducers in this type of reproduction device often leads to nonlinear frequency response and attenuation in the low-frequency region (Gutierrez et al., 2015). To mitigate the frequency response problem, a hybrid reproduction system with loudspeakers is proposed.

### LOUDSPEAKER PLAYBACK

Loudspeaker playback is broadly used in cinema production. Surround speaker systems enhance the perception of envelopment and spaciousness of a sound scene and enable designers to create an impression of movement of the sound sources around the listening space. A subwoofer speaker provides energy at low frequencies, which is especially important in cinematic sound design where emotional impact is greatly enhanced by the use of low-frequency sound effects (Whittington, 2007). The limitation of a speaker-only system is that the subjective localization of sound sources is not very accurate. Furthermore, it is hard to create a convincing virtual source positioned closer to the listener than the physical position of the loudspeaker. Besides that, in surround speaker setups, the sweet spot is usually limited to the central seating position (Rumsey, 2007).

Adding more speakers to the setup can enhance the immersion and allow for a more accurate trajectory of movement of the sound sources, but the rendering of near-field sound sources is still limited. The use of wavefield synthesis techniques would allow designers to create a very realistic sound field around the listening area, but its implementation is very expensive and requires acoustic treatment of the performance space (Boone et al., 1995).

The proposed use of a hybrid reproduction system can take advantage of both types of reproduction methods and deliver high-quality convincing and cohesive sound. Hear-through earphones enable transparent hearing and deliver an individualized mix of binaural audio to each user, providing an accurate match with the visuals – particularly important for the creation of near-field sound events, which require precise spatial alignment with visual elements. The speaker system improves the experience by providing a full frequency spectrum of sound and enhances the 3D auditory scene with far-field sounds, which can improve externalization (Mueller-Tomfelde, 2002).

### *Technical challenges*

#### DELAY

An audio signal played simultaneously through earphones and loudspeakers will not reach a listener at the same moment in time. Signals from loudspeakers arrive to a listener delayed, and the delay will depend on the distance of the listener from the speaker. This issue might be especially important if the sounds played through the device have a short temporal structure (a significant amount of transients) where the delay can be perceived by ear (Scharine et al., 1999). Delay adjustments at the binaural device, for some of the seating positions, might be necessary in larger spaces. Contrarily, this issue is not salient for sounds with longer temporal structures.

#### DISTANCE ATTENUATION

For each doubling of distance from the source, the intensity of a signal in the free field decreases by 6 dB (Shinn-Cunningham, 2000). Depending on the distance of each listener to each speaker in a chosen configuration, the signal may be attenuated to different degrees. This seat dependency must be taken into account during the mixing stage to ensure a proper sound level for each of the more sensitive positions. Special attention should be paid to the voice layer level to ensure intelligibility for all participants, which may require deviations from the 6 dB rule.

#### HRTFS RENDERING

When listening to loudspeakers, listeners perceive sound through their own natural HRTFs. The situation is different with spatial sound on nearfield devices: a listener would be usually delivered sound processed through generalized HRTF filters, which are likely non-ideally tuned to their personal spatial cues response. This might cause a problem if too similar sounds were to be played through both the earphones and the loudspeakers, as the HRTFs may color the signal spectrum and create a timbre mismatch between the two delivery methods (Takanen et al., 2012). In an ideal situation, individualized HRTF filters, measured in the listening room for each seating position, would deliver the highest possible spatial audio quality. But this is simply unfeasible, a more efficient workaround is to keep the content distinct for the two reproduction systems.

ROOM ACOUSTICS

When playing back sound on speakers, the room acoustics influence the end signal as it reaches listeners' ears. The acoustic character of the room might significantly differ from the one given to the virtual sound layer played at the earphones (for example, when using artificial reverberation). The acoustic mismatch might negatively impact the perceptual auditory integration between the two reproduction elements, affecting the smoothness and capability of immersion in the experience. For this reason, the exhibition room should be acoustically treated to reduce reflections. If this cannot be accomplished, it is desirable to adjust the reverberation of the virtual content to be similar or even slightly longer than the actual room reverberation, in order to minimize the perceived reverberation mismatch between the two reproduction systems.

### *Sound design*

The sound-design style for co-located cinematic VR is based on traditional cinematic approaches with the addition of spatial audio processing. The main stems necessary for film audio soundtracks include dialog, music, and sound effects (foley, SFX, backgrounds, and ambiances). In contrast to the stereo or surround cinematic mix, there are more audio formats available to the sound designer in shared narrative VR. The audio layers can be reproduced using different spatial audio techniques, even concurrently: as audio objects using binaural rendering, as Ambisonics files that capture the whole sphere of sound around the listener (Gerzon, 1973), as traditional stereo on headphones, as surround formats through VSS processing on headphones (Pike et al., 2014), or as a channel-based mix played back on speakers.

Each sound layer has different requirements in terms of spatial processing and diffusion (see Table 12.1). The dialogue and foley require very precise scene placement, achievable with binaural rendering. Distance cues such as level attenuation, direct sound-to-reverberation energy ratio (DRR), as well as radiation pattern, need to be added to ensure a realistic sound change during the character's movement (Shinn-Cunningham, 2000). Background and ambiance sounds are usually more diffused. They can be reproduced in Ambisonics format on headphones to allow the rotation of the sound field according to the listener's orientation, as static stereo tracks, or in surround format on speakers.

The music can be reproduced as either diegetic or non-diegetic using different spatial audio formats (Neumeyer, 2009). When using binaural techniques, the music will be perceived as coming from within the virtual space, thus diegetic. When instead using stereo or surround speaker playback, it will more likely be perceived as coming from the background.

*Table 12.1* Suggested audio techniques for audio layers

| *Layers* | *Binaural* | *Ambisonics* | *Surround* | *Stereo* |
|---|---|---|---|---|
| Dialogue | ✓ | | | |
| Foley | ✓ | | | |
| Ambiances | | ✓ | ✓ | ✓ |
| Music | ✓ | ✓ | ✓ | ✓ |

## Cave: *a case study*

An early version of the proposed sound system was implemented for a six-minute virtual reality co-located narrative piece called *Cave* (Layng et al., 2019; Herscher et al., 2019), that took place in a single, multi-user, virtual environment. The story involved one main character, one supporting character, and a virtual mammoth. The experience was prepared for a 30-member virtual audience, separated into two groups in the thrust stage format. The audience could see one another as avatars inside the virtual experience while the position and orientation of their heads were tracked using the headsets' IMUs so that the avatars' heads moved inside the virtual space accordingly. The VR experience was executed on standalone headsets for each audience member. The data and signals were sent between all devices.

### *Design choices*

The audio layers used in the experience consisted of dialogue, music, sound effects, and ambiance. The sound effects and the dialogue materials were treated as point source objects, connected to a visual component in the three-dimensional space. While the original content of these materials was in mono format, a dynamic binaural rendering engine tool transformed the sound objects into responsive stereo binaural streams, responding to the spatial relationship between characters and listeners. Additional distance cues were tuned separately for the foley, while the dialogue track had to maintain constant intensity in order to keep it intelligible. As the rendering was individual per device, each audience member was able to get a unique sound perspective into the scene.

The ambiance sounds were created from both a mix of stereo recordings and sound objects. Important and constant background sounds, such as the wind noise in the entrance to the cave or water stream, were positioned at diffuse point sources within the scene, while more general and sporadic sounds, such as drops of water, were rendered in stereo and not given specific spatial positions. The music track was exported as a non-spatialized stereo in order to give it a sense of separation from the dialogue and sound effect layers.

### *Reproduction system*

The sound reproduction system consisted of one single speaker with a subwoofer placed in the middle of the stage, and a prototype nearfield open-ear device, which allowed transparent hearing. The devices were mounted in front of the ears and were oriented to project toward the entrance of the ear canal.

The speaker unit was used mostly for the sound effects of the virtual mammoth, and it was made sure that the physical position of the speaker matched the virtual position of the mammoth's avatar as seen by all audience members. In this implementation, one speaker was sufficient because the visual object for which the sound effects were rendered was static. With more moving elements, more speakers would be necessary to ensure proper localization of the sounds chosen to come from the loudspeakers. A system-wide calibration was performed to achieve a proper blend between the nearfield devices and the loudspeakers and to ensure that the levels would be comfortable for each member of the audience.

### Discussion

In the presented case study of sound design and reproduction for immersive co-located virtual reality theatre, the cumulative effects of real-world elements (including seating, networking, and delivery of visual elements of virtual reality) and the implementation of a hybrid reproduction system delivered an effective sound experience for this shared virtual art piece.

The implementation described in the case study was limited to a single speaker and subwoofer. Adding more speakers surrounding the audience may further improve the immersion and allow the reproduction of more layers of audio other than sound effects, e.g., music or ambiance. Playing music tracks through the loudspeakers could indeed help with better separating the background music and the dialogues. We also noticed that some of the instrument tracks were perceived as coming from within the scene even though they were rendered in stereo.

Our implementation did not take into account the acoustics of the performance space due to the production limits. This resulted in sounds played through the speakers having different acoustic characteristics than the binaural layer. Furthermore, informal investigation revealed that even though the audience could see and hear one another inside the experience, their voices were not fully perceived as though they were coming from the same space as the action. Placing microphones above the audience may solve that problem. The sound from microphones would be processed in real-time through the same reverb processor used within the experience in order to ensure consistency between the sounds of the real-world natural environment and virtual scene.

Another challenge we encountered during production was the asynchronous playback between speakers and near-field devices. Even a small delay would cause perceptible distortions of those sounds played through both systems. We solved this problem by removing all of the transient sounds from the speaker playback, leaving only the sound effects of a longer temporal structure.

### Conclusion

This section presented a discussion of the principles, factors, and limitations of the sound-design theory related to the field of co-located narrative VR experiences. A first draft of this theory, reviewing technical challenges and proposing a solution based on hybrid reproduction systems, has been derived from practical experiences within prototype productions. The experience with the production of *Cave* is discussed as a platform where some of these principles were investigated and addressed to achieve insights that inform the authors' proposed framework.

Having this base to work upon, future empirical data will help validate, sharpen, and define the guidelines that may drive the creative choices of VR sound designers. It is reasonable to expect that, in the near future, technological advances are likely to affect and update the current conversation.

## References

Allen, J. B., & Berkley, D. A. (1979). Image method for efficiently simulating small-room acoustics. *Journal of the Acoustical Society of America*, 65(4), 943–950. https://doi.org/10.1121/1.382599

Andéol, G., Savel, S., & Guillaume, A. (2014). Perceptual factors contribute more than acoustic factors to sound localization abilities with virtual sources. *Frontiers in Neuroscience*, 8(December), 1–17. https://doi.org/10.3389/fnins.2014.00451

Bailey, W., & Fazenda, B. M. (2018). The effect of visual cues and binaural rendering method on plausibility in virtual environments. In *Proceedings of the 144th AES Convention.*

Barron, M., & Lee, L. J. (1988). Energy relations in concert auditoriums. I. *Journal of the Acoustical Society of America*, 84(2), 618–628. https://doi.org/10.1121/1.396840

Barron, M., & Marshall, A. H. (1981). Spatial impression due to early lateral reflections in concert halls: The derivation of a physical measure. *Journal of Sound and Vibration*, 77(2), 211–232. https://doi.org/10.1016/S0022-460X(81)80020-X

Begault, D. R. (1999). Auditory and non-auditory factors that potentially influence virtual acoustic imagery. In *Audio Engineering Society Conference: 16th International Conference: Spatial Sound Reproduction* (pp. 13–26).

Begault, D. R., Wenzel, E. M., & Anderson, M. R. (2001). Direct comparison of the impact of head tracking, reverberation, and individualized head-related transfer functions on the spatial perception of a virtual speech source. *Journal of the Audio Engineering Society*, 49(10), 904–916. https://doi.org/10.1109/ASPAA.1999.810884

Berger, C. C., Gonzalez-Franco, M., Tajadura-Jiménez, A., Florencio, D., & Zhang, Z. (2018). Generic HRTFs may be good enough in virtual reality. Improving source localization through cross-modal plasticity. *Frontiers in Neuroscience*, 12(February). https://doi.org/10.3389/fnins.2018.00021

Blauert, J., & Butler, R. A. (1985). Spatial hearing: The psychophysics of human sound localization by Jens Blauert. *The Journal of the Acoustical Society of America.* https://doi.org/10.1121/1.392109

Boone, M. M., Verheijen, E. N., & Van Tol, P. F. (1995). Spatial sound-field reproduction by wave-field synthesis. *Journal of the Audio Engineering Society*, 43(12), 1003–1012.

Brinkman, W.-P., Hoekstra, A. R. D., & van Egmond, R. (2015). The effect of 3D audio and other audio techniques on virtual reality experience. *Annual Review of Cybertherapy and Telemedicine*, 13(44).

Brungart, D. S. (1998). Control of perceived distance in virtual audio displays. In *Proceedings of the 20th Annual International Conference of the IEEE Engineering in Medicine and Biology Society. Vol. 20 Biomedical Engineering Towards the Year 2000 and Beyond (Cat. No.98CH36286)* (vol. 3(3), pp. 1101–1104). https://doi.org/10.1109/IEMBS.1998.747063

Brungart, D. S., Simpson, B. D., & Kordik, A. J. (2005). The detectability of headtracker latency in virtual audio displays. In *Proceedings of the 11th International Conference on Auditory Display (ICAD2005)* (pp. 37–42).

Brungart, D. S., Simpson, B. D., Mckinley, R. L., Kordik, A. J., Dallman, R. C., & Ovenshire, D. A. (2004). The interaction between head-tracker latency, source duration, and response time in the localization of virtual sound sources. In *Proceedings of ICAD 04-Tenth Meeting of the International Conference on Auditory Display.*

Churchill, E. F., & Snowdon, D. (1998). Collaborative virtual environments: An introductory review of issues and systems. *Virtual Reality*, 3(1), 3–15.

Colavita, F. B. (1974). Human sensory dominance. *Perception & Psychophysics*, 16(2), 409–412.

De Sena, E., Haciihabibołlu, H., Cvetković, Z., & Smith, J. O. (2015). Efficient synthesis of room acoustics via scattering delay networks. *IEEE Transactions on Audio, Speech and Language Processing*, 23(9), 1478–1492. https://doi.org/10.1109/TASLP.2015.2438547

Draper, J. V., Kaber, D. B., & Usher, J. M. (1998). Telepresence. *Human Factors*, 40(3), 354.

Engel, I., Henry, C., Garí, S. V. A., Robinson, P. W., Poirier-Quinot, D., & Picinali, L. (2019). Perceptual comparison of Ambisonics-based reverberation methods in binaural listening. *EAA Spatial Audio Signal Processing Symposium*, 121–126. https://doi.org/10.25836/sasp.2019.11

Felnhofer, A., Kothgassner, O. D., Schmidt, M., Heinzle, A. K., Beutl, L., Hlavacs, H., & Kryspin-Exner, I. (2015). Is virtual reality emotionally arousing? Investigating five emotion inducing virtual park scenarios. *International Journal of Human Computer Studies*, 82, 48–56. https://doi.org/10.1016/j.ijhcs.2015.05.004

Gerzon, M. A. (1973). Periphony: With-height sound reproduction. Journal of the Audio EngineeringSociety, 21(1), 2–10.

Gibbs, T. (2007). *The Fundamentals of Sonic Art and Sound Design*. Ava Publishing.
Gochfeld, D., Brenner, C., Layng, K., Herscher, S., Defanti, C., Olko, M., Shinn, D., Riggs, S., Fernández-Vara, C., & Perlin, K. (2018). Holojam in Wonderland: Immersive mixed reality theater. *Leonardo*, 51(4), 362–367.
Gospodarek, M. (2024). *Acoustic and Perceptual Factors Affecting Plausibility in Sound Design for Audio Augmented Reality Experiences* (Doctoral dissertation, New York University).
Gospodarek, M., Genovese, A., Dembeck, D., Brenner, C., Roginska, A., & Perlin, K. (2019). Sound design and reproduction techniques for co-located narrative VR experiences. In *Proceedings of the 147th AES Convention*.
Guezenoc, C., & Séguier, R. (2018). HRTF individualization: A survey. In *Proceedings of the 145th AES Convention*.
Gupta, R., Ranjan, R., He, J., & Gan, W.-S. (2018). On the use of closed-back headphones for active hear-through equalization in Augmented Reality applications. In *AES International Conference on Audio for Virtual and Augmented Reality*.
Gutierrez-Parera, P., Lopez, J. J., & Aguilera, E. (2015). On the influence of headphone quality in the spatial immersion produced by Binaural Recordings. In *Proceedings of the 138th AES Convention*.
Herscher, S., DeFanti, C., Vitovitch, N. G., Brenner, C., Xia, H., Layng, K., & Perlin, K., (2019). CAVRN: An exploration and evaluation of a collective audience virtual reality nexus. In *UIST 2019: 32nd ACM User Interface Software and Technology Symposium*.
Horowitz, S., & Looney, S. R. (2014). *The Essential Guide to Game Audio: The Theory and Practice of Sound for Games*. Routledge.
Jot, J.-M., & Chaigne, A. (1991). Digital delay networks for designing artificial reverberators. *Audio Engineering Society Convention*, 90.
Kaplanis, N., Bech, S., Jensen, S. H., & van Waterschoot, T. (2014). Perception of reverberation in small rooms: A literature study. In *AES 55th International Conference on Spatial Audio. Audio Engineering Society*.
Karjalainen, M., Huopaniemi, J., & Välimäki, V. (1995). Direction-dependent physical modeling of musical instruments. *International Congress on Acoustics (ICA\'95)*, 3(5), 561–563.
Kobayashi, M., & Ueno, K. (2015). The effects of spatialized sounds on the sense of presence in auditory virtual environments: A psychological and physiological study abstract. *Presence*, 24(2), 163–174. https://doi.org/10.1162/PRES_a_00226
Larsson, P., Västfjäll, D., Kleiner, M., Vastfjall, D., & Kleiner, M. (2001). Ecological acoustics and the multi-modal perception of rooms: Real and unreal experiences of auditory-visual virtual environments. *International Conference on Auditory Display*, 2014, 245–249.
Layng, K., Perlin, L., Herscher, S., Brenner, C., & Meduri, T. (2019). CAVE: Making collective virtual narrative. Best paper award. *Leonardo*, 52(4), 349–356.
Lee, K. S., Abel, J. S., Välimäki, V., Stilson, T., & Berners, D. P. (2012). The switched convolution reverberator. *AES: Journal of the Audio Engineering Society*, 60(4), 227–236.
Lindau, A., Erbes, V., Lepa, S., Maempel, H.-J., Brinkmann, F., & Weinzierl, S. (2014). A spatial audio quality inventory (SAQI). *Acta Acustica United with Acustica*, 100(5), 984–994.
Lindau, A., & Weinzierl, S. (2012). Assessing the plausibility of virtual acoustic environments. *Acta Acustica United with Acustica*, 98(5), 804–810.
Martin, A., Jin, C., & Andre, V. S. (2009). Psychoacoustic evaluation of systems for delivering spatialized augmented-reality audio. *Journal of the Audio Engineering Society*, 57(12), 1016–1027.
Martin, V., Viaud-Delmon, I., & Warusfel, O. (2022). Effect of environment-related cues on auditory distance perception in the context of audio-only augmented reality. *Applied Sciences (Switzerland)*, 12(1). https://doi.org/10.3390/app12010348
Mendonça, C., Campos, G., Dias, P., Vieira, J., Ferreira, J. P., & Santos, J. A. (2012). On the improvement of localization accuracy with non-individualized HRTF-based sounds. *AES: Journal of the Audio Engineering Society*, 60(10), 821–830.
Miles, R. N. (1984). Sound field in a rectangular enclosure with diffusely reflecting boundaries. *Journal of Sound and Vibration*, 92(2), 203–226. https://doi.org/10.1016/0022-460X(84)90556-X
Monson, B. B., Hunter, E. J., & Story, B. H. (2012). Horizontal directivity of low- and high-frequency energy in speech and singing. *The Journal of the Acoustical Society of America*, 132(1), 433–441.

Mueller-Tomfelde, C. (2002). Hybrid sound reproduction in audio-augmented reality. In *22nd International Conference on Virtual, Synthetic and Entertainment Audio*.
Neidhardt, A., & Knoop, N. (2017). Binaural walk-through scenarios with actual self-walking using an HTC Vive. In *Proceedings of the DAGA*, April (pp. 283–286).
Neidhardt, A., Tommy, A. I., & Pereppadan, A. D. (2018). Plausibility of an interactive approaching motion towards a virtual sound source based on simplified BRIR sets. In *Proceedings of the 144th AES Convention*.
Neumeyer, D. (2009). Diegetic/nondiegetic: A theoretical model. *Music and the Moving Image*, 2(1), 26–39.
Nosal, E.-M., Hodgson, M., & Ashdown, I. (2004). Improved algorithms and methods for room sound-field prediction by acoustical radiosity in arbitrary polyhedral rooms. *The Journal of the Acoustical Society of America*, 116(2), 970–980. https://doi.org/10.1121/1.1772400
Nowak, J., & Klockgether, S. (2017). Perception and prediction of apparent source width and listener envelopment in binaural spherical microphone array Auralizations. *The Journal of the Acoustical Society of America*, 142(3), 1634–1645. https://doi.org/10.1121/1.5003917
Olive, S. E., & Toole, F. E. (1989). Detection of reflections in typical rooms. *AES: Journal of the Audio Engineering Society*, 37(7–8), 539–553.
Otondo, F., & Rindel, J. H. (2005). A new method for the radiation representation of musical instruments in auralizations. *Acta Acustica United with Acustica*, 91(5), 902–906.
Paterson, J., & Kadel, O. (2019). Immersive audio post-production for 360° video: Workflow case studies. In *AES International Conference on Immersive and Interactive Audio*.
Picinali, L., Wallin, A., Levtov, Y., & Poirier-Quinot, D. (2017). Comparative perceptual evaluation between different methods for implementing reverberation in a binaural context. *Audio Engineering Society Convention*, 142.
Pike, C., Melchior, F., & Tew, T. (2014). Assessing the plausibility of non-individualised dynamic binaural synthesis in a small room. In *AES 55th International Conference: Spatial Audio* (pp. 1–8).
Pompei, A., Sumbatyan, M. A., & Todorov, N. F. (2009). Computer models in room acoustics: The ray tracing method and the auralization algorithms. *Acoustical Physics*, 55, 821–831.
Pörschmann, C., & Wiefling, S. (2015). Perceptual aspects of dynamic binaural synthesis based on measured omnidirectional room impulse responses. In *Proceedings of the International Conference on Spatial Audio – ICSA*.
Postma, B. N. J., & Katz, B. F. G. (2016). Perceptive and objective evaluation of calibrated room acoustic simulation auralizations. *The Journal of the Acoustical Society of America*, 140(6), 4326–4337. https://doi.org/10.1121/1.4971422
Postma, B. N. J., Demontis, H., & Katz, B. F. G. (2017). Subjective evaluation of dynamic voice directivity for auralizations. *Acta Acustica United with Acustica*, 103, 181–184. https://doi.org/10.3813/AAA.919045
Postma, B. N. J., & Katz, B. F. G. (2017). The influence of visual distance on the room-acoustic experience of auralizations. *The Journal of the Acoustical Society of America*, 142(5). https://doi.org/10.1121/1.5009554
Rämö, J., & Välimäki, V. (2014). An allpass hear-through headset. In *22nd European Signal Processing Conference (EUSIPCO)* (pp. 1123–1127).
Rappin, C., Palacino, J., Rueff, P., Feichter, L., & Paquier, M. (2023). Latency detection threshold of head-tracking for different head rotation speeds in binaural rendering. In *Forum Acusticum 2023*.
Rindel, J. H., Otondo, F., & Christensen, C. L. (2004). Sound source representation for auralization. In *International Symposium on Room Acoustics: Design and Science*. Hyogo, Japan.
Robotham, T., Rummukainen, O. S., & Habets, E. A. P. (2019). Towards the perception of sound source directivity inside six-degrees-of-freedom virtual reality. In *5th International Conference on Spatial Audio ICSA*, September (pp. 1–8).
Rumsey, F. (2007). Basic psychoacoustics for surround recording. In *AES 22nd UK Conference* (pp.1–9). University of Surrey, Guildford, UK.
Sandvad, J. (1996). Dynamic aspects of auditory virtual environments. In *100th Audio Engineering Society Convention*.
Savioja, L., Huopaniemi, J., Lokki, T., & Väänänen, R. (1999). Creating interactive virtual acoustic environments. *Journal of the Audio Engineering Society*, 47(9), 675–705.

Savioja, L., & Svensson, U. P. (2015). Overview of geometrical room acoustic modeling techniques. *The Journal of the Acoustical Society of America*, 138(2), 708–730. https://doi.org/10.1121/1.4926438

Scharine, A. A., Cave, K. D., & Letowski, T. R. (1999). Auditory perception and cognitiveperformance. In C. E. Rash, M. B. Russo, T. R. Letowski, & E. T. Schmeisser (Eds.), *Helmet Mounted Displays – Sensation, Perception and Cognitive Issues* (pp. 391–490). Fort Rucker, Alabama: U.S. Army Aeromedical Research Laboratory.

Serafin, S., Geronazzo, M., Erkut, C., Nilsson, N. C., & Nordahl, R. (2018). Sonic interactions in virtual reality: State of the art, current challenges, and future directions. *IEEE Computer Graphics and Applications*, 38(2), 31–43. https://doi.org/10.1109/MCG.2018.193142628

Shabtai, N. R., Behler, G., Vorländer, M., & Weinzierl, S. (2017). Generation and analysis of an acoustic radiation pattern database for forty-one musical instruments. *The Journal of the Acoustical Society of America*, 141(2), 1246–1256. https://doi.org/10.1121/1.4976071

Shinn-Cunningham, B. (2000). Distance cues for virtual auditory space. In *Proceedings of the First IEEE Pacific-Rim Conference on Multimedia*(pp. 227–230).

Slater, M. (2009). Place illusion and plausibility can lead to realistic behaviour in immersive virtual environments. *Philosophical Transactions of the Royal Society B: Biological Sciences*, 364(1535), 3549–3557.

Slater, M., & Wilbur, S. (1997). A framework for immersive virtual environments (FIVE): Speculations on the role of presence in virtual environments MEL. *Presence: Teleoperators & Virtual Environments*, 6, 606–616.

Southern, A., & Murphy, D. (2009). Low complexity directional sound sources for finite Difference Time Domain room acoustic models. *Audio Engineering Society Convention*, 126(3), 1126–1135.

Takanen, M., Hiipakka, M., & Pulkki, V. (2012). Audibility of coloration artifacts in HRTF filter designs. In *AES 45th International Conference: Applications of Time-Frequency Processing in Audio*.

Tatlow, S. (2024). Authenticity in sound design for virtual reality. In J. Cook, A. Kolassa, A. Robinson, & A. Whittaker (Eds.), *History as Fantasy in Music, Sound, Image, and Media* (1st ed., pp. 161–183). Routledge. https://doi.org/10.4324/9781003291725-11

Thery, D., Poirier-Quinot, D., Postma, B. N. J., & Katz, B. F. G. (2017). Impact of the visual rendering system on subjective auralization assessment in VR. In *Lecture Notes in Computer Science (Including Subseries Lecture Notes in Artificial Intelligence and Lecture Notes in Bioinformatics), 10700 LNCS* (pp. 105–118). https://doi.org/10.1007/978-3-319-72323-5_7

Väljamäe, A., Larsson, P., Västfjäll, D., & Kleiner, M. (2004). Auditory presence, individualized head-related transfer functions, and illusory ego-motion in virtual environments. In *Proceedings of the Seventh Annual Workshop Presence 2004, December 2013* (pp. 141–147).

Wagner, I., Mccall, R., Morrison, A., & Valle, M. (2009). On the role of presence in mixed reality. *Presence*, 18(4), 249–276.

Wallach, H. (1940). The role of head movements and vestibular and visual cues in sound localization. *Journal of Experimental Psychology*, 27(4), 339–368. https://doi.org/10.1037/h0054629

Wallach, H., Newman, E. B., & Rosenzweig, M. R. (1973). The precedence effect in sound localization (tutorial reprint). *AES: Journal of the Audio Engineering Society*, 21(10), 817–826.

Weinreich, G. (1997). Directional tone color. *The Journal of the Acoustical Society of America*, 101(4), 2338–2346.

Werner, S., Klein, F., Mayenfels, T., & Brandenburg, K. (2016). A summary on acoustic room divergence and its effect on externalization of auditory events. In *2016 8th International Conference on Quality of Multimedia Experience, QoMEX 2016*. https://doi.org/10.1109/QoMEX.2016.7498973

Werner, S., Klein, F., Neidhardt, A., Sloma, U., Schneiderwind, C., & Brandenburg, K. (2021). Creation of auditory augmented reality using a position-dynamic binaural synthesis system – technical components, psychoacoustic needs, and perceptual evaluation. *Applied Sciences (Switzerland)*, 11(3), 1–20. https://doi.org/10.3390/app11031150

Wersényi, G. (2009). Effect of emulated head-tracking for reducing localization errors in virtual audio simulation. *IEEE Transactions on Audio, Speech and Language Processing*, 17(2), 247–252. https://doi.org/10.1109/TASL.2008.2006720

Whittington, W. (2007). *Sound Design and Science Fiction*. University of Texas Press.
Wightman, F. L., & Kistler, D. J. (2018). Factors affecting the relative salience of sound localization cues. In R. H. Gilkey & T. R. Anderson (Eds.), *Binaural and Spatial Hearing in Real and Virtual Environments*(pp. 1–24). Mahwah, NJ: L. Erlbaum.
Yadav, M., Cabrera, D., Miranda, L., Martens, W. L., Lee, D., & Collins, R. (2013). Investigating auditory room size perception with autophonic stimuli. In *Proceedings of the 135th Audio Engineering Society Convention*.
Yairi, S., Iwaya, Y., & Suzuki, Y. (2007). Estimation of detection threshold of system latency of virtual auditory display. *Applied Acoustics*, 68(8), 851–863. https://doi.org/10.1016/j.apacoust.2006.12.005
Zahorik, P., Brungart, D. S., & Bronkhorst, A. W. (2005). Auditory distance perception in humans: A summary of past and present research. *Acta Acustica United with Acustica*, 91(3), 409–420.
Zhao, J., Zhang, B., Yan, Z., Wang, J., & Fei, Z. (2017). A study on the factors affecting audio-video subjective experience in virtual reality environments. In *International Conference on Virtual Reality and Visualization, ICVRV 2017* (pp. 303–306). https://doi.org/10.1109/ICVRV.2017.00070

Chapter 13

# Take me out

## Moving virtual reality outside academic laboratories to medical clinics and museums

*Rolf Nordahl and Stefania Serafin*

Virtual reality technologies are nowadays portable and mobile, in such a way that they can be taken outside laboratory settings and into medical clinics and cultural institutions. This presents several possibilities as well as challenges. In this paper, we present two research projects conducted at the Multisensory Experience Lab at Aalborg University in Copenhagen. Such projects show how virtual reality can be used in clinical and cultural settings to help people in need or to facilitate access to cultural heritage. After presenting these projects, we provide some design guidelines and learning outcomes from our experiences of using VR in the wild.

### Introduction

The use of virtual reality (VR) technologies for health and culture is certainly not new. In the context of health research, Albert Skip Rizzo, one of the pioneers on the use of VR for mental health, wonders if clinical VR is ready for prime time [1]. The article argues how, since the 1990s, several VR applications targeting cognitive, psychological, motor, and functional impairments across a wider range of clinical health conditions have been developed. Moreover, the growing recognition of the potential value of gamification in health care and the field of Games for Health is evidenced by the appearance of scientific journals and conferences focused on this topic, in addition to evolving scientific literature (see for example [2–4]).

The authors of this chapter had their first VR experience in the Fall of 2003. At that time, Professor Erik Granum from Aalborg University was coordinating a European project called BENOGO (Being There Without Going, IST-2001-39184).

The project's main goal was to investigate and develop novel camera-based technologies into an innovative mediation system that allowed close to photorealistic 3D real-time visualization of real and possibly known places for a moving observer. The project's participants captured several locations in Europe (Venice, the botanical garden in Prague, see Figure 13.1) and reproduced them in VR. The visualization provided a high degree of impressiveness and supported new empirical studies of the feeling of presence. The project acted as a melting pot for mixing novel technologies that allowed real-time visualization for a moving observer of recorded real places. The project also developed and explored new recording and visualization technologies enabling people to experience a sense of presence [5] in simulated real places without having to go there (this is where the BE-NO-GO name came from, an abbreviation of 'being there without going'). Through augmentation, visually and auditorily, a sense of life was added, including objects for interaction.

DOI: 10.4324/9780429297526-15

*Figure 13.1* Some screenshots from the Benogo project: Left: Screenshot from Venice captured with 360° cameras. Top right: Prague botanical garden. Bottom right: Sandals with pressure sensors

Projection technologies to display such captured places ranged from head-mounted displays to large screens including the six-sided CAVE present at the time at Aalborg University. Our goal in the project was to develop interactive sound effects and soundscape rendering for virtual reality and to measure the sense of presence [6, 7]. In order to ground the participants in the virtual experience, we created interactive soundscapes that also included the rendering of self sounds of the user. In order to capture the footsteps of the user in real time, we developed some sandals embedded with electronics (see Figure 13.1, bottom right). Investigations on the role of sound in creating sense of place showed that sound made the participants move and explore the spaces more [6, 7].

Almost 20 years later, VR technologies have become portable and affordable, in such a way that they can be taken outside laboratory settings and used in everyday life. In this chapter we present two case studies regarding the use of VR for health and for culture. We then discuss design principles and considerations on the use of such technologies outside laboratory settings.

## VR for health

Applications of VR in the health sector are rapidly growing, thanks to the development of user-friendly software platforms as well as the democratization of hardware solutions. VR technologies have a long history in mental health and rehabilitation [8, 9]. A common example is the use of VR for exposure therapy [10], where confronting a difficult situation in VR has been shown to help confront the same situation in real life. In the field of physical health, VR has been used for exercise, for example to motivate elderly in retirement homes to move more [11, 12], or to provide motivating physical activities during the recent pandemic [13]. In a recent collaboration started with the Copenhagen Hearing and Balance Center, part of the largest hospital in Denmark, Rigshospitalet in Copenhagen, we have been investigating how VR can be used to train spatial awareness of children with hearing impairment. Hearing technologies such as hearing aids and cochlear implants are allowing individuals with mild and severe hearing loss to be able to reacquire hearing capabilities. However, after receiving hearing aids or cochlear implants, a training program needs to be in place in order for the brain to readapt to the novel sensory capabilities achieved either by amplifying existing reduced frequencies (hearing aids) or by electrical stimulation (cochlear implants) [14].

Despite the tremendous progress, individuals with hearing loss still report difficulties in situations where several speakers are present, for example, during a cocktail party or in noisy environments [15]. VR technologies can be used to train individuals in such difficult situations [16]. This was done during the Listen Again project, a two-year partnership project started in 2021 between the Center for Hearing and Balance (CHBC) at Rigshospitalet, and our Multisensory Experience Laboratory (ME-Lab) at Aalborg University. The project aimed to use VR technologies and gamification to train hard-of-hearing children in listening and spatial awareness.

## Methods

### *Participatory design*

The VR solution was developed in collaboration with an association consulting children with hearing loss and their families, in a co-creative design approach. Specifically, four children with hearing loss were involved in the design. Following up on the co-design workshops, it was discovered that children with hearing loss perform well in relatively static scenarios such as when sitting in a classroom, but they still have difficulties in unpredictable situations such as outdoors. For these reasons two scenarios were chosen to be simulated in VR: a schoolyard and an open-school scenario simulating the National Music Museum of Denmark. In the VR experience, there are aspects of gamification. For example, trophies are earned, which optimizes the children's motivation to train. Table 13.1 shows the complete list of trophies that can be earned in the two scenarios. Figure 13.1 illustrates examples of the two scenarios. Figure 13.2 shows an overview of the virtual environments.

## The VR intervention

The application was developed using Unity 3D and Blender for the Meta Quest 2 Head Mounted Display (HMD). Table 13.1 lists all the trophies' names and the related actions in the videogame.

## Evaluating the VR solution

In total, 22 children were recruited, aged 8–12 years. Sixteen of the children were fitted with bilateral cochlear implants, five of the children were fitted with bilateral hearing aids and one child with bimodal hearing solution, i.e., one hearing aid and one cochlear implant. For practical reasons (availability of HMD displays and keeping a relatively low number of participants in the focus groups sessions), the children were divided into two groups. Group 1 represented the eldest kids and group 2 the youngest. All children were asked to train for two months three times a week. They were instructed to use

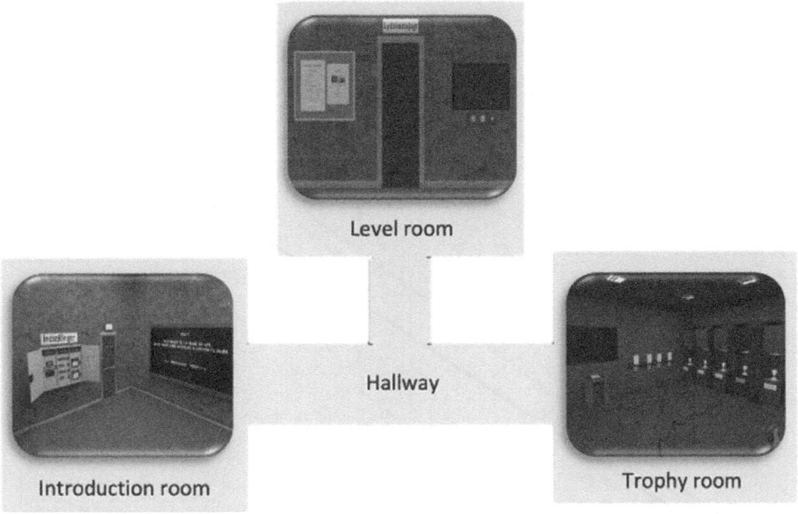


*Figure 13.2* Overview of the virtual environments: User starts in the introduction room. The level room consists of two doors, each leading to one of the two levels, an informative board, and a screen for each level. The trophy room consists of a collection of the achievements collected by the user.

*Table 13.1* Names of the 19 trophies in the game and their description

| *Trophy name* | *Description* |
|---|---|
| Quiz Master | Answers correctly to all questions at the museum |
| Placing | Places in best position at least three times in museum visit |
| Lighting Fast | Find three correct sounds in under one minute |
| Super Hearing | Selects only correct boxes in treasure hunt |
| Detective | Delivers all missing items to museum |
| The Musician | Plays all the instruments in the museum |
| Soccer Star | Scores a goal with the ball |
| Super Trophy | Achieves all trophies |
| Just Started | Plays 1 hour |
| Good Start | Plays 4 hours |
| All Nighter | Plays 8 hours |
| First Museum Visit | Completes the museum for the first time |
| First Sound Treasure Hunt | Completes the sound treasure hunt for the first time |
| Experienced Museum Visitor | Achieves difficulty level 4 in the museum |
| Experienced Treasure Hunter | Achieve difficulty level 4 in Sound Treasure Hunt |
| Museum Expert | Achieve difficulty level 7 in the museum |
| Treasure Hunter Expert | Achieve difficulty 7 in Sound Treasure Hunt |
| Museum Master | Achieve the highest difficulty in the museum |
| Treasure Hunter | Achieve the highest difficulty level in the audio treasure hunt |

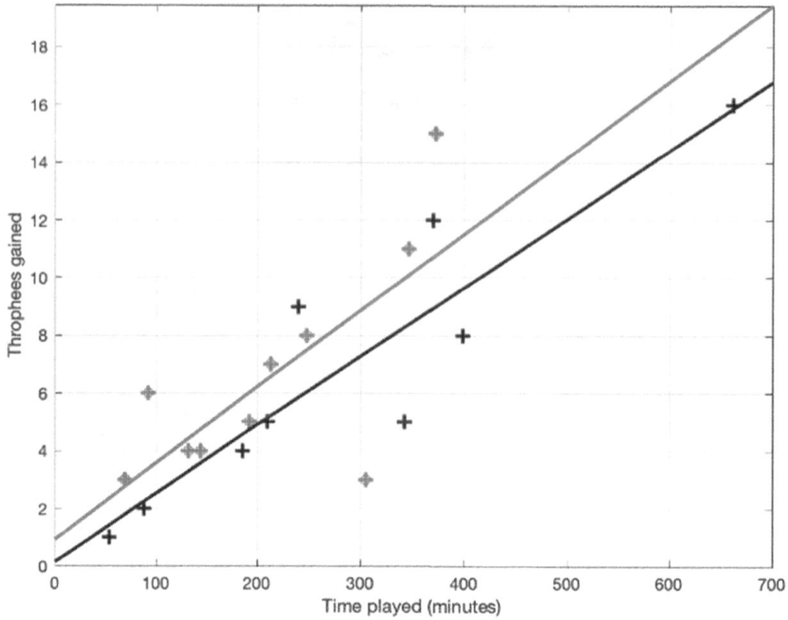


*Figure 13.3* Time used versus number of trophies gain for group 1 (black) and group 2 (grey) for the Listen Again project. As expected, the number of trophies won is correlated to the time spent playing.

approximately 30 minutes in each session and try to complete through one level each session, in order to start on a new level after every session. After the third and sixth weeks, they were sent a short questionnaire to the parents' email address, to be filled online by the children, with their thoughts regarding the game. After eight weeks of training, children met for a second focus group at the hospital and talked together about the game and their experiences since the last time. Figure 13.3 shows the amount of time played during the whole period versus the number of trophies won by the first group (black) and the second group (grey). As desired, there is a positive correlation between time spent in playing the game and amount of trophies won (correlation coefficient $r = 0.907$ for group 1 and $r = 0.7483$ for group 2).

The results from the questionnaire showed that after three weeks, 11 out of the 22 kids thought the game was fun, one thought it was not fun, and nine did not know. The situation, however, changed after six weeks, When nine kids answered it was not fun, one answered it was, and 12 answered they did not know. This clearly indicates that the training period should have been reduced to one month. In group 1, eight out of the nine children played for one hour (just started trophy), while all 12 children from group 2 played for one hour.

The best performer was in group 1 and acquired all trophies beside detective (delivers all missing items to museum) and super trophies (acquiring all trophies). He spent 11 hours and 2 minutes in VR. The worst performer completed only the first museum visit and spent overall 54 minutes in VR.

### *Discussion*

The results of our study show that children experience VR combined with gamification as a valuable technology to train their spatial awareness and hearing skills. However, gamified experiences need to become more engaging in order to keep the children motivated during the training period. The Listen Again project showed the potential of VR to improve training of hearing skills for young individuals with hearing impairment. More work is needed in several directions, for example, to understand how to quantitatively evaluate the training and how to personalize the design of the VR experience.

## VR for cultural heritage

The conservation, restoration, and maintenance of rare and precious musical instruments is currently problematic and will become increasingly so in the future.

Most instruments in museum collections are already out of playing condition, and they are deteriorating to the extent that their unique sounds will eventually be lost. To avoid any damage, most instruments at music museums are contained in glass cabinets; this aspect has the obvious disadvantage of preventing users from experiencing the feeling of touching and playing them. Curators are well aware of the problems associated with restoration and maintenance.

The emergence of digital technology has profoundly revolutionized many areas, including preservation and restoration practice [17, 18]. A defining feature that separates VR from traditional audiovisual media is the ability to give users a compelling sensation of being somewhere else than where they are physically located. This sensation is sometimes referred to as the place illusion or presence [19]. This ability to transport users to foreign and fantastic virtual places, also imply that VR can allow users to directly experience historical and cultural events, sites, and artifacts [20]. Moreover, previous work indicates that the combination of audiovisual VR and physical props serving as proxies for virtual objects can increase realism [21] and presence [22]. For that reason, the proposed work will explore how VR installations at the Danish Music Museum can be used to give audiences authentic experiences of playing instruments in correct historical contexts.

This is also the case at the Music Museum in Copenhagen, part of the Danish National Museum, which has several ancient and precious musical instruments on display. The project MOREMI (MOdelling and REconstructing Musical Instruments), supported by the Danish foundation called Augustinus (https://augustinusfonden.dk), aims to address these challenges by answering the question: How can we preserve and disseminate Danish musical cultural heritage using digital technologies? Museums usually function as places of informal learning, and attendance to this process is usually voluntary and spontaneous [23]. While traditional exhibitions often limit visitors to passive observation, VR allows for close inspection and interaction with artifacts, inviting visitors to actively participate and control their learning process [24]. The general attitude of museum professionals when considering the use of VR technology they found that this recently democratized technology had created a fundamental change, necessitating a rethinking of traditional concepts [24]. Trunfio and colleagues produced similar findings, illuminating the growing prevalence of mixed reality applications in museum settings. Their study not only suggests an overall increase in visitor satisfaction and experience but also posits the emergence of novel visitor profiles [25].

Among the different artifacts designed for the Danish Music Museum, we recently created a VR replica of a glass harmonica. As musical instruments are inherently interactive in nature and difficult to fully appreciate through visual inspection alone, this presents challenges for the museum. In this section, we provide a comprehensive overview of the museum installation project, covering all aspects of its development and execution.

Glass instruments have a rich history that can be traced back to the early days of glass-making. Among the earliest types of glass instruments were those that produced distinct tones through the striking of differently sized glass bowls. An early account of musical glasses appears in 1741 with the invention of the glass harp, an instrument invented by Richard Puckeridge [26]. The glass harp consists of stemmed glasses filled with water, altering the frequencies of the sounds produced: more water means higher frequencies, and, vice versa, less water means lower frequencies. It was not until 1762 that Benjamin Franklin introduced the glass harmonica in its modern form [27]. Franklin's design, heavily influenced by Puckeridge, uses glass bowls spun by a horizontal axle, with one side dipping into a trough of water. Sound was produced from exciting the bowls by touching the rim of the rotating glass with a wet finger [28]. The glass harmonica quickly gained popularity among artists such as Mozart and Beethoven, but association with negative events led to a decline in popularity during the 18th century [27]. In current times, what once was an instrument seemingly destined for permanence was by 1830 a forgotten instrument, a museum piece.

### *Digital technologies in museums settings*

As younger generations are more exposed to rich media and interactive content, this type of content is increasingly expected in recreational and educational facilities [29, 30]. The benefits of incorporating digital technologies into cultural heritage do not limit themselves to experiential matters. In [31], Chong et al. describe how various studies and efforts incorporate digital technologies as a preservation medium. Their findings reveal positive results for engaging and further motivating greater awareness of society's cultural content. Thus the author highlights virtual heritage as essential to reconstructing the past and safeguarding the current heritage digitally. In [24], the authors acknowledge the need to acquire new knowledge and strategies to implement and evaluate these technologies in museum settings. Their paper offers a holistic assessment of the current use of VR through interviews with museum professionals, who found that the most common use of VR is to allow visitors to experience 'impossible' spaces or time travel, providing historical context for the exhibits. This is reflected in the multitude of works describing VR experiences depicting historical sites for museums [32–35].

Although research suggests the advantages of virtual cultural heritage, the impact of VR applications on the spatial and social experience of museums remains a concern for museum professionals. As the digital space in VR remains concealed from those outside the headset, the process of people-watching suddenly lacks meaning. Symptomatic of this is reflexive concerns about how the visitor might appear to others if they inhabit this digital space themselves [24, 36]. This has sparked efforts to explore methods for facilitating co-located asymmetric interaction between head-mounted display (HMD) users and bystanders (*non-HMD users*) [37].

### *Virtual reality musical instruments*

Digital musical instruments (DMIs) are defined as systems enabling gestural control of sound production through controllers mapping gestures to sound parameters [38]. As the focus on VR systems has grown, the field of musical instruments for VR has expanded, and various authors have surveyed different aspects of this broad field. In [39], Serafin et al. surveyed VR musical instruments, distinguishing between virtual musical instruments (VMIs) and virtual reality musical instruments (VRMI), and proposed nine design principles addressing functional aspects and offering opportunities for natural and captivating experiences in VR. The importance of minimizing latency is further emphasized in a study presented in [40], which found that low latency is desirable for VRMIs, as expected. Furthermore, in [41] 3D interaction techniques for musical control in VEs are explored, and [42] provides an overview of recent musical work in VR. Certainly, the field of VRMIs has much to offer in terms of how to design well for multimodal VR, especially considering design for public spaces.

According to [31], VR has been recognized as a preservation medium. Consequently, there is an ongoing effort to create VRMIs for historical instruments. However, prior work indicates that virtual exploration of such instruments encounters challenges related to limitations of digital technology, preservation of cultural heritage, and accurate simulation of instrument acoustics [43–45]. When looking at VRMIs in museum contexts, these often suffer from a lack of formal evaluation within actual museum settings [46–49]. As discussed in [24, 31], VR's immersive capabilities effectively provide historical context to exhibited artifacts via narratives or representational virtual environments. Historical appreciation of forgotten instruments goes beyond narration, as it is closely intertwined with the interactive experience of playing them. Faithful replication of acoustic qualities, gestures, and movements can enhance understanding and appreciation.

### *Design and implementation*

Based on the notion that the glass harmonica transgresses its display case, the connection between the virtual and the real instrument plays a crucial role. The VR experience should augment the visitor experience by adding a layer of narrativity and historical context through the VE, in addition to faithfully replicating the acoustics and real interaction of playing the instrument. Taking into account the reduction in hardware maintenance and the ability to allow natural interaction, we used the hand tracking capabilities offered by the Meta Quest 2 headset, enabled through Oculus Integration for Unity.

### *Concept*

As in the case of the previously described Listen Again project, the design of the installation was collaboratively undertaken through co-design activities involving researchers and museum personnel. The research team took on various aspects of the development process, focusing on technical factors such as application development. Concurrently, the museum personnel, with their expertise in public engagement and pedagogy, provided valuable information on what would be effective within a public and educational setting. The physical installation was strategically placed facing the exhibited real glass harmonica. This arrangement ensured that, when visitors removed the HMD, they were

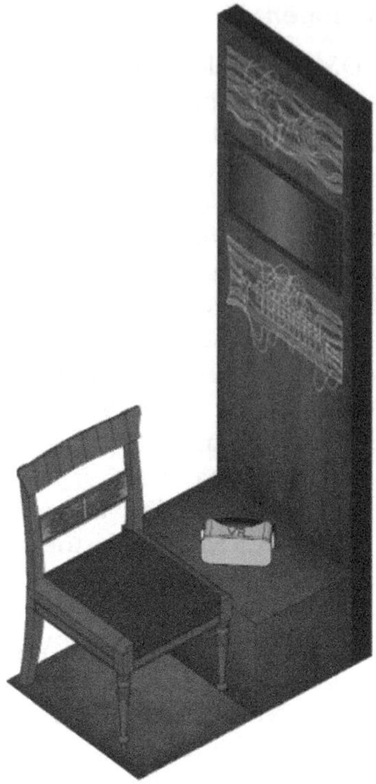


*Figure 13.4* Isometric 3D-graphic of the installation

immediately visually confronted with the real instrument. The installation was designed to aesthetically integrate with the surrounding exhibition space. A 3D sketch of the installation is shown in Figure 13.4. A screen was mounted on the wall, displaying a looped mute screen capture of the VR experience to accommodate bystanders.

The objective of the installation was to familiarize visitors with the historical background and interaction of the glass harmonica through an interactive VR experience. To achieve this, the initial step involved developing the interface and incorporating narrative elements.

### *Interface*

To facilitate an authentic gestural experience of playing the glass harmonica, the interaction has to be replicated true to its nature. First, the instrument affords control of the sound by touching the rim of the rotating glass bowls with the index finger, as shown in Figure 13.7, implemented using the *PokeInteractor* from Oculus Interaction SDK for Unity. Furthermore, the speed of rotation is set to an approximated fixed speed, opposite to user-controlled in an original instrument. The design of the virtual replica is illustrated in Figure 13.6, with the assigned notes displayed in Figure 13.5.

*Figure 13.5* Illustration of the interface with notes and corresponding frequency displayed in Hz

*Figure 13.6* Glass harmonica displayed in a virtual environment (VE) inspired by the 18th century

### *Visual feedback*

The VR experience was built using Unity version 2021.3.11f1 with the universal rendering pipeline, and the virtual glass harmonica was modeled using Blender version 3.3.1, replicating the instrument exhibited at the Danish Music Museum. The VE was designed to resemble the study room of its inventor, Benjamin Franklin, from the 18th century, according to the time of invention, as depicted in Figure 13.6. As the application was targeted to run natively on the Meta Quest 2 at the device refresh rate of 72 Hz, optimization was important to ensure a smooth experience. Three-dimensional models were optimized in Blender before being imported into Unity. As suggested in [50], to compensate for the absence of haptic cues, visual feedback was implemented by modifying the color of the bowl with which the user interacts, from dark grey to lighter grey, as shown in Figure 13.8. Furthermore, throughout the experience, the user can initiate animated elements, which were made with Blender and Cartoon Animator 5.

### *Auditory feedback*

The audio was delivered through the built-in headphones of the Meta Quest 2 headset. This decision was made to minimize the need for additional hardware within the museum setting. The immersive soundscape includes atmospheric sounds such as the wind outside and the crackling of the fireplace. The auditory feedback for the instrument was

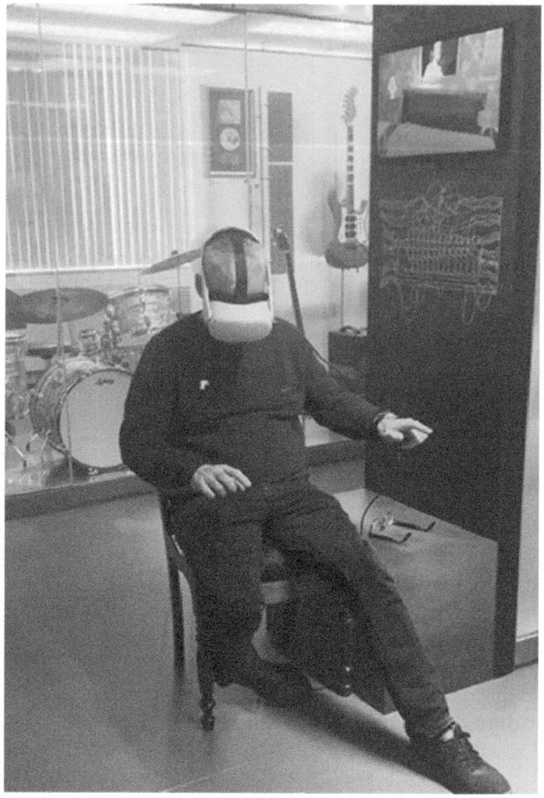

*Figure 13.7* Participant interacting with the final installation at the Danish Music Museum

synthesized using a physical model of a glass harmonica, acquired from the FAUST Physical Modelling Toolkit, which was compiled and exported as a Unity audio plug-in. The physical model produces the sound of playing glass cylinders of different sizes using a technique called 'banded waveguides', as described in [51]. Using the *FaustVariableController* script, the parameters can be modified either in the Unity inspector or by scripting using *Getters* and *Setters*. The physical model includes the following parameters:

- Basic Parameters (frequency, gain)
- Physical Parameters (base gain, bow pressure, excitation selector, integration constant)
- Nonlinear Filter Parameters (modulation frequency, nonlinearity)
- Reverberation (reverb gain, room size)
- Spat (pan angle, spatial width)

For this project, the user-controlled parameter was *bow pressure* when contact was initiated with the bowl. Addressing the challenge of creating meaningful mappings of movement to sound, as mentioned in [52], while the user kept finger contact with the bowl, the bow pressure would be mapped to a constant of 0.85. This value was estimated listening to recordings of the instrument, as the attack time, determined by the physical model, was perceived to be too long in the lower end bow pressure values. By setting a higher constant bow pressure, the sound would intrude more rapidly, resulting in a closer resemblance to the real instrument. Furthermore, adjusting the parameters of *room size* and *reverb gain* parameters to their maximum setting was found to attenuate the sound more authentically, when the user lifts the finger off the glass bowl.

### *Narrative elements*

The narrative elements of the experience extend beyond a historically relevant VE but also include a storyline narrated by an animated portrait of Benjamin Franklin, as shown in Figure 13.9. The storytelling component provides information about the origin,

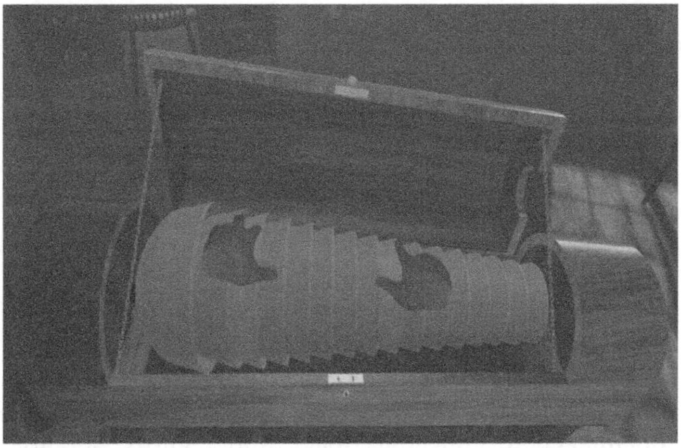

*Figure 13.8* Visual feedback when player touches the glass bowls

*Figure 13.9* Left: Button activating the narrative and tutorial. Right: Talking portrait of Benjamin Franklin

popularity, and eventual decline of the instrument, accompanied by audio clips featuring Mozart's *Adagio for Glass Harmonica* and other relevant sounds that align with the narrative content. Additionally, animated figures are incorporated to further enhance the storytelling experience. The narrator not only provides context for the user but also offers instructions, prompts them to engage with the story, compliments their playing, and serves as a guide throughout the experience. Within the VE, two labeled buttons are located near the instrument (see Figure 13.9), which allow the user to initiate the storytelling or a visual tutorial that demonstrates the interaction using animated hands. The tutorial element enables visitors to have a successful experience without requiring the assistance of museum personnel by providing clear guidance for independent navigation and interaction with the instrument.

### Methodology

This section provides an overview of the approach and strategies employed during the installation development. Collaborative design, known as co-design, was chosen as the primary method, involving active collaboration with museum personnel. The following steps outline the strategies employed throughout the process:

1. **Initial consultations:** Meetings were held with museum personnel to understand their requirements, objectives, target audience, and physical space considerations for the installation.
2. **Ideation and prototyping:** The design of the virtual space was conceptualized through brainstorming sessions and drawing inspiration from source material suggested by museum personnel. Their valuable input contributed to the refinement of ideas.
3. **Informal testing:** The prototypes were subjected to informal testing at Aalborg University in Copenhagen to collect feedback and identify potential usability flaws. Adjustments were made based on feedback and discussions were held with museum collaborators for significant design changes.
4. **Implementation and evaluation:** The final design was implemented and installed at the museum. The installation was evaluated through observations and interviews with museum guests. Subsequently, these results were qualitatively coded and condensed. The structured interview contained the questions in the next section.

### Evaluating the glass harmonica

The evaluation was carried out by museum professionals who briefly interviewed visitors during museum opening hours, covering multiple days. There were 22 participants between the ages of six and 78 (Mdn 31.0, SD 23.9). Participants had visited museums between zero and eight times a month, and 84% had previous knowledge of the Danish Music Museum before visiting that day. The study participants were selected based on their voluntary interaction with the installation. During the interactions, the museum personnel observed the participants and invited them to participate in a short interview, during which their responses to the following questions were recorded by museum personnel.

- What words can describe your experience?
- To what extent were you motivated throughout the experience?

- What did you think was good/bad?
- What do you think was difficult/easy/too easy?
- Did you experience the involvement of multiple senses during the experience?
- Did you experience the installation alone or with others?
- Did you perceive a connection between the object in the display case and the experience?
- Would you expect there to be other similar installations in the exhibition?
- We have the hypothesis that one can get closer to objects by conveying them in new ways. What do you think about this?
- If you had to give us one piece of advice the next time we build an installation at the museum, what would it be?
- Is there anything else you would like to tell us?

Participants who interacted with the VR installation were observed to spend up to 10 minutes (Mdn 4.0, SD 1.9). Initially, most participants showed hesitancy and focused attention. However, as time progressed, they became more relaxed and began to enjoy the experience; this was expressed by laughter and smiles. Most of the time, multiple individuals (up to five), often families, would participate in the experience together, actively helping each other with the HMD and interactions. Bystanders would also gather around, observing the screen and engaging in conversations with the person immersed in VR, and would frequently capture photographs of each other. While immersed, participants would exclaim words such as 'That is fun' and 'Really cool' and encourage each other to try. Additionally, it was observed that younger participants would verbally respond to the narrator in the VE.

Looking at the responses related to motivation, 68.2% explicitly stated that they were motivated or *highly* motivated throughout the whole experience, which is also reflected in the majority describing the experience as *fun*, *impressive*, and *different*. Positive feedback can be attributed in part to the fact that most participants found the interaction to be easy, which was often credited to the clarity of the instructions provided. It was observed that even the youngest participants had a successful time playing and enjoying themselves. However, a few participants did not succeed in creating sound, leading to frustration. Some participants also encountered frustration due to technical errors, particularly one of the buttons not responding correctly during the initial day of testing. This problem was evident from the significant number of participants who reported this technical problem.

The social aspect of the experience was evident as children playfully touched the fingers of the participants. However, this interaction would startle the participant who wears the HMD, disrupt hand tracking, reducing sense of presence and resulting in a poorer experience. One participant explicitly mentioned this issue, describing it as their 'hands flying around', while another participant found it very uncomfortable. The challenges with hand tracking were exacerbated by the installation's location in a heavily trafficked main hall, leading to instances of tracking loss. Participants experienced this as their virtual hands got 'stuck' or 'poorly calibrated', expressing a desire for improved hand tracking.

When asked about the perceived connection between the installation and the display case, the majority (73.6%) of the participants confirmed a clear connection between the real glass harmonica and the interactive installation upon exiting the virtual experience.

It was also observed that the participants would approach the real instrument and demonstrate how it should be played, using the correct gestures they had learnt in the VE. When asked, all participants similarly expressed that the installation brought them closer to the instrument. This was supported by statements such as, 'I absolutely think so. Touching things creates understanding and connection', 'Yes, one gets a sense of what it would do', and 'Many of the instruments hanging here are dead, so they become accessible again here'.

### *Discussion*

First, it should be noted that staff familiarity and expertise with technology are important factors when implementing and evaluating VR, as emphasized in [24]. In the case of the music museum, the staff was already used to work with VR. Although the interactive experience was novel for the participants, they exhibited familiarity with VR technology. Consequently, certain expectations arose, such as the desire to stand up and move around within the virtual environment. Although the hand tracking functionality generally functioned well, the heavily trafficked location and movement in front of the cameras on the HMD led to occasional tracking loss, rendering interaction impossible. Experiencing these latency errors has a detrimental effect on the overall experience of the VRMI of the participants, in correspondence with [39]. To solve this issue, markings on the floor could indicate the virtual space to discourage bystander interference. In addition, it is important to note that the orientation of the VE is established the moment participants mount the HMD, necessitating them to sit down and face forward to ensure the accurate positioning of the virtual instrument. Exploring technical solutions to enhance system predictability and reliability would be valuable in addressing the absence of a reset-orientation button on the Touch controllers, thus reducing the potential for user errors.

Given that the installation primarily attracted groups of people, it became apparent that when bystanders discovered that the video on the screen was not a live stream, particularly younger participants would become impatient and occasionally interact with the immersion person (such as touching their hands). According to the concepts discussed in [36], the social experience is of significance within museum settings. Implementing a live stream from the HMD would be relevant to facilitate cooperative exploration and potentially mitigate feelings of impatience. Furthermore, this observation indicates the relevance of the asymmetric interaction between HMD wearers and bystanders in museum settings. In particular, museum staff observed a considerable number of visitors photographing each other, indicating a motivation to share the experience with individuals who were not physically present at the museum.

Interestingly, around half of the participants (47.0%) explicitly stated that they believed that they experienced a sense of touch, despite the absence of haptic stimuli. They acknowledged this perceptual illusion, as they were aware of the lack of haptic feedback, but the other sensory modalities contributed to the perception of touch. The phenomenon of pseudo-haptic illusion, which is evoked by the use of cross-modal illusions, has been subject to comprehensive investigation in prior research [53]. Although it did not constitute the central area of emphasis for this project, it aligns with the museum's mission of taking music history out of the box and placing it in the *hands* of the visitors. The VR installation received overwhelming positive feedback and prompted suggestions for including additional instruments, as they wanted to experience more VRMIs. During the interviews, the participants expressed significant importance in the storytelling aspect of the experience, highlighting its role in creating a sense of presence within the VE and effectively guiding them throughout.

The clarity of the connection between the virtual and the exhibited instrument was evident among many participants as they engaged in gestures that mimic the playing while describing the instrument to their peers. This observation supports the decision to prioritize faithful replication of the interaction in the context of musical heritage and the preservation power of VR. However, it should be noted that younger participants, in particular, tended to resort to dynamic movements resembling striking piano keys if they were unable to produce sound within a short amount of time after mounting the HMD. Therefore, the static nature of the interaction may not be suitable for visitors who prefer more dynamic and active experiences. By employing virtual reality, faithful interaction replication, and storytelling, the project offers an immersive experience, bringing a historical instrument back to life, and letting visitors at Danish Music Museum touch the untouchable.

## Are we there yet?

In this section we discuss some of the learning outcomes on the use of VR outside laboratory settings. These considerations are based on the personal experience in using VR in hospitals and museums.

### *Hardware selection*

VR devices have achieved a level of democratization such that they can be purchased for relatively low cost. Nowadays it is possible to purchase devices that do not require additional hardware or a separate desktop computer and that are user friendly enough to be given to a general audience to be used at home or in public settings also by non-technical users. This has already been done in the Listen Again project, where children received an Oculus Quest 2 head-mounted display and were asked to use it twice a week at home to train. We also installed the same HMD at the Danish Music Museum, and both museum personnel and visitors can easily run the apps. The recently announced Oculus Quest Pro has also embedded eye tracking, which allows measurement in real time of what the user is looking at. This opens new possibilities for measuring behavior, obviously taking into account ethical issues and proper data management approaches. One of the drawbacks of most HMDs is the low-quality auditory display in terms of headphone quality, as discussed in the following section.

### *Auditory displays*

Regarding auditory displays, it is well-known that high-quality audio significantly contributes to an immersive experience [54]. The headphones embedded in commercially available HMDs are still rather low quality, and usually an external additional pair of headphones is used for higher fidelity auditory display. This is especially important when the user needs to be isolated from the outside world to enhance immersion in the virtual world.

From the software perspective, in recent years virtual acoustics has made tremendous progress, and VR applications include rendering of realistic auditory stimuli, including spatial rendering, room acoustics, occlusions, and personalized spatial sound. However, challenges still exist in the simulation of efficient yet accurate auralizations for applications where this is needed, such as architectural acoustics. Main issues related to the fact that accurate room simulations are extremely computationally expensive, preventing a real-time implementation. Therefore techniques to create efficient yet accurate virtual auditory environments are an active area of research [55, 56].

Moreover, the consideration that each user's individual head and pinnae shape affects the soundwave transmitted to the eardrum, or the so-called personalized head-related transfer function (HRTF), can be integrated in VR simulations [57, 58]. The question of the utility of personalized HRTFs is, however, still open. For example, [59] shows how in an interactive VR simulation where several cues are present, such as visual, motion, and proprioception cues, generalized HRTF are sufficient. This can be due to the fact that localization and externalization cues are facilitated by the other sensory modalities.

Research also shows that the use of non-individualized HRTFs when using virtualization techniques does not impair perceived auditory distance [60].

Research on interactive auralization for VR is very active. The main goal is to provide simulations that are accurate yet efficient. Understanding human perception when exposed to audio visual cues of different qualities is also an active topic of research [61].

### *Visual displays*

By replacing real-world visual information with digitally generated ones, VR-HMDs can place users inside relevant virtual environments where they can receive appropriate treatment for various mental and physical disabilities [62]. This advantage of virtual displays can also be translated into hearing impairment research. Although several tools nowadays exist to generate content for high-fidelity visual displays [63, 64], creating high-quality 3D graphics is still very time-consuming. One alternative solution is using a 360° camera that can capture real scenes to be reproduced in an HMD [65]. The advantage of using recorded footage is obviously the fact that the content is quickly and readily available and a faithful reproduction of reality. One disadvantage is the fact that it cannot be modified, and the point of view depends on the location of the camera. On the other hand, 3D-generated content is highly adaptable and modifiable but more time-consuming to create and requires skills in computer graphics in order to generate realistic simulations.

### *Natural interactions*

One key attribute of VR is its interactivity, allowing the user to take a very active role. In 1965 Sutherland envisioned the ultimate display, as 'a room within which the computer can control the existence of matter. A chair displayed in such a room would be good enough to sit in. Handcuffs displayed in such a room would be confining, and a bullet displayed in such a room would be fatal. With appropriate programming such a display could literally be the Wonderland into which Alice walked'[66].

The VR community has for several decades extensively researched interaction techniques in 3D user interfaces to approach Sutherland's dream [67]. Despite that, commercial VR devices are still based on joysticks that do not faithfully represent the complex interactions that humans can have in the real world. Some devices such as the Oculus Quest 2 have the possibility to replace joysticks with hand tracking, which allows an increase in the naturalness of the interaction. However, interacting with one's own hands removes both tactile and haptic feedback present in real-life interactions. In the physical world if we reach towards an object with our hands, we can feel the texture, shape, temperature, and other physical properties [68]. This does not happen when interacting in VR. In the case of the glass harmonica, the lack of tactile feedback was compensated by visual and auditory feedback. In [68, 69], Lederman and Klatzy present several exploratory procedures that humans use to recognize haptic properties of objects. These are typical movements that we perform to recognize several properties, such as substance-related properties such as texture, hardness, temperature, weight, and structure-related properties such as weight, volume, global and exact shape, and functional properties such as part motion and specific function. Researchers propose that such exploratory procedures should be implemented also in designing haptic interfaces in order to afford more natural interactions between humans and technology.

Among the studies examined, the interactions happen with the hands. In our literature search we found only one study that included a dual task (listening while walking) performed with the feet [70]. The objective of [70] was to investigate the effect of age-related hearing loss on word recognition accuracy in a dual-task experiment. Specifically, this research tried to cope with the limitations of highly controlled hearing tests that do not show ecological validity. This goal is achieved by creating a test scenario where HoH individuals can walk and perform a dual task.

This is certainly a very promising area for future research since it is still a challenge to understand how individuals with hearing impairment behave while performing other actions, such as walking. Overall safety measures would need to be taken to investigate how HoH individuals behave when walking in a VR, given the likely issues of stability and balance.

### *User-centered design*

As previously observed in [71–73], involving the relevant stakeholders and accounting for what they find important improves and targets the final application. To our knowledge, the first example of participatory design in the field of audiology is presented in [72]. Here a music training app is developed through a series of workshops run with the end-users. In the first workshop some mockups were created in collaboration with the end-users. This participatory design approach proved to be essential in order to create relevant solutions for the end-user. From an audio engineer point of view, it is not trivial to adopt a participatory design point of view, especially because several stakeholders are involved in the process, from the clinicians to the patients and their caregivers.

### *Gamification*

Especially when children are involved, a gamified approach such as the one used in [74–76] helps motivating training. This requires the involvement of experienced game designers. In [75] it is observed how recruitment of participants was quick and efficient, since many were intrigued by the possibility of trying a VR setup. However, many found the task boring and too long and recommended feedback after each trial to confirm that they had understood the task and to keep up motivation. One reason might be that the experience was designed by engineers without a participatory design approach involving the target group.

### Cybersickness

It is not uncommon for individuals not familiar with VR to experience cybersickness. Cybersickness symptoms cause severe discomfort and hinder the immersive VR experience. Sensory conflict theory explains that motion sickness in VR can be due to the mismatch between visual and vestibular senses [77]. Additionally, individuals with hearing loss often suffer from vestibular sensory problems [78], making this target group more prone to motion sickness. A common methodology to assess hearing loss is by using the simulator sickness questionnaire [79]. In [80] cybersickness in 360° head-mounted display VR is investigated. In traditional 360° VR experiences, translational movement in the real world is not reflected in the virtual world, and therefore self-motion information is not corroborated by matching visual and vestibular cues, which may trigger symptoms of cybersickness. New artificial intelligence software designed to supplement the 360° VR experience with artificial six-degrees-of-freedom motion was implemented, which was shown to reduce cybersickness. A study conducted by Impellizzeri et al. investigated whether cybersickness negatively affects the outcome of VR rehabilitation interventions for individuals with Parkinson's disease [81]. The results indicate that individuals with Parkinson's were not more prone to cybersickness than healthy controls. However, to our knowledge, no studies are investigating whether cybersickness can reduce the effectiveness of VR interventions for individuals with hearing disabilities.

### Addressing personal needs

The articles reviewed in this paper address several aspects of training and accessibility using VR. One important element in these situations is the fact that individuals are very different from one another, both in terms of hearing disability and in terms of preferences. From the point of view of hearing disability, each training and rehabilitation program should be designed in order to target the specific needs of the user. Moreover, when looking at musical training, it is important to observe that individuals have different preferences regarding the music they like to be exposed to, and this can highly affect their engagement in adopting the training programs [82].

### Measuring the user experience

The projects just described adopt different measurement approaches, from testing the ability of subjects to recognize speech, to performance and behavior. Measuring behavior in VR is nowadays facilitated by the inclusion of different sensors in the HMD, such as eye tracking and motion tracking.

In [83, 84] a way to measure user behavior in realistically simulated virtual environments is presented. While the solution is tested only with normal hearing individuals, the intent is to use it to monitor individuals with hearing impairment. There is evidence that head movement can affect the performance of hearing aids algorithms [85]. Head movement can be easily recorded using sensors embedded in the HMDs. Another interesting feature of novel HMDs is the fact that they include an eye tracker and pupillometry. Eye tracking and pupillometry complement other methods by providing objective physiological measures of online cognitive processing during listening. Eye tracking records the moment-to-moment direction of listeners' visual attention, which is closely time-locked to unfolding speech signals, and pupillometry measures the moment-to-moment size of listeners' pupils, which dilate in response to increased cognitive load [86]. Another objective measurement that is becoming adopted in laboratory settings is EEG [87]. In [88] EEG and VR are combined as a tool to study the neurophysiological mechanisms of everyday language comprehension in rich, ecologically valid settings, specifically a simulated restaurant in VR. Objective measurements become more challenging when the VR experiences need to be taken home for example for longitudinal training.

### *Fatigue*

Connected to the considerations regarding diverse populations, it is also important to remark that some individuals might feel fatigue when using novel technologies. This is especially the case for individuals with hearing impairment, where research has provided a lot of evidence of the connection between hearing impairment and fatigue [89, 90]. It is important to take fatigue into consideration when involving a fragile population in any sort of activity. This is especially the case when running experiments that involve the elderly. There is a need to respect their physical and mental abilities for example by ensuring that the experiences are not too tedious, too long, or too physically or mentally demanding.

### *Ethical implications*

Especially when the users are children or younger adults, it is important to study and understand the ethical aspects of using VR.

In [91] the authors call for examining the impact of VR on children's development across three domains: physical, cognitive, and psychosocial. They observe how a wide range of longitudinal impact studies is not yet matured enough to afford data for any secure conclusions. Contemporary psychology perceives child development, not as a fragmentary but as a holistic process. It is therefore essential to provide an overview of the concerns regarding the effect of VR use on overall children's development. A major worry concerns potentially harmful consequences. They suggest, therefore, a healthy Family Media Use Plan that is individualized for a specific child, teenager, and family, accordingly to each child's age, health, temperament, and developmental stage and is focused on appropriate balance between screen time/online time and other.

## Conclusion

In this paper we presented our experience on the use of VR outside laboratories, specifically in cultural and clinical settings.

In the last decade, the advances in hardware and software technologies have increased the use of VR in both laboratory and everyday settings. VR exhibits potential both in cultural heritage and in the health sector.

The audiovisual quality, ergonomics, and software tools available for VR technologies will also increase its use. There are still challenges associated with the use of VR, especially with elderly population, namely the fact that VR completely isolates the individuals from the physical world, which can be daunting. In this case, other solutions, for example based on augmented reality where the virtual world is superimposed on the virtual world, might be more suitable.

## Acknowledgments

We would like to thank the Augustinus foundation for supporting our research on virtual reality in museums, and the Velux foundation for supporting the Listen Again project. We would also like to thank all the collaborators and research assistants who participated in these projects.

## References

[1] Albert Skip Rizzo and Russell Shilling. Clinical virtual reality tools to advance the prevention, assessment, and treatment of PTSD. *European Journal of Psychotraumatology*, 8(sup5):1414560, 2017.

[2] Theresa M Fleming, Lynda Bavin, Karolina Stasiak, Eve Hermansson-Webb, Sally N Merry, Colleen Cheek, Mathijs Lucassen, Ho Ming Lau, Britta Pollmuller, and Sarah Hetrick. Serious games and gamification for mental health: Current status and promising directions. *Frontiers in Psychiatry*, 7:215, 2017.

[3] Institute of Digital Media, Child Development Working Group on Games for Health, Tom Baranowski, Fran Blumberg, Richard Buday, Ann DeSmet, Lynn E Fiellin, C Shawn Green, Pamela M. Kato, Amy Shirong Lu, Ann E. Maloney, et al. Games for health for children – current status and needed research. *Games for Health Journal*, 5(1):1–12, 2016.

[4] Jeanine Krath, Linda Schürmann, and Harald FO Von Korflesch. Revealing the theoretical basis of gamification: A systematic review and analysis of theory in research on gamification, serious games and game-based learning. *Computers in Human Behavior*, 125:106963, 2021.

[5] Mel Slater, Beau Lotto, Maria Marta Arnold, and Maria V Sanchez-Vives. How we experience immersive virtual environments: The concept of presence and its measurement. *Anuario de psicología*, 40(2):193–210, 2009.

[6] Rolf Nordahl. Self-induced footsteps sounds in virtual reality: Latency, recognition, quality and presence. In *The 8th Annual International Workshop on Presence, PRESENCE 2005, Conference Proceedings, 21–23 September 2005, London, United Kingdom*, pages 353–355. Department of Computer Science, Aalborg University, 2005.

[7] Rolf Nordahl. Increasing the motion of users in photo-realistic virtual environments by utilising auditory rendering of the environment and egomotion. In *Proceedings of the 9th International Workshop on Presence*, pages 57–63. International Society for Presence Research, Cleveland State University, 2006.

[8] Daniel Freeman, Sarah Reeve, Abi Robinson, Anke Ehlers, David Clark, Bernhard Spanlang, and Mel Slater. Virtual reality in the assessment, understanding, and treatment of mental health disorders. *Psychological medicine*, 47(14):2393–2400, 2017.

[9] Albert Skip Rizzo. Clinical virtual reality in mental health and rehabilitation: A brief review of the future! *Infrared Technology and Applications XLV*, 11002:150–158, 2019.

[10] Mark B Powers and Paul MG Emmelkamp. Virtual reality exposure therapy for anxiety disorders: A meta-analysis. *Journal of Anxiety Disorders*, 22(3):561–569, 2008.

[11] Giuseppe Riva. Virtual reality for health care: The status of research. *Cyberpsychology & Behavior*, 5(3):219–225, 2002.

[12] Emil Rosenlund Høeg, Tina Myung Povlsen, Jon Ram Bruun-Pedersen, Belinda Lange, Niels Christian Nilsson, Kristian Birkemose Haugaard, Sune Mølgard Faber, Søren Willer Hansen, Charlotte Kira Kimby, and Stefania Serafin. System immersion in virtual reality-based

rehabilitation of motor function in older adults: A systematic review and meta-analysis. *Frontiers in Virtual Reality*, 2:647993, 2021.

[13] Xiang Peng, Rashid Menhas, Jianhui Dai, and Muhammad Younas. The Covid-19 pandemic and overall wellbeing: Mediating role of virtual reality fitness for physical-psychological health and physical activity. *Psychology Research and Behavior Management*, 15:1741–1756, 2022.

[14] Colette M McKay. Brain plasticity and rehabilitation with a cochlear implant. *Advances in Hearing Rehabilitation*, 81:57–65, 2018.

[15] Andrea Ciorba, Chiara Bianchini, Stefano Pelucchi, and Antonio Pastore. The impact of hearing loss on the quality of life of elderly adults. *Clinical Interventions in Aging*, 7:159–163, 2012.

[16] Gitte Keidser, Graham Naylor, Douglas S Brungart, Andreas Caduff, Jennifer Campos, Simon Carlile, Mark G Carpenter, Giso Grimm, Volker Hohmann, Inga Holube, et al. The quest for ecological validity in hearing science: What it is, why it matters, and how to advance it. *Ear and Hearing*, 41(Suppl 1):5S, 2020.

[17] Susan S Lazinger. *Digital Preservation and Metadata: History, Theory, Practice*. ERIC, 2001.

[18] Kyong-Ho Lee, Oliver Slattery, Richang Lu, Xiao Tang, and Victor McCrary. The state of the art and practice in digital preservation. *Journal of Research of the National Institute of Standards and Technology*, 107(1):93, 2002.

[19] Mel Slater. Place illusion and plausibility can lead to realistic behaviour in immersive virtual environments. *Philosophical Transactions of the Royal Society B: Biological Sciences*, 364(1535):3549–3557, 2009.

[20] Mel Slater and Maria V Sanchez-Vives. Enhancing our lives with immersive virtual reality. *Frontiers in Robotics and AI*, 3:74, 2016.

[21] Hunter G Hoffman. Physically touching virtual objects using tactile augmentation enhances the realism of virtual environments. In *Proceedings. IEEE 1998 Virtual Reality Annual International Symposium (Cat. No. 98CB36180)*, pages 59–63. IEEE, 1998.

[22] Brent Edward Insko. *Passive Haptics Significantly Enhances Virtual Environments*. The University of North Carolina at Chapel Hill, 2001.

[23] Nikoletta Zouboula, Emmanuel Fokides, Costas Tsolakidis, and Costas Vratsalis. Virtual reality and museum: An educational application for museum education. *International Journal of Emerging Technologies in Learning (iJET)*, 3(2008), 2008.

[24] Maria Shehade and Theopisti Stylianou-Lambert. Virtual reality in museums: Exploring the experiences of museum professionals. *Applied Sciences*, 10(11):4031, 2020.

[25] Mariapina Trunfio, Maria Della Lucia, Salvatore Campana, and Adele Magnelli. Innovating the cultural heritage museum service model through virtual reality and augmented reality: The effects on the overall visitor experience and satisfaction. *Journal of Heritage Tourism*, 17(1):1–19, 2022.

[26] David A Gallo and Stanley Finger. The power of a musical instrument: Franklin, the Mozarts, Mesmer, and the glass Armonica. *History of Psychology*, 3(4):326, 2000.

[27] A Hyatt King. The musical glasses and glass harmonica. In *Proceedings of the Royal Musical Association*, volume 72, pages 97–122. Cambridge University Press, 1945.

[28] Thomas D Rossing. Acoustics of the glass harmonica. *The Journal of the Acoustical Society of America*, 95(2):1106–1111, 1994.

[29] Annamaria Recupero, Alessandra Talamo, Stefano Triberti, and Camilla Modesti. Bridging museum mission to visitors' experience: Activity, meanings, interactions, technology. *Frontiers in Psychology*, 10:2092, 2019.

[30] Arnold POS Vermeeren, Licia Calvi, Amalia Sabiescu, Raffaella Trocchianesi, Dagny Stuedahl, Elisa Giaccardi, and Sara Radice. *Future Museum Experience Design: Crowds, Ecosystems and Novel Technologies*. Springer, 2018.

[31] Hwei Teeng Chong, Chen Kim Lim, Abdul Rafi, Kian Lam Tan, and Mazlin Mokhtar. Comprehensive systematic review on virtual reality for cultural heritage practices: Coherent taxonomy and motivations. *Multimedia Systems*:1–16, 2022.

[32] Athanasios Gaitatzes, Dimitrios Christopoulos, and Maria Roussou. Reviving the past: Cultural heritage meets virtual reality. In *Proceedings of the 2001 Conference on Virtual Reality, Archeology, and Cultural Heritage*, pages 103–110, 2001.

[33] Jonna Häkkilä, Petri Hannula, Elina Luiro, Emilia Launne, Sanni Mustonen, Toni Westerlund, and Ashley Colley. Visiting a virtual graveyard: Designing virtual reality cultural heritage experiences. In *Proceedings of the 18th International Conference on Mobile and Ubiquitous Multimedia*, pages 1–4, 2019.

[34] Rafael Villena Taranilla, Ramón Cózar-Gutiérrez, José Antonio González-Calero, and Isabel López Cirugeda. Strolling through a city of the roman empire: An analysis of the potential of virtual reality to teach history in primary education. *Interactive Learning Environments*, 30(4):608–618, 2022.

[35] Gürcan Büyüksalih, Tuna Kan, Gözde Enç Özkan, Müge Meriç, Lale Isın, and Thomas P Kersten. Preserving the knowledge of the past through virtual visits: From 3d laser scanning to virtual reality visualisation at the Istanbul Çatalca Inceğiz Caves. *PFG–Journal of Photogrammetry, Remote Sensing and Geoinformation Science*, 88(2):133–146, 2020.

[36] Eryn Parker and Michael Saker. Art museums and the incorporation of virtual reality: Examining the impact of VR on spatial and social norms. *Convergence*, 26(5–6):1159–1173, 2020.

[37] Katja Rogers, Sukran Karaosmanoglu, Dennis Wolf, Frank Steinicke, and Lennart E Nacke. A best-fit framework and systematic review of asymmetric gameplay in multiplayer virtual reality games. *Frontiers in Virtual Reality*, 2:694660, 2021.

[38] Marcelo M Wanderley and Philippe Depalle. Gestural control of sound synthesis. *Proceedings of the IEEE*, 92(4):632–644, 2004.

[39] Stefania Serafin, Cumhur Erkut, Juraj Kojs, Niels C Nilsson, and Rolf Nordahl. Virtual reality musical instruments: State of the art, design principles, and future directions. *Computer Music Journal*, 40(3):22–40, 2016.

[40] Teemu Mäki-Patola, Juha Laitinen, Aki Kanerva, and Tapio Takala. Experiments with virtual reality instruments. In *Proceedings of the 2005 Conference on New Interfaces for Musical Expression*, pages 11–16, 2005.

[41] Florent Berthaut. 3d interaction techniques for musical expression. *Journal of New Music Research*, 49(1):60–72, 2020.

[42] Jack Atherton and Ge Wang. Doing vs. being: A philosophy of design for artful VR. *Journal of New Music Research*, 49(1):35–59, 2020.

[43] Antonio Roda, Giovanni de Poli, Sergio Canazza, Zezhou Sun, and Emily Whiting. 3d virtual reconstruction and sound simulation of old musical instruments. *Archeologia e Calcolatori*, 1, 2021.

[44] Angela Bellia. The virtual reconstruction of an ancient musical instrument: The aulos of Selinus. In *2015 Digital Heritage*, volume 1, pages 55–58. IEEE, 2015.

[45] Ewa Lukasik. Reconstructing musical instruments in mixed reality. In *New Trends in Audio and Video/Signal Processing Algorithms, Architectures, Arrangements, and Applications SPA 2008*, pages 201–206. IEEE, 2008.

[46] Lingyi Wu, Wei Su, Shishu Ye, and Riji Yu. Digital museum for traditional culture showcase and interactive experience based on virtual reality. In *2021 IEEE International Conference on Advances in Electrical Engineering and Computer Applications (AEECA)*, pages 218–223. IEEE, 2021.

[47] Silvin Willemsen, Razvan Paisa, and Stefania Serafin. Resurrecting the tromba marina: A bowed virtual reality instrument using haptic feedback and accurate physical modelling. In *17th Sound and Music Computing Conference*, pages 300–307. Axea sas/SMC Network, 2020.

[48] Qiuping Jia. Application of Chinese traditional musical instruments. *Scientific and Social Research*, 3(1), 2021.

[49] Abdul Syukur, Pulung Nurtantio Andono, Khafiizh Hastuti, and Arry Maulana Syarif. Immersive and challenging experiences through a virtual reality musical instruments game: An approach to gamelan preservation. *Journal of Metaverse*, 3(1):34–42, 2023.

[50] Sergi Jord'a. Interactive music systems for everyone: Exploring visual feedback as a way for creating more intuitive, efficient and learnable instruments. In *Proceedings of the Stockholm Music Acoustics Conference, August*, pages 6–9, 2003.

[51] Georg Essl, Stefania Serafin, Perry R Cook, and Julius O Smith. Musical applications of banded waveguides. *Computer Music Journal*, 28(1):51–63, 2004.

[52] Simon Holland, Tom Mudd, Katie Wilkie-McKenna, Andrew McPherson, and Marcelo M Wanderley. Understanding music interaction, and why it matters. In *New Directions in Music and Human-Computer Interaction*, pages 1–20, 2019.
[53] Karen Collins and Bill Kapralos. Pseudo-haptics: Leveraging cross-modal perception in virtual environments. *The Senses and Society*, 14(3):313–329, 2019.
[54] Willem-Paul Brinkman, Allart RD Hoekstra, and René van Egmond. The effect of 3d audio and other audio techniques on virtual reality experience. *Annual Review of Cybertherapy and Telemedicine 2015*:44–48, 2015.
[55] Stefania Serafin, Michele Geronazzo, Cumhur Erkut, Niels C Nilsson, and Rolf Nordahl. Sonic interactions in virtual reality: State of the art, current challenges, and future directions. *IEEE Computer Graphics and Applications*, 38(2):31–43, 2018.
[56] Shiguang Liu and Dinesh Manocha. Sound synthesis, propagation, and rendering: A survey. *arXiv preprint arXiv:2011.05538*, 2020.
[57] Riccardo Miccini and Simone Spagnol. HRTF individualization using deep learning. In *2020 IEEE Conference on Virtual Reality and 3D User Interfaces Abstracts and Workshops (VRW)*, pages 390–395. IEEE, 2020.
[58] Rishi Shukla, Rebecca Stewart, Mark Sandler, et al. User HRTF selection for 3d auditory mixed reality. In *Proceedings of the Sound and Music Computing Conference*, pages 84–91, 2021. doi: 10.5281/zenodo.5045168.
[59] Christopher C Berger, Mar Gonzalez-Franco, Ana Tajadura-Jimenez, Dinei Florencio, and Zhengyou Zhang. Generic HRTFS may be good enough in virtual reality. Improving source localization through cross-modal plasticity. *Frontiers in Neuroscience*: 21, 2018.
[60] Luna Prud'Homme and Mathieu Lavandier. Do we need two ears to perceive the distance of a virtual frontal sound source? *The Journal of the Acoustical Society of America*, 148(3):1614–1623, 2020.
[61] Thomas Potter, Enzo De Sena, and Zoran Cvetkovic. On the relative importance of visual and spatial audio rendering on VR immersion. In *Frontiers in Signal Processing: Audio and Acoustic Signal Processing*, 2022.
[62] Ali Adjorlu. Virtual reality therapy. In *Encyclopedia of Computer Graphics and Games (ECGG)*. Springer, 2018.
[63] Jesse Glover and Jonathan Linowes. *Complete Virtual Reality and Augmented Reality Development with Unity: Leverage the Power of Unity and Become a Pro at Creating Mixed Reality Applications*. Packt Publishing Ltd, 2019.
[64] Kevin Mack and Robert Ruud. *Unreal Engine 4 Virtual Reality Projects: Build Immersive, Real-World VR Applications Using UE4, C++, and Unreal Blueprints*. Packt Publishing Ltd, 2019.
[65] Aden Kittel, Paul Larkin, Ian Cunningham, and Michael Spittle. 360 virtual reality: A SWOT analysis in comparison to virtual reality. *Frontiers in Psychology*, 11:563474, 2020.
[66] Ivan E Sutherland et al. The ultimate display. In *Proceedings of the IFIP Congress*, volume 2, pages 506–508. New York, 1965.
[67] Joseph J LaViola Jr, Ernst Kruijff, Ryan P. McMahan, Doug Bowman, and Ivan P. Poupyrev. *3D User Interfaces: Theory and Practice*. Addison-Wesley Professional, 2017.
[68] Susan J Lederman and Roberta L Klatzky. Hand movements: A window into haptic object recognition. *Cognitive Psychology*, 19(3):342–368, 1987.
[69] S Lederman and R Klatzky. Designing haptic and multimodal interfaces: A cognitive scientist's perspective. In *Proceedings of the Workshop on Advances in Interactive Multimodal Telepresence Systems*, pages 71–80, 2001.
[70] Sin Tung Lau, M Kathleen Pichora-Fuller, Karen ZH Li, Gurjit Singh, and Jennifer L Campos. Effects of hearing loss on dual-task performance in an audiovisual virtual reality simulation of listening while walking. *Journal of the American Academy of Audiology*, 27(7):567–587, 2016.
[71] Lilit Hakobyan, Jo Lumsden, and Dympna O'Sullivan. Participatory design: How to engage older adults in participatory design activities. *International Journal of Mobile Human Computer Interaction (IJMHCI)*, 7(3):78– 92, 2015.
[72] Rachel M van Besouw, Benjamin R Oliver, Sarah M Hodkinson, Richard Polfreman, and Mary L Grasmeder. Participatory design of a music aural rehabilitation programme. *Cochlear Implants International*, 16(sup3):S39–S50, 2015.

[73] Madeline Hallewell, Davide Salanitri, Mirabelle D'Cruz, Sue Cobb, Lorenzo Picinali, Emily Frost, Stefano Tamascelli, Harshada Patel, and 3D Tune –In project Consortium. Dartanan: Prototype evaluations of a serious game to engage children in the calibration of their hearing aid functionalities. *Journal of Rehabilitation and Assistive Technologies Engineering*, 8:20556683211021527, 2021.

[74] Deborah Vickers, Marina Salorio-Corbetto, Sandra Driver, Christine Rocca, Yuli Levtov, Kevin Sum, Bhavisha Parmar, Giorgos Dritsakis, Jordi Albanell Flores, Dan Jiang, et al. Involving children and teenagers with bilateral cochlear implants in the design of the bears (both ears) virtual reality training suite improves personalization. *Frontiers in Digital Health*, 3:759723, 2021.

[75] Nille Elise Kepp, Irene Arrieta, Christina Schiøth, and Lone Percy-Smith. Virtual reality pitch ranking in children with cochlear implants, hearing aids or normal hearing. *International Journal of Pediatric Otorhinolaryngology*, 161:111241, 2022.

[76] Lone Percy Smith. Børn med høretab i hverdagen. *Audiologopædisk tidsskrift*. Audiologopædisk Forening, 2022.

[77] Adrian KT Ng, Leith KY Chan, and Henry YK Lau. A study of cybersickness and sensory conflict theory using a motion-coupled virtual reality system. *Displays*, 61:101922, 2020.

[78] Alvaro Gallego-Martinez, Juan Manuel Espinosa-Sanchez, and Jose Antonio Lopez-Escamez. Genetic contribution to vestibular diseases. *Journal of Neurology*, 265(Suppl 1):29–34, 2018.

[79] Robert S Kennedy, Norman E Lane, Kevin S Berbaum, and Michael G Lilienthal. Simulator sickness questionnaire: An enhanced method for quantifying simulator sickness. *The International Journal of Aviation Psychology*, 3(3):203–220, 1993.

[80] Iqra Arshad, Paulo De Mello, Martin Ender, Jason D McEwen, and Elisa R Ferré. Reducing cybersickness in 360-degree virtual reality. *Multisensory Research*, 35(2):203–219, 2021.

[81] Federica Impellizzeri, Antonino Naro, Giorgio Basile, Alessia Bramanti, Francesco Gazia, Francesco Galletti, David Militi, Francesco Petralito, Rocco Salvatore Calabr'o, and Demetrio Milardi. Does cybersickness affect virtual reality training using the computer assisted rehabilitation environment (CAREN)? Preliminary results from a case-control study in Parkinson's disease. *Physiotherapy Theory and Practice*, 38(13):2603–2611, 2022.

[82] Valerie Looi and Jennifer She. Music perception of cochlear implant users: A questionnaire, and its implications for a music training program. *International Journal of Audiology*, 49(2):116–128, 2010.

[83] Maartje Hendrikse, Gerard Llorach, Giso Grimm, and Volker Hohmann. *Realistic Audiovisual Listening Environments in the Lab: Analysis of Movement Behavior and Consequences for Hearing Aids*. Universitätsbibliothek der RWTH Aachen, 2019.

[84] Maartje ME Hendrikse, Gerard Llorach, Volker Hohmann, and Giso Grimm. Movement and gaze behavior in virtual audiovisual listening environments resembling everyday life. *Trends in Hearing*, 23:2331216519872362, 2019.

[85] Todd Ricketts. The impact of head angle on monaural and binaural performance with directional and omnidirectional hearing aids. *Ear and Hearing*, 21(4):318–328, 2000.

[86] Kristin J Van Engen and Drew J McLaughlin. Eyes and ears: Using eye tracking and pupillometry to understand challenges to speech recognition. *Hearing Research*, 369:56–66, 2018.

[87] Pasquale Marsella, Alessandro Scorpecci, Giulia Cartocci, Sara Giannantonio, Anton Giulio Maglione, Isotta Venuti, Ambra Brizi, and Fabio Babiloni. EEG activity as an objective measure of cognitive load during effortful listening: A study on pediatric subjects with bilateral, asymmetric sensorineural hearing loss. *International Journal of Pediatric Otorhinolaryngology*, 99:1–7, 2017.

[88] Johanne Tromp, David Peeters, Antje S Meyer, and Peter Hagoort. The combined use of virtual reality and EEG to study language processing in naturalistic environments. *Behavior Research Methods*, 50(2):862–869, 2018.

[89] Jack A Holman, Avril Drummond, Sarah E Hughes, and Graham Naylor. Hearing impairment and daily-life fatigue: A qualitative study. *International Journal of Audiology*, 58(7):408–416, 2019.

[90] Louise A Burke and Graham Naylor. Daily-life fatigue in mild to moderate hearing impairment: An ecological momentary assessment study. *Ear and Hearing*, 41(6):1518, 2020.

[91] Polyxeni Kaimara, Andreas Oikonomou, and Ioannis Deliyannis. Could virtual reality applications pose real risks to children and adolescents? A systematic review of ethical issues and concerns. *Virtual Reality*, 26(2):697–735, 2022.

Chapter 14

# Enhancing museum experiences with real-time, immersive 3D audio technology

*Kyla McMullen, Chenshen Lu, and John Glass*

Integrating immersive, 3D audio technology into museums presents a promising avenue for enhancing visitor engagement and creating a more immersive and inclusive experience. This chapter proposes a context-aware audio guide system that utilizes real-time 3D audio to complement visual exhibits and to provide visitors with personalized and interactive auditory content. The system incorporates an ultrasonic position tracking system for accurate and precise location monitoring of visitors, ensuring seamless audio updates as they move through the museum space. A pilot study conducted at the Panama Canal Society Annual Reunion yielded positive feedback, indicating high satisfaction with the system's usability and preference over traditional exhibit setups. Technical refinements, including spatial audio rendering algorithms and a user-friendly graphical user interface (GUI), are suggested to address certain limitations and enhance user experience. Furthermore, the chapter advocates for an open-source approach to the project, making the technology accessible to a broader range of museums and institutions with limited resources. The study concludes that real-time 3D audio systems can potentially transform the museum experience, fostering a deeper appreciation of art, history, and culture for all visitors. Continued research and implementation in diverse museum settings are recommended to validate and expand the benefits of this technology on a larger scale.

## Introduction

Tens of millions of people visit museums each year for varying purposes, including appreciating diverse cultures and learning about art and history via exhibitions and collections. The advent of technology has allowed curators to create more engaging and immersive experiences for their patrons. In particular, audio guide devices have been used to augment exhibits and further engage museum visitors. An audio guide device is a small, electronic system that facilitates personalized guided museum tours. They are integrated into the museum experience to supply more information about the exhibitions without a tour guide. The audio guide has many advantages over traditional means of obtaining exhibit information using labels or pamphlets. One benefit of using audio guides is that they provide accessibility for visitors with visual impairments. In addition, an audio guide can act as a dynamic interpreter to remove language barriers for the visitor. Audio tours also add rich context, background information, and present-day interpretation that may not be included using traditional means. An audio tour allows visitors to better understand the artist's goal, historical context, and significance of the artwork with additional information.

DOI: 10.4324/9780429297526-16

Including audio in the museum experience provides a more engaging and entertaining interaction that captures the attention and interest of visitors. These audio components include narrations, oral histories, sound effects, ambient sounds, environmental sounds, and music. Research has shown that people motivated by an interest in having fun spend more time in an exhibition than those less motivated by having fun. People entertained at museum exhibitions usually spend more time visiting museums than those not entertained (Falk, Moussouri & Coulson 1998; Novey & Hall 2007). The traditional museum experience involves visiting each exhibition and reading the exhibition board, which is seemingly becoming less attractive to current visitors. For example, Kim et al. (Kim, Lee & Shin 2004) reported that only 15% of the Natural History Museum visitors read the exhibition labels. Several studies have also found that people spend little time reading explanation materials in exhibitions due to accumulated fatigue or visiting time constraints (Chiozzi & Andreotti 2001, Sandifer 1997). Moreover, visitors in Kim's study spent less than 5 seconds on an exhibition if it was styled as a panel or poster. This finding would lead one to assume that a text-heavy exhibit is less desirable to museum patrons. An audio-guided tour provides a means for museum patrons to receive information about an exhibit without explicitly needing to read the information, decreasing eye fatigue. To this point, Novey and Hall (Novey & Hall 2007) found that visitors who frequently used audio-guided tours spent significantly more time engaging with each exhibit than visitors who did not use audio guides.

Even though audio content increases visitors' engagement and interest in the information offered by exhibitions, the traditional pretaped audio tour has some limitations. For example, individuals may be instructed to take a specified path for the recorded audio components playing in sequence. In other circumstances, users must manually enter a numerical code or scan a QR code (Haworth & Williams 2012; Jevremovic & Petrovski 2012; Schultz 2013) on the label to listen to the audio. Recently, visual recognition systems using cameras on portable devices have been developed as an alternative approach to detecting objects or pictures' features giving contextual information about city scenes or artworks (Bay, Fasel & Van Gool 2006; Lim, Li, You & Chevallet 2007; Skoryukina, Nikolaev & Arlazarov 2019). A study (Wein 2014) evaluating the user preference for visual recognition, QRcodes, and keypad as input method for museum guides shows that 53% of visitors prefer visual recognition, while less than 30% chose QR codes or keypads. Those scanning-based methods require manual interaction with the devices, which may interrupt the visitor's experience of engaging with the exhibition (Wacker, Kreutz, Heller & Borchers 2016) and is sometimes not usable in lowlight environments (Temmermans, Jansen, Deklerck, Schelkens & Cornelis 2011).

## Why immersive 3D audio?

Three-dimensional, or immersive, audio is a natural 'next step' to increase visitors' engagement in museum experiences. As illustrated in Figure 14.1, a 3D audio-guided tour would occur as follows. A museum patron would walk into the museum and receive a pair of wireless headphones and other wearable technology devices. As walking through the museum, the patron is able to hear sounds associated with each exhibit, co-located with the position of the exhibit. As the user moves in this environment, the sounds are updated in real time. such that as their position and orientation change, so do the sounds in the world. As the museum patron approaches an exhibit, the exhibit gets louder. Moving

*Figure 14.1* Visual representation of a museum patron wearing a pair of headphones and other wearable technology devices for the 3D audio-guided tour

among the multitude of exhibits, the museum patron experiences an auditory preview of all the exhibitions. The immersive sounds associated with each exhibit can take on a multitude of forms. For example, an exhibit can be augmented with a 3D sound oral history related to the events of the image. Alternatively, an exhibit's 3D sound could provide ambient environmental sounds to complement the display (e.g., playing the sounds of birds chirping overhead in a nature display). An exhibit can also use immersive sound to precisely represent the scene displayed in an exhibit (e.g., playing the sound of a carnival game while viewing a display of carnival games). Apart from augmenting individual exhibits, 3D sounds can be used to direct museum patrons to experience other parts of the museum. For example, if there is an exhibit on dinosaurs, a patron may hear the sound of dinosaurs roaring from the direction of the dinosaur exhibit location.

Another motivation for using 3D audio in audio-guided tours is to provide a realistic real-time environment with perceptual plausibility to orient further and increase immersive experiences for patrons with visual impairments. Museums are committed to serving the general public; however, much is left to be desired regarding the museum experience for people with visual impairments (PVIs). Museums are only recently beginning to incorporate accessibility through means such as tactile representations of art, exhibit descriptions, and specialized tours (Asakawa, Guerreiro, Sato, Takagi, Ahmetovic, Gonzalez, Kitani & Asakawa 2019; Confino-Rehder 2010; Doore, Sarrazin & Giudice 2019). Even with the use of these accessibility aids, many possibilities still exist to enable exploration of the museum prior to selecting an exhibit.

Another approach for designing a 3D audio museum guide is motivated by being aware of information regarding the context of the visitor. Some studies tracked users' gaze direction as indicators to detect the view of interest and activate the corresponding auditory component (Baldauf, Frohlich & Hutter 2010; Yang & Chan 2019). This context information could be used to indicate targets of high interest and perhaps render more content to the user describing this display or attenuate other sounds that are inconsequential to the target of interest. Other solutions render sound automatically, depending on the visitor's exact location. In this approach, visitors' engagement and sense of usability increases as there is no need for intervention between the users and devices. The interaction happens in a more 'natural' and passive manner. To achieve this interactivity, sensors embedded in the devices use several position measurement technologies to provide the users' location data. Such devices are usually realized as an indoor localization system or GPS. Abowd et al. (Abowd, Atkeson, Hong, Long, Kooper & Pinkerton 1997) developed a mobile device tour guide based on infrared (IR) and GPS for indoor and outdoor environments, respectively. In addition, a few researchers studied visitors' behaviors in museums by using Bluetooth devices to collect their location data (Yoshimura, Girardin, Carrascal, Ratti & Blat 2012; Martella, Miraglia, Frost, Cattani & van Steen 2017). The ARCHIE project (Luyten, Van Loon, Teunkens, Gabriels, Coninx & Manshoven 2006) proposed a personalized content recommendation device based on visitors' proximity to the museum artifacts. In the project, the strength of Wi-Fi signals between visitors' handheld devices and the local access points determines their position data. Recently, researchers have explored the feasibility of utilizing ultrasonic audio signals as indoor positioning systems (Lazik & Rowe 2012). To this end, some studies have implemented ultrasonic proximity detection in the context of museum guides. In one study (Bihler, Imhoff & Cremers 2011), an ultrasonic speaker was situated next to the exhibit and emitted a frequency-shifted ultrasonic signal to indicate its presence. The cell phone detected the exhibit near the visitor based on the signal strength received by its microphone and delivered relevant content associated with the indicator.

## Previous approaches

Recently, some projects have endeavored to implement interactive 3D audio augmented reality systems that incorporate the user's position, providing them with a personalized and immersive experience museum or historical heritage settings. One Such Project, CARUSO (D'Auria, DiMauro, Calandra & Cutugno 2015), is an interactive audio-augmented system that utilizes spatial sound and visual material to enhance the outdoor heritage reality experience. The project uses specially designed headphones with inertial sensors to determine the user's head rotation in 13 dimensions. A smartphone application was developed to update audio recordings related to nearby monuments by GPS data and spatializes the audio material depending on head orientation readings. Additionally, the system can recognize the user's voice commands to trigger the audio. While this system is interactive, it requires the use of GPS and cannot be used indoors, where GPS signals cannot be read.

The SoundDelta (Mariette & Katz 2009) projects proposed a distributed mobile audio-augmented system that could reproduce up to 50 sound sources to simultaneously accommodate as many as 200 visitors in a large outdoor space. The project was implemented by dividing the physical space into cells. A server encoded the sound field corresponding to each source in the active cell (the cell occupied by visitors) as an Ambisonic channel and superimposed them in the center of the cell. Using GPS data, a mobile application interpolated two Ambisonic streams from the two closest cells and binaurally decoded the streams based on the user's head rotation data. Similar to CARUSO, although this system is interactive, it cannot be used indoors.

The LISTEN (Zimmermann & Lorenz 2008) project attempted to create an invisible audio-augmented museum guide that offered visitors personalized audio information without requiring them to operate the device explicitly. The system provided personalized audio content by analyzing visitors' motions, visit history, and time spent on each exhibit. The visitor's indoor position data was measured using a set of short radio frequency sensors that utilized line-of-sight technology. The head rotation data was then calculated using the position data. The tracking system's absolute position precision can reach 10 cm. The binaural rendering was performed by a DSP processor embedded in the headphone. The LISTEN system has 5° detection resolution which is considerably higher than the minimum audible angle. Furthermore, the LISTEN system's positioning resolution is accurate to 10 cm, which is much higher than current indoor positioning systems that can obtain an accuracy within 2 cm.

In addition, the BINCI (Binaural Tools for the Creative Industries) project has been implemented within audio guides for the Opera Garnier in France as well as Die Pinakotheken, a modern art museum in Germany. BINCI uses specialized headphones for binaural listening as well as a binaural engine for rendering sound. For the museum experience, BINCI has incorporated a head-tracking device for monitoring real-time movements (Wesemann 2017). A similar solution exists within the Jabra's Intelligent Headset, a headset with multiple sensors, such as a gyroscope, accelerometer, and GPS that render 3D audio to the listener. Jabra's product allows museum curators to program their own 3D audio soundscapes for their guests (Jennings 2016). Lastly, Sennheiser employed its 3D audio technology in the Revolution exhibit at the Montreal Museum of Arts, specifically by creating a Woodstock area in which music plays from a 3D audio system (Crews 2017). While these systems and tools are highly used and interactive, their price may be cost-prohibitive for museums with limited funding.

Regardless of the complexity, current systems usually consist of two main hardware components: a tracking system that captures location and head orientation and a rendering system that binaural-synthesizes audio signals to spatial sound and delivers them to headphones. To measure head position precisely, tailored headphones are designed to be embedded with tracking devices. The spatial sound can be rendered remotely on servers and transmitted to headphones or local devices such as mobile phones or a customized DSP processor. The device can become wearable by properly configuring the hardware components and the logic to activate audio playback, eliminating the need for the user to carry or even operate it. Regardless of the system, many considerations must be attended to during design.

## Design considerations

Creators of 3D audio museum tour guides must consider many factors before implementation. These factors fall into three main categories: technical, content, and user-related. The following subsections describe the options and trade-offs for each category.

### *Technical considerations*

Our immersive audio guide for museums (SONIC) aims to optimize the following three design aspects: navigation, automatic delivery of auditory content, and visitor experience enhancement. Tracking a visitor's position and orientation continuously and precisely is necessary to accomplish these aims. Conventional implementation of technologies, such as GPS, IR, or WIFI, have drawbacks in accuracy, robustness, and ease of implementation in a large indoor environment, such as a museum. Table 14.1 summarizes the advantages, limitations, and typical features of the existing solutions for localization (Alarifi, Al-Salman, Alsaleh, Alnafessah, Al-Hadhrami, Al-Ammar & Al-Khalifa 2016; Mainetti, Patrono & Sergi 2014).

*Table 14.1* Tabular comparison of different localization technologies like GPS, IR, WIFI, and ultrasonic position systems, highlighting their advantages and limitations

| *System* | *Advantages* | *Limitations* | *Typical features* | *Accuracy (m)* |
|---|---|---|---|---|
| **GPS** | Large coverage area | Requires clear line of sight to satellites<br>Poor performance indoor | Satellite-based<br>Suitable for outdoor environment | Not applicable for indoor environment |
| **IR** | Low power consumption<br>Easy to confine the area of interest | Accuracy reduced due to multipath errors<br>Expensive if large area needs to be covered | Line-of-sight-based<br>Accuracy depending on system architectures<br>Suitable for indoor environment only | 0.01–1 |
| **Wi-Fi** | Large coverage area, low cost | Low accuracy, susceptible to signal attenuation from walls and other obstacles<br>High power consumption | WLAN-based, determine location by mapping Wi-Fi signal strength to a predefined radio map<br>Suitable for indoor for indoor/outdoor environment | 1–10 |
| **Bluetooth** | Low power consumption, low cost, high compatibility with devices | Medium accuracy, limited coverage area<br>High latency<br>Subject to complex indoor environment | Device-based, uses wireless radio signal strength to estimate locations<br>Suitable for indoor/ outdoor environment | 1–10 |
| **Ultrasonic Beacons** | High accuracy, low cost, low power consumption | Requires line of sight, limited coverage area<br>Susceptible to interference from ambient noise | Line-of-sight-based, uses sound waves and triangulation method to estimate locations<br>Suitable for indoor localization and tracking | 0.01 |

SONIC utilizes an ultrasonic position system developed by Marvelmind to track the user's location. The ultrasonic position system outperforms the previously mentioned technologies in two main aspects. First, real-time tracking is feasible due to the ultrasonic location system's excellent precision and low latency. Second, the unit installation and maintenance are low-cost, and no recalibration occurs after the system cold starts. The ultrasonic position system includes several stationary beacons and mobile beacons. The stationery beacons mounted Above Eye Level, usually on the wall of a museum. The distance between adjacent stationary beacons is restricted to 30 m for reliable system performance. A pair of mobile beacons, installed on the ends of a plastic plate, track the user's position and orientation. The position data of these mobile beacons are updated at a rate of 16 HZ through the UDP protocol over a local Wi-Fi network and transmitted to a host applicant on a Raspberry Pi 4.

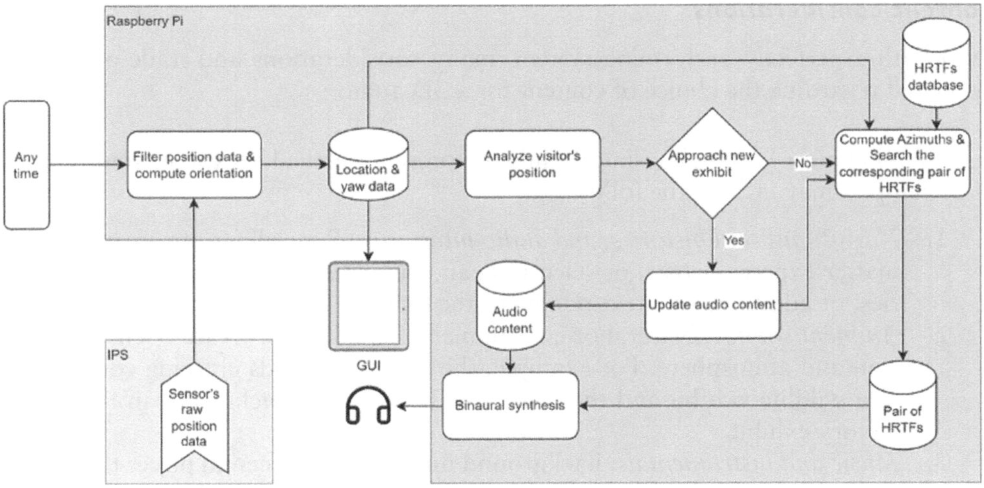


*Figure 14.2* Diagram depicting the components and data flow of the 3D audio system, including the tracking system, rendering system, and real-time audio spatialization

A customized Python application, run on a Raspberry Pi 4 receives the museum patron's position from mobile beacons, detects the exhibits near a user, spatializes the corresponding audio content for those exhibits real time and renders the spatialized audio to their headset. An overview of the functional system architecture is shown in Figure 14.2. The system uses a deque structure to hold the raw location data from the ultrasonic beacon. During the subsequent operations, the system extracts the most recent position data from the deque as the current position. To avoid the perceptual jitter of spatial audio caused by noisy or inaccurate measurements from the localization system, a simple linear Kalman filter (Bishop& Welch, 2001) is applied to smooth the trajectory of user movement. The filtered position data is then used to calculate the user's location and orientation. The user's center is first determined by locating the middle point between the two mobile beacons. Second, a vector representing the user's forward-facing direction is computed, starting at the center point and driving perpendicularly to the line connecting the two mobile beacons. The system detects the exhibit in the user's proximity by comparing the distance between the user and each exhibit location, which is predetermined and stored in the Python application. A thread control block allocates and manages threads to play audio content corresponding to the closest exhibit in the user's proximity.

We assume the sound is at the user's ear level and originates from the exhibit's center; however, sounds may be placed anywhere. The angle between the user's forward-facing vector and the vector from the user's center position to the exhibit center location may then be used to calculate the azimuth. Linear interpolation is performed on a pair of neighboring head-related impulse responses (HRIRs), which are sorted by azimuth. The HRIRs are from the CIPIC HRTF database (Algazi, Duda, Thompson & Avendano 2001). The real-time audio spatialization and rendering are performed using the sound file module.

### *Content considerations*

Even with a perfectly performing system, many considerations and trade-offs must be evaluated regarding the choice of content for a 3D audio::

- **Sound choice:** When choosing sound to represent particular exhibit, some common examples may include the following:
    1. *Narrations, oral histories, and audio interviews*: Recordings of narrators, subject matter experts, artists, or scientists can provide supplemental information, stories, or additional context related to the exhibit.
    2. *Ambient sounds*: Natural or environmental sounds can create a sense of immersion and atmosphere. For example, the sounds of birds chirping could be used in a wildlife exhibit and the sounds of heavy duty machinery can be used in a factory exhibit.
    3. *Music and instrumentals*: Background music or instrumental pieces that complement the theme of the exhibit can be used to augment an exhibit.
    4. *Sound effects*: Specific, intermittent sound effects can be used to enhance the visitor's experience and create a more immersive environment. For example, one could use the sound of a blacksmith hammering in a historical exhibit or the sounds of various exotic animals in a natural history exhibit.
    5. *Cultural sounds*: Sounds can be used to represent the culture or traditions depicted in the exhibit, such as traditional music, chants, or cultural rituals.
    6. *Historical speeches*: Recordings of famous speeches or historical moments relevant to the exhibit's content can also provide audio augmentation to an exhibit.
- **Sound representation:** Choosing the 'correct' sound to represent an exhibit can present a challenge for designers of 3D audio tours. There may be situations in which there is no direct sound representation for a visual image. For example, an abstract art piece may not have an immediate complementary sound. It is the opinion of the author that choosing a misrepresentative sound is worse than choosing silence. An ill-chosen sound may confuse the visitor, distract from the art piece, or otherwise detract from the visitor's engagement.
- **Sound location:** In a 3D audio tour guide system for museums, sounds can be placed literally anywhere. The choice of sound location is critical for immersion and interaction within the exhibit. For example, suppose that the sound of water flowing in a river is used to complement a forest painting, presenting the sound in the same location as the exhibit or at eye level may feel perceptually inaccurate. Immersion and perceptual plausibility may be increased if the sound of the river is sensed to come from the ground, where a river is likely to flow. When creating complementary audio for an exhibit, designers should test out a variety of sound locations before making a final choice.

Regardless of the content chosen to supplement the museum exhibit, the experience designers should test multiple versions of the exhibit before final deployment to determine which appeals most to the crown.

### *User-related considerations*

User-related considerations must be examined to ensure that the audio-guided museum experience is perceptible and promotes usability for museum patrons. These considerations are based on applied auditory scene analysis (ASA) principles (McGookin & Brewster 2004) and auditory masking research (Brungart 2001). Those considerations include:

- **Frequency proximity:** Frequency proximity refers to the amount of spectral similarity between two sounds in an auditory scene. For example, van Noorden (Noorden 1975) found that sounds with greater frequency distance are more likely to be perceived as separate sounds. Similarly, sounds with a smaller frequency difference are more likely to be perceived as the same sound when presented concurrently. In practice, designers of immersive audio guides for museums should ensure that sounds that are near each other and presented concurrently have significant spectral differences.
- **Temporal proximity:** Temporal proximity refers to the amount of time separating the onset of each sound in an auditory scene. For example, Darwin and Ciocca (Darwin & Ciocca 1992) found that a difference in onset time of two sounds can prevent the frequency component of a sound from contributing to the pitch of a second complex tone, even if the frequency component is present in the complex tone. In practice, designers of 3D audio guides for museums can present close, concurrent sounds offset by as few as 300 ms to increase the likelihood that the sounds are perceived individually.
- **Similarity:** Sounds with similar attributes are more likely to be grouped together; however, if each sound has a different timbre component, it is easier to distinguish the sounds as being different (Singh 1987). In practice, if it is unavoidable to have two concurrently presented sounds differ significantly in frequency, altering the timbre can increase their distinction and perfectibility.
- **Familiarity:** When presented with concurrent sounds, familiar sounds are more likely to be separated or distinguished from others. Practically, this means that the designers of 3D audio guides for museums should aim to use familiar, everyday sounds more often than unnatural, unfamiliar sounds in scenarios where sounds will likely be perceived concurrently. It should be noted that sound familiarity is context-dependent and highly subject to the culture of the museum population. Accordingly, designers should also take special efforts to learn the norms of the population(s) that will interact with the exhibits.
- **Energetic masking:** Traditional 'energetic' masking occurs when two sounds that are played simultaneously contain energy in the same critical bands, thus rendering one or both of the sounds inaudible. This kind of masking can occur if two exhibits have sounds that contain similar spectral energy that is placed near each other. This effect is particularly pronounced for speech sounds (e.g., oral histories) (Brungart 2001). It should be noted that energetic masking shares some similarities with frequency proximity since both concepts involve the interaction of sounds based on their spectral characteristics. However, the perceptual consequences of frequency proximity can include the enhanced fusion of sounds or increased interference or masking when their energies overlap. Positioning the sounds at different locations will counteract some of the effects of masking; however, the designer should consider using spectrally dissimilar sounds. In addition, the sounds should be balanced to ensure that no single

sound overpowers others, enabling visitors to appreciate each exhibit's unique auditory experience.

- **Informational masking:** Informational masking occurs when a target sound signal and a competing masking sound are both audible, but the listener cannot discern the target sound from the masking sound (Brungart 2001). Informational masking is broadly defined as a degradation of auditory detection or discrimination of a sound that is embedded in a context of other similar sounds. To avoid information masking, designers should use sounds that do not occur in the same context so that the competing or masking sound can be discernible from the target sound.
- **Managing multiple verbal stimuli:** When designing sounds to complement museum exhibits, a multitude of verbal audio augmentations, such as narrations, oral histories, and exhibit descriptions, can be used to enhance the experience. However, human listeners have audio processing limits for concurrent verbal information. For example, Treisman and Davies (Treisman & Davies 2012) point out that when a listener is presented with two sounds, detection decreases as listeners try to focus on one of the two sounds. It is suggested that there exists a modality-specific perceptual capacity that should not be exceeded.

Although these limits exist, there is research examining how to increase the discernibility of competing sounds. Many studies have shown that when two voices are presented concurrently, if the sex of the talkers is different, the impacts of masking are less prominent (Brungart 2001). In practice, if designers must present two speech sounds simultaneously, the speakers should have different sexes to enhance intelligibility.[1]

Interestingly, Gygi and Shafiro (2014) determined that reducing speech tempo by 25% improved speech perception for listeners in multi-speaker settings. In practice, designers who cannot avoid presenting multiple voices in close proximity should opt to slow down the voice of the nearest (target) sound to increase the perception.

Also, Growney and Roginska (Growney & Roginska 2022) discovered that in multi-talker scenarios, transcription performance increased when the voices of talkers were spatialized, in contrast to monaural listening. In this museum setting, this has the potential to benefit the presentation of verbal stimuli. Though design must take caution not to overload the verbal communication channel with verbal stimuli, the spatial separation of the voices may increase intelligibility. This way, users receive cohesive and personalized audio guidance without feeling overwhelmed by an excessive amount of information.

- **Balancing sound levels for the same area:** The system should dynamically adjust the volume of all presented audio based on the visitor's proximity to different exhibits. As users move closer to an exhibit, the audio associated with that exhibit should become more prominent, creating a natural and immersive experience. Conversely, audio from distant exhibits should fade into the background, reducing distractions. If many exhibits are co-located, and their associated sounds are displayed concurrently, the system gives priority to the closest system. The overall sound level should always be monitored with the addition of new sounds to avoid causing noise-induced pain for the listener.

## Initial prototype and pilot study

A prototype system, SONIC (System On Nearby Indoor audio perCeption) was created as a proof of concept to evaluate the 'look and feel' of the desired interaction. In this system, the listener used a pair of wired headphones, a laptop running MATLAB, a portable Raspberry Pi, a power source, and a lanyard with two indoor localization systems attached. The sensor detected nearby exhibits in the museum by using the data collected from the listener's location, orientation, and movement. As listeners moved around the museum, they heard real-time complementary immersive sounds for each exhibit, which were rendered, streamed, and played through their headphones. The sound reception in the headphones will reflect their movements through the room. The initial prototype did not take many of the optimal technical considerations into account, as the preliminary system was mainly focused on modeling the interaction. Nevertheless, the high-fidelity prototype was of sufficient suitability to be used in a pilot study.

### *Pilot study*

The pilot study participants consisted of attendees of the Panama Canal Society Annual Reunion in Orlando, Florida. The University of Florida has a Panama Canal Society Museum Exhibition that changes periodically throughout the year. The exhibition was titled, 'Are You Game?', which encompassed sports and recreation in the Canal Zone. The exhibit featured the sounds of various sports played in the canal zone, as well as spoken oral history quotes from various individuals recalling their experiences with the featured sports.

The aim of the pilot study was to investigate the potential impact of incorporating immersive 3D audio into the museum experience, with the goal of enhancing immersion and overall experience. Participants were recruited from the Panama Canal Society Reunion, and we employed a within-subjects design. In the two experimental conditions, half of the participants began by exploring, using the 3D auditory guide, while the other half experienced the exhibits without any audio assistance. After each person experienced the exhibit, they switched conditions and explored the exhibit again. Participants were not given a time limit during the study.

To gather data, we observed participants during their interactions and conducted a post-interaction survey. The survey consisted of structured questions, including Likert scale items and open-ended responses, to explore various aspects of their experience. These aspects included ease of use, level of distraction, preference compared to the traditional museum setup, accuracy of sound representation, and system learnability. Overall, this investigation provided valuable insights into the potential benefits of integrating 3D audio technology into the museum setting, offering a more immersive and enjoyable experience for visitors.

### *Results and analysis*

We examined the auditory guide's impact on the visit concerning its usability, sound perception, and preference/expectation. Out of 15 participants, 13 answered the survey. Although the sample size is inadequate to draw definitive conclusions, the responses may provide direction for future studies. In the pre-assessment, nine participants preferred

*Table 14.2* Tabular list of structured questions used in the pilot study survey to assess usability, sound content perception, and user preferences for the 3D audio tour guide system.

| *Usability* | *Sound content* | *User preference* |
|---|---|---|
| Q1: The audio tour was easy to use. | Q1: The sounds were representative of the images. | Q1: I prefer the audio tour over the traditional exhibition setup. |
| Q2: Most people would learn to use this system quickly. | Q2: The sounds in the audio tour were distracting. | |

*Table 14.3* Tabular representation of the average ratings and standard deviations given by participants for the usability, sound representation, and preference for the 3D audio system

| *Questions* | *Mean* | *SD* |
|---|---|---|
| The audio tour was easy to use. | 4.38 | 0.50 |
| Most people would learn to use this system quickly. | 4.38 | 0.50 |
| The sounds were representative of the images. | 4.23 | 0.83 |
| The sounds in the audio tour were distracting. | 2.23 | 1.17 |
| I prefer the audio tour over the traditional exhibition setup. | 4 | 0.58 |

audio guides while visiting museums. Two participants preferred visiting without audio guides. Following the experiments, participants rated the questions in Table 14.2 on a 5-point Likert scale, with '1' indicating 'strongly disagree' and '5' indicating' 'strongly agree'. Furthermore, we obtained feedback from participants on five open-ended questions regarding their expectations for the spatial audio guide.

Table 14.3 shows the outcomes of the questionnaires. Overall, participants were satisfied with the usability of the audio guide system, scoring 4.38 on average (SD = 0.50). Participants are also highly satisfied (average score = 4.38, SD = 0.5) with the system's ease of use, which does not place any burden on first-time users. The responses from these two questionnaires indicate that the automated auditory content playback feature is functioning as intended. Participants agreed that the auditory content was related to the exhibit image (average score = 4.23, SD = 0.83), e.g., 'the sound adds depth to photos' and 'the narration gives meaning to the exhibition'. While most participants disagreed that sound content was distracting (average score = 2.23, SD = 1.17), others had the opposite impression. One possible explanation for the finding may be the lack of an audio transition from the previous content to the next as the participant moved to a new exhibit location. The prototype device began playback automatically when the exhibit was nearby. As users cross the boundary in the physical world, the audio shifts to the most recent material. Few participants could sense that the 'soundscape didn't exactly match the photographs' as they shifted their attention to the novel exhibit while being close to the prior one. The average score for user preference is 4 (SD = 0.58). The result implied that participants tended to prefer the spatial audio tour over the traditional image and label style. An analysis of the replies to the open-ended queries discloses that the participants appreciated three aspects of the audio tour. The detailed narration enriched the reading materials and exhibition. The spatial audio augmented the visit experience and 'brought the images to life'. The ambient sound provided an enjoyable environment when reading.

## Refinement

### *Technical refinement*

While the pilot study results are promising, and the current system is stable, several limitations are currently being refined to use in future iterations of the 3D audio system. One limitation is related to sound content and its impact on user distraction. Some participants in the pilot study reported that certain sound transitions were not smooth, leading to distractions. Refining the spatial audio rendering algorithms and ensuring seamless transitions between exhibit audio content can improve the overall user experience.

### *Graphical user interface*

Another refinement in development is a user-friendly graphical user interface (GUI). Providing a GUI for museum personnel will offer a visual representation of location data being collected by the mobile beacons and will facilitate remote control over the exhibits. Using location data allows monitoring of user and exhibit positions. The system establishes a UDP connection with the custom Python application running on the Raspberry Pi and receives user location data to be displayed utilizing the Python module 'PySimpleGUI'. The GUI parses the data, enabling immersive features for the exhibits. The location data may be used to implement spatial audio and audio-guided tours. We aim for the GUI to facilitate remote control of the internals of the Raspberry Pi Python application, including manipulating exhibit location and audio data, using SSH commands. This will give operators of the GUI better and easier control over the museum, benefiting user experience. The GUI is designed to be accessible by non-technical users. An important person the design was modeled for is an exhibit curator who is proficient at using computer systems such as a mobile device but has little to no experience developing software or creating complex systems. Ensuring the GUI's usability maximizes its use outside of technical research and broadens its potential user base to the targeted demographic of real museum personnel.

To set up the audio associates with each exhibit, museum personnel would first indicate the shape of the room. Next they would place graphical markers at the locations of exhibits within the room. Next, the personnel would associate selected audio files to exhibits in the museum space shown by the GUI. The user can press a drop-down menu to select a certain exhibit based on its displayed location and a second drop-down menu that allows the user to select which audio file to match it. A button that affords 'matching' these two sounds is available. When pressed, this button action sends data which pairs exhibit and audio files on the Raspberry Pi application.

Figure 14.3 shows the wireframe of the GUI with the systems on and off. When the applications are not running, the GUI will communicate this to the user with text centered on the screen. The GUI will include a power button, indicated by the 'power' icon, to start and stop the Raspberry Pi to enable remote control of the museum exhibits. Once

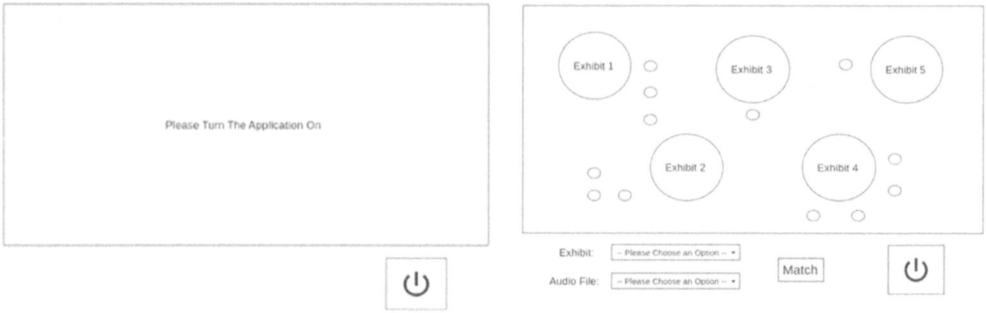


*Figure 14.3* Visual representations of the system GUI, showing the museum exhibits, visitors' locations, and controls for pairing audio files to exhibit

on, the second wireframe will be displayed. Here, the exhibits are named and distinguishable from the smaller circles representing museum visitors. Below the display of the museum exhibits and guests are the 'Exhibit' and 'Audio File' drop-down menus. They can choose from a list of selectable exhibits and files through filler text and the downward pointing arrow icon. The GUI user can click the dropdowns and select the desired exhibits and audio files to pair. The 'Match' button will be initially grayed out, implying that both an exhibit and audio file must be selected before pairing until the user has chosen a value for both fields. Then the match button will change in color and be pressable to pair the chosen values on the Raspberry Pi application.

## Discussion

The application of immersive sound into museums presents a promising avenue for enhancing visitor engagement and accessibility. The use of traditional audio guides in museums has been limited by technical constraints, lack of real-time adaptation, and user interaction barriers. The immersive 3D audio system SONIC, discussed in this chapter, overcomes these challenges by offering real-time interaction that complements the visual exhibits and creates a more immersive environment for museum visitors.

One of the key advantages of the 3D audio system is its ability to cater to visitors with visual impairments. By providing a realistic and perceptually plausible auditory environment, the technology opens new doors for including and engaging people with visual impairments in the museum experience. In addition, the pilot study conducted with the prototype system at the Panama Canal Society Annual Reunion yielded positive feedback from participants, indicating high satisfaction with the system's usability and preference over traditional exhibit setups. This result suggests that the 3D audio-guided tour has the potential to be a valuable addition to the museum experience.

The use of ultrasonic position tracking for indoor localization proved to be an effective choice for providing accurate and real-time tracking of visitors' positions and orientations. The system's low latency and excellent precision enabled seamless audio updates as visitors moved through the museum, aligning the sounds with their positions and orientations. However, it is essential to acknowledge that the pilot study had a relatively small sample size and took place in a specific museum setting. Further research with larger and more diverse participant groups in various museum environments is necessary to validate the generalizability and effectiveness of the 3D audio system. The inclusion of a graphical user interface (GUI) for setting up the 3D audio exhibit and monitoring visitor locations is a significant advancement in making the technology accessible to non-technical museum personnel. By providing an intuitive and user-friendly interface, museum curators and staff can easily manage the immersive audio system, linking audio content to specific exhibits and offering a more personalized and engaging experience for visitors. Furthermore, the system could be adapted to utilize many different sounds for the same exhibit, thus creating a different immersive sound experience for each visit. With this capability, a museum patron can visit the museum many times and have a different experience each time.

Another important consideration is the maintenance and installation of the ultrasonic position tracking beacons. While the pilot study demonstrated the feasibility and effectiveness of this technology, deploying the system on a larger scale in real-world museum settings may pose logistical challenges. Future work will focus on streamlining the installation process and ensuring the system's robustness and accuracy in diverse museum layouts.

In addition, the team aims to create an instructible, open-source user community to benefit this community on a larger scale. For example, by open-sourcing the project and providing detailed instructions, schematics, and code, the technology becomes accessible to a range of individuals and institutions. Museums, educational centers, and cultural organizations with limited resources can benefit from this technology without the need for significant financial investments or technical expertise. Open-source projects encourage collaboration and contributions from a diverse community of developers, engineers, and enthusiasts. By sharing the project's source code, others can build upon it, enhance its functionality, and address potential issues or limitations. This collaborative effort leads to continuous improvement innovation, fostering a dynamic and evolving immersive audio guide system. Different museums may have unique layouts, exhibits, and requirements. An instructible, open-source project allows individual institutions to customize and adapt the technology to suit their specific needs. They can tailor the system to include localized content, different audio cues, or even integrate it with existing museum infrastructure. Commercial 3D audio guide systems can be costly for many museums, making it challenging for smaller institutions to implement such technology. By making the project open-source, museums can leverage cost-effective solutions and utilize affordable off-the-shelf components, reducing the overall expenses associated with adopting this technology. An open-source approach fosters transparency and ethical development practices. The project's code base is visible to all, ensuring that the technology does not include hidden tracking, advertising, or other privacy concerns that may be present in commercial alternatives. This transparency builds trust and confidence among users. Proprietary technologies may face challenges if the company behind them discontinues support or goes out of business. In contrast, open-source projects tend to have a more sustainable future as they can be maintained and improved by the community, even if the original creators move on to other projects. Moreover, as technology advances, additional features could be incorporated into the system to further enhance it. For instance, integrating machine learning algorithms to analyze visitor interactions and preferences could enable the system to provide personalized audio content tailored to each visitor's interests.

In conclusion, real-time 3D audio systems for museums will transform the traditional museum experience into a more interactive and immersive journey. The pilot study provided valuable insights and positive feedback, encouraging further refinement and implementation of the technology in larger museum settings. As technology continues to evolve, the integration of immersive audio technology in museums can create a more inclusive and engaging experience for all visitors, fostering a deeper appreciation of art, history, and culture.

## Note

1 The term 'sex' is used in this context to convey the results of the study as reported. Vocal differences that are attributed to 'sex' are largely due to the spectral or pitch differences of voices. In practice, designers should opt to use voices with different pitches without referring to the gender identity of the speaker.

## References

Abowd, G. D., Atkeson, C. G., Hong, J., Long, S., Kooper, R. & Pinkerton, M. (1997). Cyberguide: A mobile context-aware tour guide, *Wireless Networks*3: 421–433.

Alarifi, A., Al-Salman, A., Alsaleh, M., Alnafessah, A., Al-Hadhrami, S., Al-Ammar, M. A. & Al-Khalifa, H. S. (2016). Ultra wideband indoor positioning technologies: Analysis and recent advances, *Sensors***16**(5): 707.

Algazi, V. R., Duda, R. O., Thompson, D. M. & Avendano, C. (2001). The CIPIC HRTF database, in *Proceedings of the 2001 IEEE Workshop on the Applications of Signal Processing to Audio and Acoustics (Cat. No. 01TH8575)*, IEEE, pp. 99–102.

Asakawa, S., Guerreiro, J., Sato, D., Takagi, H., Ahmetovic, D., Gonzalez, D., Kitani, K. M. & Asakawa, C. (2019). An independent and interactive museum experience for blind people, in *Proceedings of the 16th International Web for All Conference*, pp. 1–9.

Baldauf, M., Frohlich, P. & Hutter, S. (2010). Kibitzer: A wearable system for eye-gaze-based mobile urban exploration, in *Proceedings of the 1st Augmented Human International Conference*, pp. 1–5.

Bay, H., Fasel, B. & Van Gool, L. (2006). Interactive museum guide: Fast and robust recognition of museum objects, in *Proceedings of the First International Workshop on Mobile Vision.*

Bihler, P., Imhoff, P. & Cremers, A. B. (2011). Smartguide–a smartphone museum guide with ultrasound control, *Procedia Computer Science*5: 586–592.

Bishop, G.&Welch, G. (2001). An introduction to the Kalman filter, *Proceedings of SIGGRAPH, Course*8(27599–23175): 41.

Brungart, D. S. (2001). Informational and energetic masking effects in the perception of two simultaneous talkers, *The Journal of the Acoustical Society of America***109**(3): 1101–1109. https://doi.org/10.1121/1.1345696

Chiozzi, G. & Andreotti, L. (2001). Behavior vs. time: Understanding how visitors utilize the Milan natural history museum, *Curator: The Museum Journal***44**(2): 153–165.

Confino-Rehder, S. (2010). Art experiences for everyone: Opening the doors to the blind and people with low vision, *The International Journal of the Inclusive Museum*3(2): 27.

Crews, N. (2017). Sennheiser audio technology to power 'revolution' exhibition at Montreal museum of fine arts, *FOH Online*. **URL**: www.fohonline.com/news/16700-sennheiser-audio-technologyto-power-revolution-exhibition-at-montreal-museum-of-finearts.html

Darwin, C. & Ciocca, V. (1992). Grouping in pitch perception: Effects of onset asynchrony and ear of presentation of a mistuned component, *The Journal of the Acoustical Society of America***91**(6): 3381–3390.

D'Auria, D., Di Mauro, D., Calandra, D. M. & Cutugno, F. (2015). A 3d audio augmented reality system for a cultural heritage management and fruition, *Journal of Digital Information Management***13**(4).

Doore, S. A., Sarrazin, A. & Giudice, N. A. (2019). Natural-language scene descriptions for accessible non-visual museum exhibit exploration and engagement, in *SoL@ COSIT*, pp. 91–100.

Falk, J. H., Moussouri, T. & Coulson, D. (1998). The effect of visitors' agendas on museum learning, *Curator: The Museum Journal***41**(2): 107–120.

Growney, E. & Roginska, A. (2022). Effect of spatial audio processing on attention redirection and speech intelligibility in multi-talker communication, in *Audio Engineering Society Convention 152*, Audio Engineering Society.

Gygi, B. & Shafiro, V. (2014). Spatial and temporal modifications of multitalker speech can improve speech perception in older adults, *Hearing Research***310**: 76–86. **URL**: https://www.sciencedirect.com/science/article/pii/S0378595514000173

Haworth, A. & Williams, P. (2012). Using QR codes to aid accessibility in a museum, *Journal of Assistive Technologies***6**(4): 285–291.

Jennings, A. (2016). How audio augmented reality will transform the museum experience, *The Awe*.

Jevremovic, V. & Petrovski, S. (2012). Muzzeum – augmented reality and QR codes enabled mobile platform with digital library, used to guerrilla open the National Museum of Serbia, in *2012 18th International Conference on Virtual Systems and Multimedia*, IEEE, pp. 561–564.

Kim, C.-J., Lee, C.-Z. & Shin, M.-K. (2004). A study of visitor behavior in informal learning setting: A natural history museum, *Korean Journal of Geosciences***25**(3): 142–151.

Lazik, P. & Rowe, A. (2012). Indoor pseudo-ranging of mobile devices using ultrasonic chirps, in *Proceedings of the 10th ACM Conference on Embedded Network Sensor Systems*, pp. 99–112.

Lim, J.-H., Li, Y., You, Y. & Chevallet, J.-P. (2007). Scene recognition with camera phones for tourist information access, in *2007 IEEE International Conference on Multimedia and Expo*, IEEE, pp. 100–103.

Luyten, K., Van Loon, H., Teunkens, D., Gabriels, K., Coninx, K. & Manshoven, E. (2006). Archie: Disclosing a museum by a socially-aware mobile guide, in *7th International Symposium on Virtual Reality, Archaeology and Cultural Heritage*.

Mainetti, L., Patrono, L. & Sergi, I. (2014). A survey on indoor positioning systems, in *2014 22nd International Conference on Software, Telecommunications and Computer Networks (SoftCOM)*, IEEE, pp. 111–120.

Mariette, N. & Katz, B. (2009). Sound delta–large scale, multi-user audio augmented reality, *Proceedings of the EAA Symposium on Auralization***1**: 37–42.

Martella, C., Miraglia, A., Frost, J., Cattani, M. & van Steen, M. (2017). Visualizing, clustering, and predicting the behavior of museum visitors, *Pervasive and Mobile Computing***38**: 430–443.

McGookin, D. K. & Brewster, S. A. (2004). Understanding concurrent earcons: Applying auditory scene analysis principles to concurrent earcon recognition, *ACM Transactions on Applied Perception***1**(2): 130–155. https://doi.org/10.1145/1024083.1024087

Noorden, L. P. A. S. v. (1975). *Temporal Coherence in the Perception of Tone Sequences*, Phd thesis.Institute for Perception Research, Eindhoven. Technische Hogeschool Eindhoven.

Novey, L. T. & Hall, T. E. (2007). The effect of audio tours on learning and social interaction: An evaluation at Carlsbad Caverns National Park, *Science Education***91**(2): 260–277.

Sandifer, C. (1997). Time-based behaviors at an interactive science museum: Exploring the differences between weekday/weekend and family/nonfamily visitors, *Science Education***81**(6): 689–701.

Schultz, M. K. (2013). A case study on the appropriateness of using quick response (QR) codes in libraries and museums, *Library & Information Science Research***35**(3): 207–215.

Singh, P. G. (1987). Perceptual organization of complex-tone sequences: A tradeoff between pitch and timbre? *The Journal of the Acoustical Society of America***82**(3): 886–899.

Skoryukina, N. S., Nikolaev, D. P. & Arlazarov, V. V. (2019). 2d art recognition in uncontrolled conditions using one-shot learning, in *Eleventh International Conference on Machine Vision (ICMV 2018)*, Vol. 11041, SPIE, pp. 651–658.

Temmermans, F., Jansen, B., Deklerck, R., Schelkens, P. & Cornelis, J. (2011). The mobile museum guide: Artwork recognition with eigenpaintings and SURF, in *Proceedings of the 12th International Workshop on Image Analysis for Multimedia Interactive Services*.

Treisman, A. M. & Davies, A. (2012). Divided attention to ear and eye, in *From Perception to Consciousness: Searching with Anne Treisman*, Oxford University Press, Chapter 2, pp. 24–31.

Wacker, P., Kreutz, K., Heller, F. & Borchers, J. (2016). Maps and location: Acceptance of modern interaction techniques for audio guides, in *Proceedings of the 2016 CHI Conference on Human Factors in Computing Systems*, pp. 1067–1071.

Wein, L. (2014). Visual recognition in museum guide apps: Do visitors want it?, in *Proceedings of the SIGCHI Conference on Human Factors in Computing Systems*, pp. 635–638.

Wesemann, E. (2017). Listen up, 3d audio is about to reshape your museum experience, *Tech Guru Daily*. APA Citation.

Yang, J. & Chan, C. Y. (2019). Audio-augmented museum experiences with gaze tracking, in *Proceedings of the 18th International Conference on Mobile and Ubiquitous Multimedia*, pp. 1–5.

Yoshimura, Y., Girardin, F., Carrascal, J. P., Ratti, C. & Blat, J. (2012). New tools for studying visitor behaviours in museums: A case study at the Louvre, in *Information and Communication Technologies in Tourism 2012*, Springer, pp. 391–402.

Zimmermann, A. & Lorenz, A. (2008). Listen: A user-adaptive audio-augmented museum guide, *User Modeling and User-Adapted Interaction*18(5): 389–416.

Chapter 15

# Immersive technologies and artificial intelligence

## Therapeutic tools for the treatment of neurocognitive diseases and functional diversity

*Nuria Lloret Romero and Jorge Sastre Martinez*

In recent years we have witnessed the increase in the use of immersive technologies in various fields of application, entertainment, education, industry, culture. In the field of health there have also been great advances, especially in the fields of surgery and telecare. But what happens with the application of these technologies in areas related to functional diversity, mental illnesses such as ADHD, Asperger, or autism? Can immersive technologies and AI help patients in their therapies? Can they improve their quality of life and expand their integration into today's society? In this chapter we want to present the results of the TDHAEXPRESION research project carried out thanks to the award granted by the Takeda Pharmaceutical Company. For this project, the design, development, and implementation of immersive technologies, VR, AR, immersive sound, and AI have been carried out for the treatment of neurocognitive diseases in treatment programs for said diseases. Data have been collected in real time for the development of a customizable and adaptable immersive space that would allow the treatment of diseases through different critical variables such as intensity, duration of therapies, or frequency of use, among others. Once these characteristics were assessed, associated designs and developments have been carried out, and a system with approved therapies has been implemented. This system has allowed a personalized application of these therapies for each patient and contributed to the study of possible efficient and safe treatment of neurocognitive diseases using these technologies. We are not going to present the project data in an aseptic or technological way. We want to present what the experience of working with neurologists and therapists specialized in these pathologies and neophytes in the use of immersive technologies has been like. What has been the experience in dealing with patients and patients' families? Pros and cons of use and how we believe the treatment of some of these diseases can change with the use of these technologies. We want to offer a realistic vision of the experience carried out and what future we believe it may have in the application for the field of functional diversity. The success of the project implies achieving an improvement in the quality of life of these patients, especially their incorporation into their jobs, in addition to cost savings in the work, family, and social spheres and significant savings to the Spanish health system. The project has been developed with various patient associations and with neurologist specialists, psychiatrists, and psychologists from the mental health areas of several Spanish hospitals.

DOI: 10.4324/9780429297526-17

## Theoretical framework

This project involved the design, development, and implementation of new immersive technologies and artificial intelligence for treating neurocognitive diseases in therapeutic programs. To achieve this, we designed, developed, and implemented new technologies based on virtual reality (VR), augmented reality (AR), immersive sound and artificial intelligence (AI), and real-time data collection to create a customizable and adaptable immersive environment for treating neurocognitive diseases. These technologies considered critical variables such as therapy intensity, duration, and usage frequency, among others(Arino etal., 2014). After assessing these characteristics, associated designs and developments were carried out, and a system of standardized therapies was implemented. This system has enabled a broader application of these therapies and contributes to the efficient and safe treatment of neurocognitive diseases through new technologies.

We focus our research in three areas:

1. **Analysis and research of new therapies for neurocognitive and functional diversity diseases**
   - Mainly, these analyses have been carried out at three fundamental levels: identification of knowledge, data, algorithms and therapies associated with the solution(Adams et al., 2009).
   - An in-depth analysis of the main neurocognitive disorders has been carried out, identifying their characteristics and symptoms. These disorders include cognitive problems such as memory disturbances, behavioral changes, communication difficulties, and problems with daily activities.
   - The causes and risk factors that can trigger these disorders have also been examined, such as traumatic brain injury, stroke, brain tumors and substance abuse.
   - A detailed analysis of the data, patterns, and algorithms necessary to develop personalized therapies has been carried out. This includes the identification of behavioral patterns in patients with neurocognitive diseases and the selection of algorithms that allow the adaptation of therapies according to individual needs.
2. **Analysis of current conventional treatments**
   - Second, the main conventional treatments and current challenges have been analyzed and identified to understand where a treatment using virtual reality (VR) could have a place. Regarding the analysis of existing conventional treatments, a comprehensive evaluation has been carried out of the therapies and approaches currently used to treat neurocognitive disorders (Arildskov et al.,2022). This analysis has been conducted to fully understand the current treatment landscape and determine in which areas the implementation of virtual reality (VR)–based therapies could be relevant.
3. **Analysis of currently available virtual reality technologies and immersive sound to treat neurocognitive disorders**

In this section, the main technologies to treat neurocognitive disorders (neurofeedback, virtual reality, and augmented reality and immersive sound have been analyzed and identified, taking as reference results from previous studies and the benefits they provide in this field, with the main focus being in virtual and augmented reality in gamified environments.

## Methodology

Following is the description of the project's development, the software, and the study phases carried out toward achieving the project's objectives.

### *Phase 1: analysis and research of new therapies for neurocognitive diseases*

In the first phase of this project, an analysis and investigation of the different critical variables of the project at all levels of its development were conducted. Primarily, these analyses were carried out at two fundamental levels:

1. **Identification of knowledge, data, algorithms, and therapies related to the solution:**
   a. An in-depth analysis of major neurocognitive disorders was conducted, identifying their characteristics and symptoms. These disorders include cognitive problems such as memory disturbances, behavioral changes, communication difficulties, and issues with daily activities.
   b. The relationship between neurocognitive disorders and neurodegenerative diseases such as Alzheimer's and dementia was studied. Additionally, the causes and risk factors that may trigger these disorders, such as traumatic brain injury, stroke, brain tumors, and substance abuse, were examined.
   c. A detailed analysis of the data, patterns, and algorithms necessary for developing personalized therapies was conducted. This includes identifying behavioral patterns in patients with neurocognitive diseases and selecting algorithms that allow therapy adaptation to individual needs.
2. **Identification of entities, means, and resources required for the development of the project and the initial training related to functional testing in phase 3:**
   a. A process of identifying and recruiting testing centers was carried out, which will be fundamental for validating and testing the solutions developed in later project phases.
   b. An analysis of the necessary means and resources for implementing the proposed therapies was conducted, ensuring that the appropriate tools were available for testing and developing the technology.
   c. Training was provided to the identified centers to ensure they are prepared to use the new technologies and provide valuable feedback during the tests.
3. **Analysis and identification of major neurocognitive disorders:** As part of the project, an in-depth analysis and identification of the main neurocognitive disorders that could benefit from the proposed therapies were conducted. Neurocognitive disorders represent a set of conditions that often result in impairment of mental functions. Although these disorders are more common in older adults, younger

people can also be affected. Cognitive functions that may be altered in these disorders include:

- Memory disturbances: Patients may experience difficulties remembering past events, close individuals, or important information.
- Behavioral changes: Neurocognitive disorders may manifest in personality and behavioral changes, including irritability, aggression, or apathy.
- Communication difficulties: The ability to communicate effectively may be compromised, making social interaction and expressing thoughts and feelings more challenging.
- Problems with daily activities: Tasks such as dressing, cooking, or performing household activities may become difficult for those suffering from neurocognitive disorders.

These disorders can have various causes, which were considered in the analysis. Some may be related to neurodegenerative diseases such as Alzheimer's or dementia(Clay et al.,2020). These conditions cause a progressive deterioration of the brain and nerves over time, resulting in gradual loss of cognitive function and reduced quality of life.

In addition to neurodegenerative diseases, neurocognitive disorders may also develop as a result of brain injuries, such as traumatic brain injury (TBI), strokes, brain tumors, or substance abuse. The underlying cause and severity of these disorders can vary significantly from one patient to another, making identification and personalized treatment essential for effectively addressing each case(Ghai & Ghai, 2019).

4. **Analysis of conventional treatments for neurocognitive disorders:** In addition, the main conventional treatments and current challenges were analyzed to understand where VR-based treatments could be implemented. An exhaustive evaluation of current therapies and approaches for treating neurocognitive disorders was carried out to fully understand the existing treatment landscape and determine areas where virtual reality (VR)–based therapies could be relevant.

   The treatments for neurocognitive disorders vary greatly depending on the underlying cause of the disorder and the individual needs of each patient.

   Here are some of the conventional treatments analyzed:

- Neurology: In cases where the neurocognitive disorder is caused by an illness like meningitis, analgesics may be prescribed to alleviate headaches and antibiotics to eliminate the infection. However, these treatments do not directly address the associated cognitive dysfunctions.
- Neurosurgery: When there is severe brain damage due to traumatic brain injury or a brain tumor, neurosurgery may be necessary to repair or remove the lesion. While this approach addresses the physical problem, it does not directly treat cognitive impairments.
- Neuropsychology: Neuropsychology focuses on evaluating and treating cognitive, emotional, and behavioral problems that may arise due to neurocognitive disorders. This approach is crucial for addressing cognitive difficulties, although its effectiveness may vary depending on the severity of the disorder and the patient's individual response.

- Speech therapy: Speech therapy is used to improve communication in patients with neurocognitive disorders who experience difficulties in this area. This may include therapies for speech recovery and improving language comprehension.
- Occupational therapy: Occupational therapy helps patients develop daily skills that may have been affected by neurocognitive disorders. This may include training in activities such as dressing, personal hygiene, and household tasks.
- Physical therapy: Physical therapy improves the strength, coordination, balance, and flexibility of patients. Although it focuses on physical aspects, it can be relevant in cases where neurocognitive disorders affect mobility and physical function.

It is important to highlight that these conventional treatments have their own value but also present limitations and challenges, particularly in terms of therapy personalization and patient motivation. The implementation of VR-based therapies addresses some of these challenges by providing a highly customizable and engaging approach to treating neurocognitive disorders.

5. **Analysis of virtual reality technologies and immersive sound technologies currently available for neurocognitive disorder treatment:** This section analyzed and identified the main technologies used to treat neurocognitive disorders (neurofeedback, virtual reality, immersive sound, and augmented reality), referencing previous studies and the benefits they offer in this field, with a focus on augmented reality in gamified environments.

- Benefits of VR in treating neurocognitive disorders: The use of VR technology for treating psychological disorders began in 1994 at the Human–Computer Interaction Group at Clark University in Atlanta, where the term 'virtual reality exposure therapy' was coined. This technique was first applied to treat a case of aerophobia. VR technology offers new opportunities for developing innovative tools in neuropsychological evaluation and rehabilitation through the creation of scenarios that would be difficult or impossible to achieve using conventional neuropsychological methods. Regarding its intervention in patients with dementia, VR allows for the development of cognitive evaluation methodologies, rehabilitation strategies, and therapeutic activities.
- Advances in psychological treatments: The field of psychological treatments has advanced significantly in recent years. Since the beginning of the evidence-based psychology movement, numerous treatment protocols have proven effective and efficient. In this regard, the role of information and communication technologies should be emphasized. The use of VR has garnered interest from clinicians and researchers for over a decade, as it allows users to face feared situations through standardized environments that emulate reality. These environments are capable of evoking emotional states such as anxiety, cognitions, and behaviors that are very similar to those experienced in everyday life. This makes VR a potential tool for functional rehabilitation, not only in the neuropsychological sphere but also in the motor aspects of various types of pathologies, as it enables precise control in presenting complex stimuli, allowing for the highly accurate evaluation of cognitive and functional performance(Goharinejad et al.,2022).

- Research other therapeutic games using immersive sound: Several therapeutic games and applications demonstrate the effective use of immersive sound in promoting cognitive and emotional health:

  i. Virtual Reality Classroom: This game uses spatial audio to simulate real-world distractions, helping ADHD patients develop focus and task persistence in controlled environments.
  ii. Aquarium Mixed-Reality: Designed for children with autism spectrum disorder, this therapeutic game uses ambient underwater sounds and interactive auditory feedback to create a calming and engaging environment that aids cognitive training(Huang & Liao, 2017).
  iii. Emotion Regulation Games: Immersive sound in these VR games for psychosis patients helps modulate emotional states, providing calming effects during therapy sessions.
  iv. FitMi Therapy: This rehabilitation game integrates rhythmic and auditory feedback to improve motor coordination in stroke patients, leveraging sound to enhance focus and precision.
  v. Jewel Mine: This ADHD therapy game uses auditory signals to guide children through attention and memory tasks, reinforcing positive behavior with reward sounds.

In summary, VR and immersive sound and AI research has increased the knowledge associated with technology in the treatment of neurocognitive disorders by providing a highly personalized and realistic therapeutic approach(Alcañiz-Raya et al.,2020). The ability to recreate specific environments and situations in a virtual setting makes it a valuable tool for neuropsychological evaluation, cognitive rehabilitation, and improving functionality in areas such as motor skills.

### *Phase 2: design and development of personalized therapies for neurocognitive diseases*

After identifying the needs for non-pharmacological therapies, the development of specific games using virtual reality (VR) began.

The development process was divided into the following phases:

1. **Design and development of virtual reality technologies – conceptual design:** Based on the defined requirements, initial concepts of the games were created. Sketches, diagrams, and descriptions of the user interface, virtual environments, visual stimuli, and expected interactions were generated.
2. **Prototype design:** Photoshop was used for rapid design prototyping of the user interface sketches. An aesthetic inspired by Ancient Greece was chosen due to the affinity of the target audience (older adults) for history and also for the potential appeal to a younger audience.
3. **3D modeling and rendering tools:** Tools like Blender were employed to create 3D models of virtual objects to be used in the therapies. These models were designed with attention to aesthetics, usability, and ease of interaction for the patients.

4. **Interface design – using unity for interface design:** Unity was the primary tool used to design immersive virtual environments and to define interactions with virtual elements and sensory feedback.
5. **Immersive sound:** Development of immersive sound design for each virtual space according to the therapist´s instructions regarding the patients´ needs.
6. **Development and testing of immersive spaces:** The development of spaces such as the game rooms continued, as captured in the code documentation and interface designs.

### *Phase 4: integration functionalities of artificial intelligence*

Simultaneously with the development of the games, functionalities of artificial intelligence (AI) are integrated to enhance the user experience within the immersive environment. To achieve this, methods for detecting machine learning in games were established.

The following steps were undertaken:

1. **Data model design:** The design of the data model was a fundamental stage in the development of our AI solutions for the treatment of neurocognitive disorders. During this phase, several key tasks were carried out to ensure the appropriate collection, labeling, and management of the datasets necessary for training and validating our AI models. The most relevant aspects of this phase follow:
    - Collection of relevant data: The first step involved identifying and collecting the necessary data to train our AI models. This included acquiring information from patients participating in therapies and utilizing our solutions. The collected data encompassed patients' responses to specific questions designed to assess their cognitive, emotional, and functional states. Interaction records from patients with our virtual reality (VR) and augmented reality (AR) applications were also gathered, allowing us to track how they interacted with the therapies and which areas needed attention.
    - Data labeling: Once the data was collected, it was labeled. This involved categorizing and annotating the data in a manner comprehensible to our AI models. For instance, patients' responses to specific questions were labeled based on their content and relevance to assess therapeutic progress. Interaction records were labeled to identify patterns of behavior and cognitive performance.
    - Design of data structure: An efficient data structure was designed to store and manage the collected datasets. This included the creation of databases and storage systems that enabled quick and secure access to information. The data structure was tailored to meet the needs of our AI models and ensure the integrity and confidentiality of patient data.
    - Implementation of analytical tools: Analytical tools were implemented to allow our team to examine and understand the collected information. These tools facilitated the identification of trends, patterns, and relationships within the data, which proved essential for the continuous improvement of our therapeutic solutions.
    - Assurance of data privacy and security: Special attention was given to the privacy and security of patient data. Robust security measures were implemented to protect confidential information, and applicable regulations and guidelines concerning health data protection were followed.

In summary, the design of the data model was a thorough process that allowed for the effective collection, labeling, and management of essential datasets for the development of our AI solutions. This phase was crucial to ensure that we had the necessary information to train and validate our models and to guarantee the privacy and security of the data of the patients involved in our therapies.

2. **Integration of the model into the games:** The integration stage of the artificial intelligence (AI) model into the games was a crucial step in the development of therapeutic solutions for virtual reality (VR) and augmented reality (AR). This phase is described as follows:

   - Choice of development platform: For the integration of the AI model, Unity was chosen as the primary platform. Unity is a widely used game development engine that offers a flexible and powerful development environment. Its capability to support VR and AR applications made it the appropriate choice for our implementation, as it is designed to create interactive real-time experiences.
   - Preparation of the AI model: Before integration, the AI model was prepared. This involved adapting the pretrained model to be compatible with Unity and to operate within the VR and AR environments. The model was configured to interact and provide real-time feedback based on the actions and responses of the patient.
   - Development of the user interface (UI): An intuitive and engaging user interface was designed to enable patients to interact with the therapies effectively. This UI was tailored to be compatible with the VR or AR experience, which involved creating visual elements and controls that adapted to the virtual or augmented reality.

3. **Integration of AI into the game environment:** The AI model was integrated into the game environment developed in Unity. This was accomplished through the programming of scripts and components that allowed the model to interact with game elements and respond to patient actions. For instance, if a patient was engaged in an activity within the game that required cognitive decision-making, the AI model could evaluate those decisions and provide immediate feedback.

   - Testing and adjustments: Following the initial integration, extensive testing was conducted to ensure that the AI model functioned effectively and provided the desired therapeutic feedback. During this phase, adjustments and refinements were made to the interaction between the patient and the AI system to optimize the therapeutic experience.
   - Performance assurance: Performance testing was carried out to ensure that the integration of the AI model did not negatively impact game performance in terms of speed and fluidity. Resources were optimized to guarantee a seamless experience.

In summary, the integration of the AI model into VR and AR games was a critical technical phase that enabled real-time interaction between the patient and our artificial intelligence system. Unity was the chosen platform for executing this integration, with particular attention given to the user interface and testing to ensure an effective and satisfactory therapeutic experience for the patients.

4. **Integration of motion detection:** The VR games already incorporated motion detection functionalities through the device used for gaming, specifically the VR Quest 2 headset. However, to enhance the experience for the elderly audience, the XR team implemented improvements in motion detection for this demographic, as testing revealed that this group faces motor difficulties.

    - Real-time data collection development: The entire gaming ecosystem was created to collect real-time data from users who tried them. For this purpose, a user registration gateway was created to identify the movements of each patient. The data that was collected depended on each game. The following is a summary of each of the data to be collected:

        i. Game: *Triton's Treasure* – Time to perform the open lock action

        - Time to perform the action trying to open the padlock with the wrong key
        - Time to perform the action trying to open the padlock with the correct key, but the wrong color
        - Total game time

        ii. Game: *Zeus's Beads*

        - Number of coins moved correctly
        - Number of coins moved incorrectly
        - Number of lives gained
        - Total points

        iii. Game: *Mirage Room* – Number of objects placed correctly

        - Number of objects placed incorrectly.
        - Time of each round
        - Total game time

        iv. Game: *The Mysterious Door* – The player has passed the round.

        - The time the round lasted
        - The player has lost a life.
        - The total time of the game (the time of all rounds)
        - Indicator if you are impatient (select a rock before the series is finished indicating)

        v. Game: *The Wishing Well*

        - The player has picked up the correct item.
        - The player has lost a life.
        - The total time of the game (the time of all rounds)
        - Indicator if you are impatient (select a rock before the series is finished indicating).

5. **Motion detection integration:** The VR games already featured motion detection functionalities through the VR Quest 2 headset. However, enhancements were made to better suit the older adult population, addressing motor difficulties identified during testing.

The entire game ecosystem was designed to collect real-time data from users during gameplay. A user registration gateway was created to track patient movements and actions. The data collected varied depending on the game.

### *Phase 5: integration of immersive sound*

Sound plays a crucial role in developing and deploying immersive technologies, offering opportunities to enhance therapeutic interventions. Immersive sound, including spatial audio, binaural beats, dynamic soundscapes, and interactive auditory feedback, creates a realistic and engaging auditory environment that complements visual and haptic feedback(Perez-Marcos et al., 2018). This multisensory approach is particularly beneficial for patients with attention deficit hyperactivity disorder (ADHD), as it aids focus, reduces hyperactivity, and promotes relaxation during therapy.

Specific techniques and technologies must be utilized to integrate immersive sound into our therapeutic games. The following outlines the key steps involved in programming immersive sound design and asset creation:

- Sound design and asset creation: Create dynamic soundscapes through layered audio tracks, such as ambient, background music, and real-time sound effects.
- **Spatial audio:** We used binaural recording techniques or 3D audio engines to simulate realistic sound positioning in our games in order to give a realistic experience to our patients.
- **Interactive sound effects:** Design sounds that react to in-game actions, such as a key clicking when inserted into a lock or a chime playing when a task was completed successfully.
- **Integration with game engines:** We utilized Unity and Unreal Engine to integrate audio assets into our games. We need to implement spatial audio libraries or plug-ins, such as FMOD, to add 3D sound capabilities. These libraries allow us to attach audio sources to in-game objects, creating a realistic auditory experience in our games.
- **Event-driven sound triggers:** We programmed sound effects to trigger based on specific player actions or game events supervised by the therapist. We used scripting languages, C# scripting in Unity and Blueprint in Unreal Engine to define these triggers(Liu et al.,2021).

   i. In *Triton's Treasure*, the sound of unlocking a chest plays when the correct key is used
   ii. In *The Mysterious Door*, a celebratory tone sounds when a sequence is completed correctly

- **Adaptive soundscapes:** We need to integrate AI to modify audio based on gameplay or patient responses. For example:

   i. Adjust the tempo or volume of background music in response to patient stress levels, detected through wearable biometric devices used in our games.

ii. Alter soundscapes dynamically as the patient progresses through levels to maintain engagement.

- **Testing and iteration:** Conduct user testing to ensure that soundscapes are neither overstimulating nor too subtle. Iterate based on feedback to refine the balance and timing of auditory elements.
- **Customizable settings:** We programmed the games to allow therapists to adjust sound parameters, such as volume, type of background music, or intensity of sound effects to meet individual patient needs.
- **Current applications and challenges of immersive sound:** Immersive sound addresses several therapeutic objectives for ADHD, but challenges, such as overstimulation risks and dependency on sound cues, must be considered with the therapist experience.

Examples of sound implementation in current games include:

i. *Triton's Treasure*: The creaking of the chest and the jingling of keys are timed to match player actions.
ii. *The Wishing Well*: Flowing water sounds and gentle chimes are used to create a calming environment.
iii. *Zeus's Coins*: Real-time clinking sounds signal successful actions, reinforcing focus and motivation.

We are working now on future developments in immersive sound technology integrated in our games:

- **Real-time biometric integration:** We want to develop other games where sounds adapt to the player's physiological state (e.g., stress or focus) using data from heart rate monitors or EEG devices.
- **Procedural audio:** We need to generate soundscapes dynamically based on game environments or patient actions, offering unique auditory experiences in every session.
- **Natural language interaction:** Future research includes enabling voice recognition to let patients interact with the game through spoken commands, enhancing engagement and accessibility.
- **Emotion-responsive sound modulation:** Finally, we will use AI to detect emotions from facial expressions or vocal tone and adjust soundscapes to provide calming or energizing effects as needed.

### *Phase 6: development and testing of immersive spaces*

Once the games were developed, the revision and testing phases began. Additionally, a dissemination phase took place, during which the development team presented the project and shared results.

1. **Game review:** A review phase was conducted by the XR team to document all games and validate the functionalities related to AI, motion detection, and real-time data collection.

2. **Project dissemination:** The development team was invited to present the project in various forums at the Polytechnic University of Valencia.
3. **Game descriptions and specifications:** Each of the developed games is described next with its technical specifications:

   - *Triton's Treasure* objective: The user must open a treasure chest by unlocking locks with the correct keys. Both the shape and color of the key must match the lock. The game collects real-time data such as time spent performing specific actions, and offers four levels of difficulty.
   - *Zeus's Coins***objective:** The user captures as many coins and hearts as possible within a limited time. Real-time data includes the number of correctly/incorrectly moved coins and lives gained.
   - *Mirage Room***objective:** The user places objects in their correct position, with different levels of difficulty determining how objects are disordered and where they need to be placed.
   - *The Mysterious Door***objective:** The user must press the sequence of stones indicated. The game collects data such as whether the player progresses to the next round, loses lives, and more.
   - *The Wishing Well***objective:** The user selects the correct objects from a series presented, avoiding penalties for wrong choices. The game collects real-time data on the user's performance.

## Results

The design, development, and implementation of new audiovisual technologies, metaverse, and artificial intelligence for the treatment of neurocognitive diseases in treatment programs for these diseases have been successfully completed through the design, development, and implementation of new technologies based on virtual reality (VR), augmented reality (AR), artificial intelligence (AI), and real-time data collection for the development of a customizable and adaptable metaverse that allows the treatment of diseases through different critical variables such as intensity, duration of therapies, or frequency of use, among others. Once these characteristics have been assessed, associated designs and developments will be carried out and a system with approved therapies will be implemented.

The tool developed is new and innovative due to the applications and functions the project encompasses and does not represent an improvement on a previously existing tool.

## Discussion and conclusions

Once the data has been presented objectively and in detail, it is crucial to interpret its significance. We need to explore the underlying meanings of the statistics, linking figures and values – down to seconds and hundredths – with the signs, symptoms, challenges, or strengths of the participants. This process involves integrating technology with pathology to correlate changes in performance with tangible benefits or drawbacks for patients, their disease management, and their daily lives.

At this stage, we can address the primary research question: whether these technologies serve as effective tools for neurodevelopmental disorders, including ADHD, Asperger's, and autism. Patients with Alzheimer's disease, other dementias, Parkinson's disease, multiple sclerosis, stroke sequelae, ADHD, autism, and Asperger's have particularly benefited from this project. Successful implementation implies an improved quality of life for these patients, especially regarding workforce reintegration, while also achieving cost savings for families, workplaces, and the Spanish healthcare system.

The introduction of immersive sound in the escape room as well as in each of the games has been decisive in giving users a feeling of reality in the game, which allows the data obtained in the game to be much more accurate since the users believe they are playing in a real analog environment. Without the inclusion of immersive sound, this feeling of reality would be impossible since, for users with these pathologies, the auditory sense is important for their ability to interact with these tools.

We must independently analyze variations in results, considering factors such as age, association, gender, and medication intake, both globally and in segregated groups. Our aim is to identify patterns reflecting the internal changes patients experience during therapy sessions and to investigate potential external factors that might impact performance. This analysis will guide adjustments and enhancements for future iterations of the developed spaces and research, optimizing the project's efficacy.

Ultimately, we seek to determine whether therapies utilizing immersive technologies and AI can become valuable allies in the future treatment of these pathologies. The expertise gained from previous projects and their success has garnered support from CEAFA, FEVAFA, and Valencia General Hospital.

## References


Adams, R., Finn, P., Moes, E., Flannery, K., & Rizzo, A. (2009). Distractibility in Attention/Deficit/ Hyperactivity Disorder (ADHD): The Virtual Reality Classroom. *Child Neuropsychology*, *15*(2), 120–135. https://doi.org/10.1080/09297040802169077

Alcañiz-Raya, M., Marín-Morales, J., Minissi, M. E., Teruel-Garcia, G., Abad, L., & Chicchi Giglioli, I. A. (2020). Machine Learning and Virtual Reality on Body Movements' Behaviors to Classify Children with Autism Spectrum Disorder. *Journal of Clinical Medicine*, *9*(5), 1260. https://doi.org/10.3390/JCM9051260

Arildskov, T. W., Virring, A., Thomsen, P. H., & Østergaard, S. D. (2022). Testing the Evolutionary Advantage Theory of Attention-Deficit/Hyperactivity Disorder Traits. *European Child and Adolescent Psychiatry*, *31*(2), 337–348. https://doi.org/10.1007/S00787-020-01692-4/METRICS

Arino, J. J., Juan, M. C., Gil-Gómez, J. A., & Mollá, R. (2014). A Comparative Study Using an Autostereoscopic Display with Augmented and Virtual Reality. *Behaviour & Information Technology*, *33*(6), 646–655. https://doi.org/10.1080/0144929X.2013.815277

Bioulac, S., Micoulaud-Franchi, J. A., Maire, J., Bouvard, M. P., Rizzo, A. A., Sagaspe, P., & Philip, P. (2018). Virtual Remediation Versus Methylphenidate to Improve Distractibility in Children with ADHD: A Controlled Randomized Clinical Trial Study. *Journal of Attention Disorders*, *24*(2), 326–335. https://doi.org/10.1177/1087054718759751

Clay, F., Howett, D., Fitzgerald, J., Fletcher, P., Chan, D., & Price, A. (2020). Use of Immersive Virtual Reality in the Assessment and Treatment of Alzheimer's Disease: A Systematic Review. *Journal of Alzheimer's Disease*, *75*(1), 23–43. https://doi.org/10.3233/JAD-191218

Dragusanu, M., Troisi, D., Villani, A., Prattichizzo, D., & Malvezzi, M. (2022). Design and Prototyping of an Underactuated Hand Exoskeleton with Fingers Coupled by a Gear-Based Differential. *Frontiers in Robotics and AI*, *9*, 862340. https://doi.org/10.3389/FROBT.2022.862340/ BIBTEX

Freeman, D., Yu, L. M., Kabir, T., Martin, J., Craven, M., Leal, J., Lambe, S., Brown, S., Morrison, A., Chapman, K., Dudley, R., O'Regan, E., Rovira, A., Goodsell, A., Rosebrock, L., Bergin, A., Cryer, T. L., Robotham, D., Andleeb, H., . . . & Waite, F. (2019). Automated Virtual Reality

(VR) Cognitive Therapy for Patients with Psychosis: Study Protocol for a Single-Blind Parallel Group Randomised Controlled Trial (gameChange). *BMJ Open*, *9*(8), e031606. https://doi.org/10.1136/BMJOPEN-2019-031606

Ghai, S., & Ghai, I. (2019). Virtual Reality Enhances Gait in Cerebral Palsy: A Training Dose-Response Meta-Analysis. *Frontiers in Neurology*, *10*, 421514. https://doi.org/10.3389/FNEUR.2019.00236/BIBTEX

Goharinejad, S., Goharinejad, S., Hajesmaeel-Gohari, S., & Bahaadinbeigy, K. (2022). The Usefulness of Virtual, Augmented, and Mixed Reality Technologies in the Diagnosis and Treatment of Attention Deficit Hyperactivity Disorder in Children: An Overview of Relevant Studies. *BMC Psychiatry*, *22*(1), 1–13. https://doi.org/10.1186/S12888-021-03632-1/FIGURES/4

Huang, T. L., & Liao, S. L. (2017). Creating E-Shopping Multisensory Flow Experience Through Augmented-Reality Interactive Technology. *Internet Research*, *27*(2), 449–475. https://doi.org/10.1108/INTR-11-2015-0321/FULL/XML

Johansen, T., Strøm, V., Simic, J., & Rike, P. O. (2020). Effectiveness of Training with Motion-Controlled Commercial Video Games for Hand and Arm Function in People with Cerebral Palsy: A Systematic Review and Meta-Analysis. *Journal of Rehabilitation Medicine*, *52*(1), 0–10. https://doi.org/10.2340/16501977-2633

Lehne, M., Sass, J., Essenwanger, A., Schepers, J., & Thun, S. (2019). Why Digital Medicine Depends on Interoperability. *NPJ Digital Medicine*, *2*(1), 1–5. https://doi.org/10.1038/s41746-019-0158-1

Liu, J., Bian, Y., Yuan, Y., Xi, Y., Geng, W., Jin, X., Gai, W., Fan, X., Tian, F., Meng, X., & Yang, C. (2021). Designing and Deploying a Mixed-Reality Aquarium for Cognitive Training of Young Children with Autism Spectrum Disorder. *Science China Information Sciences*, *64*(5), 1–3. https://doi.org/10.1007/S11432-020-2941-7/METRICS

North, M. M., North, S. M., & Coble, J. R. (1998). Clark Atlanta. Virtual Reality Therapy: An Effective Treatment for the Fear of Public Speaking. *The International Journal of Virtual Reality*. University Atlanta, Georgia, USA Editor: Han Sieburg.

Paro, M. R., Hersh, D. S., & Bulsara, K. R. (2022). History of Virtual Reality and Augmented Reality in Neurosurgical Training. *World Neurosurgery*, *167*, 37–43. https://doi.org/10.1016/J.WNEU.2022.08.042

Perez-Marcos, D., Bieler-Aeschlimann, M., & Serino, A. (2018). Virtual Reality as a Vehicle to Empower Motor-Cognitive Neurorehabilitation. *Frontiers in Psychology*, *9*(October), 409091. https://doi.org/10.3389/FPSYG.2018.02120/BIBTEX

Part II

# Production and creative practice

Chapter 16

# Sound installations

## The art and implementation

*Charlie Morrow*

### How I got started

Immersive audio, in my definition, is created physically and/or electronically. In both instances, it immerses the listener in an audio experience that is a step beyond a frontal or single source experience. Physically immersive experiences include out-door sonic events and events in resonant environments such as underwater or very reverberant spaces. In these instances, the listener is immersed and at times saturated with sound. A loud explosion is an example. Robert Thurman says, 'Death is the ultimate immersion'. Ambiences are immersive in the real world. I have made my work around them, usually building the ambience first and working in added layers of sound.

My work in immersivity starts with what you'd call *physical immersivity*. I come from a medical family. I've actually had the experience of doing autopsies. I was going to be a doctor, but I took a left turn. My whole life has been as a creative person, mostly working in sound and music and creating immersive experiences. In the 1950s, I began creating physically immersive experiences for music performance, sound works, language/poetry, performance actions, and conceptual events.

This grew from my love of pageants. The first pageants I experienced were celebrations in grade school as well as ceremonies of local Passaic, New Jersey, Catholic Churches and synagogues. I think the deepest experience that I had in terms of immersion as a child was attending six o'clock mass and actually somehow feeling that the Holy Spirit had come through my body and through the whole place. It was like a wave that traveled through the congregation. You would see people reacting to this spiritual wave. A wave that was unseen but totally felt as a group. And for me, these are the immersions.

In an even earlier one, I remember, the sounds I heard before being born. That waking up to sound identified: what's me, and what's the rest of the world? The main thing was being able to reach beyond myself and to understand that I am a self. Without that, I was part of my mom's body. So these very fundamental organic processes inform my idea of immersion.

As a fetus, I was immersed and heard sounds from the world beyond myself. I heard and identified immersive spaces by their audible boundaries. This led me to later develop a method that made those boundaries vanish and to replace them with electronic sonic illusions. In the 1950s I studied electronics on my own. I became a radio amateur, building gear as an operator of communication systems with Morse code and voice. I installed antennas and RF ground systems. I learned later that humans hear ground reflections, starting in our crawling stage and then throughout our lives. The ground reflections are

DOI: 10.4324/9780429297526-19

*Figure 16.1* Morrow Sound cube



specific to height above ground. We identify sound location in the sound ambiences of the world with planar, spherical, and height coordinates. Ground reflections shaped my thinking and the logic of my patented immersive sound method for physical and virtual spaces.

The Morrow Sound cube, the building block of my 8.1 modular design, has four floor- or ground-level speakers and four overhead speakers, avoiding ear level diffusion. These cubic units can be built out in any direction to create a unified sonic illusion space which backgrounds the physical architectural acoustics. Subwoofers, center point, and other mono sources, as well as directional speakers, effectively edge blend beams within the 3D sound field. As well, our software scales sound objects.

I planned to be a medical doctor. I studied chemistry and anatomy. Wave motion and sensation of sound captured my imagination. Calculus and topology studies opened the door to macro-shaping of music, sound works, language/poetry, performance actions, and conceptual events. The first events I created were humorous musical and sonic moments. This work included super slow and spatial performance of known music works, mime performances of music, many without sound, and pranks.

My immersive events started with sound games and were influenced by parades and mass events, indoors, outdoors and underwater. Spectacle and mixed live and electronic shows. Years later, my electronic immersive MorrowSound works would be called Sonic Puppetry. They derive from these roots.

Figure 16.2 shows images from my event composition Copenhagen Waves (1985) with several thousand performers, boats of performers, many churches bells and lights, a helicopter with the lead singer Festival Fool, as well as radio and TV live reportage and transmission.

*Figure 16.2* Pictures of Copenhagen Waves, Fools Festival 1985

Photo ©1985 Copenhagen International Festival

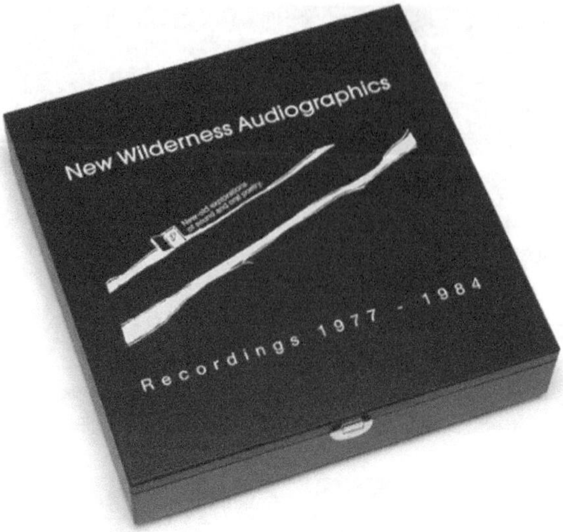


*Figure 16.3* New Wilderness Audiographics – Recordings 1977–1984 (11LP Box + Book)

*Published by Soundohm, 2024*

In the 1960s. I built and operated professional audio and broadcast studios, starting in my 20s, to produce an income from audio, music, and media production. As well, I produced and published other artist's works. Vinyl-on-Demand published a box set of some of our best releases, 11 LPs and my book (shown in Figure 16.3).

As an official event maker for New York City, I created celebrations in public spaces and on the airwaves. All were immersive. I built very loud and very quiet audio environments, environments with motion, gesture, and sometimes without physical sound. I worked with loudspeaker builders and circuit benders. I crafted and combined branded sound and purely artistic sound. The immersivity was that of sound in physical spaces, not of real or virtual 3D electronic sound formats.

In my essays on sound art, I say that sound art is the original practice. When people began to find a place to live and negotiate the world as early human beings, they relied on their wisdom, about sensations and opportunities and data and handling danger through their senses. We can look for the roots of sound art in the way that people use sound and seeing and hearing. Looking and listening. Both. Being two different things. The bandwidth of perception versus the human mental apparatus.

In discussing sound art in the 20th century, I was faced with the issue that people, particularly in Europe, used to say in order to be a qualified sound artist you couldn't be a composer. That reminded me of the *Light Poem for Iris Lezak* that Jackson Maclow wrote, which said 'the sun must meet the following requirements in order to be a star'.

I believe that sound art is the original art and from that come talking and speaking and so forth. We are basically going back to rudiments when we talk about sound art. Similarly, immersivity grows out of the same bedrock. All sensory environments are immersive. And so the question then is what are the ways in which we're able to perceive them? What's the bandwidth of perception? And then what can we do with that? The earliest needs for having awareness of this are in order to eat, reproduce, and defend. In sensing of place and in place making. Negotiating space sonically. In coordination with seeing, thinking, and doing. All very fundamental. And so immersive in the sense of being enveloped by external things or immersive as far as being in yourself. As Robert Thurman says, 'The ultimate immersion is death'. The poet Jerome Rothenberg said that the ultimate immersion for a writer is being able to sit alone, imagine a universe, and put it down on the page. So when we're talking about *immersive*, we're talking really about the marshaling of our senses and our thoughts. In order to do functional things – either to think or create.

I was creating musical events and creating a lot of jokes. I was a jokester as a teenager. One of my earliest pieces was to perform Gabrielli's *Sonata pian e forte* for brass choir walking in slow mime. Slowly crawling up to the stage. It was good music and it was fun. But it was also the beginning of ultra-slow things, with tempo change. It was the beginning of the idea of macro-composing – which is taking something and then spinning it all. The other early one was an 'arrangement' I did in a rehearsal of an orchestra in which I played. I arranged with all the orchestra that, at a certain time, everybody would stop playing and belch and cough. The conductor stopped and laughed. understood immediately what was going on. Everybody got it. It was a moment that made me really interested in spending my life doing this sort of thing. What proceeded from that was developing multiple skills in different areas.

I earned my living as an arranger and composer in art projects and in commercial areas, as a jingle writer, sound designer, and producer. I felt that having a studio of my own was important. Therefore I built one in my then New York City flat and have always had a studio since.

With poet Jerome Rothenberg and Rip Hayman, I formed a non-profit organization, the New Wilderness Foundation Inc., for art events/celebrations; publications (*EAR Magazine* and *New Wilderness Letter*); broadcasts on public radio, TV, and cable; the Audiographics cassette label and a hive of bands: the Ocarina Orchestra, New Wind Band, Grand Conch Chorus, and the New Wilderness Preservation Band. We specialized in environmental performance, site-specific events, outside the concert hall with its controlled acoustics. Immersion and physical context figure throughout.

New Wilderness Audiographics is one of the primary labels of sound art in the world, where artists co-produced the sessions. That was possible by having people come in to learn to use the equipment because that was the beginning of combining this kind of philosophical interest in sound with my technical background,

As a kid ham radio operator, I built my own radio (RF) and audio (AF) systems. I built a Klipsch speaker, and I became particularly interested in antennas and ground wires. My strongest influence in creating what is MorrowSound was my work in a field called QRP. In the ham radio world, QRP means to talk with a radio system at the lowest possible power level to the largest possible part of the world. QRP could best be done on an island where you would have your long wire antennas at full wavelength (or multiples of full wavelength). You'd have similar ground wire patterns. You would put ground wires into the saltwater and use a metal stake underwater. My friend Art Kaye (who later on put the TV broadcast antenna on the Empire State Building) was a fellow radio amateur. He put in this huge ground wire system, which made that antenna activate. I understood later on that sound works in some of the same ways as RF.

When I was a kid crawling around, I heard sound reflections. As I stood up, I continued to hear the sound relative to the position of my head but at different heights and angles. My invention is to create a believable audio field, that you can walk around in, using the fewest speakers possible. By matching the ground reflections to reflections in the atmosphere, we found that there was an area, a ratio, where you get the immersive illusion. We found that you can dial into it, and dial out of it, and that you actually can get partial immersivity. It is a nonlinear but continuous sound pressure spectrum – where you can go from non-immersivity to full immersivity and out of it again.

We humans have a differential operating system that knows when it's on the beam. That is to say that it's easier to hear when we're going from one perceptual image to a different perceptual image than it is to understand subtle changes in a continuous experience. Often, people can't hear a very slow crescendo. It's almost like we need steps, and our brains have thresholds of comparison.

A scientist friend of mine was studying an Indian Yogi's audio perception. His study found that the most highly developed Indian Yogi could hear the same sound repeated hundreds of times, and each instance would register the same brainwave experience as the initial experience. So what we're dealing with is the combination of when you are able to differentiate something from what came before and how much you're flowing with it. It's an immersive experience. Right now you don't doubt that the sky's up. We have constructed this illusory world. In reality, none of this exists as we perceive it. It's all the synthesis of our inputs in an agreed-upon physical and mental language. That's what reality is. Reality for audio, then, is when you feel like whatever you're listening to sounds believable. If it was slightly off, you would doubt it. And if it was way off, you would understand that you weren't there. The only way to do it is as simple as possible. So I use an 8.1 sound cube system with parallel sound fields. With the right ratio of parallel sound fields, you get the immersive experience. It's the kind of thing that happened with the RF fields and ground wires in Art Kaye's ham radio QRP rig when the antenna and the ground wires were in harmony. Suddenly, it was able to reach 100 times further around the world.

In the late 1990s, I met acoustician Brian Katz who interested me in Ambisonics. I had previously met Michael Gerzon through participation in English guitarist Derek Bailey's Company Week, London, in the early 1980s. It was inspiring to learn that Gerzon had created Ambisonics.

I changed the elements of Ambisonic speaker placement and created MorrowSound as an 8.1 modular system. My team and I, including sound designer Jeff Aaron Bryant, found ways to make multi-zone spaces, some huge, and created a manipulatable control for immersivity itself. A goal for these electronic MorrowSound fields was overwhelming the physical environment so that it seems to disappear. Walls, ceiling, and floor, at correct immersivity, sonically vanish, so a virtual sound world can be built.

In 2018 I partnered with William David Fastenow and his company Park Boulevard Productions where we continued to hone this concept and develop MorrowSound products and content as a co-venture. Together we found new ways to create multiple planes of sonic information derived from multiple points both above and below listening height. The idea was that by designing the way a space will *sound*, we can affect the way a space might *feel*. Fastenow's work with 'subject-oriented' sound helped us to consider the listeners' unique vantage points as a key ingredient of our affective design process.

Our work with believable ambiences gave way to circadian sound masking. All of this work continues with Fastenow and Bryant in the MorrowSound co-venture.

## Working on a project

I approach every project in a unique way. Every project is different. Most of my work is based on really listening to what people want. Really figuring out what it is actually about. Typically, everybody comes ready to tell me how to do the piece. They know better than me, but they're not sure how to get there. I don't ever have a situation where someone says, 'Do whatever you want'. I am usually asked to join a team one way or another – even if it's only one other person. The first step is really hearing what they want. The major ingredients of a project are the spec, the schedule, and the budget. It is very important to find out about those things, because there's only so much you can do.

The first step is to find out what are the constraints of the project and the logistics. I begin every project by worrying about what can go wrong. The spec is very important. We don't live just in a silo of sound. You have to be very precise about the delivery system. For example, I am consulting on a project now for a sculptor working on a sculpture with 15 sonic segments. It's a giant dinosaur. I arranged to provide the speakers for her. We met with the speaker company and decided that she'd be well off to have a system using active speakers and to use Dante. The question came up, what audio player? From the standpoint of a professional sound person, you cannot take a basic Mac Mini or laptop and install it in a museum. Also, if you're in any kind of situation where the fire department has to do an inspection, you're suddenly confronted with the fire laws and fire rating of the parts.

Another example is a project that I was working on in Alaska, an installation at the Smithsonian's Arctic Studies Center exhibition in Anchorage. It turns out they use the chase over the ceiling for the air conditioning. This is where I had put my speakers, which were not UL rated for that. I learned about this after everything was installed and the museum was built. It meant finding an alternative device after the whole job was done.

The next question was how do you run it? You can't run it off a CPU because CPUs run hot. CPUs also get regular operating system updates. If your system is online, that may change while you're operating your system. This means you cannot have remote maintenance unless it is on demand, which means that somebody authorized has to go and hook you up to the internet and supervise your remote maintenance. You have to watch that your operating system is not changed automatically, effecting the program. These are elements and complexities we are planning for before we're discussing content.

The next element in the project is doing a preliminary walk-through. I have seldom produced an event for a space that I couldn't see. Every space has its own feeling, and you'll always notice something. You begin to get a sense of the place and how people will come and go in it, and so forth. It's a flow. The flow is interesting, as are the constraints. For example, one of the biggest factors in any group setting is the sound of footfalls. In some cases, such as a hospital setting, people are told what shoes they can wear. On a construction site, helmets must be worn. All of that becomes very important. You sound design for a place, and then it's filled with people chattering on phones and the sound of feet. So all of that is the prevailing environment in which your sound idea has to live. In immersive experiences you must consider all these elements: what's in the environment? What do you want to do in it? How long will it be operational? And so on.

When I work on a project I often don't go into the project knowing what the configuration of the loudspeakers will be. This is because most people don't want an immersive installation, they want an audio production company. Generally speaking, immersion is very hard to sell because it uses more speakers. When you come right down to the budget cutting moment, you either have to have a cheaper solution or have no sound. Even in very large projects, the first thing to go is the immersive sound. This is not a pleasant part of the process. You can work on a project for two years and then find yourself with no installation.

Right now one of the real futures for our business is the growing construction of human habitats. I believe in controlling the environment of apartments and living spaces. The really interesting aspects of immersive sound are the architectural, electronic, and hybrid components, working on standards – whether it's the noise of the city or in working spaces. Open plan working spaces are very common now. One of the things that my company does is natural sound masking. The reason for this is that we have found that studies of various forms of flavored noise (white noise, pink noise, etc.) have definite degenerative aspects for both human physiology and psychology. We work with what we feel is a *healthy sound.*

Let's start with the fact that while I have a tool called the MorrowSound cube, this is just one of the tools in the kit that my company offers as designers and installers. It's similar to saying that I have a special paintbrush, but I also work with pencil, etc. So I have gone away from referring to it as Morrow Sound, and now I refer to it as *healthy sound.*

We have different kinds of immersivity for different areas. Part of the immersivity is the use of noise masking of various degrees, for people who are working and want total privacy – or no privacy. All of these things are programmable. In some installations, the soundscaping systems are programmed over days and months and so forth. For example, in one of the hospitals where we did an installation, they wanted it to always be spring. Within that environment the trajectory of bird flight in the soundscape mimics the natural movement of birds.

The granularity of my work is that the spatial 8.1 system is just one piece of it. We work with it because we have a library of 8.1 experiences, and the library makes it cheaper to deliver. Otherwise, if you have large installations, you're often building the library up from new elements. In that case you have to think about the acquisition rights – which is either the cost of recording or the cost of licensing particular sounds. When I worked for a game company with the remit to use all of their sounds, that division of the company had to license the use of the game sounds from the central part of the company (because usually big companies charge themselves for using their assets).

### *Bread and Puppet*

In the Bread and Puppet Theater, all the sounds are going to be played by people. Their major pieces are going to be done outdoors, except mine which is going to be in a dirt floor barn cathedral. It starts from the performers and the environments that they're going to perform in. For Bread and Puppet, I took Beethoven's *Grosse Fuge* – which was a favorite of Bread and Puppet's director Peter Schumann – and we imagined Beethoven being deaf and writing *Grosse Fuga*, one of his very last pieces. I went in and stretched seven sections of the piece – going from real time into ten times slower. The experience is a sense of Beethoven's mind drifting and his thinking about the next world. (You can see how this is related to my experience in the church when I was a kid, when the Holy Spirit came through on a Sunday . . .). The string players didn't want to play a written part that suddenly showed a much slower tempo. I couldn't just write $t = 60$ and then $t = 6$. They wanted to see all the notes written out in a familiar tempo. A big issue is notation so that the musicians can comprehend the beat, the number of notes, and enjoy playing.

I determined as a young jingle writer that I would bring a magnetized combination of conceptual thinking and Hollywood sound. I thought it was very important to be able to have the edge on the sound when you needed it. The degree of 'bite' is something that you would think of consciously, as the maker of sounds. You want to have total control over whether you sound like a documentary, or a telephone on location, or a really intense experience made to knock your socks off. For example, I made a life-size train, made just of 3D sound. It won the TEA Grand Award for exhibits. It is in the Great Platte River museum celebrating this history location in the prairies in Nebraska. It was all about transportation. We had buffalo herds and the sounds of moccasins. We had the sounds of an 1860s Rocky Mountain train as an overhead experience. People would hear (not see) the train. We worked with the input of the Smithsonian Institution transportation historians. They unfortunately did not have the sounds of the whistle so we had a whistle builder in Denver build the correct period whistle. The train sounds came from a recording by an accountant in Pennsylvania whose hobby was recording trains. He had a wire recorder which he used to make the recordings in 1938–1939. He used a wire recorder because the dynamic range is greater than that of magnetic tape. The recordings were approved by the Smithsonian consultants.

## Audubon's Aviary

My favorite project is called Audubon's Aviary, which was a multi-year project for the New York Historical Society. It was done in the Great Hall. They have the Audubon original paintings, prints, and sculptures. They can only show them for a couple of months

*Figure 16.4* Audubon's Aviary show with MorrowSound 16.2 system and sensor speakers. New York Historical Society 2006

Photo ©2006 Maija-Leena Remes

due to light sensitivity before they have to go back to storage. They're all regulated by conservatory rules about light on old paintings. The idea of the project was to develop a very intense audio experience that would be an attractor for people. I put up multiple Morrow Cubes there and built these devices that I called 'sensor speakers'. It was a gimbal-like pipe with two boxes on it: one was the speaker that could be aimed up the wall, and the other had a motion sensor on it. We could regulate at what proximity it would trigger when kids got close to it. We also tied it to circuits that would decide how much of a bird song to pick off each play so that you wouldn't ever hear the same thing repeat like a cuckoo clock.

So there were two experiences. One activated four big zones in the big exhibition room, which we filled with the sound of a summer day because it was generally cold in that room. (We've often used that trick of an ambience that sounds warm.) We then built a sonic moving flock of cranes. When the system works right, the 3D sound illusion overwhelms the physical reality. This is how you know the immersivity is correct – when the walls, the ceilings, and the floor sonically disappear. Then we could fly in the sonic cranes. It sounded like they flew in from distant West 42nd Street, through the walls of the West 77th Street Historical Society. They flew into and completely through the Great Hall. Then they took a turn east and they flew through the walls out over Central Park. We had to decide with the curators how often the dramatic crane experience would occur and for how long. Part of that experience design was that the geese will have trajectories and a dwell time while in the exhibition space. We had a focus group to listen to the experience and talk about the travel time of the cranes, and how fast the cranes would disappear. The end result was that the goose fly-by would happen three times an hour.

Our sensor speakers were mounted below many of the bird paintings. When the bird sounds were triggered by visitors' motion in the vicinity of each painting, there was the illusion that the sounds were from the birds in the actual Audubon painting of that species. Some of those painting birds would occasionally seem to acoustically jump out from the sensor speaker and come onto our Great Hall 3D system which carried the flock of geese experience. You'd hear a painting bird come out and move around in its real-life normal pattern. We had to devise rules for the bird sound triggers so that the experience seemed natural. Also to prevent visitors from playing with the sound triggers for fun.

I did several shows where I used that trick to create a warm atmosphere in a room. Another example of this was for the Microtropolis show for Microsoft. It was November, and it was cold in the pier building on the shore of the Hudson River. The idea was to play a recording of an obviously summer day and somehow, for a lot of people, it makes them feel warmer. You have to choose the right sonic illusion to warm the place and the right feeling of warmth. Sometimes I think of my work as sound illusions. Maybe I'm a sonic illusionist rather than a sound designer because we're fooling the brain and in the most pleasant ways.

## Main lessons learned

The most important lesson is to keep track of the details and descriptions of your work, label all your sound files, keep notes and photos. You do this to be able to get back to anything that you've done to make changes and new thoughts about it. You have to back it up twice, at least. Imagine if you're in a mood of creativity and you sort of just do things and don't label properly. You can lose track of your work, with hundreds of (and sometimes more) files, operations and decisions. It could suddenly all fall apart and would need to be restored, or you could get a macro request for a change. If you don't label things properly and don't keep track of your work, the consequences of that can be most horrible.

We generally find that once an environmental installation is in, clients want touch-ups about a month or so into it. This is in our contracts, based on experience. That grew out of the responses of the focus groups involved beforehand.

We find ironically that many of the most interesting sounds in soundscapes are the ones that have the fastest burnout for listeners. A month or two after you have created your favorite sound, people will have heard it too many times. If you're creating a sound environment, think about the idea that people should live with it. Any natural environment is constantly changing. When you build something, you have to come up to that standard. Otherwise, people will notice. People absolutely notice there's something mechanical about an environment that's mechanically repetitious. It was Milton Babbitt who said, 'Nothing grows older, faster than a new sound'. He was thinking of people who are going out of their way to find something really unique. I knew Milton. He was a man with a great sense of humor.

If you're working with a set of ingredients, and the ingredients are heard identically every time, this becomes repetitive and unnatural. For example, if you're working with the sound of a bird, it's very important for the particular bird to be heard differently at different distances. In nature, that bird is never the same twice. Developing this skill involves spending time listening to nature. You have to be careful about not making things sound artificial. Paradoxically, perfect artificiality is what show business is about. When you think about the most dramatic scenes in any film – the scenes that people want to see over and over again. On one hand, you have the need for variety, and on the other hand a need for an absolutely unique, iconic experience. When making the work, you have to understand whether you're making an icon, or a varying perceptual palette. It's the duality and juxtaposition of the constant and the changing. It's the duality and juxtaposition of the constant and the changing.

Chapter 17

# Perspectives on multi-channel compositions

*Michael J. Schumacher*

## Multi-channel sound requires a 'new wine, new bottles' approach

I've been composing with multi-channel sound systems for over 30 years. From the outset, my creative process has focused not on adapting stereo to multi-channel formats but on composing directly onto an immersive setup of my own design, one that began with 18 speakers and evolved into a 'standard' 8.1 setup. The computer has been my medium, initially utilizing synthesizers (through MIDI) and later transitioning to self-contained software that could play, process, and spatialize sound. This software has employed algorithms capable of generating varying outcomes with each presentation of a piece. Additionally, the manner of presentation, the rooms in which the pieces are heard, as well as the social conditions – seating, lighting, the conduct of the audience – have all been conceived as integral to the works.

Including my early encounters with quad, I've seen a number of efforts by the industry to promote multi-channel sound as a way to experience music (as opposed to movies or games) to the public. Each of these has had their challenges, certainly with regard to personal or home listening but even in the adoption of immersive systems to enhance live concerts. This history falls notably short of the pioneering spirit and ambitious innovations characteristic of the early experimental phase in the 1950s and '60s by composers like Amacher, Cage, Lucier, Oliveros, Stockhausen, and Xenakis, among many others. In my view, the primary reasons for these shortcomings are twofold: first, the failure to cultivate a niche market by developing a repertoire that truly leverages the capabilities and special attributes of immersive sound; and second, the industry's economic model, which necessitates selling millions of units of a particular format to justify the investment.

## Encounters

My first encounter with surround sound was when my sister purchased a Superscope QA-420 quadraphonic integrated amplifier in 1972. It had four discrete inputs and outputs; when she wasn't using it, I'd experiment by plugging in cassette decks, portable radios. and a record player, add two of my speakers to hers, and make four channel audio collages. (Much later on, I used this same amp in my first 18-channel setup.)

In the early 1970s, the 'hifi' stores were pushing quad, but it didn't go anywhere. My sister, for example, used the Superscope to listen to stereo (or mono) records. There were quad LPs, but you needed a special needle. There was also the possibility to 'decode'

DOI: 10.4324/9780429297526-20

stereo records through a matrix to get quad, but it didn't seem worth it, or even appropriate; the music we listened to, mostly pop, rock, and mainstream classical, sounded like it was coming from a stage, so what was the point of the rear channels? Quad tape decks were probably the way to go but were expensive and more challenging for the casual user than LPs. So in the case of my sister and me, you had, in miniature, the situation that has seemed to repeat itself every so often with regard to surround or immersive formats and music: on the one hand you have the average consumer, satisfied with stereo and resistant to paying for the new technology, and on the other hand you have the 'enthusiast' who, as often as not, is also a content creator.

Reflecting on the 1970s, often considered a 'golden age' of hi-fi, the prevalent listening practice of that era was, in its own way, an immersive experience reminiscent of attending a live concert. My father, for instance, had a respectable audio setup at home featuring Wharfedale speakers, a Garrard turntable, and a Sherwood receiver, arranged in our living room with the speakers positioned symmetrically around his preferred chair. Every so often, he would bring home a new LP from one of the two local record stores in Elizabeth, NJ – Alwilk and Vogel's – or occasionally from Sam Goody in New York City. He would then sit back in his chair and devote his full attention to the album from start to finish, much like being in the audience of a live performance. This setup essentially turned the living room into a virtual concert hall. Advertisements from that period for hi-fi systems often depicted a man, sitting in a chair, in 'deep listening' mode, with his 'significant other' standing admiringly or proudly at his side. Some ads might not feature people at all, instead showcasing an upscale designer chair, symbolizing the listener's physical and mental engagement with the music, alongside the sound system.

Another formative experience in immersive sound for me came from the cinema, with films like '*Nashville*', '*Star Wars*', '*Apocalypse Now*', and '*The Exorcist*', to name a few – not only because of the surround sound systems (when used) but also due to the comprehensive immersion into their meticulously designed worlds. '*Star Wars*' was my first introduction to the particular emotional impact that sound originating from behind the audience can have, triggering an instinctual 'fight or flight' reaction. The auditory experience in '*The Exorcist*', which utilized elements of Musique Concrète, was so intense and terrifying that I found myself needing to step out of the theater momentarily to regain my composure. Films like '*Nashville*' (and later, for me, those of Jean-Luc Godard) demonstrated the impact of complex, multilayered soundtracks and suggested the potential in using multi-channel systems to enhance separation. These experiences in movie theaters made clear the importance of Chion's idea of the 'added value' of image coupled with sound and begged the question, what to do when there's no image?

## IU, first electronic compositions

While doing my undergraduate work at Indiana University, I composed some pieces in the electronic music studio. The studio had an Emu modular synthesizer, Eventide H910, Lexicon PCM 42, a Sony 16 track and a couple of Ampex 1/2″' mastering decks, along with EQs and compressors that, unfortunately, didn't interest me at the time. I decided to present one of these compositions at my senior recital and told my teacher, Bernhard Heiden, with whom I'd focused on acoustic pieces and hadn't, until then, shared the electronic stuff. He said, in his inimitable, growly way, 'The only thing I like about electronic music concerts is when something goes wrong'. Reflecting on this led me to consider the theatrical and visual dimensions of live performances, including something as apparently straightforward as a piano recital. The idea of an audience simply gazing at a pair of speakers on stage struck me as potentially lacking, both visually and in terms of expectations regarding musical structure. To address this in my electronic composition, I opted to darken the room and added an additional stereo pair at the back, limiting visual stimuli and creating an immersive sound environment.

## 1980s: Juilliard, Cunningham, etc.

While at Juilliard in the 1980s, I went to the Cunningham shows at City Center, with Cage, Tudor, Kosugi, Behrman, King, and others performing the music, and, as I recall, they would use surround systems at least some of the time. These concerts were what got me into experimental music. There was also '*Roaratorio*', with musicians and speakers placed all around the audience, Tudor's '*Rainforest*' at Lincoln Center, many concerts at Phill Niblock's Experimental Intermedia space, which had Klipschorn and Cornwall speakers surrounding the audience, and, perhaps most important of all in terms of creating my own listening space, La Monte Young and Marian Zazeela's '*Dream House*'. Later, in 1992–1993, spending a year in Berlin, I heard a number of concerts presented by Folkmar Hein, including Xenakis's Bohor in eight channels. There were also immersive events produced by Freunde Guter Musik (Matthias Osterwold). While in Berlin, I visited Klangraum Krems in Austria, another inspirational space for multi-channel and immersive sound, run by the late Jo Aichinger.

## Sullivan street

In 1989 I moved into a one-room studio apartment and began building a multi-channel sound system. This was built around the synthesizers I was using at the time, an Oberheim Xpander with six discrete outputs and a Yamaha TX-802 with eight. A piano 'module' and stereo cassette deck brought the total number of channels to 18. I found some old amplifiers around my parent's house and bought more at garage sales and used music stores, especially Rogue Music (developing a particular fondness for the Crown DC series amps, eventually acquiring six of them). I also gathered a diverse collection of speakers, varying in brand and quality, some home-made. These speakers were distributed throughout the apartment at different elevations, without regard to creating a 'sweet spot'.

The synthesizer outputs were connected directly to the amps/preamps/receivers; no mixer was used; it was purely 'point source'. (In any case, panning sound didn't interest me. At this time, though I had heard the music of composers such as Bayle and Parmegiani, I was not aware of GRM's Acousmonium, but, in any case, the idea of performing the movement of sound in space was antithetical to my approach.) A computer, via MIDI, controlled the synthesizers. Using system exclusive messages, I could change synth parameters as part of a composition. I used programs like Mark of the Unicorn's Performer and Opcode's Max.

## Metaphor for multi-channel sound

At this juncture, an idea struck me that would guide both the design of my multi-channel systems and my approach to composition. This insight drew heavily from John Cage's concept of rhythmic structures that serve as frameworks to be filled, using the fundamental interplay between sound and silence as the cornerstone of compositional structure. I began to see the 18 speakers in my apartment as individual 'portals' to distinct sonic activities. When a speaker – or 'window' – was closed, the sound it held was silenced; as it opened, the sound became audible, revealing a slice of an ongoing sonic process, much like the continuous sounds one encounters in everyday life, such as conversations, traffic, birds, radios or TVs, car alarms, sirens, airplanes, or someone practicing an instrument.

I envisioned an automated mechanism to control these windows, mixing and matching these sounds in varying combinations and intensities. This setup allowed for the auditory landscape to shift dynamically, where the interplay of different sounds could lead to a change in perception. Similar to how harmonization can transform the interpretation of a melody, the contextual changes brought about by these combinations of sounds would suggest new meanings, by either highlighting or concealing certain elements and drawing connections between their pitches or rhythms.

Initially, my sound palette primarily consisted of synthesizer-based sounds. The goal was to craft sounds that were absorbing on their own yet could seamlessly blend with others to form complex textures of varying densities. By around 1996, computers became capable of playing back sound files, marking a significant turning point. I then adopted a modular approach to composition, designing discrete, stand-alone 'objects' that could either synthesize or playback sounds. These modules were capable of manipulating various parameters in real-time, governed by algorithms that also managed the timing for activating or deactivating each module. This method laid the groundwork for a series of compositions I referred to as *Room Pieces*.

## Studio 5 Beekman

The *Room Pieces* not only demanded an extensive setup of speakers but were also of long duration. Observing the execution of the algorithmic processes playing out over weeks or months in my apartment, I became attuned to the transient quality of the sounds. The awareness that the specific sound combination I was hearing at any moment was unique, ephemeral, contributed to a distinct state of perception. Additionally, I started to incorporate the spatial positioning of sounds as a fundamental element of the composition's structure. Neither of these important aspects of the music could be captured on a stereo recording.

In 1996, after my partner at the time, Ursula Scherrer, moved into my apartment, I rented a studio space on Beekman Street near City Hall in downtown Manhattan. This studio was situated in a charming old 'Queen Anne–'style building, primarily occupied by accountants and lawyers. The building's robust wall construction gave me the confidence that I could make sound without causing inconvenience to or being disturbed by the neighboring offices. The studio featured a compact foyer leading into three connected rooms, which allowed for distinct areas dedicated to equipment, listening, and relaxation.

However, the sound insulation was not as effective as I had anticipated, and I had to construct an additional wall lined with lead sheets to mute the constant ringing of telephones from the adjacent office. Following this modification, along with sealing the windows tightly and installing wall-to-wall carpeting, I managed to create a highly private listening environment, short of constructing a decoupled, floating room.

The intimate listening area, measuring approximately 10 feet by 16 feet, was outfitted with ten speakers: six mounted on the walls and four positioned on stands. The adjoining room housed the equipment, including amplifiers, a matrix mixer, and a computer, with cables neatly routed through the walls. A mobile work desk equipped with a computer screen and keyboard could be moved easily in and out of the listening room as needed.

Inviting friends and fellow composers to experience and share their work led to the birth of Studio 5 Beekman, a gallery that quickly became a focal point for experimental sound artists. The gallery, operational on Thursdays and Saturdays – the same schedule as the Dream House – eschewed traditional openings to preserve the integrity of the sound pieces, emphasizing quiet engagement over social chatter. An adjacent 'chill' room served as a salon for post-listening discussions, often with the artists themselves. The inaugural exhibition in January 1997, a collaboration between David First and Patricia Smith, set the stage for a series of notable shows by artists like Ellen Band, David Behrman, Maria Blondeel, Tom Hamilton, John Hudak, Ron Kuivila, Christof Migone, Lee Ranaldo, and Leah Singer, Steve Roden, Stephen Tunney, and Achim Wollscheid, all embracing multi-channel sound. Over four years, this space not only hosted these exhibitions but also served as my personal studio for composing over 30 *Room Pieces*, leveraging the studio's unique capabilities to explore the ramifications of computer-generated, immersive soundscapes.

## My thoughts on multi-channel music composition, ca. 2000

Over the decade-plus of working with multi-channel sound, I'd developed specific notions about what kinds of music suit this format best and how it should be presented to audiences. Essentially, I saw it as a case of 'new wine, new bottles'. My focus wasn't on using multi-channel systems to mimic environments captured with Ambisonic microphones or in reimagining existing musical forms, from strophic to symphonic, with instruments or voices placed arbitrarily in an exploded but arbitrary spatial arrangement. Instead, I believed that the work should be rooted in process, characterized by nonlinearity, openness, generativity, and algorithmic composition.

Regarding the implementation of multi-channel sound, I felt then, and still do, that it should be point-source, engaging directly with a space's existing architecture rather than trying to simulate an alternative environment, as Ambisonic, or other virtual techniques attempt. I've observed that point-source works are versatile, fitting well into various spaces without being tied to a specific site – contrary to the misleading flexibility often attributed to Ambisonic systems.

The materials used, which could range from simple loudspeakers and amplifiers to specially modified speakers, sound-producing gadgets, reflectors, resonators, or containers, should be essential to the sound work itself, not extraneous visual fluff made for the benefit of potential investors. The work is envisioned to be non-performative, as the theatrical aspect of performance can detract from engaging with the musical relationships at the heart of the piece. Additionally, the work is not designed to be interactive in the sense that audience participation goes beyond listening.

The strategic placement of speakers to spread the composition's sounds throughout a space is a critical aspect of the work's form, not merely a secondary consideration. Incorporating sound from behind the listener is vital for tapping into profound emotional responses linked to the brain's 'fight or flight' mechanism. While a stage full of speakers may be visually impressive, the absence of rear channels reduces it to a mere electronic orchestra, fostering a conventional listening experience akin to that of a concert hall. At the same time, in immersive musical compositions, rear channels aren't limited to producing startling effects; instead, they enhance the listener's sense of presence, attentiveness, and spatial awareness. This dynamic interplay between the composition's structure and the listening experience encourages an openness to fluid 'moment' forms, freeing composers from the conventional formal restrictions typical of concert music performances.

## Diapason

In 2000, choreographer Liz Gerring, whom I've collaborated with since our student days in the 1980s, along with her husband, attorney, and arts patron Kirk Radke, offered me an opportunity to co-manage a space in midtown Manhattan. Liz planned to use this space for rehearsals throughout the week, while I would organize sound installations and various forms of sound art on Saturdays. Kirk generously covered all associated costs, including rent, artist fees, technical assistants, and grant writing services.

The space we leased was L-shaped, featuring a main area measuring approximately 60 feet by 30 feet and a smaller, secluded room at the back, sized about 20 feet by 12 feet, accessible via a short corridor. This layout allowed for the installation of versatile multi-channel sound systems in both rooms. The primary area could support up to 24 speakers, while the backroom was equipped with an 8.1 channel system, adjustable according to the needs of specific installations. Video projectors were available in both rooms. The larger space was furnished with four expansive area rugs and numerous pillows, creating a comfortable 'Dream House'–style environment for immersive listening experiences. The back room featured wall-to-wall carpeting. Facing Sixth Avenue, the main area's windows were fitted with sliding metal shutters, effective at muting city traffic. The rear room offered complete seclusion, devoid of windows, and the lighting throughout was kept simple.

I named the sound gallery 'Diapason', and over the following decade, across its original Sixth Avenue location and two subsequent sites in Brooklyn, I curated approximately 100 sound and intermedia installations. Like Studio Five Beekman, exhibitions ran for a month, with no openings. Silence was enforced, just as it would be in a concert hall, a movie theater, or a library. The setup with carpets and pillows encouraged visitors to stay and immerse themselves; people often lingered listening for 2–3 hours.

Diapason featured multi-channel sound installations, including notable works specifically created for the gallery by distinguished artists such as John Luther Adams, Bruce Andrews, Olivia Block, Kabir Carter, Richard Chartier, Tony Conrad, Evelina Domnitch and Dmitry Gelfand, Judy Dunaway, John Duncan, Bernhard Günter, Jim Haynes, Catherine Christer Hennix, Leif Inge, Tetsu Inuoue, Jason Kahn, Ron Kuivila, Alvin Lucier, Al Margolis, Charlie Morrow, Erik Nauman, Phill Niblock, Matthew Ostrowski, Andrea Parkins, Zeena Parkins, Eliane Radigue, Matt Rogalsky, Ursula Scherrer, Lary Seven and Fabio Roberti, Carl Stone, Fred Szymanski, Ed Tomney, Stephen Vitiello, Barry Weisblat, Monika Weiss, and Amnon Wolman. The gallery not only showcased these installations but also hosted regular performances that either extended over long durations or made use of the multi-channel sound system. Additionally, Diapason was home to a series, *OptoSonic Tea*, curated by Katherine Liberovskaya and Ursula Scherrer, which blended visual arts and sound in unique gatherings, and a 'house band' named Analogos.

This band, comprising James Fei, David Galbraith, Tom Hamilton, Kato Hideki, Sergei Tcherepnin, Stefan Tcherepnin, Ed Tomney, and myself, specialized in performances on analog, modular synthesizers.

## Thoughts on creating a public 'listening space'

Diapason and Studio 5 Beekman specialized in showcasing immersive sound and multimedia installations, setting them apart as unique 'sound galleries'. Unlike more or less conventional venues for auditory experiences – such as concert halls, churches, art galleries, museums, theaters, pop-up spaces, salons, nightclubs, recording studios, living rooms, or personal listening through earphones – sound galleries offer a distinct environment. These spaces are designed to integrate the listener's social interaction, physical orientation, and focused attention with the structure of the auditory works presented. As the repertoire for immersive sound art grows, the demand for specialized listening environments increases, potentially driving further innovation in this field. The financial barrier to creating such spaces has significantly decreased, primarily due to technological advancements like digital multi-channel amplifiers. While many immersive auditory spaces exist within academic and well-funded cultural institutions, the majority of grassroots music venues still rely on traditional stage setups with stereo sound systems, missing out on the rich potential of immersive sound experiences.

The concept of a 'sound gallery' emerges as a solution when traditional music outgrows the social and acoustic confines of the concert hall. This shift can be traced back to Arnold Schoenberg's pioneering work in serial music and process composition, notably seen in his Opus 19 piano pieces. Schoenberg aimed for a 'complete liberation from form and symbols, cohesion and logic', challenging the conventional structures of musical presentation. The quest for alternative spaces capable of accommodating the evolving forms of modern music led to innovative experiments like the Philips Pavilion and Stockhausen's '*In den Höhlen von Jeita*' – to cite two specific examples – the emergence of avant-garde jazz and minimalism' within New York's loft scene, and the proliferation of radio studio concerts across Europe during the 1950s and '60s.

Diapason is a direct descendant of this movement, seeking to enhance the symbiotic relationship between music and its architectural and social contexts. It integrates elements from various settings such as salons, lofts, art galleries, and recording studios to offer a unique environment for experiencing sound art.

At Diapason, we showcased a broad array of immersive 'sound art', without the intent to strictly define the genre but rather to provide artists with a platform to share their work with an engaged audience. When contemplating the characteristics of what could be considered an 'ideal' sound space, albeit cautiously using the term, several key features come to mind:

1. **Control over external sound:** The ability to manage outside noise is crucial, not necessarily to eliminate it but to have the option to do so. Distractions such as a receptionist taking calls should be avoided to maintain a connection to the auditory experience.
2. **Volume flexibility:** The space should allow for the sound to be played at high volumes without the concern of disturbing neighbors, accommodating a wide range of dynamics.

3. **High-quality equipment:** The quality of the audio equipment should be the best the budget allows, ensuring an optimal listening experience.
4. **Multiple isolated rooms:** If possible, the venue should feature several acoustically separate spaces, allowing for the simultaneous presentation of multiple pieces.
5. **Communal areas:** Designated spaces for social interaction outside the listening areas encourage discussions among listeners and interactions with the artists.
6. **Adjustable ambience and lighting:** The atmosphere of the space should be modifiable, with a preference for dim lighting to enhance auditory focus. Minor adjustments toward darker lighting can significantly heighten the sense of hearing.
7. **Minimalist aesthetic:** The visual design should avoid an overly technical appearance.
8. **Unobtrusive climate control:** The heating, ventilation, and air conditioning systems should be discreet and effective, preventing discomfort from extreme temperatures.
9. **Non-directional sound:** Instead of a singular optimal listening position, the space should encourage a feeling of envelopment, with sound coming from all directions, with noticeable shifts in perspective as one changes position. This setup fosters a sense of discovery and enhances auditory awareness by not having a fixed focal point (though I've always discouraged people wandering around a space during a presentation, as it creates a distraction for other listeners).

These guidelines aim to create a space that not only facilitates immersive sound experiences but also fosters deeper connections among the listener, the sound, and the surrounding environment.

## Compositional approach

In my approach to composition, I work within structured frameworks governed by specific principles, akin to classical forms such as 'Sonata-Allegro' or 'Rondo', but specific to the unique context of computer-based and multi-channel sound environments. One example of such a structure is *Room Pieces*, characterized by their long-duration, generative nature. Each version of *Room Pieces* is created from a selection of sound and process modules from a collection developed over years.

A 'module' within these pieces is essentially an algorithm that generates a particular sound, either through software instruments or by manipulating existing recordings. These algorithms introduce variability and unpredictability, actively engaging with spatial dynamics. My library, comprising about 500 modules – ranging from instrumental improvisations and field recordings to spoken word and ambient sounds, often heavily processed – serves as the foundation for these compositions.

For any given *Room Piece*, I might use between 20 and 50 modules. The operational core of the piece involves cycling modules on and off, with the ratio of active to inactive phases dictating both the likelihood of each module's activation and the overall texture of the piece. I utilize a system based on six prime numbers (17, 23, 29, 31, 37, and 43) to determine the timing of all compositional events, from module activation cycles to the duration of sound events. This creates a vast range of timing possibilities, from milliseconds to several minutes, ensuring a rhythmic cohesion without resorting to simple, repetitive patterns.

The duration of each module's active and inactive phases is chosen from one of the six primes, and the relationship between these phases is defined by specific multipliers. For example, a multiplier difference might result in a sound being more likely on than off, with durations varying significantly.

Spatialization has been a central element of *Room Pieces*. The specific spatial positioning of each module is critical, influencing its interaction with other sounds within the piece. A sound's placement in space becomes a structural aspect, a basis for variation and development. The interaction of the modules, how they continuously (re)contextualize each other, is the way meaning is engendered.

Echoing Roger Ebert's sentiment regarding the image in film, I believe sounds achieve their fullest expression when allowed the freedom to evoke a broad spectrum of associations, unconfined to narrow interpretations. In other words, sounds should not be chained to ideas but should be allowed to resonate within the structural framework of a given work as well as the personal associative world of the listener. Contrary to John Cage's perspective that musical relationships hinder the act of listening, my work embraces the dynamic relational interplay of sounds. However, these relationships are not prescriptive but are designed to be flexible and open to interpretation, offering a suggestive rather than definitive experience.

## Description of the spatial algorithm used in *Room Pieces*

The original sound source is typically mono, but it undergoes a series of processing steps:

1. The sound is divided into two separate channels.
2. Each channel passes through independent high- and low-pass filters, with specific frequencies for each. For instance, channel 1 might have frequencies set at 320 Hz and 5,120 Hz, while channel 2 could be set at 1,280 Hz and 20,480 Hz. Available frequencies for the high-pass filter are 80, 160, 320, 640, and 1,280 Hz, and for the low-pass filter, they are 20,480, 10,240, 5,120, 2,560, 1,280, and 640 Hz. These filters can be optionally bypassed.
3. A delay effect is applied to the second channel, adjustable from 0 to 373 ms, based on the sound's characteristics. Delay options include 0, 0.1, 0.5, 1, 2, 3, 7, 13, 23, 37, 53, 67, 91, 123, 153, 203, 271, and 373 ms. The maximum delay is set by a variable in the Max patch.
4. Amplitude modulation (AM) is applied to each channel, with variable depth (0.1 to 1, in 0.1 increments) and speed (1 to 25 Hz, in 1 Hz increments). This effect can be turned off; if enabled, there's a 25% chance it will affect a channel.
5. Distortion, specifically bit-crushing, can be applied to each channel. This effect can be turned off or, if enabled, has a 20% chance of affecting a channel. Options include bit depths of 10, 12, or 14 and sample rates between 8,000 and 22,100 Hz. A variable allows for an 'auto-off' feature that stops the distortion shortly after the sound starts, creating a burst of distortion at the beginning.
6. The channels are then processed for spatialization, treating eight loudspeakers (or any number up to 24) as points in a circle, with each channel panning between two adjacent speakers but selecting different pairs for each. This results in the mono signal potentially expanding to four speakers, each with unique EQ, delay, AM, and distortion effects.
7. In the final output stage using a spatialization tool, each channel goes through a wide bandpass filter to give each speaker a distinct sonic character, and compression to adjust for room acoustics. These effects can be bypassed. The setup may include up

to four different stereo reverbs (e.g., concert, plate, room), with reverb characteristics changing in real time, whenever a momentary silence allows. Changes can include source and room presence, filter frequencies, distance, and other parameters.

This process transforms a monophonic sound into a complex spatial audio experience, moving from behind to in front of the listener, emphasizing spatial awareness and the listener's physical relationship to the unfolding soundscape.

### Case study: portable multi-channel sound system

The portable multi-channel sound system (PMCSS) is a compact setup I designed in 2013 for easy transport, fitting into a suitcase and weighing under 40 pounds. This system enables me to present my multi-channel compositions in diverse venues, from galleries and living rooms to concert halls, theaters, libraries, classrooms, and clubs.

Originally, the system featured 12 channels, utilizing an assortment of small speakers with 2-, 3-, and 4-inch drivers housed in various containers – boxes, glass jars, plastic food containers, plumbing pipes, and other resonant objects – either brought along or found at the performance site. Each speaker was engineered to produce its own distinctive sound. While I still possess this initial setup, my current preference is for an eight-channel system equipped with high-quality, full-range 4-inch drivers, all housed in yogurt containers to enhance their bass response by approximately 30 Hz. I request that venues provide a subwoofer to complement the system. The setup includes a Crown eight-channel amplifier and a compact audio interface, with the addition of the computer's headphone outputs to independently manage the subwoofer channel. This configuration suffices for spaces up to 2,500 square feet. Utilizing 20-foot, tangle-resistant speaker cables with RCA connectors enables swift speaker placement throughout the venue. RCA couplers extend cable reach, facilitating speaker distribution over significant distances. Setting up typically takes about 30 minutes, with speakers positioned on various surfaces or suspended to achieve a balanced yet non-symmetrical sound dispersion.

My performances range from solo sets to collaborations with instrumentalists, poets, dancers, singers, and actors, employing interactive improvisational software developed in Max, which draws on the algorithms from my generative works. The PMCSS is also used for presenting both fixed-duration pieces and long-duration installations. My catalogue includes over 60 fixed compositions, spanning 2–80 minutes, alongside roughly a dozen generative installations. The *Room Pieces* concept, a staple of my work since the late 1980s, now encompasses around 500 modules, offering limitless possibilities for new installations tailored to each unique auditory space.

### Case study: living room pieces

*Living Room Pieces* is a sound installation that reimagines the domestic environment as a space for active engagement and discovery through listening. Once activated, it sporadically produces sounds, each occurrence unique, drawing from a vast library of audio elements. This unpredictability enriches the listening experience, altering one's connection to the surrounding physical, architectural, social, and sonic environments. The sounds generated blend with and enhance the room's natural ambience, emphasizing the listener's presence in the moment – alert, focused, and adaptable.

This piece is a computer-generated, multi-channel composition of extended duration, structured around specific rules for timing, signal processing, and spatialization, all executed in real time. This framework ensures continual variation while projecting a coherent, overarching form. Each instance of sound is singular and ephemeral, yet the algorithms provide a comforting sense of deliberate structure.

The composition cycles through a seven-day period, established each time the software is restarted. Ideally, after its initial activation, it runs indefinitely as long as the system remains powered.

The first installation of *Living Room Pieces* took place in 2005 at the Chelsea Hotel in the home of artist Antoine Laval, utilizing 12 speakers distributed throughout his apartment, where it operated continuously for a year. The programming adapted to the time of day, becoming quieter overnight.

Another notable installation occurred at EMPAC in Troy, New York (curated by Micah Silver), employing the venue's PA system with around 30 speakers across different areas, running non-stop for a year. In Berlin, Carsten Seiffarth's Singuhr Gallery offered a version in a hotel room during the summer of 2007, allowing visitors to immerse themselves in the piece over a day or two.

The version of the piece launched in March 2021 aims to bring this immersive experience into more homes, making it more widely accessible. To this point, the 2021 edition has been experienced by about 30 households, through lending programs offered by ESS in Chicago, Black Hole in LA and Harvestworks in New York City. Another edition was released in November 2024, through Brian Chase's Chaikin Records. Both these editions comprise a compact, self-contained sound system with two speakers and a Raspberry Pi computer, capable of playing over 7,000 sound files. The playback is determined in real time, ensuring a richly diverse and unpredictable auditory experience, yet maintains a coherent structure that goes beyond randomness. This approach underscores sound's omnipresence and the challenge of disregarding it, ensuring that a sound installation designed for living spaces reflects the variability and unpredictability of everyday life while sustaining an underlying order.

## Conclusion

After dedicating over three decades to the field of surround, multi-channel, and immersive sound, my perspective on its future is ambivalent. On one side, the advancements in technology – both software and hardware – are unprecedentedly accessible and affordable, making the creation and dissemination of new works more feasible than ever for both artists and audiences. However, the pop music industry's overwhelming influence has led to a creative environment that is often homogeneous and predictable, with a tendency to replicate past successes in pursuit of social media validation rather than genuine innovation.

Despite these challenges, I remain optimistic about a shift in how immersive music is marketed. While immersive audio is likely to continue its growth in gaming, film, and television – increasingly becoming a staple in home entertainment systems – I'm hopeful for a paradigm shift toward targeting smaller, specialized audiences for music in this format. This approach would leverage the internet for broader outreach without relying on large-scale sales to deem projects financially viable.

Additionally, I observe an expanding global network of DIY venues offering immersive sound to the public. I envision these grassroots spaces forming alliances with more established, financially secure institutions, facilitating an exchange of resources and repertoires. This collaboration could significantly enrich the immersive sound landscape, encouraging innovation and diversity in the music we experience.

Chapter 18

# Spatial sound composition

## Concept and practice

*Daniel Neumann*

With this essay I aim to introduce the concept of Non-Representational Spatial Sound Composition (NRSSC) and illustrate it by means of four case studies from my own artistic practice. To distinguish between the conceptual framing and my artistic considerations I'm choosing to write in first person when talking about my artistic practice and in third person when presenting definitions of theoretical concepts.

### Non-Representational Spatial Sound Composition

This concept provides a framework for spatial sound works that are composed in dialogue with and are considered for a site – a specific physical location. Works that engage acoustic features and form unique temporary sound spaces have a different approach to spatialization than surround sound studio compositions, cinema sound, works in Ambisonics, etc. The latter three attempt to represent a certain preexisting real or studio-made spatiality, aiming for spatial transferability from one space to another.

Helga De La Motte-Haber describes two different artistic intentions in composition:

> [O]n the one hand, sound is represented in space, and on the other, space is presented in sound.
>
> (De La Motte-Haber, 2015, p. 172)

This essay makes a case for works of the second approach. These are works that engage in the creation of unique tonalities, situating a space through sound, articulating and foregrounding spatial conditions of the room and making them felt. *Tonality* here is referencing an open tuning, derived in part from the architecture, not in reference to a key or scale in any specific Western tuning system like equal temperament, just intonation, etc.

My practice in general is centered around how sound interacts with space and how spatial perception can be shaped by sound. The works are deeply rooted in the process of real-time listening, often culminating in hybrid installation-performance formats, challenging the notion of fixed and final works, and understanding sound as an intersubjective field enabled and expanded by audio procedures (Lähdeoja & Montes De Oca, 2021).

#### *Sound space, folded*

Traditionally, the act of composing and performing/listening happens in separate spaces: composition takes place in a studio, while performance takes place in a concert hall

DOI: 10.4324/9780429297526-21

or theater. Non-Representational Spatial Sound Composition folds them together. Here, works are made for and with specific locations, engaging with unique characteristics, temporality, and context – all interconnected with the physical architecture.

### *Worksample 01::* From Within a Field

*From Within a Field* was a series of works in formats between electroacoustic live performance and installation for multi-channel speaker systems. In the series, which ran 2013–2017, I used loudspeakers in interaction with architectural or environmental space as my instrument to live-compose a spatial sound field. It was adapted to many different situations: a five-story atrium, a techno club, gallery spaces, a villa in Southern Italy, a DIY music venue, an outdoor cinema, etc. The work was developed on systems ranging from six guitar amps to a 24.4 3D-speaker array. Live diffusion and a Continual Concrète compositional method were other characteristics of this work. I'll step through some of the iterations next to illustrate the evolution.

## Continual Concrète

A given sound material is applied, diffused, processed and re-recorded over multiple iterations. It is therefore continually shifting, morphing, accumulating, and developing. Traces and imprints of different times and spaces are layered, staggered, cut, and merged. The use of audio tools, audio processing, and the ways of using sounds are derived from the historical Musique Concrète, where composing starts with concrete sounds, not with an abstract idea.

In contrast to the fixed media compositions of Musique Concrète, *From Within a Field* did not result in a finished, hermetic work. The sound material – over time, not for one piece – was unfolded, processed, recorded again, modified, and reapplied to ever new listening situations. The sound materials were ambiguous electronic sounds and textures, recordings taken in the spaces (a nighttime security guard using the elevator in the Pratt Atrium for example), recordings of the various iterations themselves (the snippet of the elevator ride was also in the room recording of the performance, which is then used during future iterations) and other random sonic occurrences (electromagnetic interference in one of the mic channels). With the process-based Continual Concrète composition method in *From Within a Field*, the architectural space became a generative force.

There's a lineage of working with sound in this way – *space presented in sound* – from prehistoric rituals in caves to church organ music to the Audium in San Francisco (Blesser & Salter, 2006a). There is also a lineage of individual works and artists who articulate spaces: Maryanne Amacher, Mark Bain, those who use live-diffusion approaches, *Ricochet Lady* and *I'm Sitting in a Room* by Alvin Lucier, *One*[3] by John Cage, and many more. See the accompanying webpage for a running list of works in this lineage. The approach is framed in Amacher's own words as: 'Working with the space and its existent architecture is the only way I'm able to create the experiences I wish to make with my music. . . . The music I compose takes on its 'life' only when it becomes PART OF THE ARCHITECTURE IT SOUNDS IN' (Amacher, 1983).

Often these works exist in a blurry Venn diagram overlap area between formats of electroacoustic, multi-channel music performance, and sound installation with loudspeakers.

In folding together the composed space and listening space, there is generative interaction – a dialog, neither separation nor (modernist) domination or overpowering control. It encourages a form of open active listening, in the following fold:

**Studio/the Composing Space:** The space where composing happens, where the composer makes the work, which traditionally is happening in a studio, a separate, specific space.

**Listening Space:** The listening space is the setting in which the work is heard: living room, concert hall, desk, headphones.

**Composed Space:** Composed space is a term coined by Dennis Smalley and derived from Michel Chion's notion of *espace interne*, referring to the internal organization of compositional elements in the work as they are recorded onto a fixed medium. Frank Ekeberg Henriksen added a helpful distinction to this by speaking of intrinsic (magnitude, shape, density), spectral (low, high), and extrinsic (direction, distance, movement) spatial properties for individual sounds. These individual sounds with their intrinsic, spectral and extrinsic properties are arranged into a broader (musical) system, which create fields, streams, and events in spatial interrelations.

**Sound Space:** The space where composing happens and the listening space fold together. Composed Space will only be developed in dialogue with the listening space. Works are composed in an architectural listening space and do not contain any precomposed spatiality that needs to be represented in playback. An obvious example for this representative approach would be the helicopter made to virtually fly over the audience.

The series *From Within a Field* was built on that fold of Sound Space, turning an architectural situation into a process of forming, of generating a Sound Space. The *#02 lvl UPLOAD* iteration, which took place at the Pratt Institute in New York City, involved distributing speakers into a vertical array over the four floors of the Pratt Atrium. Each speaker was placed in a way to help activate the architecture.

I operated with the idea that the dialogue with the space starts with where to place the loudspeakers and how to arrange things in space, which requires a sort of physical or architectural mode of listening to the space. The Sound Space unit becomes the instrument that's being played live, building on the following thought by Francois Bayle about the Acousmonium: 'A loudspeaker in a certain space becomes an instrument. The speaker is the resonating exciter – like a string; and the space is the resonator – like the wooden body of a violin. Every space and every speaker system will act differently' (Bayle, 2017).

In two performances in Spain in November 2016 – subtitled *Widening the Field* – the sound system was a fairly standard eight-channel horizontal ellipsis. Because my set was part of a festival with other works on the program, I could only rearrange a few of the speakers. I pointed one into a large empty gallery and two others towards the walls. I used sounds that had a fairly diffuse intrinsic quality for the remaining speakers and created a slowly developing very wide and spacious field. The sonic material functioned indeed like *what* is played on Bayle's violin and thus Sound Space was formed.

An important aspect here, of course, is the loudspeaker system itself, overtly at the center of these works. Surround or immersive-representational techniques typically require standardized, very high-quality speakers in standardized configurations, and therefore are often extremely expensive technologies. Bernhard Leitner in his artist talks mentions this inverse proportional relationship between site specificity and transferability. Yet I would argue that another factor is in play, which is access and affordability. Works in standard formats, while building on standardization and transferability, actually become less transferable because they're much less accessible, behind tall commercial or ivory walls. In contrast to such approaches, I hold that NRSSC does not require expensive loudspeaker arrays: the technology required is very flexible and very open. One can make a beautiful work with a pile of discarded hi-fi speakers and some used amplifiers from a flea market, or guitar amplifiers, or studio monitors, or a mix of it all. Differing from standardized 3D-surround sound practices, here the speaker configurations are idiosyncratic and vary every time. The loudspeaker, together with its surrounding space, becomes the instrument. With every space and every speaker system acting differently, exploring these variations was a key concern of *From Within a Field*.

To me, a more accessible and inclusive approach seems aesthetically more desirable. That became the foundation I laid in this series of pieces, a set of realizations that continue to ground my work to this day. The restriction in transferability is outweighed by the flexibility and openness of the process. Process versus fixed object is part of this aesthetic choice. NRSSC is a very process-based approach – an open work in which technology is secondary. If foregrounded, the discourse takes too much of a part in the consumer ideology of our late capitalist society.

Openness is also key in selecting the architecture of the physical space for displayed work opposed to the one-track-purpose of a concert hall. A gallery, or abandoned industrial space, e.g., is a more open context as it has no built-in stage or seats and is fluid: 'The contextual situation of the gallery and the effects of the art audiences' consumption of music within the exhibition space have generated new approaches to music performance and new musical subgenres'(Kelly, 2017). I'm fascinated by concerts in gallery spaces because they require a very sensible dialogue between the musician/performer and the acoustically difficult space. It foregrounds the spatial aspect versus the internal components of the sound. It is fragile. There is no ignoring the space in these scenarios, only engaging with the space; which again, to me is aesthetically more desirable. I believe this practice of comprehensive engagement is biocentric as opposed to extractive; it is crucial in how humans treat the environment and our planet as a whole.

Performances act as temporary formations where the different spatio-temporal fragments, including recordings from previous iterations and layers of different materials strung and woven together in the present moment. The listeners and myself the composer/performer are *Within a Sound Field*. This series of works foregrounds listening as a process. Sounds are chosen to encourage an attentiveness to the sound field – or Sound Space – in which one sits. The term 'Sound Field' became a very technical term with the popularization of Ambisonics and other 3D formats over the last decade or so, which is why I've chosen to use 'Sound Space'. The terms seem somewhat interchangeable, with Sound Field being more centered on the propagation of a sound, which is another factor that lets me lean toward Sound Space. Rather than following Pierre Schaefer's call for a 'reduced listening', I invite a spatially expanded listening and *reduction* becomes my responsibility as the composer – that is, the reduction of narrative and semantic content to encourage spatially expanded, environmental listening. In the growing fields of tones listening from within becomes more important than listening to something.

Walter Branchi notes:

> The effect of a musical composition as such on the listener is secondary to the way in which it harmonizes in a set of systems of which the listener is part. The pulsing of a composition interacts with the surrounding space and perception is of the global interaction of all the entities that partake in the musical event.
>
> (Branchi, 2012)

## Site specificity

For the compositional approach laid out in this essay, it is important to address the term 'site-specific', since it has been in use since the 1960s – and has had its own inflation – to describe artworks that operate in and address the relation to specific sites. Over time the term's usefulness became questionable (Poole, 2015). This realist approach, which assumes a preexisting site (as a container) for putting something special into it, seems much less relevant and engaging for NRSSC. Instead, these artworks themselves can be understood as sites that are being produced or generated. The installation/performance – the Sound Space – as a temporary formation is its own site.

### *Temporary formations as aural architecture*

At first, I thought of NRSSC from a fairly technical background. The representation of composed sonic spaces through super-exclusive systems, and questions of transferability were not much of an artistic interest for me. Then I came across Marianne Amacher's ideas and approaches: more phenomenological, embodied, opposed to others that strive for objectification, confinement, or dominance of spaces. Those represent something that *is*, not *what is*, therefore lowering what could be perceived, obstructing it, like Baudrillard's simulacrum or Debord's spectacle. Our world, our so-called culture does that a lot, plastering our minds with media distractions, producing strings of situations where essence precedes existence. So I think art, and especially an ephemeral medium like sound with such a high potential for fluidity and presence, does not need to reproduce those conditions.

As an artist, I regard composition, and in that sense also curation/organizing, as creating temporary formations where 'Formation' means a thing that is conditioned by different kinds of elements' (Hanh, 2008) and all of the elements in play come together – space, loudspeaker system, sonic material, listeners – for a certain moment in time.

#### *Worksample 02::* Resounding the Tower

For this installation-performance (which was part of *Newark Rhythms*), I distributed eight loudspeakers throughout the Brutalist monument designed by Robert L. Geddes–Hill Hall at the Rutgers University–Newark campus.

The 'tower' of this building is currently sitting there as a layered, inaccessible enigma; only functional as an alarmed emergency staircase. From its initiation, this now-closed external stairwell invited speculation and mystification. I aimed to sonify its interior; using the tower as a large vertical filter, playing sounds into it, while simultaneously recording the acoustic responses with various microphones at different heights and placements.

The counterpart to this tall vertical structure, and another prominent structural feature of Hill Hall, are internal horizontal ramps. These long concrete ramps became the listening space where sounds from the tower dynamically diffused through the remaining six speaker-channels placed in close relationships to the architectural structure. All audible sounds were sounding through the tower – either live or during the preparation period. The sonic material was mostly electronic sounds and recordings, both generated on-site.

During the performance, the audience moved through the Sound Space, experiencing the aural architecture formation with the central question of whether the verticality of the tower – that itself was off limits – would become audible. Could this verticality be felt on the horizontal ramps?

## Aural architecture

A temporary **formation** includes, or might be based on, spatial perception and felt space. The listeners are in it, it is shared, yet subjective; it includes spatial, architectural, and emotional resonances conditioned by our subjective memories. It is a social space; it is fluctuation 'that includes but exceeds human contributions' (Cox, 2018) what Amacher calls *perceptual geometries*. Hence the Sound Space is generative – 'a continual production that involves the listener as intersubjectively constituted in perception, while producing the very thing he perceives' (Voegelin, 2010). Sound Space is the material condition for this: Human-made and pertaining to the ear, to hearing, and to the vibratory perception of our bodies. Sound Space and Aural Architecture are related terms, the former weighted toward physical conditions. Aural Architecture 'is that aspect of real and virtual spaces that produces an emotional, behavioral, and visceral response in inhabitants' (Blesser & Salter, 2006a) and centers the *experience* of the Sound Space. It is more fluid, grounded in the sonorous. According to François J. Bonnet, 'every audible is, a priori, sonorous . . .. The passage from the sonorous world to the audible world unfolds across a veritable genealogy of the sensible' (Bonnet, 2016). Aural Architecture is expanded, grown toward perception, and dynamically situated. It could be understood as the aural aspect of Gernot Böhme's *Felt Space.* He expresses the difference comparing felt space and metrical space: 'Felt space and metrical space are situated on different ontological strata; namely, the one on the stratum of being, the other on the stratum of representation' (Böhme, 2021).

Formations are aesthetically interesting if they imbue a uniqueness, a concrete here and now. What distinguishes that from aesthetics geared toward illusions? The recipients are active participants within this singular formation, not consumers of a spectacle, kept in awe by a mystified, unreachable apparatus. Listening is understood here as an active production, as active becoming-being, as making sense, as being within a field. 'The performer is thus not a creator but a node through which the sonic flux passes' (Cox, 2018).

### *Thresholds of decentering*

These discoveries led me to pursue the development of an instrument for this exact purpose – to make tones for multi-channel speaker systems in situ. In 2020 I leaned into further work on the software instrument Room Tone Generator [RTG]. Until then, RTG was a simple Max patch for my colleague Axel Töpfer, who wanted to generate subtle background tones for his installations. Tones that were based on the room resonances of the spaces the installations were shown in. Using room resonances vitalizes and animates the sound field when visitors are moving through it.

### *Worksample 03::* Room tone generator

*Room tone generator* (RTG) became a more complex software instrument intended to produce finely tuned, complex background tones to alter the room atmosphere. In this instrument, everything revolves around one center frequency [THE Frequency], the frequency of its main oscillator. This frequency gets tuned to a subjectively intriguing resonance in the physical space. All other parameters and voices of the RTG are brought into play in sequential steps, and are in variable mathematical ratios to THE Frequency. That way, all harmonics, partials, and overtones of a sound are related, and as sound waves with various wavelengths, they are in a specific and unique relationship to the architectural space. This allows for the creation of spatial tonalities, which merge with the architecture and situate an Aural Architecture Formation through sound, instead of using sound as a means of representing a preexisting spatiality.

In November 2022 I presented a work for the festival Espacios Resonantes in Santiago de Chile for an abandoned underground cistern. For this work, I decided, for the first time, to exclusively use the RTG to create all of the sounds. I devised a compositional structure from having eight loudspeakers available: there are 56 different triangles in an octagon, and this gave the basic spatial form. I created 56 triphonic (three-channel) tones. Each of these three-channel tones were placed in three of the eight speakers. For the concert I mixed the 56 triphonic tones live, shaping them into an expansive dialog with the unique acoustic space of the cistern: 17 m in diameter, 5 m tall, made out of concrete, round, with parallel surfaces of floor and ceiling.

The title of the work became *Das Schwellenstück/Thresholds of Decentering*. For this series I usually title the works also in situ, once I'm in the process of making the work, once I get a feeling for what the work wants to be. I wanted to perform this work with decentered listening as the driving force for me as the performer, trying not to think about wanting to entertain the audience but inviting them to become part of my Aural Decentering. I used an iPad mixer controlling a 56 × three-channel playback. I sat in the middle of the audience, the audience on scattered chairs, loose, chaotic, in all directions – no rows, forced geometric shapes, or other authoritarian gestures.

## Aural decentering

The concept of 'decentering' is borrowed from psychology and applied to a mode of listening. Decentering means taking on multitudes of perspectives, consciously leaving our own center, and taking on different standpoints. When applied to listening, this could mean guiding one's attention to various points and peripherals of the always dynamic acoustic horizon – a form of active spatial listening. '

> The concept of virtual sonic boundaries leads to a new abstraction, acoustic horizon, the maximum distance between a listener and a source of sound where the sonic event can still be heard. Beyond this horizon, the sound of a sonic event is too weak relative to the masking power of other sounds to be audible or intelligible. The

acoustic horizon is thus the experiential boundary that delineated an acoustic arena, a region where listeners are part of a community that shares an ability to hear a sonic event. . . . The concepts of area, horizon and channel originated from the language of soundscapes (Truax, 2001) but they are especially relevant to the analysis of aural architecture'.

(Blesser & Salter, 2006b)

When listening to spatial sound, the wandering of our attention is activated and engaged. We travel around the room with our attention, and when listening to a changing soundscape, as is the case with listening to a musical work, we encounter certain thresholds that will make our listening direction change – fast, or very slowly, sometimes even going back in time! The term 'acoustic horizon' is a very useful term, and it defines the acoustic field around a listener, the subjective perspective of what and how far one can hear. Since sound is ephemeral and changing, the three-dimensional acoustic horizon also constantly morphs and changes, contracting and expanding. With the spatial and dynamic shaping that a performance of live mixing offers, I can play, morph, and shape these listening thresholds.

To summarize, the Sound Space is built as a temporary formation – a subjective dialogue between physical properties and generated tonalities. Thus it shapes and alters perceptive qualities of a space to form an aural architecture which encourages a decentering of the audience's listening foci: an Aural Decentering. This doesn't mean not paying attention; on the contrary, it means establishing a focus that affords one to move the attention to different places and peripheries within the aural architecture.

### *Tone Field Horizon*

This past summer, I was able to work on a new performance-installation for the Wave Farm Art Park, as well as a radio transmission mixed live from the various points that made up the field. It's using the Art Park as a spatial instrument in a sense. It was my first time using the RTG for an outdoor site.

#### *Worksample 04::* Tone Field Horizon

Five remote stations, each consisting of a loudspeaker and a microphone, were installed around the premises and connected back to the transmission studio. I produced tones for each of the stations in response (and with respect) to the nearfield environment around them using the RTG. The project culminated with a twofold public performance: one part for a local audience who freely moved between the different acoustic arenas yet were unable to listen to all the sites at once. Microphones picked up the altered environment and were fed live to the radio broadcast studio, where I was mixing those different streams into a compounded broadcast. The second audience, removed from the local sites, were listening to the mix of the dislocated five feeds broadcasted over WGXC-90, yet they weren't able to access individual locations. It was a play with the differences, the beauty and the limitations of both – our subjective, one-point-in-space perspective and the broadening yet disconnecting technological answer to that.

## Acoustic horizon/acoustic arena

The terms 'acoustic horizon' (a) and 'acoustic arena' (b) by Barry Truax, quoted in (Blesser & Salter, 2006a) define (1) the acoustic field around a listener, the subjective perspective of what and how far one can hear or (2) the area a certain sound is audible in. Transmission technology expands and complicates them. It could be considered that in an acoustic horizon, human listening is the locality of perceived existence.

This horizon in *Tone Field Horizon* split onto very different ontological planes for the two audiences – in loco and tele.

For the on-site audience, the tones in the environment (RTG within an ecosystem) expand perceived spatiality. The tones are placed into the environment in order to direct listening by focusing attention. When listeners walk around Wave Farm Art Park, they experience the shifting acoustic horizon in relation to how close or far they are from the various RTG speakers: the aforementioned wandering of attention (and the body), activated and engaged. The process of decentering is evoked by being in a field that can only be grasped in parts, never all at once, as is the case with listening to a living ecosystem.

The tele-audience experiences a very different acoustic horizon. They hear an all-at-once conglomerate from the different locations. Their horizon is expanded to the Art Park, yet it is also out of focus in a sense. Their process of decentering is the task of imagining different individual locations. Through the radio broadcast, I aimed to highlight sound as an intersubjective biocentric field that enables and is expanded by an extended universe of interference with equal value placed onto all human and nonhuman contributors, inspiring ecological connectivity.

Furthermore, as the broadcast is a creation of new sites for listening, tele-participants in the greater Hudson Valley had control of the mixed soundscape tele-arena from the Art Park. And like the local engagement, remote listeners are not passive spectators: they too, are a part of and affect the sonic ecosystems within their horizons/arenas. The site of each Wave Farm broadcast receiver (in a car with the windows down, out of a speaker on a front porch, etc.) marks centers of new circumferences. The previously introduced Continual Concrète composition method is here spatially expanded, happening simultaneously in many places. All over the Hudson Valley, new tone fields – new acoustic arenas – expand like many ripples in a pond. Temporary formations feeding into other temporary formations. In flux: when we respect the experience of sound as an equally participatory act by all environmental elements.

## Conclusion

Non-Representational Spatial Sound Composition offers a compelling framework that emphasizes the dynamic relationships among sound, space, aesthetics, and listening. By folding together the processes of composition and performance within specific physical environments, this approach transcends traditional paradigms of sound presentation and representation. Through case studies of *From Within a Field*, *Resounding the Tower*, *Das Schwellenstück/Thresholds of Decentering*, and *Tone Field Horizon*, I have illustrated how sound can be used to articulate and engage with the unique acoustic features of a space, creating temporary formations that invite active listening in the form of Aural Decentering.

This methodology not only challenges the notion of sound works as fixed tokens or as a mere representation of space but also redefines it as a lived experience that fosters a deeper connection between listeners and their surroundings. In a world increasingly dominated by commercialized sound technologies, the flexibility and openness inherent in NRSSC provide an inclusive alternative that prioritizes process over product. Ultimately, I hope this exploration encourages a more biocentric and participatory approach to sound art, one that resonates with the intricacies of our shared environments, cultivates a heightened awareness of the sonic landscapes we inhabit, and inspires art as practice of sensing, not extraction.

## References

Amacher, M. (1983). *The Amacher Notebook: About the Big Waves of Structure Borne Sound* [Unpublished manuscript].

Bayle, F. (2017, August 8). *François Bayle about the Acousmonium, Paris* [Video]. YouTube. https://youtu.be/1tsEdyw0s8c

Blesser, B., & Salter, L. (2006a). *Q+A Dialogue Handout* [Unpublished manuscript].

Blesser, B., & Salter, L. (2006b). *Spaces Speak, Are You Listening? Experiencing Aural Architecture*. MIT Press.

Böhme, G. (2021). Felt spaces. In S. J. R. de la Cruz (Ed.), *The Oxford Handbook of Sound Art*. Oxford University Press.

Bonnet, F. J. (2016). *The Order of Sounds: A Sonorous Archipelago*. MIT Press.

Branchi, W. (2012). *Canto Infinito: Thinking Music Environmentally*. Open Space.

Cox, C. (2018). *Sonic Flux*. MIT Press.

De La Motte-Haber, H. (2015). Space impression mediated by sound. In *Compositions for Audible Space*. Transcript Verlag.

Hanh, T. N. (2008). *Awakening of the Heart*. Parallax Press.

Kelly, C. (2017). *Gallery Sound*. University of California Press.

Lähdeoja, O., & Montes De Oca, A. (2021). Co-sounding: Fostering intersubjectivity in electronic music improvisation. *Organised Sound*, *26*(2), 123–135.

Poole, M. (2015). Specificities of sitedness. In *When Site Lost the Plot*. Urbanomic Media.

Truax, Barry. (2001). *Acoustic Communication* (2nd edition). Ablex Publishing/Greenwood Press.

Voegelin, S. (2010). *Listening to Noise and Silence*. Continuum.

Chapter 19

# Immersive theatre

## The emergent edge of storytelling with sound and spectacle

*Rhiannon Catalyst*

On any given day, we can hear the words 'immersive experience' tossed around by acousticians, artists, technologists, transmedia storytellers, marketing executives, investors, theme parks, and museums, meaning a litany of things on the morphing linguistic and imagination highways (or 'electronic superhighways' as coined by Nam June Paik) of our increasingly interconnected and interdisciplinary world. Countless academic courses and degrees from Extended Reality to Immersive Media Communications are offered, with industry-leading experts teaching the next generation of spatial sound creators in world-class immersive sound labs like those at McGill University, Georgia Tech, New York University, and Rensselaer Polytechnic Institute's expanded immersive production complex that includes a 500-channel Wavefield Synthesis System. It's important to discuss the context of this global explosion of focus on immersion arts and techniques. The various sectors driving the surge in experiential and immersive creativity are diverse yet deeply intertwined, part of a broader shift toward participatory culture and what economists Pine and Gilmore coined the 'Experience Economy' in 1998 (Pine & Gilmore, 1998). The economists predicted it would follow the agrarian, industrial, and service economies and cite *Future Shock*, in which Alvin and Heidi Toffler explore themes of change and adaptation in a rapidly evolving society including shifts in societal focus toward quality of life and psychic gratification. They discuss how 'experience industries' will rise to create 'pre-programmed experiences, including simulated environments that offer customers a taste of adventure, danger, or other pleasure' (Toffler & Toffler, 1970). In this chapter, we'll explore the modern explosion of immersivity in sound and experiential spectacle while examining its broadly impactful and intentionally transformative roots. We'll look to the future as well as journey into the past to understand how history and culture have long foreshadowed today's virtual and immersive art forms – long before the age of *Future Shock* and both inside, and decidedly outside of consumer culture.

As someone working at the intersection of art and research, I have found both experiential and empirical evidence that immersive experience design is not only transformative but also central to a broader cultural shift – one that includes creators of varying backgrounds, disciplines, and eras. As experiential culture continues to rise and humanity faces unprecedented challenges, this is an ideal moment to reflect on its history and future potential. We have seen the growing integration of holistic and interdisciplinary practices mirror developments in business, design, spirituality, and academia. Meanwhile, the history of grassroots immersive theater and sonic creativity, which we will explore with ardor, offers both perspective and inspiration for expanding access and inclusion in immersive thinking. First, let's consider definitions.

DOI: 10.4324/9780429297526-22

It's exciting to be part of the global influx of immersive art and culture creators taking our senses by storm. The word *immersive* has major buzz, and significant confusion can arise alongside excitement when a concept grows to such commercially compelling status. As Gregory Hansen of *Sleep No More* shared in our conversation for this book, 'Everything is an immersive experience now. Opening your *mail* is an immersive experience' (Hansen, 2024). Massive cultural trending of a medium and concept is, naturally, a double-edged sword: while it can become easier to pique interest, we also experience increased dilution and, at times, public cynicism through overuse of the terms.

Let's discuss just a few of the multifaceted definitions and dizzying array of theatrical immersive experiences utilizing the power of immersive sound in creative and experimental ways. Some can be literally dizzying, leading to techniques like teleportation in VR or anchoring of sound objects, such as a main vocal, while other objects, frequency bands, or channels move or are widened and spatialized in varied ways in a mix. Anchoring can help audiences stay grounded in ever-shifting, richly imagined worlds.

I want to underscore from the start how *none* of the projects or spaces we'll discuss would be possible without deeply and remarkably complex collaborative processes between people working in many disciplines, with many perspectives and backgrounds. That is the only way endeavors of this magnitude and complexity are possible. I believe this is one of the reasons it holds power far beyond any singular production or event – power to drive social change, shift culture, and transform *how* we work together. It fosters a culture of enhanced cooperation while continually pushing the boundaries of the technology and art forms we use. As we'll see, my collaborators and generations of our predecessors have used immersive theater quite deliberately to do just that.

This is also part of a greater cultural trend toward interdisciplinary thinking and collaboration across disciplines that we can see reflected in art, business, diplomacy, and higher education. Of course, this is one creator's lived experience and research, by no means exhaustive in scope. The field is beautifully wide and multilayered, and this chapter is intended to open exploration of the subject including elements infrequently covered in other publications, aiming to provide a starting context while encouraging dialogue, experimentation, further research, and redefinition.

Culturally speaking, immersion may be as old as language and ritual. It's arguably one of our first modes of communication, tied directly to performance, ceremony, and spiritual practice across cultures. Perhaps our drive toward immersion in any era is an attempt to recreate our earliest sensory immersions. As Charlie Morrow posits, 'We are born immersed' (Morrow, 2022). I think our drive toward immersivity, or being reimmersed, is a shared human drive toward a visceral connection to that which is greater than ourselves. Or, rather, our true and intrinsically connected nature – to viscerally remember our divinity and wholeness, our connection to source and universe. Many indigenous cultures have sacred ceremonial practices and plants used to aid in the erasure of a sense of separation from the universe, from our ancestors, our Goddesses, Gods, and spiritual dimensions. A study in Peru with 74 individuals attending Amazonian healing ceremonies including entheogenic uses of ayahuasca found subsequent common threads of 'personal shifts towards relationality' as well as 'boundary dissolution' and the 'blurring of perceived boundaries between themselves and nature' (Manuel-Navarrete et al., 2024). As discussed later in this chapter, we began using immersion techniques to access the sacred, divine, and spiritual realms very early in human cultural history. Contemporary artists, healing practitioners, spiritual leaders, Indigenous scientists and artists, scientists studying consciousness – many now draw from ancient immersive traditions through immersive performances, rituals, sound baths, and more. Collectively, humans use the power of immersion to achieve many aims – creative, commercial, educational, medical, and beyond. We've created a *culture* of immersion.

I think it's no coincidence and quite profound that this explosion in immersive fixation arrived in an age saturated with battlegrounds for our attention, where there's naturally a resulting cultural pendulum swing toward seeking full presence and deeper sensory engagement. As the age of information, targeted ads, screens and scrolling compete for our focus, people crave going deeper, finding authentic and full engagement, especially with experiences that can create visceral embodiment while also allowing for rich conversations and social connections to be sparked. Immersive theater clearly dovetails with our collective focus on wellness, mindfulness, and presence as it envelops and even attunes our senses more deeply, especially if we practice things like deep listening. I think this is a big part of its success and widespread embrace. It has the power to create a feeling of full saturation and engagement of the senses, even wholeness and meditative calm, heightening awareness, even inspiring transformation on personal and collective levels that we'll explore clear examples of. It's profound to reflect on how Pauline Oliveros, one of the most influential composers and mentors of the 20th century, talked about her Deep Listening practice, which began with physical movement and breathwork to get centered in the body before the listening begins:

> Deep Listening represents a heightened state of awareness and connects to all that there is.
>
> (Oliveros, 1971)

Deeply tuning in, to sound and to our own senses, to reach a meditative state of heightened awareness and sense the interconnectedness of all . . . this is the language both of immersion and of spiritual practice. Personally, as an artist often thinking and dreaming in immersion and spatial composing, I find these are powerfully different ways we wire (or rewire) our creative mind to work, to train and become attuned to dream in orchestration of many senses – and sometimes many scales – at once. Part of the training is deep humility, learning to resist falling under any illusion that we can control these senses, anyone's unique experience and way of interacting with the art, or that it's our place to even try. For me, this is a creative, spiritual, and healing practice. For Oliveros, when she published '*Sonic Meditations*' in 1971, she penned goals including 'expanded consciousness' and 'humanitarian purposes; specifically healing' (Oliveros, 1971).

From reverberant caves and petroglyphs, to cathedrals and stained glass bathing us in light and spatial sound with long tails and pictorial tales, history suggests the urge toward simulated immersion and immersive storytelling began as an attempt to grow closer to the divine, to thin the veils between life and death, and even, as stated by Morrow, to return to the natural immersion of prebirth. Later we began using immersion to create spaces of our own free expression, for social impact, and to intentionally shift perceptual experience and consciousness. We began using it to turn within, to create new worlds, to heal, and to feel a visceral connection to our planet and the vast and unseen universe, which can perhaps bring us full circle to the earliest origins of creative immersion.

## What is immersive theatre?

Immersive theatre is a form of socially engaged art that can blend spiritual and cultural traditions with modern technology, driven by the universal human desire to play, connect, and explore the unknown. Writer Suzy Woltmann calls immersive theatre a 'performance art movement that aims to transform audience members from passive recipients to active participants' (Woltmann, 2023). Immersive theatrical practice takes many forms including the elimination of a traditional stage, breaking the fourth wall, engaging multiple senses, creating elaborate explorable sensory environments, and transforming audiences into active cocreators. Because it centers on shifting audiences from passive observers to active participants, immersive practice demands a critical understanding of the historical roots of experiential and participatory culture to appreciate its current trajectory and cultural significance.

> [It's] the theater of intimacy. . . . I think audiences want this because it flings the challenge of creating meaning and interpretation back on the audience members. It's the way we've always imagined the virtual world, but it's alive.
>
> (Gantner & Lee, 2010)

Common devices employed by creators include telling stories through activation of many senses – embodied storytelling, Ambisonic and binaural sound, 360 projection, and elaborate environments with props designed to be interacted with to encourage playfulness and curiosity, and individually or collaboratively solving puzzles. Props can include 'easter eggs' hidden through spaces to create treasure hunts for information and story threads, as in Meow Wolf's *House of Eternal Return*, where visitors can pour over physical items in the house and work to piece together what happened to the family that's nowhere to be seen. Meow Wolf is one of the most successful examples of large-scale commercial story-driven immersive installations, employing thousands of local artists, first in Santa Fe, New Mexico. Local artists created elaborate thematic rooms in a remarkable model that became franchised and helped transform the world of immersive. Les Stuck, Senior Sound Technologist for Meow Wolf, has an extensive history in immersive sound dating back to the 1980s including work with IRCAM, Max MSP, Ableton, Cycle74, and Frank Zappa. He brought all this experience to their interactive spatial sound for physical environments.

Immersive theatre sometimes uses blindfolds to bring participants to secret locations and to usher us into a new world, or masks, like the hit production *Sleep No More*. The use of masks – which has a strong connection to historical cultural practices – can heighten sensory focus, create mystery and anonymity, ritualize the performance, create liberation from inhibition, and offer a sense of initiation into a secret social group. The tradition of secret clandestine locations connects to guerrilla street theatre and the underground rave scene, with map points and other tactics to earn access to information and protect its secrecy. This further enhances the feeling of being transported to another world, ritualizes the process, and carries the exciting charge of prohibition-era rebellion – even if it is just pretend, a playful masquerade. Originally, as in Prohibition speakeasies, these happenings were sometimes illegal or borderline legal in spaces that needed safeguards to keep from being shut down. Now it's more often simply a fun psychological trick to build mystique.

The curation of the journey and entrance into an immersive narrative world is crucial for establishing a sense of stepping into the story. Immersive sound plays a significant role in creating this transition. The art of crafting entrances is a discipline in itself, one I've passionately honed over years across various spaces. The practice of immersive theatre is a form of worldbuilding. The story and its world can be designed for either linear or nonlinear exploration. While participants often have the freedom to move about, some experiences place them in fixed seats, offering interactivity and agency through means other than physical movement.

It's important to note that we're defining a form driven by many creators and theorists who intentionally strive to subvert previous notions of how we perceive and interact with theatre, space, and our place within it. While I don't believe this has to mean it's esoteric, it *is* a moving target that takes many shapes, and it continues to be defined as to what qualifies as immersive theatre, as well as what we wish it to become. It's open for reinvention – while its roots are very old, it's just beginning to come fully into its own as a form of collective expression and social artistic inquiry. It's a rebellious medium by nature asking to be challenged and transformed with its core transformative nature.

Immersive theater and storytelling can also transport us to new places and invite us to walk in someone else's shoes – it can activate empathic responses by placing us in new perspectives. Transmedia Producer and Narrative Designer Caitlin Burns, whose work has been seen by millions in Avatar and Halo, talks about a piece she cocreated with Future Colossal:

> [Participants are] transported through a year in the life of the Arctic while lighting, wind, fog and spatialized audio effects play with perception of temperature and time. Beyond creating a sense of wonder, sensory immersion leads to active learning and engagement. Panoramas of a vibrant living world are suddenly covered by oil; the sounds of oil-drilling, gas exploitation and mining prevail. You then hear the voices of the legislators who threaten the Refuge and soon after, in stark contrast, the hopeful voices of the local Gwich'in people describe thousands of years of culture you can help preserve. You are no longer just transported to a new, fascinating place: you now have the opportunity to alter its history.
>
> (Burns, 2020)

This directly echoes the aims of award-winning immersive theatre production Bella Gaia, created to simulate the Overview Effect and inspire planetary stewardship. We have found – and later I'll share evidence to back this up – just as Caitlin concludes, that:

> They find people connect with places they've never visited and feel prompted to take action to protect them.
>
> (Burns, 2020)

We will discuss more examples and go in-depth into many of the spatial sound modalities employed in immersive theater. Regardless of format, in the most successful immersive experiences, the path one takes is a choice – we are given agency in our experiential travels. This means we become more deeply invested in a story. Naturally, this is complex to build with many more variables, possible paths, and mixes than a single storyline or sonic experience tracing linearly from A to B to C. At the time of this writing in

Shanghai, highly skilled immersive experts including Jeanne Angel, Digital Designer with Disney's Imagineers, have been busy building one of the most ambitious immersive rides the planet has yet seen. This requires skilled and adventurous minds – minds who thrive on solving complex puzzles and the excitement of seeing a multifaceted creation come to life. It also usually means many things will break and many sleepless nights, as the risks grow exponentially higher with more pathways through a carefully imagined and rendered world. With those many paths, we create many experiential avenues, so audiences can return again and again, having new experiences every time.

Unlike the world of rides and interactive media design, immersive theater doesn't necessarily require a massive amount of technology and speakers to be considered immersive. Just like some of the most successful 3D sound systems I've heard have been beautifully tuned and elegantly simple quad or eight-channel cube systems, some of the most powerful immersive theater sound I've experienced has been relatively low tech, sometimes created with sound sources like vocalists placed strategically in space without any amplification. Perhaps quad is like the Occam's Razor of immersive sound in some scenarios. Of course, the simplest solution isn't always best, but it's great to consider the principle of parsimony when designing sound solutions and theatrical experience design.

In 2018, The 'Mile Long Opera' Highline Project recruited a thousand singers from choirs across New York City to perform texts written based on interviews with New Yorkers about what seven o'clock means to them. People could wander the Highline Park immersed in a thousand voices placed carefully in space by two architects, singing different pieces of true personal experience written by a poet and based on interviews with diverse voices. You could also experience the project digitally in 360° on a mobile device. The highest-tech element built for this highly immersive narrative and sonic experience was the app. A huge part of the effort was in the massive coordinated community engagement project to recruit and train performers across the five New York City boroughs along with the interview process, composition, staging, and app development. This is often the case – that some of the highest degree of complexity required is related to human team building and communications, along with technology. In all of these adjoining worlds, building soft skills is as important as building technological skills.

I share this to underscore the fact that immersive theater and experience building is accessible to everyone with a passion for the medium and bold creativity, and beautiful immersive sound experiences can be achieved without necessarily having access to large-scale immersive systems or labs. The highest tech labs are playgrounds of endless wonder and inspiration, but we can also create immersivity with what we've got around us, and we often do. There's a long history in grassroots and DIY spaces of doing exactly that, and the things created there have been hugely influential on the high-tech world as well. Just look at the trajectory of the *House of Yes*, co-founded by powerhouse duo circus artists and entrepreneurs Kae Burke and Anya Sapozhnikova, which started with the performance ensemble *The Lady Circus* and two DIY House of Yes venues. *HOY*, as it's known colloquially, now exists as one of the most successful commercial immersive nightlife venues in New York, a high-production projection-mapped immersive theater space also booked to curate their signature brand of experiences around the world. That is no accident – they are some of the hardest working people I've ever met, and build brilliant teams that bring many perspectives and skills into the mix. Speaking to diversity and inclusion, Kae Burke talks about the power of different perspectives:

> Parties are more fun, meetings are more fun, and art projects are more fun when there's people with different ideas and you're not just screaming into your own echo chamber. That's why diversity is important. It's not just because it's a socially-conscious and a socially-impactful decision – it's actually just better.
>
> (Burke, 2019)

Immersive sound in an immersive experience can involve live sonic elements placed in space, a binaural headphone mix, or a multi-channel system utilized for participants encouraged to roam and find sweet spots that shift the experience of a narrative. Participants are invited to influence the experience in some way as they are given agency to alter the story and/or paths of travel. Note, I use the word 'participants' rather than audience members with intention, since it's inherent to true immersive theatre that people have agency. Thus, if you're producing immersive inclusively, there is no longer an audience, only participants.

## Sonic effects in immersive theatre

### *The pied piper*

The *pied piper* effect is a sonic device we employ to guide or entice participants to follow an actor or path in an immersive narrative using localized sound sources, like a human voice, flutist, or amplified point source open-air sound, to guide people on a desired path through an immersive experience. Environmental sound is often created and used at an entrance or between chapters of a story unfolding through different spaces. The audience is encouraged to arrive somewhere at a particular time, or drawn there through spectacle, and then a *pied piper* leads them to another location, a person acting as a bridge and sonic connective tissue.

### *Localized sounds*

Spatial sound experiences can be achieved with multiple physically placed and localizable sound sources – stationary or in motion – such as musical instruments, localized amplified sounds, noisemakers, and voices. It may also involve a narrative built around local, naturally occurring sounds.

### *Non-localized sound*

Non-localizable sound can be strategically placed and used spatially to become an integrated part of an immersive, natural sonic environment. Participants can move freely within or between spaces, creating their own spatial mix while experiencing different elements of the environment. Distant sound events can be designed to overlap just enough to maintain the illusion, ensuring the immersive 'spell' is never broken. Often, a continuous compositional thread or ambient drone ties the spatial nodes together – functioning like a film's room tone – to unify the experience. A notable example of this approach can be found in the *Sleep No More* case study.

### *Virtual acoustic*

Virtual acoustics can replicate and transport participants to another place in space or time using a surround sound system. In the Bilge Siren Project, we used the popped balloon technique to take Ambisonic impulses in the belly of a sinking ship, created a 'bilge verb' virtual acoustic IR, and used that in many immersive theatrical experiences designed to evoke nautical or otherworldly presence.

### *Group binaural experiences*

Group wireless headphone experiences such as silent discos and binaural theatrical experiences can very effectively fuse real-time and pre-recorded worlds. *The Encounter*, a theater piece using a dummy head system live on stage to capture pre-recorded sounds is an excellent example of binaural theatre. One of the best examples I've experienced of the silent disco model separated the low-end frequencies from the highs, with the treble in wireless headphones with radio transmission and the bass in subs placed at the four corners of a pop-up dance floor experience in a gallery garage. Sometimes, silent discos will offer multiple radio channels for participants to choose from along with volume level further allowing for user customization, limiting sensory overload, and making immersive experiences more inclusive. It can also allow experiences to happen at places and times that noise ordinances otherwise wouldn't allow.

### *Sound clash*

The sound clash is a key part of a sound system culture and of participatory culture history and practice as a whole. Global sound system culture originated in 1950s Jamaica, in the post-war Kingston inner city, with heavy bass, dubplate riddims, and trucks carrying speaker stacks powered by generators to create community street parties called sound clashes (Sparks, 2025). Dubplates were often cut for live testing in street dances and sound clashes with audiences and MCs creating interactive musical exchange with producers (Mann, 2022). They all competed for popularity with words, exclusive records, and the power of their low-end bass frequencies. This culture traveled to the birth of hip-hop at the Park Jams of the Bronx with Jamaican-born Kool Herc and his system the Herculords, to Bristol, and to the streets of present-day New York City where DJs load trucks with techno and jungle music devotees, creating an intentional friendly sound clash in the NYC Dance Parade and beyond. It has powerful effect, with organic spatial sound distribution and overlapping multi-channel or stereo images reverberating through the streets, sometimes for miles, sometimes from the back of moving flatbed trucks and buses, or decentralized events like the Mutek and Teknival experimental avant-garde festivals started in Montreal in 2000 that spread across the globe. Many of these events are built with multiple 'mutant' or 'Frankenstein' sound systems (see Figure 19.1). Sound clash occurs organically at many festivals and venues with multiple stages without intentional competition as well, creating a distinctive and evocative immersive sonic experience.

*Figure 19.1* The *Frankenstein* sound system wall built by underground electronic music crews Renegade Virus (NYC) and Havok (SF) at the Northtek Teknival Festival in the freetekno scene.

Ontario, 2005

### *Haptics and vibrotactile, Audio Description (AD), and sign dancing*

Exploring ways to indicate and experience sound and vibration without hearing can be considered powerful immersive sound devices in immersive storytelling when used to create an inclusive immersive sonic experience. We'll discuss examples of all three in project samples. It's an intriguing perceptual exercise to consider how to describe sound for people who can't hear it, which we do daily in evolving creative ways for virtual broadcast production, Audio Description, VR, and XR. I hope this will soon increase in immersive theatre as well.

### *Soundscapes*

Naturally occurring environmental sounds – whether experienced though natural listening, in binaural, through multi-channel sound systems, or by some combination of these techniques – often play a starring role in immersive theatre and experience design. They can provide a sense of place, time, an environmental wash evoking emotion that changes over time, or tell a complete narrative. There are many elements of a soundscape, as discussed in the seminal work *The Soundscape*, by R. Murray Schafer, very useful to immersive producers. Developing a practice of field recording over decades, starting when I was 15 years old, helped me develop critical listening, my own recording practice, and environmental sensitivity that aided me in growing toward creating environments with sound and performance.

In conclusion, I've learned both as a creator and someone experiencing many forms of immersive sound that subtlety can be one of the most critical effects. Too much movement (of sound objects, vibration in vibrotactile composition, or visuals) can be disorienting and distracting (or even nauseating) from a true immersion state, pulling me out of an experience as the brain knows physical authenticity on visceral levels. One of the keys is to create an experience we can sink into, lose and find ourselves in, a multisensory creative meditation that saves space for our personal perceptual journey.

## Origins of immersive theatre

The history of immersive theater is long, rich, and multifaceted, shaped by innovators from diverse disciplines, generations, and cultures that continue to influence immersive practices today. While specific immersive sound technologies are thoroughly covered in the editors' first book, here we focus on immersive theater's evolution and function with select innovations and applications of immersive sound discussed along the way.

Stained glass cathedrals in Europe built in the Middle Ages bathed worshippers in light, creating enveloping stories through reflections of light and sound. The cathedral room reflections and long-tail reverberations of these sacred spaces amplified the power of music and mystical experience, such as Gregorian chants and the more complex liturgical compositions of Hildegard of Bingen, fostering a heightened sense of mystery and visceral connection to sacred stories.

Many scholars of archaeoacoustics and architectural history suggest a long continuum linking prehistoric ritual caves to medieval cathedrals and theorize that both types of spaces were intentionally designed or selected for their ability to produce transcendent sonic immersion. The environments evoked awe and reverence, enriching sacred spaces and experiences where non-localized sound enhanced a sense of omnipresence and mystery. Work by Reznikoff and Dauvois (1988) established that Paleolithic cave paintings consistently cluster at zones of strong resonance, indicating that sound was central to ritual experience. They proposed the 'astonishing echo-effects' and likelihood of people using chanting and sound in ritual, making it quite likely that they chose locations with the best acoustics (Reznikoff, 2002). Subsequent quantitative analysis confirmed this statistical likelihood of alignment of 'acoustic hot spots' with visual motifs. In the Christian tradition, similar priorities can be seen in subterranean worship sites and later in monumental cathedrals, where long reverberation and diffuse sound fields supported liturgical chant and mystical atmosphere. Conceptually, Boren (2021) situates these sacred soundscapes along a spectrum between intelligible 'free-field' acoustics and cave-like 'diffuse field' envelopment, the latter theoretically used as a conscious strategy for evoking awe and sacred presence. Regardless of how direct the lineage of connection is between the two types of spaces, both environments demonstrate a deep human impulse to use resonant enclosures to envelop the body in sound, facilitating experiences of awe, mystery, and the sacred.

Our roots in immersive ceremonial storytelling arguably stretch back to our earliest known artistic and symbolic expressions. While traveling across Japan with an intercultural peace delegation, I had the honor of sitting in ceremony with a remarkable group of global Indigenous leaders including Aboriginal Human Rights Leader and Unity Earth Global Ambassador Jason Kelly, also a direct descendant of the oldest known human lineage in Australia, tracing his ancestry back 45,000 years. Mungo Lady and Mungo Man were discovered by geologist Jim Bowler in 1968 and 1974, respectively, widely accepted as the oldest human remains found anywhere in Australia, and some of the oldest anatomically modern remains found anywhere on the planet (Allbrook, 2024). They were buried ritually, adding to evidence of how far back we've been holding ceremonies. As the moon rose over Mt. Fuji, Kelly shared origin stories passed down through his ancestors. These stories were a beautiful and poignant reminder that the histories I and many cite are, in fact, relatively young. It's critical to recognize and acknowledge that much of the documented research available on immersive traditions centers primarily on Western cultural practices. As an artist and researcher, I'm compelled to explore beyond the lens we're given. Especially as it's increasingly clear to me that the impulse toward immersion may be as ancient as art and culture itself. Aboriginal art is often regarded as the world's oldest *living* artistic tradition. A charcoal kangaroo in the Kimberly region has been firmly dated as 17,300 years old, and researchers think a rock painting in Nawarla Gabarnmang may date back 28,000 years (David et al., 2013). In South Africa, researchers found a rock art etching dated back 73,000 years (Henshilwood et al., 2018). Given this incredible legacy of human art and continual lineage of Aboriginal art and culture interweaving ancestral Dreamtime stories, ceremonial practice, and symbolic thinking brought to life in many artistic ways alongside sophisticated Indigenous science – I wonder if Aboriginal storytelling may also hold keys to some of the earliest human expressions of intentional immersive experience.

Festivals across cultures and centuries demonstrate a shared human impulse to transform public space into a stage for deeply symbolic, multisensory storytelling unfolding through streets and rivers with costume, ceremony, and music. For example, Nowruz, meaning 'New Day', has been celebrated for over 3,000 years in Persia (now known as Iran) and marks both the vernal equinox and renewal of nature with the triumph of light over darkness. Celebrations vary across regions but often include symbolic preparations of fire and water, ritual dancing and performative processions, and filling vessels with water to bring abundance to the home (UNESCO, n.d.). In Mexico, Día de los Muertos also reflects this shared human impulse, filling streets and cemeteries with parades, music, costumes, and altars that blend Indigenous and Catholic traditions into a multisensory communion with the dead (UNESCO, 2008).

Around the globe we see narrative experiences designed for theatrical interactivity with processions marking the passage of important events and seasons. Multisensory experience and storytelling are often used as a device for collective transcendence, for exploration of the lines between life and death, to teach moral lessons, to celebrate important figures, and for the ancient universal drive to honor our ancestors. The Great Dionysia in Athens marked Spring and featured dramatic playwright competitions, a significant part of the institutional development of Greek Theatre where tragedians such as Aeschylus, Sophocles, and Euripides first staged their works (Csapo & Slater, 1995). We see this across continents and time, cross-cultural practices with a long lineage suggesting very old immersive roots.

In the mid-17th century, 'illusionists' began experimenting with theatre through elaborate optical and sonic devices, most often themed around raising the dead. Rasmussen Walgenstein's *lanterna magica* projected haunting images of a corpse for the court of King Frederik III (Grau, 2007), creating immersive experiences with storm sound effects in a predecessor of cinema.

> The evening's entertainment began with a simulated storm complete with thunderclaps, wind, hail, and rain. The dramatic climax was the conjuration of spirits.
>
> (Grau, 2007)

*Magic lantern illusionists* were early immersive theater artists. Occultist Johann Georg Schropfer took this further with:

> projection of ghostly apparitions onto smoke using a concealed magic lantern. The images produced by this technique were flickering and ephemeral. . . . Schropfer used a whole suite of tricks including projection with mirrors, hollow voices spoken through concealed tubes, assistants dressed as ghosts, and thunder sound effects. To this arsenal of illusions Paul Philidor added the recently invented Argand lamp, which produced a much stronger light and thus enabled larger audiences to see the images – this was the birth of the phantasmagoria.
>
> (Grau, 2007)

We see early origins of immersive theatre in participatory pantomime performances of ancient Rome, in ancient Greek theater, the tradition of Commedia dell'Arte, and performance art traditions like *tableaux vivant*. Exploration of the evolution of early immersive theatre also brings us to Antonin Artaud, the French Surrealists, and Theatre of Cruelty. We can see correlations with the Dadaists, Surrealists, Letterists, and Absurdist movements.

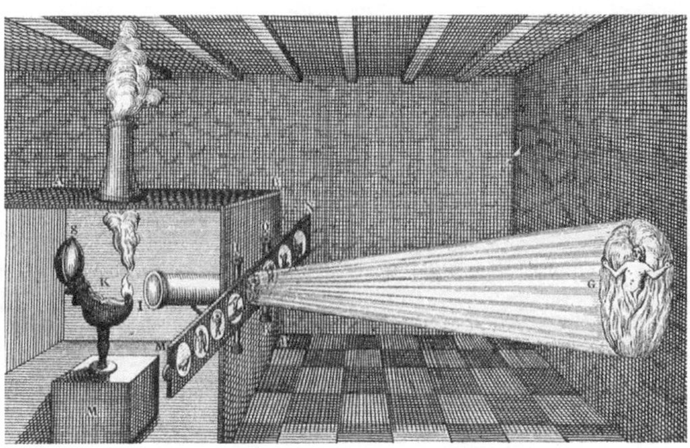

*Figure 19.2* Illustration of a magic lantern (or *laterna magica*) contained in the work '*Ars Magna Lucis et Umbrae*' (1671) by the mathematician Athanius Kircher

Connecting pre-modern and modern histories, immersive street theatre created within festivals, protests, and parades often employs grassroots spatial sound techniques together with embodied experiential elements, archetypal characters, dance battles, and giant puppets that can clearly be seen above the heads of mass rivers of people. In the Czech Republic, master puppeteers Leah Gaffen and Mirek Trejtnar teach the next generation of puppet carvers. They and their daughter Johana taught me how puppet theatre helped save the Czech language and culture during the Habsburg Empire. Jacklyn Janeksela in the BBC 'Koha Ditore' broadcasts writes extensively about this history and speaks of:

> evidence that the act of creating dolls has long been a form of protest for the Czech people. Seventeenth-century wood carvers . . . began making puppets for actors in Bohemia soon after Ferdinand II came to power. . . . The puppets were the only remaining entities that had the right to speak the Czech language in public.
>
> (Janeksela, 2018)

The historic and hugely influential work of the Bread and Puppet Theater – founded in New York in 1963 by Peter Schumann and vibrantly active for decades – exemplifies modern immersion, puppetry interwoven with participatory spectacle and ceremony. At

*Figure 19.3* Schooner Apollonia transformed into a 64-foot sturgeon puppet with moving eye and fins by OBE Award-winning artist Gregory Corbino for Riverkeeper

the time of this writing, puppeteers and mission-driven immersive producers have joined forces with scientists, a flotilla of historic ships, and Riverkeeper (the Hudson River's oldest environmental organization) under the visionary leadership of Tracy Brown, to inspire stewardship and actively protect the waters through distributed and connected narrative spectacles activated on both ships and land. Along with producing the flotilla, I developed a participatory sound project around soniferous migrating fish of the Hudson River, and a collaborative storytelling process designed to unfold in concert with lobbying efforts and a theatre-driven curriculum led by Superhero Clubhouse. The soundscape of the event on land included the Funk Rust Brass Band, Indigenous speakers, and songs written by students of the NY Harbor School amplified in public parks. From the ships, Misha Mayeur and I performed and transmitted live on Wave Farm Radio, while the flotilla connected with multiple destinations activated by roving musicians and puppet parades built with Heather Henson and the Jim Henson Creature Shop. The ambient sounds of the water with the ship's diverse whistles and bells provided a beautiful nautical sonic backdrop along with the chatter of marine radios and crews. For the soniferous fish project, sound artist Aine Pearson and I researched and publicly presented a range of hydrophone recordings capturing fish sounds including the low frequency rumble dubbed 'Sturgeon Thunder' by sound artist and scientist Chrys Bocast. Many branches of the *Migration Celebration* team worked together beautifully to directly link the impact of immersive experience with action. These new studies of the transformative history and practice of immersive theatre are evolving on stages, on rivers, in cinema, and in parks all around us, often mirroring society's most urgent needs for change and collective expression.

## Participation and transformation

The roots of contemporary immersive theater are historically transformative and avant-garde. These roots are deeply intertwined with those of sound art, electronic dance music, puppet theatre, and participatory culture as a whole. At times, this culture has sought to disrupt and transform, embracing radical self-expression and creative resistance culture; at others, it focuses purely on aesthetics. Regardless of whether it aims to catalyze change or its art for art's sake, it challenges conventions of proscenium stage theatre, thus transforming the traditional audience perspective sound experience by eliminating divisions, asserting that anywhere can be a stage or ceremonial space, and turns passive receivers into active participants. As discussed in this chapter, the origins of ancient and modern immersive theatre are also historically interwoven with ceremonial practice, spiritual practice, embodied experience, movements for social change, free expression by marginalized communities, and cultural preservation, while challenging the core definition of theatre. This evolution has always gone hand in hand with explorations and advances in technology and spatial sound, at times deeply embracing analog forms like vinyl and cassette culture.

The Living Theatre, the oldest experimental theater group in the USA, was founded in New York City in 1947 and contributed to an increasing shift toward immersivity, interactivity, and direct spectacle in theatre. They were committed to reshaping societal power dynamics by shifting from a competitive, hierarchical model to one rooted in cooperation and communal expression. The troupe sought to achieve this by disrupting audience complacency with engaging performances and spectacle.

It's not hard to see the direct correlation with the essential aims of immersive theatre, to utilize theatre of all kinds for transformative and socially engaged arts, a creative questioning of status quo culture and power structures.

The language of spectacle is one of the great traditions of immersivity. Julian Beck, co-founder with actress Judith Malina, said The Living Theatre's mission is:

> To call into question who we are to each other in the social environment of the theatre, . . . to set ourselves in motion like a vortex that pulls the spectator into action, to fire the body's secret engines, . . . to move from the theater to the street and from the street to the theater.
>
> (Beck, n.d.)

Taking action was sometimes but not necessarily connected to political activism – in many cases the medium is the message – activating community members to become engaged and participatory in co-creation and community is, in itself, the aim – to resist complacency.

Participatory culture is a direct response to consumer culture, a rebellion and shifting from viewing citizens of the state as passive consumers to active producers of everything from media to theatre. Flash mobs, fan fiction, cosplay, and social media are all part of the participatory cultural shift that gave rise to the rising tide of immersive theatre and immersive excitement. The emergent field of Transmedia Storytelling is a perfect illustration of how this can be utilized by design on very sophisticated levels, reaching many industries at once. Transmedia Storytelling is a powerfully potent tool set with the truth of immersion and participatory power at its core. Jeff Gomez, co-founder of Starlight Runner Entertainment and an early mentor, reframes Joseph Campbell's classic *Hero's Journey* for our next era of storytellers. Many of our most beloved narratives, from *The Wizard of Oz* to *Star Wars* to *Harry Potter*, have used the same narrative formula of a singular protagonist who saves all after first refusing the call. Gomez rewrites this tired adage with what he calls the Collective Journey Model[1] – a narrative practice that disrupts the story status quo and truly reflects the sociological shift we're living while revealing the power of collaborative, participatory creative agency, a narrative structure that moves beyond individual heroism toward stories animated by ensemble perspectives and shared creative power. This mirrors the growing scholarly acknowledgment of what we already know – that storytelling today is increasingly immersive and collaborative, with audiences and creators co-authoring the narrative experience.

A lineage of avant-garde theorists connects key points in the story of avant-guard theatre, one of which is the Letterist International (LI), founded by Guy Debord (1952–1957). This self-proclaimed motley crew of writers, sound poets, painters, and bohemians – culturally connected to the Beat Generation in the USA – laid the groundwork for the subsequent formation of the Situationist International (SI), active in Europe from 1957 to 1972. The SI published *Internationale Situationniste* (1958–1969), a journal that articulated its core ideas. During its early years (1957–1962), the SI focused on psychogeography and the *dérive*, creating situations in urban environments to foster creativity and critically engage with daily life. Another key concept, *détournement*, involved creatively reusing existing media to critique consumer culture. This approach anticipated modern practices such as appropriation and sampling, as well as protest art and culture jamming techniques like *billboard liberation*, which began in 1977 in San Francisco and became popularized by internationally renowned artists and groups such as the Billboard Liberation Front (BLF), Shepard Fairey, and Ron English.

In 1992, Brazilian dramatist and director Augusto Boal announced he was running for commissioner (mayor) of Rio de Janeiro, and he won with the campaign slogan, 'Have the courage to be happy'. Boal employed actors to perform interactive street theatre in low-income and marginalized working-class neighborhoods, using performances to communicate political policies and uplift morale in communities often excluded from the arts and policy messaging. He believed in engaging audiences directly, encouraging them to participate, propose solutions, and reshape the narrative on stage. Boal's approach, which he termed 'legislative theater', aimed to influence politics through art rather than letting politics reshape his vision. This method transformed Brazilian political engagement and inspired social advocacy groups worldwide.

> We use theatre to discuss problems of communities, workers, Blacks, women, street children, the unemployed, the homeless, etc. We don't want a passive audience, simply watching. We propose, on the contrary, that the public participate, interfere, enter on to the stage and propose alternatives for the plot: create a new story. . . . Theatre is political and politics is theatre.
>
> (Heritage, Boal, 1994)

Boal's Legislative Theatre is rooted in Forum Theatre, a participatory form of performance that addresses community experiences with specific issues. Its goal is to actively involve the audience in understanding a chosen issue and collaboratively discovering creative solutions. He created the Forum Theatre practice in 1973 as part of the Theatre of the Oppressed (TI). Examples of this practice are now used across the globe including in Africa where Forum Theatre serves as a tool to foster unity and help heal the emotional scars left by colonialism, civil wars, and modern slavery.

> Throughout the 20th-Century, avant-garde theatre has been concerned with politics or activism in one way or another. From the actions of the Futurist Movement through to the groundswell in political theatre after the 1960s, radical thought has always been applied to theatrical forms. Over the past four decades, activist groups have used ritual concepts, often including participation of the spectator, taken to the streets, and organized festivals and spectacles presenting ideas marked by a considerable departure from the usual or traditional.
>
> (Sant, 2012)[2]

With immersive theatre, stories unfold outside traditional stages – in public spaces, winding through multiple locations, the whole city and world can become a canvas for creation and transformation, with viewers becoming active participants with the power to express and affect dramatic narratives and paths. This can be used to galvanize participation and change on a larger social level.

Around the globe, starting in the mid-1990s, organizations such as Reclaim the Streets (aka RTS) used giant puppets, celebratory parades, and peaceful direct action framed as giant free street parties to gather masses and encourage collective cultural use of public space. This was a direct part of the culture our storytelling lineage, from which early modern immersive theatre culture was born. The ethos of RTS came from ideals of collective ownership of public space aka The Commons and from the idea that culture is up to all of us to co-create. The organization was first formed in London by the environmental advocacy group *Earth First*, and the first event in the UK in 1995 featured rave culture secret location tactics, dramatic street theatre, free food tables, and dancing in the middle of a busy London intersection set to a bicycle-powered sound system. Subsequent RTS events drew as many as 8,000 people and influenced art culture around the globe, including many of the early immersive theatre groups we co-created with during the late '90s and early aughts.

In 1997, brothers Geoff and Aaron Taylor Kuffner co-founded the artists' collective Ransom Corp with dancer/filmmaker Kathi Von Koerber Kaesha. The group produced ritual immersive performances on many scales in the USA and Europe, including in public spaces, in abandoned buildings, and in nature, and toured Europe with programs featuring video installation, interactive musical instruments, butoh-inspired dance, and aerial performance. Geoff Kuffner calls what they create *audience immersion in service to emergent ritual.*

> We sought to scramble and unscramble the senses, to explore spatial and perceptual possibilities in order to give birth to new and sometimes very ancient experiences. We cultivated personal and collective mythologies to enable and inspire understanding and transformation within our species.
>
> (Kuffner, 2002)

All experiences were designed to follow a set of guiding principles, in a nod to seminal groups like the Living Theatre, and a significant part of our collectively emerging modern immersive culture. The principles include creating a festive, carnival-like atmosphere that infuses performances and gatherings with participatory and transformative energy, empowering the audience to internalize and embody this vitality. (Kuffner, 2002) They treated every element, including fliers and entrances, as part of a total experience and environment.

> (They) programmed the nights to move in waves, to have silences and crescendos, like a jazz score where all players will step forward to solo and to recede to the back.

The study of crowd psychology, and intentionally acting as architects of experience on levels that focus on dynamic narrative arcs, responsiveness to collective crowd energy, and giving individual collaborating artists moments to create solo pieces within an overarching narrative container – this is all key to experiential design. It's also telling that it's described in terms of music and sound. Sound is central to every spatial and immersive experience we create, even if the sound is silence or natural ambient environmental sound, this is always considered. There is a live feedback process that happens where artists respond to sonic, environmental, and participatory elements that can't be anticipated. So we must become or work with skilled improvisers that can activate and hold and sculpt these spaces while responding in real time to the audience and to space, our cocreators. Sound and compositional arcs are also a metaphor for understanding how to move space and crowds, how to shape participatory experience, and how to move emotion. Many people whom I've seen become absolute masters of this craft; for example, multidisciplinary artist Akim Funk Buddha developed a great deal of this strength first as a street performer. Amanda Palmer goes into great depth about this in her book *The Art of the Ask*. Acting as an architect of experience is an emergent art form that calls on the ancient, invites the future, and demands total presence in the present, all at once.

After returning from the SKY PLOTS European Tour in 2000, the Ransom Corp founded 23 Windows in Bushwick, Brooklyn. The old Brewery served as a home for many experimental artists until 2021. Taylor Kuffner went on to create the Gamelatron, award-winning sonic kinetic sculptures fusing the thousand-year-old tradition of Indonesian Gamelon music with robotics to create intentional sanctuaries of immersive sound and light driven by his compositions that are often site-responsive. The Gamelatron has been featured around the globe, including multiple times at Burning Man, where it formed the base of the interactive center temple in 2017. He built the prototype piece while working in residency with celebrated engineer and technologist Eric Singer at the League of Electronic Musical Urban Robots.

Around' the world, especially in cultural hubs like London, Brazil, Tokyo, and New York City, artists evolved dramatic immersive traditions that originated in ancient rituals and practices, gaining powerful momentum through various eras, including countercultural and activist art movements of the 1940s, '50s, '60s, and '70s.

## NYC: underground culture of immersion

In NYC during the late 1990s and early 2000s, this global lineage continued to evolve while a freshly collaborative era of boundary-breaking creativity was unleashed at a creative crossroads. A perfect storm of influences ignited a bold new era of experiential art, transforming the city into a living lab for immersive creation. Beginning in 1995, Anita Durst's ChaShaMa began offering temporarily unused real estate to artists for creative activations, transforming the cultural landscape by creating increased access to space – the one thing New York artists needed most as gentrification and skyrocketing rent prices began displacing and forcing more and more of the creative class to migrate. The turn of the millennium, coupled with the politics of the post-9/11 period, saw a vibrant new era of grassroots artistic expression fueled by a deep determination to create a more peaceful and inclusive future while aiming to liberate free expression and advance social justice through the arts. Wonderfully strange theatrical spectacles happened in storefront windows and historical theatres on 42nd St., including TeleTwinTowers for Peace, produced by Missy Galore with a 50-hour nonstop lineup in Times Square. I joined other performers in the 42nd St. storefront windows with speakers spilling avant-garde sound and experimental theatre onto the street for bewildered tourists with baby carriages. The world of DIY culture lit up with elaborate co-created experiential lineups and environments, in now legendary events hosted by key artist collectives, including Rubulad, Ransom Corp, Blackkat, Complacent Nation, Madagascar Institute (pioneers of the NYC hacker and maker communities), Dumboluna, Amoeba Technology, Wanderlust, The Lady Circus (founders of the House of Yes), and Burning Man–born organizations like SEAL Arts and 'street theatre club' Kostume Kult.

Together, all became part of a renaissance and intentional shift from competition to collaboration among key culture 'Makers'. As cooperative planning and mutual aid became more clearly necessary for the survival of individuals, organizations, and the culture at large, much of the NYC nightlife culture shifted toward more collaborative cultural strategizing and co-producing with shared interests and advocacy in the face of mounting political pressures, draconian Cabaret Laws, and the incessant battle for affordable space. In the post-9/11 landscape, working together wasn't just admirable; it was essential for survival and cultural rebuilding. The infamous 'dog-eat-dog' ethos of New York City made way for more strategic alliances and creative mutual aid, including increased community building and shared harm reduction resources. It was an empowering and shimmering silver lining in a deeply vulnerable time.

This is often a truth that – in the midst of tragedy and deep uncertainty, with focused work, collective crisis, and trauma – can reveal a new unity and determination. In truth, we didn't have a choice but to learn and get better at working together to survive and support new growth. The entire city became a canvas for reimagination.

Our imagination was definitely not limited to land – the waterways became a vast networked playground for mischief, spectacle, and costumed theatrical battles, often led by the immutable artistic force that was Orien McNeill. McNeill was famously baptized by William S. Burroughs in his mother's sink with a bottle of tequila, an act which artist Duke Riley astutely observed properly 'foreshadowed an exceptional life of unparalleled adventures'. Those adventures included countless vast and absurdly ambitious collaboratively created immersive spectacles that snaked through the waterways of New York City for decades with McNeill as captain, pied piper, mentor, and devotee of artistic chaos. Jeff Stark rallied a crackerjack theatre troupe of talented city misfits to co-create his original work *The Dreary Coast* on the city's most polluted canal, complete with boats, high priestesses, and a full score in an epic work *The New York Times* called 'reeking but rhapsodic' (Soloski, 2014). Artists with roots in street art, public art, and even deep sculptural studies, like Tom Beale, ESPO, and Caledonia Curry (popularly known as Swoon), did brilliant things to both well-lit galleries and shadowy industrial alleyways. Five years after being one of the first artists to ever transform the walls and corners of the Lunatarium, Swoon designed and led, together with McNeill and 50-plus other collaborating artists, an almost unimaginable series of 'Swimming Cities' projects – sound system–laden sculptural junk rafts, floating works of art they lived on while wafting art and mystery down rivers including the Mississippi and Mahicantuck. This series culminated in the Swimming Cities of Serenissima, where the brilliant and brazen shipped the handmade boats in pieces to Slovenia, then reassembled and floated their way to Italy. They invited themselves to the Venice Biennale and were met with critical acclaim (and doubtless some light legal *kerfuffles*).

Back on dry land, site-specific productions by theatre and opera directors like Anne Bogart expanded classic forms into unconventional spaces with blurred lines. It was in this same era when De La Guarda shook up off-Broadway New York with an Argentinian theatre troupe fusing aerials, multisensory avant-garde theatre, an omnipresent stage, and thumping percussion-heavy techno. Eight cast members on bungee cords flew above audience heads, walked on walls, played with space and water, and, yes, you got wet. Occasionally, they plucked a wide-eyed audience member from the floor, flying them up into the sky to further remind us that there was no fourth wall. It was brilliant, and it took the city by surprise, partly because it was packaged as an evening theatre experience while it was in fact a visceral immersive theatre experience capped with a brief dance party you wished would keep going into the night. For that, you had to go to the Lunatarium.

The Lunatarium emerged as a landmark immersive arts space in the post-9/11 era, a towering multimedia cathedral on the Brooklyn waterfront. Housed on the top floor of an old sugar refinery framed by the Manhattan Bridge and a power plant, its 20,000 square feet of soaring 30-foot ceilings, columns, and vaulted windows offered breathtaking views of the Manhattan skyline – a skyline forever changed just weeks before the doors officially opened. The venue's significance in advancing participatory and immersive arts stemmed from its iconic location and deeply significant time for culture, combined with management by a collective of artists who embraced bold, often risky experimentation to push creative boundaries and foster free expression.[3] We had a vast, humbling amount of space to fill for each uniquely themed event and opened at a time when the city and world were in collective mourning, deeply needing community and healing through art and culture. We found ourselves at the forefront of a new cultural era as the next millennium had just begun, and then the world as we knew it disappeared overnight. Arts and immersion have always played a key role in transformation and collective healing, as discussed and shown by countless studies and organizations, including the CAIR Lab and US Surgeon General (CAIR Lab, 2024). We opened when nearly everything else was shut down because people need community. We had work to do.

As Music Director and Stage Manager of the Lunatarium, I focused heavily on the sonic experience, the overall event flow, and crafting collective moments of experiential crescendo. We curated with a core team and with co-producers, and being a small grassroots team we all wore many hats. The space also hosted outside producers while working with hundreds of artists and volunteers. The lineups I personally designed were crafted intentionally to move between diverse genres to tell a story and to draw together diverse communities.

This was an intentional playground for immersive and experiential R&D. Along with the intrinsic spatial nature of warehouses adding a natural height channel to all sound experienced, the soaring ceilings provided a perfect context for fire performers, aerialists, inflatables, and circus arts. All were curated to explore and push the experience of space and to fill the immense expanse with creativity and wonder. It was a dream for many to have that vertical axis to play with. Few knew how to play with the art of spectacle or needed massive space for artistic experiments more than aerialists and Burners – those devoted to poi, fire arts as a whole, and the Burning Man Festival with its signature celebration of large-scale interactive art and participatory culture. Relationships were built, and some crews became resident performance troupes along with breakdance crews, fashion and costume designers, VJs, punk rock circuses, human-powered carnival ride builders, propane sculptors, sound artists, installation artists, performance artists, inflatables artists, DJs, live bands, and beyond. The space became a vital hub for creators of many disciplines as we hosted and co-produced with countless groups over four years, like Anakin Koenig Airways, who built massive inflatable explorable lounge environments for events like the *Airport* Ambient Festival.[4]

The Lunatarium belonged to an expanding lineage of venues with an unabashed grassroots immersive ethos – places that were maximalist mazes of found objects and whimsical experiments not yet safe for mass consumption but offered (sometimes to thousands of revelers a night) anyway, perhaps paving the way for the above-board more polished commercial enterprises we are seeing now. There's no doubt we were all inspired by the iconic Rubulad, where energy wound through fever dream rooms in roaming currents, a decentralized river with estuaries of attention, drawing all to roam, connect, and compare notes about what they'd encountered. A main stage still existed, but performers could and would appear anywhere, including balconies, suspended from the ceiling,

*Figure 19.4* Artistic rendering of the cavernous and architecturally resonant Lunatariam in DUMBO, Brooklyn, circa 2001 with 30-foot ceilings, multiple sound systems, and fire dancers on a floating platform only accessible by climbing.

and spinning fire from a floating second-level platform accessible only by climbing a telephone pole-style column. Sound was inherently spatial, shaped by the architecture and surrounding decor. The space itself became an instrument played by artists of many mediums. At Rubulad, every inch of the maximalist dream was layered with handmade flotsam, a constantly evolving kaleidoscope and textural playground.

Door Manager and founding member Micha Waschke welcomed guests as he played flute, an enchanting literal pied piper creating an unexpectedly mythic entrance as people waited for a freight elevator to transport them into another world, immediately seeing that this wasn't a standard nightclub. Entrance experiences are key. In immersive world-building, entrances are not mere doors; they are portals. In her book *Parades, Parties, and Protests,* Sarah Sparkles describes this curated portal at the Lunatarium:

> We ventured inside and we escorted into an intergalactic freight elevator, decked out with metallic foil, blue lights, ambient music, and space alien operator when we arrived at the top floor the doors of the mothership open to reveal a cavernous 20,000-square-foot warehouse that had been transformed into a realm beyond my wildest imagination.
>
> (2013)

The Lunatarium (aka Luna) served as a canvas for some of the most ambitious and large-scale experiments I've ever witnessed. Experimental, interactive, and immersive arts unfolded in many forms. Madagascar Institute built human and jet-powered carnival rides that looked like they could eat you alive (and in fact might if you looked at them sideways). A larger-than-life robotic game of Simon made by sound artist and metalworking maker Eric Singer spewed flame jets you had to feverishly mimic with a four-way controller built from drum pads. Playing that game was truly one of the most thrilling moments in my young and sometimes dangerous life. A floating perch in the middle of the space, allowing fire spinning and projecting in any direction with video crews including Honeygun Labs, Funktaxi 1533, VJ Mojo, Dorian Orange, and Metal Tiger. Events featured diverse genres of live and electronic music, sound installations, and live legends like Run DMC, Afrika Bambaataa, and DJ Craze. Local and internationally acclaimed artists like Christina Wheeler, DB, DJ Olive, Miss Kittin, Rasputina, fflood, and countless more artists performed there along with legendary pioneering dance music lables such as Breakbeat Science, Sonic Groove, and Liquid Sky. The space created cultural shockwaves. We were nominated for Best Club by the US Dancestar Awards in 2003, Best Alternative/Unusual venue by *URB Magazine* in 2002, and multiple Best of NY awards from the NY Press, all while striving to walk a comically thin tightrope of keeping things quiet and promoting a massive venue producing – at its height – three events a week with unique themes, decor, and co-producers.

Simultaneously, large-scale participatory spectacles took place in other industrial spaces, in crumbling old manors, and on the city streets. In the spring following 9/11, the epic-scale Madagascar Institute event *The Hurting* facilitated collective catharsis and mass processing of pain inside a 60,000-square-foot warehouse animated by an abundance of fire, music, four distinct sound systems, and cacophonous artful demolition. Headliners included 'The Smasher', a machine built by founder Chris Hackett whose sole purpose was destruction via good old-fashioned jet-powered smashing. Only in New York, as they say.

Together, we honed the craft of large-scale immersive spectacles and crescendo moments, using crowd energy as a medium to build moments of curated collective awe. This relied on curating multidisciplinary lineups that build with dynamic and intentional arcs in the experiential flow, with seamless transitions between sets and key moments, while studying and developing intuitions about crowd psychology. We are always thinking about spatial and temporal flow through any large-scale immersive experience. The experience of space and spatial sound, whether amplified or open-air, is critically considered, in terms of the macro-level of the entire event, the buildup to the grand spectacle, and the apex experience itself.

The key is building and carrying energy in a narrative sense, with dynamic arcs, just as we consider in compositional song structure. You can't drop that energy, can't miss a beat. We curated many disciplines to come together to create a gestalt moment that's more than the sum of the parts. Crafting 'midnight moments', as we often called them, is an art form in itself, and deeply fulfilling and exciting when you pull it off, especially with hundreds or thousands of people. These crescendos could be brass bands paired with fire and dance or a culminating theatrical tableau of characters you've seen separately throughout the night. At one event in my *Nebula* series, I worked with a DJ to cut the amplified sound at the height of the party when the dance floor was at its peak. It's brazen and rare to do that on purpose, but we had cleared space on the dance floor for a Wing Chun Kung Fu School to surprise the crowd with martial arts demonstrations set to live hand percussionists. Everyone worked together to present something beautifully unexpected, transportive, and powerfully alive. Cueing the DJ to hit back in with a high energy record at exactly the moment the live elements were complete, guiding the experiential choreography without missing a beat, makes the crowd go wild.

A few of my personal favorite immersive performances artists created at Luna included an ensemble from Philadelphia that danced with fire while simultaneously carving massive ice sculptures with chainsaws. At a weekly event co-produced with DJ Spooky, a ghostly performance art troupe dressed in all-white Victorian garb and face paint created a striking interactive installation. They silently handed strings to participants across the expansive venue, weaving massive human webs as partygoers stood still, holding space and forming intricate shapes. Another group guided blindfolded guests with great care, one at a time, through tactile and auditory sensations before being awarded the prize of a plate of spaghetti under the towering Manhattan Bridge. At an Alice in Wonderland–themed New Year's Eve event attended by 2,000 revelers, artists created a white paper rose garden repainted red by frantic playing cards in costume during the event. A driving creative force behind the venues decor and aesthetics, Artistic Director and VP Mark Winkel led a huge volunteer team to paint the dance floor in a chessboard pattern, and guests danced so hard they erased all traces of it by the end of the night (Green, 2010).

We built a cast of characters and dramatic vignettes played out weaving through the space and involving surprised guests in the story. I played the Queen of Hearts. The Wedding of Chunk and Chove was another unforgettable immersive spectacle, this absurdist story having first begun with theatrical episodes played out on the streets of New York by improv artist Kurt Braunohler, inspired by Shepard Fairey's Obey street art campaign and Andy Kaufman, created collaboratively with the NYC improv community. A clandestine message went out, in the same spirit as it often did:

> WEAR WHITE IF YOU'RE A CHENGWIN.
> WEAR BLACK IF YOU'RE A CHUNK.
> WINNER GETS CHOVE.
> EVERYBODY ELSE GETS A PARTY.[5]

Kurt the Inevitable (2002)

The aforementioned text comes from the public call to rally participants for the flash mob–style comedic bird-battle stories that unfolded in public spaces. These kinds of artist calls are intrinsic to how participatory dramatic arts are made. It's critical to take a careful tone whether it be comedic, inspiring, mythic, or somber, and guerrilla marketing is always elemental. In this case, the eight-foot-tall, portly costumed hybrid chicken characters played out a love triangle complete with feathery bird battles over months in the middle of Manhattan among confused and amused tourists and participants, until finally ending up at The Lunatarium for an immersive wedding reception no one would forget (Callahan, 2005).

This underground culture carries a pronounced and profound sense of humor, ripe with absurdist and dadaist nods. From the more recent group Shadow Traffic's 'Competitive Winter Picnicking' to Brooklyn's 'Condiment Wars', 'Idiotarod', and 'Battle for Mau Mau Island', humor and improv comedy have long been a staple in this world, and thanks to a next generation of makers and the unceasing need for humor, it certainly lives on.

In the post-9/11 art party and creative resistance culture, the use of ritualistic and ceremonial elements was key in this emerging immersive movement. This revival reconnected us to our experiential roots while also foreshadowing our immersive future. Artists and producers transformed industrial sites, public green spaces, and other unconventional locations into labyrinths of collective immersive rituals – many culminating in poi and fire dancing.

Alongside the post-rave culture of multimedia art parties, which served as a testing ground for experimental and immersive projects, some of the most electrifying spatial sonic landscapes emerged from free subway train parties, mass street gatherings, and large-scale spectacles that drew thousands to participate. These narrative-driven, roving, and meticulously planned episodic events and parades centered on sound, art, and collective spectacle. Their spirit and lineage were deeply rooted in sound system culture – a direct homage to the dub sound system movement of Jamaica, blended with the unmistakable grit and flair of New York City. The entire city became a canvas, while every event featured massive sound systems, many of which mobile rigs carried by biodiesel buses converted to run on vegetable oil and retrofitted with turntables. Artists left beauty in their wake as art parades transformed the streets into living galleries.

A key part of this lineage was New York City's punk and squat scenes, which visibly merged with rave culture at Blackkat's annual *Mayday Festival* in Tompkins Square Park – a free, multigenerational event commemorating historic workers' rights riots that took place there in 1874 and 1988. Co-producer Arrow Chrome later converted an old school bus to run on biodiesel and took the *Biotour Bus* on the road, traversing the USA on a dance music and French fry oil–fueled eco-tour, using sound and spectacle to educate people on the sustainable benefits of biodiesel while hosting impromptu DJ sets and parties to foster connection and awareness. The 1990s and early 2000s were an era of boombox brigades and brass bands, of spontaneous sonic uprisings in the streets.

One now-legendary co-creative triptych of renegade street parties, named for its timing on the first warm night of the year, took this ethos to another level. Thousands of people showed up to the mysterious starting point of the roving fable that was collectively organized inside dimly lit vintage wooden kosher wine casks underneath avant-guard LES club *Tonic*, one of John Zorn's sonic homes. When Akim Funk Buddha famously repelled off the Brooklyn Bridge surrounded by stiltwalkers, brass bands, waving flags, and the setting sun, he shouted, 'We're going to reclaim the ritual!' In order to keep things peaceful and moving, a strategic plan of working with law enforcement was designed and enacted with a combination of clear point people, fairy-winged guides dubbed 'coaxers', carefully establishing trust and no overt political agenda, and at times asking for forgiveness rather than permission. In the words of William Etundi, a veteran activator of mythic, boundary-testing public celebrations and rituals cocreated with many of NYC's boldest producers – events that continue to whisper inspiration to more entrepreneurial immersive productions including his own membership model social club CX – 'We simply explained to the cops that we aren't anti-anything; we're pro-everything'. Deep in Brooklyn, DJ Chrome outfitted a flatbed truck with turntables on a suspension rig, allowing vinyl to be played in the streets as a parade followed behind. Skeptics asked, 'Why not just use CDJs?' But that wouldn't have been brazenly, absurdly ambitious to the point of being comical, which was a key motivation. Naturally, the needle skipped a few times. But that day, we danced to vinyl in the streets, and the event became legend.

## Spatial sound and sound system culture

This growing mass immersion culture in NYC, heavily tinged with homespun punk rock ethos, was also deeply intertwined with the lineage of rave and club culture along with their predecessors, Jamaica's sound system culture. Some raves gave sound system crews and speaker companies top billing alongside headlining DJs, while notable clubs and selectors became renowned for their systems and spatial innovations. Beginning at the *Experimental Soundfield* at Glastonbury in 1992, Funktion-One began installing four-point and six-point speaker systems, experimenting with innovative left and right stereo placements. Their explorations of spatial sound built on live traditions established by the classic quadraphonic systems of the 1970s. Launched in Manhattan in 1970, the pioneering invite-only *Loft* parties of David Mancuso, working in collaboration with sonic architect Alex Rosner, were as celebrated for their impeccable sound clarity as they were for their music and transcendent atmosphere. Rosner's setup – four Klipschorn speakers with false corners – set a gold standard for audiophile experiences in the underground club world. Fifty years later, at the time of this writing, I attended the still-ongoing Mancuso Loft party, where the sound system remains quad enveloping the dance floor with incredibly lush high-fidelity sound sourced from live vinyl DJs and VJs using analog overhead projectors – an audible through line of care, intention, and acoustic ethics carried across generations. Similarly, sound engineer Steve Dash played a key role in defining the legendary *Paradise Garage* and later carried its legacy of sonic excellence to *Sound Factory and Twilo*, which became renowned for their clear and thunderously powerful sound.

> The sound which is extremely intelligible is entirely concentrated on the 60 × 60 dance floor and there is a decibel drop as you walk off. We have four speaker stacks in each corner and on the side of the floor, blowing in, are two hanging side-fill custom-made boxes. There are no other speakers anywhere else, they're set back and pointed towards the floor and are designed to fill the corners. Delay lines ensure that the audio hits the floor at the same point in time.
>
> (Dash, 1994, as cited in Fikentscher, 2000)

Rave and club culture, with their inherently immersive nature – multimedia environments, elaborate themes, and at times groundbreaking sound innovations – served as significant test beds and chapters in the evolution of immersive theatre.

The early immersive New York culture, like immersive theatre as a whole, brought together inspiration and learning from many music traditions – Jamaican Sound System culture, the birth of hip-hop and cyphers, lyrical and movement improvisation in public spaces, the downtown experimental scene, and street theatre used as social commentary and collective change agent. Added to all that was the power of the dance music and DJ culture from Detroit techno and Chicago House to New York City's legendary Studio 54 and Phazon Sound eras, the UK and stateside jungle and breakbeat scenes, and the multimedia festival traditions of Europe and Canada like MUTEK. Much of this emerged from the urgent need for welcoming spaces that empower free expression – centering Black, Brown, and queer identities and culture, and embracing people of all generations drawn to the rhythm and energy of those records. At the same time, the tapestry of global cultures that defined New York as a cultural Mecca also played a vital role. All these traditions coalesced to inform the experimental immersive sandbox of New

York City's millennial warehouse culture. I'd argue that the post-rave experiential culture forged there became one of the Americas' most distinctive and globally recognized cultural exports – now emblematic of the Brooklyn brand discussed in Gregory Hanson's case study in this book.

Rooted in festival and rave culture, street theater, and participatory culture, the experiential boom of the early 2000s laid part of the foundation for today's immersive theater explosion. The culture developed in many places globally, and New York City became a significant nerve center. Decades of emergent culture and storytelling with thousands of participants, hundreds of large-scale art crews, organizations, and Burning Man camps were co-produced on the global stage of post-millennial NYC, and joined by crews from California, Black Rock City, and beyond. Together, this was a significant part of the birthplace of modern experiential culture. These kinds of spaces are important to the continued development of emerging arts and immersive culture as they've historically served as studio space, performance space, and training grounds for artists who've gone on to work in museums, Disney Imagineering, and in high-profile commercial interactive agencies. Academia is one clear path and research ground for developing these practices and mastering skills, and the streets and DIY spaces are often another, creating additional accessible and diverse worlds for emerging creators.

## A chorus of forms

Modern immersive theater and sound experience takes many forms, from massive experiential playgrounds and explorable inflatable sonified environments designed for free roaming, to a single space where intimate interactive multisensory experience unfolds, and you're instructed to stand in a sweet spot, to multimillion-dollar complexes with state-of-the-art 360 projection and dome arrays designed with rotating exhibits open for performers to occasionally 'activate' and theatrically enhance installations. Or we can wander into a multiverse sea of box trucks, each containing an interactive, whimsical world of its own – as we do in *The Lost Horizon Night Market*. This series, replicated virtually during the COVID-19 pandemic, stands as an example of absurdist immersive theatre and participatory culture at its gritty best. Or it can be an immersive sound play like the triptych *A Thousand Ways* by Abigail Browde, where you bring your own theatre and chair.

Stories can unfold throughout many rooms or even multiple locations, as in the 'choose-your-own-adventure'–style events groups like Dumboluna, Complacent, and *thedanger* began producing at the turn of the millennium where one ticket gained you entry to many locations with a connected storyline. Roving brass bands often linked them all and led paraders to discover new chapters of the night, sonic experience being connective tissue between them all, both amplified and open air. Later, inspired by and working in collaboration with these earlier experiential event producers, The JunXion NYC[6] used the 'choose-your-own-adventure'model, linking their locations with buses and an ice cream truck retrofitted to carry DJ decks where the ice cream peddler used to stand. They became part of the next generation defining the crossroads of underground nightlife culture, festival culture, and high production immersive events at spaces like their own X Marks the Lot and later Arts District Brooklyn, featuring cutting-edge immersion with the heart and soul of grassroots culture.

Many forms of spatial sound are highlighted in varied ways within immersive theatre. We see spatial relations, crowd flow patterns, and performance-based spatial arts explored. Some manifestations occasionally do occur within the context of a static stage – this is the case in the full-scale ensemble touring immersive theatre production Bella Gaia. More often, we design and encounter a variety of roving experiences that grant participants the agency to navigate the space, whether freely, as in *Sleep No More*, or in a more structured linear format like *Then She Fell*. These experiences may unfold in virtual or public spaces or span across multiple boroughs or even cities. Scale, space, and sound are integral elements of these creations, continuously explored, and reimagined to enhance immersivity.

As more and more people have the opportunity to explore intentional immersive sound and theatre experiences in powerfully evocative ways, we've seen an explosion of pop-up immersive brand experiences and public spaces set up where guests pay to take photos wowing their friends on social media in themed rooms with little or no narrative component beyond decor and props. Nothing wrong with that, but if someone *only* has this kind of branded commercial interaction with 'immersive experiences', they may form a strong and limited view of what they are. Hopefully some who find these kinds of hyper-commercial spaces first will later find more artistic, richly narrative, and sonic immersion, drawn in by the excitement of their next best Instagram shot and accidentally coming upon a *Dream Think Speak* production or Victoria Vesna's next immersive Sci-Art experiment. It's also key to note that there is a vast industry of immersive branding that is rich with opportunity and creativity, and many experiences that are commercial also involve deep artistry by some of the best creators and producers in the business.

Some of the world's biggest entertainment brands are breaking new ground with new kinds of immersive theme parks designed around transmedia storytelling and deeper fan engagement. These creative and technologically innovative theme parks employing thousands include the new Universal Epic Universe and Netflix House. There's also a trend of a lot of sizable-budget commercial immersive experiences that are elaborate light and sound shows, most involving loosely tied story threads, art installations, and spatial elements you wander through. It's hard to count how many 'Immersive Van Gogh' productions there have been now – at least ten different groups globally have replicated this money maker. They are not immersive theater, but they are certainly part of the immersive experiential culture explosion. Some things in this category are particularly beautiful and innovative along with often supporting huge numbers of local artists as in the case of Meow Wolf, Arcadia Earth, and teamLabs.

The Astra Lumina Experience is another example of this trend. It was not focused on storytelling per se but was created for all generations while beautifully celebrating the natural environment it's set within. Sound was distributed throughout the walk-through experience with local speakers nicely camouflaged in moss at specific stations. You could hear both local and more distant ambient sounds creating an otherworldly but very natural listening experience as you walked along the path flanked with light, sculpture, and architectural installations. In all, they created the illusion of a true multilayered environment.

## Co-created works

In this section I will share a few projects that illustrate some of the concepts previously discussed in the chapter. I founded Catalyst Culture Labs as a co-creative incubator and production house at the emergent edge of immersive and socially engaged arts. We collaboratively create experiences, environments, and community alongside leaders from diverse fields driving social change across disciplines, cultures, and generations. As an actively producing R&D collective, we focus on experimental, experiential work – often shared through immersive theatre, hybrid events, and global broadcasts, with a strong emphasis on immersivity and *sciart*. Next are some categories of immersive theatre and sound projects we've been part of over the last two and a half decades and how they directly relate to what we've discussed thus far. We learn so much from and with every creator and collaborator we've ever worked with, there are too many to share here. Deep gratitude to all who came before us and work to advance these worlds.

When creating the character of the 'Bilge Siren', the original goal was testing water quality and capturing impulse responses to create custom convolution reverb of the gorgeously resonant rusting hull of a sinking ship – a decommissioned commercial ferry that came to serve as a secret immersive theater venue and floating immersive arts incubator for 15 years in the heart of industrial Brooklyn. Inspired by the beauty and unique nature of the ships' sonic signature (or voice/acoustic environment), this began a project that unfolded in many forms over three years. This began a focus for me of creating much of my immersive theatre with water and boats. I created the Bilge Siren as a whimsical nod to the sonic device that alerts you loudly if a ship is filling with water, as well as the mythical siren that sings to doomed sailors. I used the character to creatively alarm people that we were in a true climate emergency as well as the ship herself actively sinking. Toxic water was pumped out constantly – our team was given permission to turn off the pumps just long enough to record clean takes in the hull, which I did numerous times with different instrumentalists. This project focused first on virtual acoustic as we used the popped balloon technique to create our signature 'bilge reverb',[7] which was then used in many performances, including immersive plays I wrote and performed. For *Homesick: An Immersive Journey to Earth, Space, and Sea in 3 Movements* at Harvestworks,[8] we collaboratively created improvisational electroacoustic music with live visual projections, the spatial bilge reverb, and a drum machine we built out of environmental samples from the ship together with NASA recordings released in celebration of the 50th Anniversary of the Apollo Moon landing. Another piece created for the Multi-channel Music Festival, called *The Ninth Channel*, used conceptual interactive components.

The incredible multidisciplinary artist community working together over the years on this ship created a series of immersive theatre events involving elaborate narratives told through interactive roving performances, a mix of scripted and improv acting, mysteries to solve, fire dancing and performances on additional vessels and floating rafts, various sound sources placed throughout the ship to create multi-channel amplified environmental experiences designed be discovered and mixed in open air with the natural sounds of the creaking ship and steel reflections, all of which supported the layered narratives unfolding on four levels of the vessel.

I frequently acted as the 'pied piper', asked to create an immersive ceremony and bridge by drawing and holding attention with an amplified experiential performance at the helm of the ship, then a moment of shared intimacy (like moving around in silence anointing each person with rose water), then compelling participants to follow my voice, a moving localized sound source snaking through levels and story chapters carrying a lantern to end up at the top where they are delivered for a culminating experience. In one culmination, this was a live performance by Filistine, who have been creating albums at sea for years. That set was one of the most sonically and visually compelling I've ever experienced. Grey Filistine created effected polyrhythmic drum patterns furiously on a miked Home Depot shopping cart while Nova sang through a megaphone with visuals they created themselves projected onto the ship's smokestack. Just brilliant. People were told to meet at a location nearby and brought blindfolded to the ship. So the pre-entrance and entrance experiences are carefully designed theatre as well and begin to focus revelers on their senses of sound and touch while also protecting the secrecy of the location. I've used the 'pied piper' method both amplified and open air – with a body pack and just the voice depending on how quiet an audience you can command (working with brilliant immersive storyteller Joro Boro on the ship, you could hear a pin drop), how well you can project and utilize the acoustics of the space, and what other spaces are activated at the same time to create sound coupling or soundclash.

*'Ancestors and Glaciers'*, an immersive ice theatre experience, revolves around a multichannel recording of a melting iceberg made with hydrophones at Diamond Beach next to the Jökulsárlón Glacier Lagoon in Iceland with Paul Geluso. We've performed this evolving piece in a number of spaces including experimental electronic Festival *Woodstockhausen*, for the first time adding collaborator Candy Brigham who climbed inside and moved artfully in a full body balloon. It was a shock to the audience when she appeared completely encased in this balloon up to her neck, and it looked eerie and beautiful as we created a dance battle and played with the sound of the inflatable through an effected mic. Every immersive theatre piece I write includes spatial sound, costume, archetypal characters, self-created scent, visuals and light, narrative and music that are partially precomposed and partially improvised, and interactive elements to create a whole sensory story experience told through many senses. In this case, the interactive piece involved a costumed Ice Queen giving ice to participants to create a moment of connection and a visceral sensation of melting ice.[9] As a side note, Jaanika Perna creates powerful interactive climate art involving ice, beautiful graphite drawings, movement, and multisensory experience. I find it powerful and liberating for collaborating artists to create a sort of creative 'container' – a well-defined narrative, leading theme and character that is both compelling and archetypal and therefore universally connective, and a compositional foundation within which collaborators are given wide berth and space to create and improvise so the piece is a true collaboration while also defining clear direction. I'm excited to see where this piece can go when we keep developing the composition and the visuals to accompany a piece that is completely driven by the beauty of the sound. The sound is incredible, a very clear recording that feels wide and very listenable with percussive moments, crackling, a soft background of rushing water created by the sea and occasional squeaks of escaping air together with spatial reverbs.

Developed in ongoing collaboration with NASA, Bella Gaia originated from a NASA educational grant and evolved into a full-dome planetarium film with a 5.1 surround sound mix, created by Kenji Williams in partnership with the University of Colorado, Boulder.[10] By blending supercomputer data visualizations and orbital imagery from Hubble and other space missions together with music and dance performances, the production offers a visceral simulation of the *Overview Effect* – the profound shift in consciousness astronauts often describe after viewing Earth from space, as coined by Frank White.

This visceral experience of Earth from space aims to activate empathy and inspire planetary stewardship by catalyzing a transformative sense of interconnectedness to Earth, humanity, and the cosmos. By design, it is immersion as medicine, working to heal a sense of personal relationship to nature. Bella Gaia (which means 'Beautiful Earth') uses immersion techniques and neuroscience to explore the impacts of the Anthropogenic Age, spark empathy, and ultimately create impact. Neuroscience and communication studies show that learning is enhanced when both the right and left brain are activated, making Bella Gaia's combination of music and science particularly powerful. A NASA-funded study demonstrated this impact, with 89% of viewers reporting a stronger connection and knowledge of the Earth after attending.[11]

Bella Gaia is a powerful example of immersive experiences that are scalable, as it occurs in many forms, and continually evolves as a living piece responding to changes in environment and culture. We customize the full-scale touring ensemble theatre production with great care for every theatre we perform in. Each show is a storytelling journey from the outer reaches of the galaxy into our solar system, where we experience the Earth from space and then zoom in to celebrate different cultures of the planet. At the same time, the full-dome and 360° video experiences continue to be shared around the world, curriculum-based workshops are taught with NASA educational partners, and intimate immersive salon experiences have been successfully piloted as well. A full-scale vision is in the works that involves Web3 technology and immersive storytelling.

I've toured with Bella Gaia for eight years as Immersive Tour Producer and Technical Director, along with Directing VR Engagements. One recent milestone was a collaboration with Earth Harp for the 100th Anniversary of the Miller Outdoor Theatre in Houston, Texas.[12] Two Earth Harps, built with strings extending from the stage to the back of the 10,000-seat venue, transformed the theater into a giant resonant instrument. In rehearsals with William Close, I suggested letting the strings vibrate without any other accompaniment for a moment in the show where Close talks about how the Earth Harps are built and played. The space hummed with an immersive, haptic resonance that was unforgettable, and we were allowed to be completely enveloped by the sound and nothing else.

Another collaboration with Purpose Earth featured a live Bella Gaia performance with violin, alongside nine interactive workshop stations led by me and NASA educator and ICESat-2 outreach lead Valerie Casasanto. Hands-on workshops included tactile educational books like *Getting a Feel for Eclipses*, designed for blind and low-vision students, and the NASA Merge Cube AR app, which transforms a paper cube into a handheld Earth displaying live satellite data. We hosted an event for 700 New York City public school students during the worst wildfire smoke of 2023, poignantly allowing participants to view the smoke from space on the AR globe – an unforgettable, immersive experience.

*Music: Not Impossible* (MNI) began with a different approach to sound and the importance of shared experience, asking how they could take the visceral and social power of sharing a concert and make that more accessible and inclusive of audience members who are deaf or hard of hearing. I was part of a group of the first group of artists[13] to ever create vibrotactile compositions with a haptic suit invented by visionary composer and technologist Daniel Belquer, controlled with transmitters and programmable through Max MSP and a custom-coded platform that continues evolving with each release. Paul Geluso and I had the opportunity to work directly with ASL interpreters and disability advocacy experts like Amber Galloway, who became famous for her incredible sign dancing skills that keep up with beatboxers and MCs. This was one of the most exciting projects of my career and set me on a new path as both an immersive artist and a broadcast producer. We now have an ongoing artistic collaboration with MNI and Daniel Belquer, and I also now work professionally in the field of accessibility education

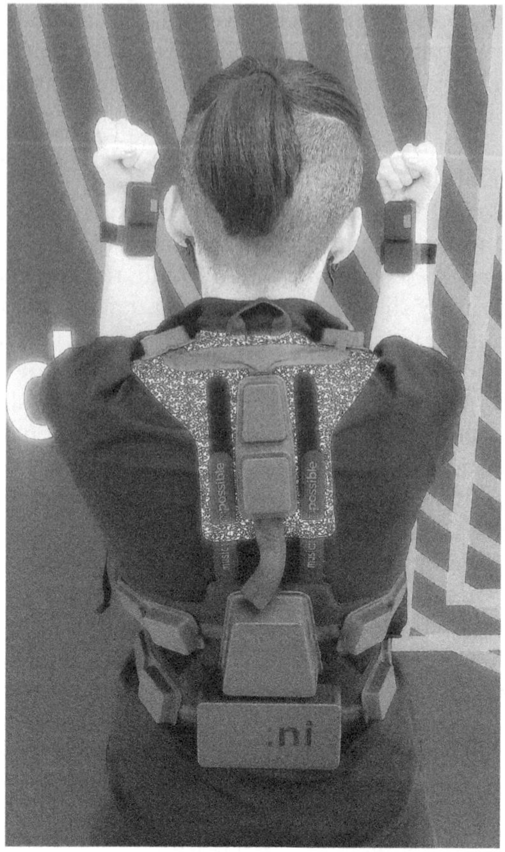


*Figure 19.5* Daniel Belquer's haptic suit invented for *Music: Not Impossible*, with dynamically assignable actuators throughout the vest, wrist, and ankle vibrotextiles first created for deaf concertgoers. Worn by the author during her haptics project *To Sense the Sea*, commissioned by MNI in 2020.

and technology, producing broadcasts and video content, which are evolving rapidly with key partners in the field.

Considering the suit has 24 points of vibration, including a subwoofer at the base of the spine, it's a straightforward application to map the points to frequency bands or (as we did) to use the suit for live input. We tried many techniques, but ultimately I started thinking about how powerful reverb is to individual and shared experience. Of course, it gives us information about spaces, and also the ephemeral – the transcendent experience of sensing space and vastness and feeling connected to that vastness. I wanted to explore making reverb accessible to people without the sense of hearing, so we mapped the suit in a way to attempt to simulate a sense of reverb in haptics, to create haptic reflections and echoes. What that translated to, at least for version one, was a kind of cascading domino effect chain of haptic firing that grew less intense as it traveled across the body points, as that seemed maybe a metaphor for the sensory experience of reverb, though not for the physics of it. We presented the final piece *To Sense the Sea*, which included a short film, live performance with spatial reverb and composition, vibrotactile composition, scent, and story at the Museum of the Moving Image alongside a group of creators who did completely different things with the technology.

Catalyst Culture Labs (CCL) worked with Bridge Multimediaand Beautiful Earth Foundation to create an experimental and inclusive 360 video experience for those with visual impairment and sighted audiences that simulates the Overview Effect with an accessible immersive connection to the planet and cosmos. The core of the piece was the Bella Gaia VR headset experience created by Kenji Williams. I worked with the Bridge Multimedia team,[14] who wrote and recorded a beautifully poetic and artful audio description script, we spatialized the audio using GRM Tools Spaces, and that spatialization involved an artistic process of considering spatial sonic experience for both sighted and visually impaired participants, which fundamentally changed the way I think about spatialization and extended reality. Of course, the only way we can truly know what an experience is for disabled communities is to have disabled advisors in the project – to consult experts with lived experience, gather more and more qualitative analysis, build diverse and inclusive collaborative working and testing groups, and to reduce barriers for disabled creators to lead the next generation of immersive experience design and to be included in dialogues about what inclusive immersive experience can mean. This is always considered by Bridge Multimedia, which works with many industry-leading professionals with disabilities. Once complete, the piece was shared with educational leaders in Washington, D.C., as well as with sight-impaired blind members of a board of advisors who gave us positive and helpful feedback.

Just as we know we can create inclusive immersive sound experiences with deaf communities, we can create visual immersive experiences with blind communities, and many can benefit from incredible work-disabled creators and scientists who are creating that push and break boundaries while shifting the way we think about sensory experience and immersion.

In 2018 the EDM pop music duo Memba approached me and the late performance artist and culture champion Ming Lui with a unique challenge – to create a fully immersive EP box release and event built around an elaborate cosmological narrative. They invented a planet they called Saga-I, complete with its own language and lifeforms. We released a limited-edition multisensory box for the heads of key music industry leaders to experience the planet and music, and we introduced the box to the masses with a body-painted alien visibly overwhelmed by the hyperstimulating sound and visual landscape of Times Square to create quite a stir among tourists. The singles from the album were all released in conjunction with imaginary relics of the planet. It was part guerrilla marketing campaign, part immersive theatrical concert, and because I was the one producing this, we took very direct inspiration from my years of producing in New York City's grassroots underground immersive world covered earlier in the chapter. Roving performers included Akim Funk Buddha creating beautiful acrobatic and ritual sonic spectacles with dance, beatboxing, and Tuvan throat singing. 'Vocal acrobat' Zero Boy and sharp comedic voice actor Steel Filipek wandered with megaphones improvising and engaging the crowd, while a scent station allowed people to smell the planet, and Memba performed with their homemade inflatable weather balloon MIDI controllers. Along with producing the multisensory box release and event, I agreed to take on the task of being the olfactory artist to create the scent of Saga-I. It was part clove to evoke the warmth of

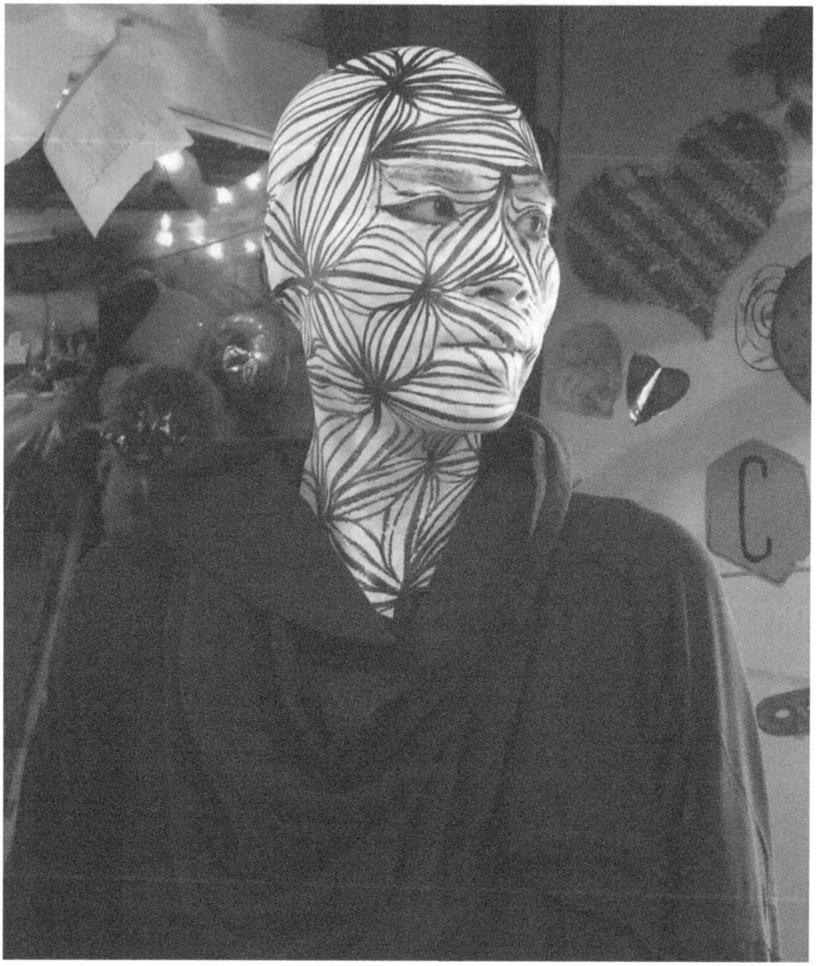

*Figure 19.6* Ming Liu as a body-painted alien at Memba's immersive EP release party, New York City, 2018

the fiery world, part dirt, and part ozone (astronauts report the scent of space is slightly metallic, which ozone also is – if you've ever gone outside right after a thunderstorm strikes lightning and smelled slightly metallic air, that could be the unstable atoms that were temporary split). This was an impressive vision from the young artists that elucidates the richly narrative possibilities in the world of commercial releases and branding that can go far beyond simple prop-driven Instagram pop-up shops and employ stables of authentic artists while they do.

The immersive, sound-centered theatrical pieces I've co-directed and performed range widely in scale – from expansive multi-venue stories unfolding over boroughs and even years to deeply personal sonic experiences tailored to a single participant. Just as we co-create mass spectacles, we create intimate one-on-one moments. These moments are especially impactful when interwoven into a broader narrative, forming a cohesive, interactive tapestry. My role has often involved crafting these intimate experiences while also being one of the core writers collaboratively developing overarching narratives that other artists can seamlessly integrate into, creating a dynamic container for interactive storytelling. For example, in the 'Reclamation' experience, conceived with headdress artist and performer Michelle Palmer and producer Misha Mayeur, we transformed a historic Soho social club into an immersive Goddess-centered narrative unfolding across five floors with many interactive experiences and performers. Key immersive experiences included a tableau vivant drawing salon within Taylor Kuffner's mesmeric Gameletron sonic

*Figure 19.7* Author performing while holding a book written collaboratively onsite with participants in the multilevel immersive piece 'Reclamation', featuring open air spatial sound from Taylor Kuffner's Gamelatron installation. Headdress by designer and collaborating performer Michelle Palmer. Curated by Misha Mayeur and William Etundi.

installation, roving one-on-one interactive storytelling, and a crescendo at the height of the event involving a costumed vocal performance with an original soundscape and song I completed during the event by weaving in names and stories attendees shared through the earlier acts. This approach blends transmedia storytelling with immersive theater, creating collaborative, multilayered environments where artists and audiences co-create meaningful experiences. Sound becomes a central element, shaping not only individual moments but the entire narrative flow across interconnected spaces.

We've also been producing many hybrid experiences designed for simultaneous virtual and physical audiences together with immersive theater and spatial sound elements. This includes events and global broadcasts created for participants both in person and distributed globally and virtually, like the CoCreation Festival, which is built by teams working together in Australia, Vancouver, and New York. For World Unity Week, Unity Earth joined hundreds of organizations to build a virtually explorable immersive world we dubbed 'One World', which included multi-channel soundscapes to create non-localized environmental sound online while roaming the interactive world map created in partnership with Sparkle, a company born first at Burning Man. This was another interesting intersectional moment – innovators born in a participatory culture stepped up to try solving collective business problems faced when creating compelling virtual gatherings, as well as cultural challenges and opportunities we all faced across the globe, to find new ways to gather and create shared experiences while physically and socially distanced. While our physical events and tours were halted, this collective cultural shift created rippling waves of innovation in the hybrid and immersive space. From distributed performances using high-quality real-time web-based audio, to experiencing *The Encounter* remotely with people across the globe, to regular virtual broadcasts by circus and cabaret stars like Bindlestiff Family Circus and Charming Disaster, these experiences were truly collaborative and deeply intimate, which helped ease the distance as people experimented with new ways to share and co-create. Victoria Vesna's *Alien Star Dust* project exemplifies this era of innovation, bringing artists together remotely under her direction to create a shared virtual experience. This immersive and intentionally challenging multimedia meditation centers on micrometeorites and incorporates a spatialized drone machine, sonified COVID-19 data from Johns Hopkins, ethereal improvisational vocals I create, dynamic visuals, and an augmented reality experience accessed through a custom app.

One of my favorite contemporary examples of immersive storytelling through sound that I've experienced without having any involvement in the piece is *The Gauntlet* by composer Sxip Shirey and choreographer Coco Karol. The Gauntlet is an immersive choral experience that guides listeners through 'sonic architecture', corridors crafted from the human voice.[15] A libretto is drawn from 'movement interviews' developed and conducted by Karol with members of the community in somatic practice that in itself sounds like an intimate one-on-one immersive storytelling experience. The piece is then performed by locally assembled choirs, making each iteration unique and site-specific, as the choir moves like a river through a tunnel created by listening participants. We are also invited to experience a version created for headphones mixed spatially with SPAT. The piece explores how we listen and engage with sound differently when experiencing it in a dynamic social setting – a 'sonic social space'. Its language of listening and immersion is itself spatial, embodied, interactive, and challenging.

Above all, the story, goals, and potential for shared experience must inform the technology, not the other way around. This is one of the most powerful things about the MNI

Haptics project – it was born out of a desire to create a more inclusive and embodied visceral experience. It's one of the best examples of immersive from inception, and the potential for customized and inclusive shared experience drove the entire design process, which was also informed by feedback from people with diverse sensory experiences of the world. The most exciting things I've seen happen are when the doors are thrown open and diverse artists with very different visions for tech are given the opportunity to use it in different ways.

## Accessible and inclusive immersion

Among the most exciting and transformative frontiers are inclusive immersive design and experiential accessibility. Immersive narratives, by nature, are designed to engage multiple senses. We seek to tell stories in varied, flexible, and participant-driven ways. This creates added opportunity for diverse sensory experiences and needs, particularly when we design for accessibility from the start. With over 25% of US adults[16] and 16% of the global population[17] living with disabilities, failing to prioritize accessibility means excluding vast numbers of potential participants, creators, and shared experiences as well as lost business opportunities. When we prioritize accessibility, the whole of society benefits in a multitude of ways, including increased innovation and development of our social and creative capital.

Education and representation are a big part of this growing movement, with organizations like *Immersive and Inclusive Audio* advancing education and mentorship opportunities for underrepresented groups in the audio industry. In immersive theatre, groups like Oily Cart have spent decades pioneering and advancing inclusive *sensory theatre*, pushing progress for theatre arts alongside our fundamental understanding of disability, to shift attitudes from a medical to a social model. As understanding and appreciation for disabilities and neurodiversity grows, so does the recognition that disability is part of the healthy biodiversity of the human species rather than something to be remedied. We are collectively advancing our means to design experiences and fields that accommodate diverse sensory needs, preferences, and forms of communication. This can have vast and far-reaching implications for society as a whole. As Dr. Shelley Moore shared at the 2024 NYU PINE Summit:[18]

> Biodiversity, like disability diversity, strengthens and is necessary for the health of the whole.

Moore goes on to speak about shifting from the medical model and colonial mindset, where disability and anything that doesn't fit into a majority status quo is treated as something to be fixed, and embracing a social model where we recognize that disability is biodiversity – just as we see in nature – creating a healthy and whole ecosystem. Moore talks about the importance of listening to those with lived experience and putting resources into improving and removing barriers in environments and attitudes, often in ways that will support everyone (Moore, 2024).

As we continue getting better at creating accessible technology and arts including audio description (AD), captioning, haptics, screen reader technology and compatibility, assistive devices and more, we also recognize that all people experience and create in different ways and that's a positive thing that pushes society to think and grow in new expansive directions. It's exciting and telling that Audio Description is also now elevated to an art form itself, with an annual Awards Gala founded by The ACB honoring the best of the best.

Collectively we create more and more customizable immersive experiences including adjustable intensity levels on haptic devices, volume, EQ, captioning and speed controls on VR, and multiple ways for people with communication differences to interact with immersive theatre – to allow participants to tailor sensory elements for different levels of personal sensitivity, and our vastly different sensory and expressive experiences of the world. It's exciting to expand immersive experiences to incorporate additional sensory layers along with visual and sonic, such as proprioception, vestibular, taste, and scent, which are less understood and closely linked to memory and sense of place.

Programs that blend sound and taste, like Piet Goddard's *Krug Echoes*, pair sensory elements to unlock deeper immersion. Bridge Multimedia's *What Does Audio Description Taste Like?* brings together disabled and non-disabled accessibility leaders, all blindfolded, to collectively experience jazz and chocolate pairings, sparking discussions on sensory diversity and its impact on shared experience.

It's critical for the advancement of immersive design as a whole to encourage diverse creators and center the lived experience of disabled participants. This will be a key part of broadening what's possible in immersive arts as inclusion not only serves social justice but also fuels innovation, bringing fresh perspectives to shared experiences. Accessibility isn't just beneficial –it's essential for a thriving creative culture and society.

Dr. Wendy Sapp, Senior Producer with accessibility leaders Bridge Multimedia, notes that innovation often arises when new perspectives enter a field, challenging previous traditions or boundaries. She discusses Bridge Multimedia working on emergent projects, and how exciting and challenging it is to keep doing things you've never done before.

> Innovation often comes from people who aren't in the field. I've been in the field of educating kids who are blind for 30 years. I've done so much, and have this great wealth of knowledge. But it can be hard to think completely outside the box, because I know what all the boxes have *been* for the past 30 years. Bridge has been doing Audio Description for 15 years, and we're very good at it. But to come up with new and innovative things that are outside the box of the framework you work within, it's key to keep doing new things, and work with people new to the field.
>
> (Sapp, 2024)

Along with making accessible and customizable immersive experiences for participants, it is key to make immersive design accessible to creators with diverse sensory experiences of the world, and we will see and hear things emerge that we literally couldn't have imagined. Just as we heard Kae Burke share about inclusion in immersive theatre and clubs, this isn't just a moral imperative – it means more innovation, fresh perspectives, more richly and widely shared experiences. It means we make better things. Accessibility and inclusion benefit all of society. I'm excited to hear more immersive sound and haptic experiences created by people with different experiences of sound than I have. To do this, we need to challenge our own biases about who has the right senses to create or experience with, and be open to not knowing. Just as the Cadillac of immersive sound is pieces created for spatial sound from the beginning, the same goes for accessibility. We can up-mix and retrofit when needed, but it's much more powerful when things are created with immersion and accessibility in mind from inception.

At NYU Steinhardt's Iris Cantor Theatre during the inaugural 'Immersive Mixtape' Festival, the boundaries between performer and environment dissolved. David Rothenberg, renowned for his pioneering interspecies musical collaborations with birds, insects, and whales, created a captivating piece celebrating a song composed by a Humpback Whale off the coast of Maui. I vocally improvised with David and the Whale, which was mixed to envelop the audience in a 56-channel spatialized soundscape. Eventide's Blackhole reverb for the vocals, together with Rothenberg's resonant and beautifully recorded bass clarinet, enhanced the immersivity of this environment where human and whale song intertwined, inviting listeners into a shared sonic experience.

As part of the Immersive Mixtape event, I participated in a roundtable discussion moderated by Leila Adu-Gilmore, focusing on strategies to increase access to these technologies for marginalized communities and their music creators. Composer, engineer, and educator Delia Beatriz, aka Debit, shared that even though she's a music technologist familiar with multi-channel systems and related concepts, it wasn't until she was at a sound clash – a Bristol dub show – that she fully grasped immersive sound because the experience became visceral. She goes on to speak to the power of breaking down the model of the sitting audience, stage, and stereo image, and how the way she thinks about advanced music technology has shifted:

> In my research, I started to see the parallel with cumbias, trabajaras, and sonideras in Latin America, building cabinets and systems. A lot of these practices are from socioeconomically marginalized communities, and sometimes that's part of the inventiveness and improvisation, a scrappiness through the development of the systems that then become these majestic things. So, departing from the idea that immersiveness is necessarily a high-tech thing, and something we can only learn in school, was a big conceptual shift for me. And with pre-Hispanic research, learning that there are echoes and delays built right into the design of the pyramids (was a revelation).
>
> (Beatriz, 2024)

Jamal Mims, aka MC Tingbudong, points out that in making immersive sound more inclusive, we are confronting structures and systems that fundamentally reject the full humanity of individuals in their wholeness. He goes on to talk about how important a role sound, music, and immersion can play in creating shared and uniting experiences.

> So how do we use sound to connect cultures? If you're looking at the global lingua franca, that language is hip hop. . . . But it has to be marshaled towards that greater good of figuring out how we make the world more accessible. How do we overthrow those systems that are in the way of people being able to see and interact with each other? Human beings, no matter where they're born, living a life worthy of a human being?
>
> (Mims, 2024)

We have an unprecedented opportunity as immersive artists, engineers, educators, and advocates. Because of exponentially improving technology, dovetailed with shifting societal attitudes and priorities, we have the chance to make immersive theatre and sound more accessible and inclusive than ever before. The explosive popularity and interest in immersive theater can be another gateway for many to experience firsthand the incredible power of spatial sound. I believe deeply in this beauty and power, and the power of creating shared experiences and creative tools that are made more and more inclusive is tapping into the full power of immersion that we're still just beginning to uncover.

## Immersion as medicine

Notable institutions are working to collect data linking immersive experience with healing, which will likely increase support for continued research and treatment in this direction. Research has shown promising results in brain mapping, indicating the power of immersion to reduce both anxiety and pain in patients undergoing medical treatments.

> The Virtual Reality Fine Art Program at Montefiore seeks to diminish anxiety, pain and opioid addiction through stimulus-rich and curated artistic environments. These experiences will serve as *immersive analgesics* allowing physicians to treat their patients more effectively by improving patient's health and hospital experiences and reducing reliance on pain medication, especially opioids. Since this collaboration proved successful, we are now planning on expanding our efforts to use fine art and advanced technologies for pain remediation in our Pediatric Sickle Cell, Hematology, Oncology and Bone Marrow Transplant patients.
>
> (Davis, 2018)

Another organization utilizing the power of sound and immersion for healing is Sound Mind, founded by Chris Bullard as a community hub and multi-stage festival working with over thirty mental health partners to destigmatize mental health with the power of music and immersive experience. The festival expanded to become the Sound Mind Center, a music and wellness center which offers experiences from sound baths and sound beds, to workshops and concerts designed to create connecting and healing experiences.

Decades after I was first transported by the transcendent open-air spatial sound of Paul Winter wafting into the catwalks I raptly listened from at ten years old in the Cathedral of St. John the Divine, Sound Mind hosted a gathering in the historic institution devoted to interfaith peace and non-denominational sacred experience. Bullard shared that he learned only afterwards that the cathedral is built on grounds that once held a youth asylum.

> We've come a long way, he says, in reference to the way we collectively deal with mental health.
>
> (C. Bullard, personal communication, March 15th, 2024)

Bullard is part of a next generation of producers using sound and immersion to catalyze change, partially in response to a personal journey with bipolar disorder and also in response to statistics that Gen Z, one of the youngest generations at the time of this writing, is about twice as likely to report feelings of depression than older generations.[19]

## The power of collective immersive experiences

Simulating, creating, or recreating an experience of immersion may be one of our oldest human impulses since we began painting in resonant caves with long-tail reverb. We encounter the frequent misconception that immersive theater, as well as immersive sound and virtual reality, are recent inventions. However, the roots of immersive storytelling date back hundreds or, arguably, thousands of years. Immersion may be our first and most natural state of consciousness. I would posit that the rising interest in immersive experience is due in large part not only to advances in technology but also deeply entangled with broader cultural impulses toward mindfulness, embodied presence, participation, and visceral connection. I wonder if our modern fixation on immersivity, along with being a calling to the future, is also a reclamation of a past when we lived in immersed embodied experiences, with sacred ceremony being central in our daily lives, and without the constant competition for our attention we experience now. Immersive arts also hold the potential to help us heal, at times offering connection, a sense of belonging, relief from pain and anxiety, and bringing us into something larger than ourselves.

The ancient roots of immersive storytelling are revealed through archeoacoustic studies of caves, petroglyphs, and medieval cathedrals. Across eras and cultures, practices of intentional immersion have been intertwined with spirituality, ceremonial rituals, and reaching out to ancestors and deities. Contemporary immersive experiences – whether in Brooklyn or Japan – continue these traditions. From the Burning Man Temple, to the sprawling 100,000-square-foot maze of temples at City of Gods in Industry City, dotted by hundreds of artists like Michelle Palmer offering her elaborate headdress art and archetypal ritual activations, *temple theatre* and Goddess-centered ritual immersion remain integral to our modern immersive repertoire.

Theater has long been a catalyst for transformation, and immersive theater amplifies that impact by centering transformation and direct audience engagement. We also know that increasing dynamic range and spatialization enhances the emotional impact of sound. A 2021 academic study titled 'Attending Live Theatre Improves Empathy, Changes Attitudes, and Leads to Pro-Social Behavior' found that seeing theater can increase empathy, shift socio-political beliefs, and boost altruistic behavior – effects that were strongly correlated with how transported participants felt by plays (Rathje et al., 2021). Immersive theater is explicitly designed to transport its audiences, activating multiple senses and inviting them into the story, thereby heightening empathic response.

Some say the most powerful aspect of immersive experience is the ability to foster shared, meaningful encounters. Technology is a brilliant and beautiful tool, but it shouldn't lead experiential design. What drives truly memorable experiences is a strong narrative and deep emotional resonance. Prioritizing storytelling and social psychology before or on par with creative technology is essential to create impactful and community-building experiences. A huge part of why immersive theatre and culture has such power is the hunger for community and connection, and all kinds of immersive creators, including those making VR and binaural experiences designed for private listening, can benefit by considering how are discussions and connections encouraged around this work? And, furthermore, how can those dialogues include diverse voices? Like the history of dance music, immersive theater and sound have the potential to bring together diverse communities, fostering unity by highlighting our shared humanity. Projects like *Music: Not Impossible*, producing global broadcasts and events with Unity Earth and Purpose Earth, and countless dance floors worldwide have shown me firsthand how shared emotional experiences can transcend language and cultural barriers, creating ripples with lasting impact.

Emmy award–winning composer Gary Malkin describes immersive arts as 'catalysts for coherence', while posing the critical question, 'How could these art forms play a role in fostering more peaceful and collaborative ways of engaging with each other, or even with ourselves?'

> Visualize city councils invoking shared purpose through multi-sensory experiences, or corporate boardrooms reconnecting to humanity before making decisions. Envision classrooms worldwide beginning with videos that foster mindful collaboration and curiosity. This is the immersive arts' magic: recalibrating our collective intentions.
>
> (Malkin, 2023)

For emerging creators, the ability to collaborate effectively in diverse, cross-disciplinary teams and to communicate a shared vision is essential. Balancing technology, art, and storytelling relies heavily on teamwork, and immersive experiences come to life through the collective effort of passionate innovators.

One of the greatest powers of immersive experience building is this: if we as creators can design experiences that are scalable – that evoke a sense of spaciousness, of connection to a richly represented or imagined world greater than the sum of their parts, and also create a sense of intimacy at the same time – that's where the deepest impact and engagement lies. Regardless of scale, the strength and success of projects like Bella Gaia – and the ability to realize wildly ambitious visions or sustain long-term collaborations with theatre ensembles and organizations like NASA – all stem from relationship building.

For young, emerging talent and producers, learning to effectively collaborate in cross-disciplinary teams, communicate a vision, let go of ego, and adapt within diverse groups are some of the most vital skills, not only for this work but for any endeavor. Along with technological skills and balancing art with science, technology with storytelling, a big part of the art of the work is communication and rallying people to do monumental things together that we can *never* do alone.

I became a producer, artist, and curator when I was very young. Even as a child, I was captivated by the alchemy of communal multimedia experiences – bearing witness to immense productions that soared, faltered, and sometimes completely unraveled. That early awe never left me. Decades later, I still carry the wide-eyed wonder of the child I was in the cathedral catwalks of St. John the Divine. If you're fortunate and smart, you surround yourself with people more experienced than yourself – people who may dazzle you with their insight and vision. Immersive theater and sound spectacle are still emerging into their modern expressions. It's a boundless unfolding: an ever-evolving constellation of ideas, voices, and innovations. And there is much space for new voices to emerge and weave together.

All of this is timelessly emergent and exciting for storytellers, producers, and participants as it encourages us to create elaborate works involving more of the senses and more of our creative and strategic faculties, bringing more intimacy and connection into focus. With collective immersive experiences, we can explore stories that are different every time, depending on who is perceiving and completing the puzzle each day or night.

It challenges, activates curiosity and wonder, and requires bravery as it involves more risk, intimacy, and vulnerability, and that vulnerability is shared between core performers and visiting participants. Naturally, as the risks rise, so can the rewards. It's a big roll of the dice we take together. The form also means creating on a holistic environmental level, which in itself I think shifts the way our creative minds think and work. Creating immersive theater imbued with thoughtful spatial sound is a practice of collaborative world-building or of fostering a deeper sense of connection to the infinitely beautiful and irreplaceable worlds all around us. Creating new worlds or a new sense of place within a rendering of a spatial setting we already inhabit, while activating participation and our place within that world, it is possible to induce an embodied sense of belonging and community. It can also be a powerful tool to explore a *lack* of embodied belonging and to reclaim identity in the face of cultural erasure. As an artist, one of the most exciting aspects of working in immersion for me personally is its power to transport people in a deeply visceral way to beautifully unique and threatened environments. By engaging the senses and activating empathic, embodied connections, many climate artists seek to inspire stewardship and a profound sense of responsibility for these fragile ecosystems. Through the fusion of technology and the arts, we can bring audiences into these environments sonically, cinematically, or virtually, allowing them to experience their beauty, urgency, and vulnerability firsthand. An embodied sense of place and passion can become a sense of responsibility to nurture that place.

Perhaps in a sense, we can consider immersion a state of consciousness. Like the Overview Effect, we are working to create an all-encompassing embodied sense of presence within the world we wish to conjure or sustain, which can be Earth. Immersive theater and spatial sound can connect powerfully with stewardship of our planet. Immersion not only invites the creation of new worlds but also encourages creative attempts to evoke even a fraction of the rich, multidimensional beauty of the natural world, both around and within us. Whether one views the world, as some quantum scientists and spiritual leaders do, as having a holographic nature, it's certain that a mere stereo image falls short of capturing the full, sensory experience of being alive on this pale blue dot we call Earth. From using HRTFs to craft realistic binaural soundscapes to imagining spatial sonic environments to engaging senses like scent and proprioception, immersion is far from a new concept. Yet today, there's a heightened sensitivity to its techniques and potential, and an industry rapidly growing to explore it – one that also introduces important social and ethical considerations for the future.

The co-evolving fields of immersive theater and spatial sound are central to the aptly named Experience Economy and shift toward participatory culture. Alongside sociological phenomena like Burning Man, transmedia storytelling, and DIY culture, immersive theater and sound are transformative in nature and historically demonstrate power to inspire intentional shifts in our cultural fabric. They can act as R&D labs and sandboxes for broader cultural shifts, extending beyond individual productions. This process is both organic and purposeful. This includes changes in how we enact healing, both in clinical and personal settings, along with entertainment, education, business, and many other aspects of life. Likewise, the recent surge in immersive experiences is no coincidence. Amid the battle for attention in our screen-focused, digital lifestyle, there's a growing emphasis on mindfulness, meditation, and grounding physical experiences, in part a clear response to the demands of contemporary existence. At the same time we see the power of virtual experience can be manifold, and directly part of healing and mindful practices as well as fostering connection and belonging as in social VR, XR, and MR platforms.

Alongside phenomena like Burning Man and DIY culture, immersive labs both commercial and grassroots have functioned as R&D sandboxes for advancement along with broader cultural shifts beyond individual productions, shifts like the increase of interdisciplinary collaboration and spatial thinking. Advances in immersive sound have often gone hand in hand with moments of historic change. Part of this may be because of larger cultural shifts and also because engineering minds like Blumlein responsible for leaps forward in recording, playback, and spatial technology are also tasked with engineering solutions for communications, war, and peace.

More fields directly intertwine with immersive theatre than we can go into here, from immersive theme parks rides to escape rooms, 4D and 8D cinema (and, of course, cinema as a whole is one of our oldest immersive forms), metaverse MMORPGs, domes, XR, ASMR, and beyond. People around the globe in many industries are doing amazing work in these fields and others. The industry group *No Procenium* does a wonderful job of fostering community and collaboration between veteran and emerging creators in the immersive theatre world, and many organizations and symposiums like the Future of Storytelling Festival and Creative Tech Week have delved deeper into the subjects herein for those eager to grow and share in this arena.

Particle Ink, Third Rail, PunchDrunk, Meow Wolf, teamLab, The Sphere, Amoeba Technology, Indexical, Shadow Traffic, House of Yes, the Without Walls Festival, countless Burning Man artists, and Starlight Runner Entertainment, among many others, have been profound innovators creating beautiful spaces for meaningful exploration. New York City events communities like The JunXion and CX have done interesting things with immersive theatre crossing over from the underground to the commercial. During the global pandemic of 2020, we collectively created developments in hybrid and remote experiences with embedded spatial sound that have changed the face of the events and experience world. Catalyst Culture Labs CoCreation Festival was created to explore this edge with the power of globally shared experience in collaboration with Unity Earth TV. Countless other groups and artists, like Swoon, Orien McNeill, Jeff Stark (also responsible for the tour de force Nonsense NYC, which arguably has done more for New York City DIY culture than any other publication), and Jacklyn Atkinson among many others, have created beautiful innovative public works that have inspired me and many others to envision what can collectively be created with some brilliant and brazen inspiration coupled with a healthy dash of disregard for common sense. Industry groups and symposiums centered on immersive arts – like No Proscenium, NYU's Immersive Mixtape Festival, ITCH, IEI, The Big Thing, Immersive Acts, and many others, are intentionally expanding access to spatial technology, creating collaborative immersive sandboxes, and space for dialogues on innovation and ethics. All of this is testimony to the power of immersive sound and arts to help dismantle barriers and make diverse voices heard, continuing the inherently transformative nature and tradition of immersive arts as part of a continuum of connection and social change.

As immersive experiences shift from grassroots origins to mainstream and niche markets, tension has arisen between exclusivity and inclusivity. While exclusivity can be a potent marketing tool, it can also hinder progress if it restricts access. Building inclusive, accessible spaces for diverse audiences and creators is essential for the future of these mediums. Some of the most innovative work has come from creators with limited technological resources, demonstrating that it certainly can be *the wizard, not the wand*. Much innovation has emerged in underground spaces where creativity thrives amid constraints along with the free exchange of ideas on the emergent edge, which is heavily populated with creative technologists looking to collaborate across disciplines. These advanced spaces are invaluable for pushing artistic boundaries and the overall field and technology forward.

Around the globe, we collectively experiment, push, explore, collaborate, fail, test further, succeed, upgrade, and repeat as we passionately collectively push forward what is possible in immersive sound, theatre, and experience. Risks have to be taken (by creators, curators, companies, and investors), more and more doors and barriers broken down, and respect and education built to share a long legacy of countless innovators contributing to the greater whole.

The possibilities for immersive experience and spatial sound are vast, stars glowing blue-hot and moving at the speed of light while part of a larger constellation of shifting orbits. Great immersive storytelling blends senses of spaciousness and intimacy, creating something larger than the sum of its parts with a powerful emotional pull. The drive toward inclusivity, storytelling, and mindful design continues to shape a vibrant, interconnected world of immersive arts.

As an artist and a passionate advocate for humanity's potential to evolve through collaboration guided by heart, art, and purpose, I am continually inspired by the transformative power of immersive arts to connect, educate, and ignite change.

This is not only an exciting moment to be part of the field – alive with innovation, discovery, and collective wonder – but also part of a much older story. Immersion is a timeless force within our cultural evolution, woven into the fabric of how we tell our most self-reflective stories and bring them vividly to life.

## Notes

1 Authored by Jeff Gomez of Starlight Runner: https://blog.collectivejourney.com/why-is-this-happening-d1287d5ee4ee
2 Gratitude to Dr. Toni Sant at the University of Salford for creating this Applied & Interactive Theatre Guide, an excellent resource for further research. https://www.tonisant.com/aitg/Radical_Theatre/index.shtml
3 Led by Founder and President Sebastian Holzmeister, VP and Artistic Director Mark Winkel, Door Manager and Founding member Micha Waschke, Music Director Rhiannon Catalyst, Director of Marketing Jacky Tran, and Street Team Director Steve B/Courage Productions. Main sound and A1s: Music First Production and David Junglez with Opus Audio. https://en.wikipedia.org/wiki/Lunatarium
4 Akairways Sketches for the Lunatarium venue: https://www.akairways.com/lunatarium_sketches.shtml
5 Direct text from Kurt the Inevitable's public invitation on the NYC Improv forum, spread via web and word-of-mouth. No actual birds were bred or hurt in the making of this event.
6 The JunXion NYC was created by Myk Tummulo with support from a beautifully strong community and influential producers including Lindsay Arden Cooper, Misha Mayeur, Elena Barbara, and Joey Angel Bonilla.
7 Instrumentalists included composer and trumpetist Takuya Nakamura. Bilge reverb impulses and recordings made and engineered with Zhenya Warshavsky and Paul Geluso.
8 Homesick was created with Asukaya Bailey, Kevin Ramsey, Jeremy D. Slater, and Paul Geluso. Special thanks to Lindsay Arden Cooper, The Queen of Cups, and to Harvestworks Digital Media Arts Center and Carol Parkinson for immeasurable support and influence of experimental arts. https://www.harvestworks.org/
9 Thanks to The Turquoise Barn, Michael Milton, and Indexical, a group dedicated to experimentation through music that includes Jean Marc Jot, creator of the Spat software library for immersive music creation and performance. https://www.indexical.org/

10 Full-scale ensemble theatre show created in collaboration with NASA, Kristin Hoffmann, Choreographer Irina Akulenko, and additional collaborators. https://www.bellagaia.com/the-experience/bella-gaia-live
11 NASA Study details: Reporting Period: March 26, 2014–July 1, 2015 Principal Investigator, Valerie Casasanto; Funded by NASA Grant #NNX11AH30GNASA Research Opportunities in Space and Earth Sciences (ROSES 2010): Opportunities in Education and Public Outreach for Earth and Space Sciences (EPOESS) *Beautiful Earth: Experiencing and Learning Science in a New and Engaging Way*
12 Earth Harp Collective created by William Close. Collective artists Adrienne Byrne, Shawn Barry, Rich Sherwood, and Luna Achiary.
13 Special thanks to fellow MNI artist, brilliant photographer and immersive creator Ashley Garner.
14 Project AD and narration by Kara Bonensteil, John Cavanagh, and Thomas Reid under the guidance of founder Matt Kaplowitz and Senior Producer Wendy Sapp. https://bridgemultimedia.com/
15 Phrases quoted from Sxip Shirrey, composer of the Gauntlet, from: https://www.sxipshireymusic.com/the-gauntlet
16 Center for Disease Control.
17 World Health Organization.
18 The NYU PINE Summit is an annual virtual conference focusing on inclusion and neurodiversity education, which I produce with the PINE team and Bridge Multimedia. https://www.pineprogram.org/
19 Research from the Walton Family Foundation and Murmuration (2022), conducted by John Della Volpe and Social Sphere found that Gen Z (42%) is about twice as likely as Americans over 25 (23%) to battle depression and feelings of hopelessness.

## References


Allbrook, M. (2024). Life stories: Mungo Lady and Mungo Man. In *Reframing Indigenous Biography* (pp. 18–21). Taylor & Francis.

Beatriz, D. (2024, October 11). *Immersive Mixtape Festival Roundtable* [Conference presentation]. NYU Immersive Mixtape Festival, Iris Cantor Theatre, New York City, NY.

Beck, J. (n.d.). *History*. The Living Theatre. https://livingtheatre.org/history

Boal, A., as quoted by Heritage, P. (1994). The courage to be happy: Augusto Boal, legislative theatre, and the 7th International Festival of the Theatre of the Oppressed. *The Drama Review, 38*(3), 25–34. Cambridge University Press. https://www.jstor.org/stable/1146376

Boren, B. (2021). Word and mystery: The acoustics of cultural transmission during the Protestant Reformation. *Frontiers in Psychology, 12*, 564542. https://doi.org/10.3389/fpsyg.2021.564542

Burke, K. (2019, April 2). Inside the immersive events created by House of Yes and Caravan Gitane. *Forbes*. https://www.forbes.com/sites/lisakocay/2019/04/02/house-of-yes-caravan-gitane/

Burns, C. (2020, April 24). Design trends to love: Sensory immersion. *LinkedIn Pulse*. https://www.linkedin.com/pulse/design-trends-love-sensory-immersion-caitlin-burns

CAIR Lab. (2024). *Healing, Bridging, Thriving: A Reflection by CAIR Lab*. National Endowment for the Arts Blog Series. https://www.arts.gov/stories/blog/2024/healing-bridging-thriving-reflection-cair-lab

Callahan, M. (2005, December 6). Pranks a million–Merry groups pull massive jokes on N.Y. *New York Post*. https://nypost.com/2005/12/06/pranks-a-million-merry-groups-pull-massive-jokes-on-n-y/

Csapo, E., & Slater, W. J. (1995). *The Context of Ancient Drama*. University of Michigan Press.

Dash, S. (1994), as cited by Fikentscher, K. (2000). *You Better Work! Underground Dance Music in New York City*. Wesleyan University Press.

David, B., Barker, B., Petchey, F., Delannoy, J.-J., Geneste, J.-M., Rowe, C., Eccleston, M., Lamb, L., & Whear, R. (2013). A 28,000 year old excavated painted rock from Nawarla Gabarnmang, northern Australia. *Journal of Archaeological Science, 40*(5).

Davis, O. (2018, June 22). VR art reduces need for opioids in NY hospital. *Forbes*. https://www.forbes.com/sites/charliefink/2018/06/22/vr-art-reduces-need-for-opiods-in-ny-hospital/

Gantner, V., as quoted by Lee, F. R. (2010, July 27). Theater for audiences of one. *The New York Times*. https://www.nytimes.com/2010/07/28/theater/28one.html
Grau, O. (Ed.). (2007). *MediaArtHistories: Remember the Phantasmagoria! Illusion Politics of the Eighteenth Century and Its Multimedial Afterlife* (pp. 136–161). MIT Press. https://doi.org/10.7551/mitpress/4530.001.0001
Green, P. (2010, February 17). Decorating the night in Brooklyn. *The New York Times*.https://www.nytimes.com/2010/02/18/garden/18party.html
Hansen, G. (2024). *Personal communication with Rhiannon Catalyst.*
Henshilwood, C. S., d'Errico, F., van Niekerk, K. L., Dayet, L., Queffélec, A., & Pollarolo, L. (2018). An abstract drawing from the 73,000-year-old levels at Blombos Cave, South Africa.*Nature*, 562, 115–118.
Inevitable, K. (2002). Chengwin – This Saturday. *Improv Resource Center Forums*. https://improvresourcecenter.com/forums/index.php?threads/chengwin-this-saturday.30967/
Janeksela, J. (2018, August 21). Why Czechs Don't Speak German. *BBC Travel*. https://www.bbc.com/travel/article/20180820-why-czechs-dont-speak-german
Kuffner, G. (2002). *Performances*. Ransom Corp. https://ransomcorp.org/perf.htm
Malkin, G. (2023). The power of the immersive arts to catalyze a more peaceful world. *Kosmos Journal*. https://www.kosmosjournal.org/kj_article/the-power-of-the-immersive-arts-to-catalyze-a-more-beautiful-world/
Mann, L. K. (2022). *Rude Citizenship: Jamaican Popular Music, Copyright, and the Reverberations of Colonial Power*. University of North Carolina Press.
Manuel-Navarrete, D., et al. (2024). Ayahuasca ceremonies, relationality, and inner transformation. *Journal of Psychedelic Studies*. Advance online publication. Taylor & Francis Online
Mims, J. (2024, October 11). *Immersive Mixtape Festival Roundtable* [Conference presentation]. NYU Immersive Mixtape Festival, Iris Cantor Theatre, New York City, NY.
Moore, S. (2024). Supporting One, Supporting All: Collaborative Classroom Support Models for Inclusive Classrooms [Webinar]. NYU PINE Summit: Co-Creating Inclusion. https://www.pineprogram.org/summit
Morrow, C. (Host), interviewing Fitzhugh, W. (2022–present). *Immersed in the Arctic: Immerse! Book and Podcast Series*. https://charliemorrow.com/pdfs/immerse-billfitzhugh-transcript.pdf
Oliveros, P. (1971). *Sonic Meditations*. https://www.soundportraits.info/wp-content/uploads/2019/06/Oliveros_Pauline_Sonic_Meditations_1974.pdf
Pine, J. B., & Gilmore, J. H. (1998). Welcome to the experience economy. *Harvard Business Review*, July–August Issue.
Rathje, S., Hackel, L., & Zaki, J. (2021). Attending live theatre improves empathy, changes attitudes, and leads to pro-social behavior. *Journal of Experimental Social Psychology*, *95*, Article 104138. https://doi.org/10.1016/j.jesp.2021.104138
Reznikoff, I. (2002). Prehistoric paintings, sound & rocks. In E. Hickmann (Ed.), *Studien zur Musikarchäologie III* (pp. 39–56). Rahden (Westf.).
Reznikoff, I., & Dauvois, M. (1988). La dimension sonore des grottes ornées. *Bulletin de la Société Préhistorique Française*, *85*(8), 238–246. https://doi.org/10.3406/bspf.1988.9349
Sant, T. (2012). *Applied & Interactive Theatre Guide*. University of Salford. https://www.tonisant.com/aitg/Radical_Theatre/index.shtml
Sapp, W. (2024). Personal communication with Rhiannon Catalyst.
Soloski, A. (2014, October 31). 'The Dreary Coast,' on the banks of the Gowanus Canal. *The New York Times*. https://www.nytimes.com/2014/10/31/theater/the-dreary-coast-on-the-banks-of-the-gowanus-canal.html
Sparkles, S. (2013). *Parades, Parties, and Protests: Creative Resistance Culture*. 218 Press.
Sparks, M. (2025). A history of sound system and emcee culture. In R. Bramwell & A. de Lacey (Eds.), *The Cambridge Companion to Global Rap* (pp. 34–46). Cambridge University Press.
Toffler, A., & Toffler, H. (1970). *Future Shock*. Random House.
UNESCO. (2008). *Day of the Dead – Mexico*. Intangible Cultural Heritage. https://ich.unesco.org/en/RL/day-of-the-dead-00054
UNESCO. (n.d.). *Nowruz: Celebrating the New Year on the Silk Roads*. UNESCO Silk Roads Programme. https://en.unesco.org/silkroad/content/nowruz-celebrating-new-year-silk-roads
Woltmann, S. (2023, May 8). What is immersive theater? *Backstage*. https://www.backstage.com/magazine/article/immersive-theatre-explained-75850/

Chapter 20

# Sound for immersive theater

## Audio engineering for *Sleep No More*

*Gregory Hanson*

*Introduction and interview by Rhiannon Catalyst*

*Sleep No More* has captivated attention and admiration around the globe, earning its place as one of the most successful immersive theater experiences in history.

Punchdrunk's *Sleep No More* transforms traditional theater by dropping audiences into a sprawling, maze-like hotel where Shakespeare's *Macbeth* unfolds around them in real time. Rather than sitting in seats, masked visitors wander through dozens of intricately designed rooms across multiple floors, following unmasked performers who tell the story through dance, movement, and atmospheric encounters rather than traditional dialogue. The production weaves together the Scottish play with influences from Hitchcock's psychological thrillers and historical witch trials, creating a noir-tinged world where guests become detectives piecing together fragments of narrative. The experience emphasizes physical storytelling and environmental immersion, with individual audience members crafting their own adventures by choosing which characters to follow and which secrets to uncover in the darkly theatrical landscape. Every performance becomes a personalized mystery where the boundary between observer and participant dissolves, making each visit a completely different journey through the same story.

For the purpose of this discussion, immersive theater is defined and explored in detail in my contribution within this book. The performance *Sleep No More* took place at the iconic McKittrick Hotel – a venue whose name is often used interchangeably with the show itself in casual conversation. As of this writing, the legendary New York venue has just officially shuttered its doors, but *Sleep No More* continues on in Shanghai, and Punchdrunk continues producing other productions like *Viola's Room* in London.

In this contribution, we delve into the unique artistry of immersive theater through the lens of Gregory Hanson, Audio Supervisor and In-House Sound Designer for *Sleep No More*. Hanson shares his insights into the challenges and triumphs of working in this genre and offers his vision for the future of immersive theatrical experiences.

The success of *Sleep No More* can be attributed not only to its innovative format but also to the meticulous attention to detail by its creative team, with Hanson's contributions standing out as a key factor in its enduring appeal. His work exemplifies how sound design can elevate immersive theater from a visual and physical experience to a deeply emotional and sensory journey.

As the landscape of immersive theater continues to evolve, productions like *Sleep No More* serve as a benchmark for what is possible when artistry and technology converge. Gregory Hanson's work offers a glimpse into the future of the medium, where sound

DOI: 10.4324/9780429297526-23

becomes an even more integral part of the storytelling process, enabling audiences to lose themselves completely in the worlds they inhabit.

## Background

*Sleep No More* is a well-oiled machine. Part of the reason so many people love to go back is that you can't experience all of it in one go – it's not possible – so it's just as magical the fourth, fifth, or sixth time as it is the first. I still remember my first experience seeing it, and I was completely blown away. It was a combination of not knowing where I was physically, losing that sense of direction, and also 'What did I just see??'

As an audio engineer, New York City is a great place to work because you get a little bit of everything, and it can help you find your direction. So I got thrown a few system optimization projects and many utilitarian things, like migrating to a modern Show Control System. One of the first creative projects I was assigned and stuck with is parties that the venue holds for the bigger holidays. They involve everything – immersive performances, bands, theatrical things; it's a massive transformation of the space that guests are allowed to explore. It's sort of a cross between immersive theater and a nightclub. If you've ever been to the House of Yes in Brooklyn, it's that sort of vibe. Each of those venues has its own kind of space and identity, they all do their own thing, but they fit into a larger culture that's developed by many people creating and seeking experiential environments. Anyone on the production crew knows from experience that these environments offer something for everyone – from a professional or an audience perspective – making them incredibly fun.

In 2016 we did a series of parties, one a month, and they were all themed after cool movies, movies with a cult following. We did *Romeo and Juliet*, *Alice in Wonderland*, *The Wizard of Oz*, *Clue*, the *Nightmare Before Christmas*, etc. They were really fun, but they were in rapid succession, and we needed a team to do dedicated work just for that. So I was named the Associate Sound Designer for all those events. And that meant any number of things from making immersive soundscapes, designing song and dance numbers, producing tracks, cleaning up tracks produced by someone else, and all of the above. I worked in conjunction with a counterpart who designed all the system work, and someone who did the staffing, and we were a core dedicated team just handling those events because they were such massive recurrences. After that, an opportunity arose for me to become the head of the audio team. I took over that role formally in 2017 and held it until 2022.

Like all theaters during the COVID-19 pandemic, we shut down for a period of time, but we survived – we came back, we got everything going again. My official title was Audio Supervisor and In-House Sound Designer. In a sense, it was two roles. For one I was doing the administrative work of staffing and running it, technical maintenance, all the things you would expect from a theatrical audio supervisor. And then there was the in-house sound designer piece, which meant anything we wanted to produce in-house, I was handling the sound design for it, or working in collaboration with the show sound designer to make it fit the venue.

Everything else produced at the McKittrick Hotel happens in repertory – *Sleep No More* doesn't stop while other things are happening. That means really intense coordination. A lot of compromises, cleverness, and rehearsal from the technical side of things. The double-edged sword and the interesting thing is, nothing is really impossible, right? From the dreaming perspective, nobody wants to eliminate something because it might be technically challenging. The team there is full of incredible people who have such a high aptitude for leaning in to make something a reality. It's the actual epitome of teamwork; it's interdepartmental, everybody working together toward what you'd expect in a high-level theatrical production. It's almost on the level of a Super Bowl half-time show transition – no one can do anything until the stage gets there, then we can set up phase 1, but another person can't set up their pieces until we finish setting up ours – etc. It's an intensely concentrated team effort. You have to be prepared to set up a lot of elements very fast, prep infrastructure, and hide a lot of things really well. It's a lot of cable drops to make things both accessible and hidden, setting up portable speakers and portable mixing consoles, and so on.

### *Audience perspective*

From the audience's side, it's perceived as different systems creating a whole. To keep the events interesting, we try not to repeat them wherever possible. You aim for a balance of innovation and time-tested elements. As in, 'This worked really well last time, and we're also really interested in doing something different'. Or conversely, 'This didn't work at all last time. So let's not do that again. But what do we replace that element with?' It's a lot of soft changes between events to make it function, with the one consistent thing being *Sleep No More*. And that core show, from a technical perspective, is its own beast and has to work every time. You don't mess with *Sleep No More*.

One interesting aspect of these parties is how the audience enters a space they were previously familiar with, only to find it completely transformed. That may mean they're greeted by a 16-piece brass band that wasn't there ten minutes ago. The team scrambles and everyone makes it happen. You're working together to do what may seem to be impossible. And it's not without conflict, of course. Like any project where people are emotionally invested in the work, you're going to have conflict and struggle to try to get things right, but at the end of the day, everybody remembers it's for a common goal. Everyone's working toward a successful performance. When you have a moment of feeling defeated from a technical perspective, there are so many good resources to turn to. And when you do something as niche as immersive theater and do it full-time, you get really, really good at it. It starts to become muscle memory. Not even because of repetition – you start to remember: 'We *know* this. We *got* this.'

For instance, putting a singer in the middle of the room with all the loudspeakers pointed at her might seem crazy to most people, but we've done that 100 times, and it works every time. You figure it out once, have the same team do it, and every time it gets better with refining. The process involves a *lot* of trial and error and carefully weighed trade-offs. Everything is connected to everything else, and a lot of things don't work. But you will never know what *will happen* if you don't try. You try to be respectful of other people's ideas, and stay in your lane. But if something's interesting, nobody's going to tell you not to participate in helping to figure that out. For instance, I worked with the team on a visual effect we were trying to accomplish – a smoke tornado – we wanted to have smoke swirling around with confetti, and we tried so many different things. We rented fans, we bought fog machines, we were cutting up cardboard, it literally felt like we were doing an arts and crafts project. We never could get it to work – the physics were just not there for us. But we spent the time and we had a vision and we got somewhere with it.

When it comes to the sound of an immersive show, it's crucial not only to the audience but also to the actual structure of the show. The original Sound Designer, Stephen Dobby, is an incredible person. He's been a huge influence and a mentor at times, and he did an amazing job with the sound design. The beauty of *Sleep No More* from an audience perspective is that it's what *you* make of it. You are the director and camera person of your own movie as an audience member. Having the sound support that, regardless of the path you choose to take, is crucial. It's important to have all the same feelings that you get from a film score, except here you can move through the score in a multitude of different ways. The music is the underscore of the story. As you're exploring and watching the show, it feels right that what you hear matches where you are physically and what you're seeing.

Anybody who's created a score knows how hard it can be. When you finally see it locked into place, that's a unique and interesting feeling. If you can translate that into multiple people experiencing that same thing, then you know it's been achieved. And I think Stephen did that. What you're seeing matches what you're hearing and feeling, physically and emotionally, and that's pretty interesting. There are a lot of places where you can see different things and hear different things, depending on how someone personally explores, and they all fit together. At the end, people walk away singing the music. That's how you know it really worked. When you hear a lot of people leaving and whistling a little piece of the music then you know it will stay with them. It's so interesting to witness the impact as people are leaving. The most common thing I hear is, 'What did I just experience?' or 'What just *happened*?' You hear people try to break it down together to figure it out. 'Did you see this?' 'No! What do you mean, that happened?' Or, 'No, that's not what happened to me at all!' People talk about it together to try to piece the puzzle together. That's part of how it has social and connecting power.

### *Mentorship*

Mentor relationships are crucial to the development of immersive experience and immersive sound design. I didn't know what immersive theater was before *Sleep No More*. Going to a theme park like Disneyland is probably the first experiential, immersive-type experience a lot of people have. If you've been to Galaxy's Edge – the Star Wars ride – it's a multimedia experience, and it's so much more than just a theme park ride. You are fully put into the Star Wars world. You don't have to be a Star Wars fan to be captivated and appreciate the technology and effort, and how seamless the whole experience is. Perhaps most people's closest experience to something like immersive theater early on is a theme park or something like Rainforest Cafe. There's nothing else like an immersive experience.

In regard to immersive theater and immersive sound, some problems you solve and practices you develop are very complex, and some are surprisingly simple. For example, it may seem pretty obvious but if you can't see where the sound is coming from, it's going to be a better experience. Sound enthusiasts are excited by the sight of loudspeakers, but nobody else pays attention to it. That can really work to your benefit when you're trying to immerse someone. In a movie you have diegetic versus non-diegetic sound. Unless you're trying to produce a diegetic sound, why would you want to be able to tell where that sound comes from? Immersive experiences are all about transporting people out of their reality and into the new world you're creating. You always have to ask yourself, how can we help someone fully suspend their disbelief and leave their current reality behind? One key is eliminating the localization of sound. A lot of companies are now producing site-specific and smaller-scale immersive theatrical performances. When you walk into a site-specific piece, I find nothing disrupts the illusion of a new world more than walking into a space and immediately spotting a speaker where the driver and sound source is easily localized. It breaks the immersion and undermines the atmosphere.

Something interesting in the sound design for *Sleep No More* is the specific drone sound that is always present. It's such a low frequency that it becomes largely omnidirectional, simply because of the nature of the content. And that omnidirectionality helps you remove yourself from hearing the scuffle of other people in the stairwell, or hearing when a door closes that you didn't open. It's just there, like the underscore of the movie. It's a non-diegetic sound supporting the world, and that can be true anywhere you go.

In my opinion, there are two ways that you can think about immersive theater. There's the forced perspective – for instance, you're sitting down watching a movie with surround sound, or you're in a Broadway theater seated in the orchestra and there're surround speakers and effects. That's immersive because you hear sound coming from above and all around you. Then there's this sort of free-roaming opportunity created by *Sleep No More* and other experiences popping up all over. They're both immersive, but they're very different ways to think about it. In the former, the forced perspective, it's all about using localization to reinforce the story. The Broadway show '*Six*' is a pretty good example of using the d&b Audiotechnik Soundscape and performers moving across the stage for tracking, and you really can feel it – you can be sitting house left and feel where that singer is, localizing to that. That's a very interesting way to use that technology and the direction the technology has gone. The technology is largely moving towards the concept of 'Let's figure out how to localize – let's move sound, have jets fly overhead in a movie, have sounds come from behind you'. And so on. That sort of localization is what you get with binaural audio, that intense localization, and that is an interesting approach. If you're sitting there, closing your eyes, listening to one of those famous binaural ASMR audio clips on YouTube where someone is cutting your hair, you get that effect. It does feel real and almost gives you a sense of unease. But as soon as you open your eyes, you've lost it, right? You're not imagining it anymore. You're suspending your disbelief while your eyes are closed, immersing yourself and playing into it, but as soon as you open your eyes and remove yourself from that position, suddenly it's not real.

On the other hand, in a big environment where you have the opportunity to remove the localization, you can't really escape it. There's sound everywhere, you can't control where it's coming from, and it can be a very effective tool. It wasn't until many years of doing this work that I was able to put a pin in what that was. During the COVID-19 pandemic in 2020, the only show I worked on was *NightWood*, a show by a small New York City–based company called Clerestory Light. There isn't really a narrative to it, it's outside at the estate of Edith Wharton in the wintertime, so it's cold. And it's just an experience, kind of like a light show with macro scenes. In some of them, you're in the environment. In some of the larger scenes, you watch events unfold in the woods from a distance, with loudspeakers positioned far off in the forest. So you lose localization and it's a free field, or it's as close to free and open as possible, give or take some trees and hills. In that open free field with big, wide, covered loudspeakers far off in the distance, where you can't perceive the origin of the sound, it feels like you're looking at a painting with sound playing in your head. It's such an interesting way to build an immersive experience. Your entire field of vision is consumed by what you're seeing, while the score envelops you completely. You're fully immersed. No matter which direction you move, you remain fully immersed in both the visual and auditory elements. It's kind of the opposite of moving sounds around you – the sound is just everywhere. That's my perspective on one way to make it work.

There are so many ways to transport people with immersive sound, even back to their present reality. For instance, nothing grounds you more than silence. Like that tense moment when you're at a concert and the PA cuts out – one second the kick was punching you in the chest and you're surrounded by people screaming. Then, all of a sudden, it's silent and you remember, 'Oh, I'm here, and 10,000 people are experiencing the same thing at the same time'. That's an interesting thing. Those are accidental moments, but you can use that same tool effectively.

We had a really good run with *The Woman in Black*, which we co-produced for quite a while. There's a scene where there's no sound over the PA. The actors weren't amplified – the PA was only used for sound effects. That dynamic range is just incredible – going from silence, 50 dB SPL, to all of a sudden there's a door slamming that's 90 dB from the back quarter, then back to silence. Using that tool really grounds people. There are also some interesting moments where things like that happen in *Sleep No More*. It can be a powerful way to shape experience.

## Lessons learned

By the time I got there, *Sleep No More* had been going for about three years. I can tell you the biggest mistake you can make is changing things too fast. For instance, I'm not a dancer, but I know that sound plays such a crucial role in knowing how to react and what to do when, and that everyone has a different system for listening. I can listen to it from a sound person perspective, or from an audience perspective, or from an interpreted audience perspective. However, that might not be how other people, who are using the sound performatively, hear it.

I can remember going in and changing the settings for a loudspeaker processor and thinking to myself that even putting this in for one moment, I've got to limit it more, lower this compression threshold just to get it to kick in a little bit more. But that was disastrous because it ruined a key point for a dancer. I had no idea the effect that something so small could have. It reminds me of fidelity and sound, and why fidelity is important. The analogy I always came up with is that, if you took a small pixelated image and blew it up, then you would see all the defects. You would see where the pixels end, and start. If you zoomed out, it would look clean. If you zoomed in, it would look wrong. The same thing is true about sound. When you start amplifying steps at a high volume level, you start to hear those imperfections. The difference between sound and anything else is that it's really hard to quantify what sounds good and what sounds bad. You can objectively measure audio compression loss – analyze what happens when you compress a sound file and quantify the data loss. But when you actually play that compressed audio, the perception becomes subjective. But if you change something without telling anyone, as I did, I discovered I could unknowingly influence how people responded to the sound. This taught me an important lesson: never make changes too quickly, and always know how to revert them. What seemed incorrect to me at the moment was actually valuable to someone else. This is crucial for any theatrical sound designer to understand because it's easy to second-guess yourself as a designer and think, 'That's not how I intended it to sound', or 'That's not the effect I was going for'. But if you're saying that without consideration for the way other people may be currently reacting to it, then it's a little misguided. It's a word of caution to anybody out there who's changing work that's affecting the way other people are responding. It took a while for me to learn that. I realized, 'Oh, I get it, it's ME. I'M the problem'. There are a lot of humbling lessons to learn. That one was a game-changer.

Another important factor is being open to experimenting. You're emotionally invested, you want it to be huge and successful, and you want it to live for a long time. If that's the goal, then you have to remember that this is going to be bigger than your own emotions right now. It's going to turn into something else. What's the piece today? It's a living, breathing thing that continues to develop. I haven't been part of *Sleep No More* full-time for about a year now, so if I went back, things are going to be different than when I left. And they're not necessarily huge things. But once you're there for eight or nine years as I was, you see many changes. You have to be willing and open to change. Nothing's worse than being the person in the room who is not willing to compromise. That is true in all trades, but it is perhaps a little more emphasized when you're creating art. You have to be willing to take a step back and say, 'Okay, I wonder why that's not working'. A lot of times, you can tell by the gut reaction of people. If people are reacting to something, and it's not the way that you expected, it's probably not working. And you don't have to take that personally, you just have to recognize it and say 'Okay, I'm gonna try something else'. Or, 'I'll save that for later – maybe it'll work better in a different place or a different piece'.

The foundation of any sound design is absolute technical reliability. Every system must be consistent, stable, and systematic. When you press go, you expect the cue to fire and it needs to fire, without exception. This is a fact in regular theater as well, but in immersive theater, there's even more at stake. When something goes wrong, you can't rely on conventional recovery mechanisms. If performers aren't in the right place at the right time, the entire experience falls apart. You can't get by without that tech perspective of reliability and redundancy. If you don't have redundancy in your system, then you have to take a step back and reexamine things. When you're creating works, if it's not in three places, it doesn't exist. Tools like Dante provide essential technology for building this redundancy, ensuring that technical failures never become showstoppers. In immersive theater, where the audience experience depends on seamless integration of all elements, technical reliability isn't just important – it's everything.

Like any production, there's a balance of budget versus tactical aspirations. This has played a lot into my philosophy in general on creative immersive experiences. It's very different from that forced perspective approach to sound design because I don't need to localize to one point. Therefore, I don't need giant line arrays helping me get coverage across the whole room and localized into the stage. There is a place for that design, and plenty of such places out there. But here, the key is lots of coverage, getting the sound everywhere possible. In my opinion, that's the key to success. And that is the reason it works so well because there are so many places to discover, and sound is *everywhere*.

Dividing up what sound goes where is a math game based on resources, available content, and the resolution you want to have. The challenge lies in recognizing that you're placing sound throughout an interconnected space. Unlike traditional theater with acoustically isolated rooms, immersive environments require you to consider how every audio element will interact with every other element. Nothing's worse than going to a music festival and the other stage is clashing with the sound of the one you're trying to focus on. In immersive theater, this problem multiplies across every room, corridor, and transition space. Success requires treating the entire venue as a single, complex instrument where each speaker placement, volume level, and content choice affects the whole. Every decision must serve both the local experience and the broader sonic ecosystem.

The craft lies in mastering the acoustical response of each speaker placement. You're solving a three-dimensional puzzle: hiding speakers while achieving optimal room coverage, all without overpowering adjacent audio sources. It's a balance that demands both technical precision and artistic sensibility. That is hard. The Massachusetts version of Punchdrunk's *Sleep No More*was set in a big open field. We approached this challenge compositionally – writing all the music in the same key so that any combination of sources would create harmony rather than chaos. Whether you're standing next to a speaker or catching distant sounds across the field, the audio layers blend naturally together. Even if you're hearing something else from far off in the distance, it has a way of seeming to fit together. That's one of the things we worked on while I was there – improving that response. Sometimes the solution was as simple as repositioning a speaker by a few feet; other times it meant optimizing our use of the technology itself. We treated the entire field as a single, complex acoustic instrument that needed careful tuning. The technology and the affordability of the technology has become a lot more accessible over the past decade. This technological evolution hasn't just lowered barriers – it's fundamentally expanded what's possible in experiential storytelling. It's phenomenal. It's been really enabling.

Talking about moving something from an input to an analog mixer, then straight out to power amplifiers. Moving from that to a proper loudspeaker processor where we have the ability to delay speakers, change their tonal response, their phase response, all of the actual tools you have that might seem tedious on paper can make a really big difference. For example, if you've got a hallway full of speakers, where do you time align to or do your timeline at all? How do you get any sort of coherence in that space from a phase perspective – you put a bunch of speakers down the line, but what's the right answer? Sometimes there isn't a 'right' answer, and sometimes there's a way to do it that results in a compelling effect that serves the artistic vision better than textbook implementation. That's one of the interesting things about immersive sound in theater specifically – we don't have to adhere to any specific format. We don't have to adhere to 7.1, or 7.2, or 11.4.2. You can do whatever you want, as long as you've got the tools. This is interesting because I don't feel the software we typically use is really set up for that. For example, if you go to set up a surround bus in your Digital Audio Workstation, you're limited to 5.1 or 7.1 or whatever it is you want to do. But say my budget allowed me to have six speakers, and I'm going to put them in a circle. How am I going to do that? It would be interesting to get some flexibility from some of the software to allow us to do that. The divide between live sound capabilities and recorded sound workflows continues to hinder immersive production. What we need is unified software that treats spatial audio as a flexible canvas rather than as a collection of presets. That would be great to see.

In immersive sound design, every decision must serve the story. This principle holds true across all applications we've explored, but the depth of this relationship reveals something profound about the medium itself.

While content selection – what audiences hear – represents the most obvious narrative tool, the delivery method – how they hear it – carries equal storytelling weight. The spatial positioning of a whisper, the decay time of footsteps in an empty corridor, the way voices blend and separate as characters move through space – these technical choices become narrative choices.

Consider how a voice coming from behind an audience member creates different emotional meaning than the same voice broadcast from overhead speakers. The content remains identical, but the delivery method changes the story being told. This is where immersive sound design goes above and beyond traditional audio work: the medium itself becomes part of the message. And, of course, the sound has a massive impact on what the message is along with setting the mood, it's the vibe that sets everything up.

## Impact and accessibility

I think talking about the impact that spatial sound and immersive theater can have is an important question, and very relevant. When you look at the communities that are involved with theater and arts, people are no longer creating art just for art's sake. There's a healthy relationship of saying something about the world and the way things are going in the world. Art is a very good outlet for that process. For one, because you can get your message out to a lot of people who are willing to listen to it and you give people the opportunity to interpret it for themselves. I think the biggest impact these kinds of things can have is making that opportunity accessible to other people. That's also going to be the biggest challenge for technology. Because this stuff is not inexpensive. It's more affordable than it was, but it's not inexpensive. Also, immersive theater has to be discovered. These days, everything's immersive, right? Websites are immersive. Opening your mail is an immersive experience. Not to sound cynical, but it's true. Everybody's branding, everything is immersive. And that's fine. It's a buzzword, that's fine. But there is something unique when you get down to the real application and heart of what it means to make something that's truly immersive. So in what direction is that going to go? And where is immersive now? What impact could it have? It's a good question – I think the technology is going to keep moving forward. So how do we make it more accessible?

We're seeing a lot of emphasis on locating sound in places, moving them and giving a 360° perspective from your living room. But I don't know that it has to be that way, and I'd be interested to see different interpretations of immersive sound coming from different people. The only way for that to happen is for different people to be exposed, discover it, learn about it, and try it in different ways. Then we can fully see what impact it can have on the world, as the tech helps us innovate art, in the same way that recording technology innovated music – this is the next generation. It's exciting how hard it is to predict.

I'll close with one other experience we produced at the McKittrick – a show that will stay with me for a very long time. In 2018 we did a co-production with the Scottish theater company Vox Motus, called *Flight*. It was a really impactful narrative, told like a mashup of an audiobook and a graphic novel, all told through a physical model set up on a revolve. It was like watching a storyboard. It scrolled past you as it was on this revolving stage that moves very slowly, with little vignettes that would light up. It was the definition of a technological challenge. By no means am I taking credit for the design of the show, they had already been on a run before, but we worked to redesign the system a bit to optimize it with integrated systems they could take on the road with them. That was a very interesting and unique challenge. It was on a revolving stage, and you were meant to watch it by yourself using headphones. You sat in a booth and watched the revolve move in front of you. Two minutes later, the person sitting to your left would get the same scene you saw. There were 24 seats, and we had a bass shaker on the bottom of the chairs. It was a monster of a show to get right from a technological perspective.

The highest achievement in immersive technology is invisibility. When audio, visual, and physical elements merge seamlessly, audiences experience something that feels effortless – even inevitable. The most sophisticated technical orchestration disappears entirely, leaving only the story and the emotion it carries. It's really remarkable when it's done so well, that it seemed simple. I think we're going to see really interesting things leaning into that kind of multimedia experience. The future belongs to creators who ignore the limitations others have accepted. Every 'impossible' idea deserves exploration. Every unconventional approach might unlock new forms of expression. The most transformative immersive experiences will come from those willing to experiment beyond established boundaries. If you've got an idea, I think you should try it.

Chapter 21

# Cinematic sound immersion

*Steve Schatz*

## Immersion

What does it feel like to be 'immersed'? More specifically, what do you think being immersed means in an auditory sense? It could describe the frantic sounds of the hustle and bustle of city life on the way to your office, a busy kitchen in the midst of a hurried frenzy preparing for the lunch rush, or, contrary to the noise – the few rare moments you have to yourself in a quiet room, hearing your own breaths, bathing in bare silence. I believe immersion means to feel fully encapsulated by your environment, whether it is quiet or noisy, somewhere in the middle, or nothing at all.

If you pause for a moment and look at the world around you . . . it is changing. Nothing remains static, even if you are in a motionless space. The 'steady' hum of the room you're in is anything but steady; it's oscillating. The people around you are in perpetual motion, moving in every direction. It is something that is being *presented* to you, an ambient world in constant flux, almost as if to keep you entertained. You are the audience, taking it all in.

When we go to the cinema, it is the same – a world is being presented to you. It is carefully imagined, and in some cases agonizingly crafted and evaluated until an emotional evocation has taken place. In cinema, that world can be invigorated by the power of immersive audio. Pure cinema is living within the world the filmmakers have created, without noticing that any of it exists.

The first example of immersion is inconspicuous. You are absorbing the environment in your current space and time, treating yourself to a unique experience. The second example is more deliberate – it is the purposeful art of filmmakers creating the space *for you*, using cutting-edge technology and practices and expressed through a wide range of mediums.

Our goal as filmmakers is to try and mimic our world to give us a sense of familiarity but also to enhance that sense of realism. The possibilities are endless. Sound can come through the secret passageways of our subconscious mind and affect us emotionally and effortlessly.

## Humble beginnings

I would like to take a quick visit back to the evolution of sound in cinema and show you where it began, before we can see into the beyond. It is a riveting journey that parallels the history of filmmaking, a true testament to human ingenuity and innovation. In the

DOI: 10.4324/9780429297526-24

early days of cinema, we had silent movies. The audience could enjoy a visual spectacle but lacked the auditory dimension we sometimes take for granted. Live musical accompaniment, often by pianists or orchestras, provided some of the ambiance in theaters, but it was a far cry from the synchronized sound we know today.

If we're going to talk about the dawn of cinema sound, we have to give credit to the landmark film *The Jazz Singer* from 1927, which marked the transition from silent film to sound in the cinema. It utilized Vitaphone technology, created by Western Electric (and later purchased by Warner Brothers in 1925) to synchronize sound recordings with visuals. This breakthrough spurred a race among movie studios to integrate sound into their films, leading to the development of sound-on-disc and sound-on-film technologies.

By the 1930s, the use of microphones, amplifiers, and ***multi-channel audio*** allowed for the creation of more intricate soundscapes, enabling filmmakers to incorporate dialogue, music, and sound effects seamlessly, further enhancing the immersive experience. The 1940s brought advancements in ***stereo*** sound, adding depth and dimension to movie audio. This was further refined in the '50s with the introduction of widescreen formats like Cinemascope, which enabled filmmakers to match expansive visuals with immersive sound. The next few decades gave rise to surround systems like Dolby Stereo and THX, enhancing the audience's engagement by enveloping them in a three-dimensional auditory experience.

By the time we reached the 21st century, we witnessed a revolution in digital cinema, with Dolby Digital and DTS becoming the standard. The use of computer-generated imagery (CGI) in films opened up many avenues for sound design, allowing for the manipulation of sounds in ways previously unimaginable. This era also saw the emergence of object-based audio, which treats sound elements as individual objects, enabling more precise placement within a three-dimensional field. It took us to new auditory 'heights' – for the first time we were able to envelop the audience from directly above, whereas before that was limited to the side and rear speaker arrays only. When you can place sounds in specific locations or even individual speakers in a home setup, audiences are immersed in a 360° sphere that heightens emotional engagement and realism. These newer forms of ***immersive audio***, combined with advancements in visual effects, have ushered in a golden age of filmmaking, where storytelling is not limited by technical constraints but enhanced by them.

The era you currently are living in is the era of ***modern*** immersive audio, whose primary goal is to more accurately recreate and reimagine how we hear things in the natural world. By the time you read this, there may be something completely different out there than the formats I've mentioned, which speaks to how quickly technology evolves. Through these advances, we can conclude that immersion can come in many forms and impact our senses in varying degrees, and that it will continue to change well into our future.

I did my best to get the jump on emerging technologies, studying every format that was available and reaching out to those who were at the forefront of new developments. I started in immersive audio around 2012. At that time, only a handful of movies were given the greenlight by studios to deliver films with immersive formats, with Dolby Atmos being the leading cinema audio technology. There were other formats then too, such as Auro-3D and later DTS:X (2015) to name a few.

The first movie to be mixed in *native* 'object-based' audio was ***Oblivion***(Kosinski, 2013), starring Tom Cruise as Jack Harper, 'a veteran drone repairman [who] uncovers a crashed spacecraft with contents that question his mission and his own identity'. Sounds like a great opportunity to use immersive audio, if you ask me. It's amazing how art can lend itself to pushing technology to its limits, simply by the nature of the content.

The term *native* refers to the mix being created and developed from the ground up in its original format. Using the workflows that require it, the result is a seamless monitoring environment from the digital audio workstation (DAW) to the speaker system. The two re-recording mixers on this film are close friends and colleagues of mine, and they are two of the best in the business. I've sat alongside them and many other sound designers, mixers, and editors for the better part of two decades, absorbing their knowledge into the designs I work with today. I will share with you the constructs that I feel are most important when creating an immersive audio experience.

## Immersive design

Immersive audio design is an art and a science that takes us beyond mere sound. It is a philosophy that encapsulates the essence of storytelling, connecting audiences with narratives on a profound level. At its core, it seeks to envelop the audience in a way that transcends the boundaries of the screen or speakers. It's about making viewers not just hear but feel the world of the story. This philosophy embraces the idea that sound is vital in conveying emotions, creating tension, and building empathy with characters.

To achieve this, immersive sound designers employ a range of techniques and concepts. Sounds are meticulously placed around the audience to ensure that every rustle, whisper, or nuance feels as impactful as any larger counterpart. Immersive audio design values subtlety. It recognizes that sometimes the absence of sound can be just as meaningful as its presence. Silence becomes a main character. It is a tool for creating tension and anticipation, letting viewers' imaginations fill in the blanks, fostering a deeper connection with what they are seeing and hearing.

All of this champions the idea that sound should be adaptive. Just as a film's visuals change, so should its audio. In a way, this can be considered ***interactive storytelling***, as seen in video games, audio should respond to a player's actions – but in cinema, it is your imagination that is responding to what is presented to you. Immersive audio for cinema strives to make you not just a spectator but an active participant.

Creating a memorable audio presentation in cinema is done with a polished sense of taste and technical chops. I will discuss a few points that I believe are integral to achieving both the artistic endeavor of the craft, in addition to mastering its scientific approaches.

- **Serving the narrative:** Immersive audio should always serve the narrative. Start by thoroughly engaging with the story, finding the emotional beats of the characters, and the nuances that their world is set in. This will help guide your decisions in crafting their environment, helping you emphasize key moments and create a cohesive soundscape.
- **Dialog clarity:** While immersion is key, if you are working with material that has dialog as the primary storytelling element, remember that dialog clarity is paramount. Ensure that dialog remains intelligible. Position dialog in the most appropriate locations while maintaining focus on the characters during each and every scene. Enhance realism by paying attention to perspective changes – utilize delays, reverb,

and other time-based effects to invoke depth of field when a character's voice is near or far away from us.

- **Subtlety:** Immersive audio can be about subtlety as much as it can be about power. Consider the small things – a creaky upstairs floorboard, a callout from distant voices, a bird flapping its wings, an indiscernible radio call. These details can add depth and authenticity. Using immersive audio allows you to place and move sound objects overhead, or pinpoint specific areas of the theater. Use this capability to fully immerse your audience. In a rainstorm scene, you can make raindrops appear to fall from above, creating a sense of depth and realism. You can also carefully place unique 'rain on object' sounds, like rain falling on the broken wing of an airplane, and pan it across the screen, as our hero walks up to it and observes the wing's damage.
- **Dynamics:** Take advantage of dynamic mixing capabilities. Use volume, panning, and depth to guide the audience's attention and emotions. Create tension by gradually moving a sound closer or build excitement by tastefully panning sounds around the theater during an action sequence.
- **Music:** You must also recognize the power of music. Integrate the score seamlessly into the immersive soundscape. Look for moments to highlight (or not highlight) the music over other elements like dialog or sound effects in order to enhance emotional impact. Use surround channels and a bit of height to give the music a spherical quality. You can also place certain orchestral elements and anchor them to specific speakers in the room to add width and detail to your sound space.
- **Audience's perspective:** Think about how the audience will react to what is presented. Consider the placement of speakers in a theater, whether it's a commercial or home theater, and how different elements will be perceived from various seating positions. Take into account your audience's listening environment and how your design choices will translate to their medium. Test your designs to ensure it works for all viewers, and create your work in a space that you can trust.
- **Collaboration:** Filmmaking and immersive audio design is a collaborative effort. There are many collaborators on a project, from pre-production to on-set, all the way to post-production and marketing. Everyone will have an opinion one way or another. Ultimately, you will work closely with a director, film editor, producer, and other members of the creative team. Effective communication is essential throughout the entire length of the film in order to align the audio design with the overall vision of the filmmakers. Discussions begin early in the process, sometimes months before shooting begins. Great screenwriters incorporate sound cues in their scripts that bring the story to life. As sound designers, our job is to carry that through every stage of the production to the final product.
- **Emerging technology:** As you aim to explore the future of sound, keep an eye on emerging technologies and trends in immersive audio. The field is constantly evolving, and staying informed will help you push the limits of what's possible. Never stop learning and experimenting. The more you explore and innovate, the more enthralling your immersive audio designs will become.

To summarize, designing an immersive experience at the cinema is about combining technical expertise with storytelling. It's about using the tools at your disposal to transport the audience into the world of the film, enhancing their emotional connection to the narrative.

## Do's and don'ts

When actually sitting down to do the work, it may seem quite intimidating, whether you're a seasoned veteran or a beginner. So many tools and methods are at your disposal. You might find yourself asking, 'Where do I start?' In the context of cinema or television, here are a few good starting points on what to focus on in the early stages of developing the soundtrack.

If you are doing a feature, episodic, or video games, I will again stress that the dialog will be the anchor of your soundtrack. It will dictate where all the other elements like music, ambience, or 'hard' effects (sweeteners), and *foley* will be placed and balanced as you begin to craft your mix. It is imperative that every syllable of every line of dialog is heard. Most cinema or television eventually make its way into the home where most people will hear it, and we cannot rely on all of those environments being set up correctly, minus a few audiophiles. We have to rely on a system that is not calibrated perfectly and other distracting noises competing for our attention, like a dishwasher running, dogs barking, or a pickup truck bumping loud music on your street.

How dialog relates to immersive audio starts with my pertinent question of, 'What is immersion?' First we have to understand that it is being *immersed* in the story that will captivate you, and if you cannot understand the dialog, you are not part of the narrative, and your attention has been lost. It won't matter too much how cool the sound effects or music are at this point. Dialog intelligibility can be a contested topic, especially during scenes that involve a lot of dynamics and density. Some filmmakers may believe it's the unintelligibility of the dialog that draws you in deeper because in the real world, we won't hear every syllable people utter. As a dialog editor or mixer, you may disagree! Above all else, it will be up to the filmmakers on how a scene plays, and if they have an idea of how a scene should sound, your job has become easier. You have less to figure out because they have laid it out for you.

Now that we understand the fundamentals of dialog as the central part of the immersive experience, we can begin to discuss how it can be used in the **immersive audio construct**. Can we pan it around the room? Should we place it overhead to mimic a character calling out from above? These are the challenges you will face when creating your mix. Sometimes placing dialog overhead works; other times it may seem gimmicky. Sometimes music coming from overhead works, and sometimes it doesn't. Instead of feeling like a 'sphere' that envelops you, placing things in an array overhead can feel oppressive or enclosed, as if too much content is coming from above. Instead, you may choose to place a specific element in a pinpoint location (individual object-based speaker) as opposed to a static speaker array, which is the default behavior of some overhead digital signal processing these days. Experimentation is key. What works best for the scene and how it affects you, the characters, and your audience, will be a fun game of trial and error.

The potential limits of your soundtrack will also depend heavily on the willingness of your filmmakers to experiment. What if the movie *The Birds* (1963) were remixed in immersive audio today? That sounds like a really cool thing to try, right Alfred? One would think that movie lends itself to a great deal of immersive audio possibilities. It would be frightening to reach that level of detail for such a classic film by the Master of Suspense, Alfred Hitchcock.

If he were alive today, I would try to convince him that an immersive audio mix for *The Birds* would elevate his movie to a whole new level and that bird flaps swooping from all directions would be absolutely terrifying. He may agree wholeheartedly, or on the contrary – he may disagree, as he takes a more classic approach to cinema. He may be completely satisfied with his mono mix from 1963 because it centers the audience to one location – the screen – and it doesn't detract from that anchor point. Yes, even a single-channel mix can be enough, and we can conclude that the term **immersive** is not dictated by how many speakers you have but by your own emotional response to the content. It is up to you.

There are many more pieces of advice to give as far as what works and what doesn't. Some of it is personal taste, some are rather rigid guidelines. If something is emotionally successful or unsuccessful, it is up to the individual viewer. Particular filmmakers and collaborators get hired because they know the difference between the two. We'd like to believe we bring experience to the perception of what works and what doesn't work. It can apply to directors as well as sound designers, mixers, or any other filmmaking crew. Again, I'd like to emphasize that all of this is subjective, but I recommend understanding the fundamental concepts before attempting to go outside the box and break the rules. Your experience with different types of projects and their unique requirements, crew personalities, and challenges will develop your taste and what you believe will be creatively abundant and technically sound.

We'll shift gears for a moment and get into the nitty-gritty. I have strong feelings about this next bit. You might be tempted to propagate a sound into multiple speakers or 'diverge' it, to give something size and power. Divergence is splitting the energy of one audio signal into multiple speakers or source points. Without proper usage or knowledge of time-alignment and phase, diverging audio into multiple locations can prove to be monotone and phase-incoherent. Even if the surrounds are properly time-aligned, you will most likely run into phase issues either at the center position or at different locations around the theater. An optimal choice may be to layer your original sound with similar ones and place them around the sound field to create a more natural illusion of size.

Now that we took care of that, let's go back to a couple of legacy formats for a bit. You have heard about 5.1 and 7.1 mixes. In a typical theater you have multiple speakers or **speaker arrays**, placed on the side and rear walls. You will notice that the surround speakers in a movie theater are quite small compared to what is behind the screen, which you actually cannot see. Take my word for it, they are much larger than the surround speakers! That is by design – the soundtrack is meant to be favored by the front channels to keep you engaged with the story, and all the other speakers around the theater are the supporting cast.

You may have also noticed that when something gets panned off the screen into the side or rear channels, it loses some of its power and low-frequency significance, which results in a slight tonal change if you are listening carefully enough. Every theater engineer will strive to maintain tonal balance throughout their theaters; however, the surround speaker size and frequency response will vary, and they will not be capable of the power that the screen channels and low-frequency effects (LFE) channel can deliver.

Fortunately, in the new age of emerging immersive audio formats, this limitation has been lifted and grants the usage of **full-range surrounds**. By way of bass management (Hirvonen & Robinson, 2016) and rear subwoofers, this enables full-frequency content to move around the theater without losing its punch and low extension. Many audio designers prefer this feature over the usage of height channels, as it has quite an advantage over traditional channel-based sound reproduction. As a result, we are capable of creating the most consistent and impactful immersive experience possible.

I'd like to touch on audience translation once again. The content we create is not only for ourselves, but for thousands if not millions of people. So, how will we manage to dumb down what we have created in a calibrated environment into a tiny phone speaker? In my experience, the fewer speakers that need to translate to a smaller system, the better. By this I mean, the fewer speakers a sound 'occupies' at the time of design and mixing, the more likely it is to translate correctly to a pair of earbuds. We've now touched on the topic of **down-mixing**.

In an immersive audio system with a high channel count (e.g., 72 or more), say I choose to pull a single rain ambience effect from the walls into the center of the room in order to 'fill' it. The impression of size can be misleading. It may sound impressive in the theater, as your rain is emanating from almost every speaker in the room, but what is actually happening is that you are 'mono-izing' that rain. The same sound playing in many speakers won't make it sound bigger. It will sound flat and one-dimensional. To make matters worse, it could translate to phase cancellation on smaller consumer playback devices. This draws similarities to the overhead array example I mentioned earlier. Once more, it would make sense to use similar types of effects but with different frequency responses to add variety and layers. At this point you can confidently experiment with filling the room from above and every direction.

Be vigilant about checking how a mix will translate down to the lowest-common denominator or perform a separate mix and monitor in that denominator to begin with. The general piece of advice is to keep your main elements on the screen and walls, and use ancillary speaker locations, like overheads and **prosceniums**,[1] sparingly to enhance special cinematic moments.

## Fin

I would like to close and reassure you that the future of immersive audio is a thrilling prospect. We can anticipate refinement and wider accessibility of current technologies. The potential for new forms of media and storytelling is boundless, and your role as a creative sound professional positions you at the forefront of this exciting future. It's a journey into uncharted territory, where the only limit is our creativity.

Thank you for joining me on this adventure through time and sound.

## Note

1 In modern immersive cinema (and some audiophile home theater setups), prosceniums are 'wide' speakers that represent the angle viewed from the audience for a traditional soundstage. They were introduced in immersive audio to widen the front image, extending beyond typical LCR. The goal in using them is to widen the stereo spectrum beyond the edge of the screen frame while keeping that sound in front of the audience and not allowing it to be perceived as coming from any of the surround speakers.

## References

Hirvonen, Toni; Robinson, Charles Q. (2016) *Extended Bass Management Methods for Cost-Efficient Immersive Audio Reproduction in Digital Cinema [PDF]*. Dolby Laboratories, Stockholm, Sweden; Dolby Laboratories, San Francisco, CA; Paper 9595.

Kosinski, J. (Director) (2013) *Oblivion*. Universal Pictures.

Chapter 22

# Make the loudspeakers disappear

*Nathaniel Reichman*

When I was very young, the Merce Cunningham Dance Company came to a performing arts center in Anchorage, Alaska. I sat in the fourth row, watching and listening as composer David Tudor performed a live electronic score for the dancers. The sounds seemed to come from in front of me, from the ceiling above me, from behind me, and at other times as though they were emanating from behind the curtains backstage. It's likely that his method was simpler than I remember. But the goal was quite clear: he was shaping an ephemeral, three-dimensional space for all of these sounds, and the very locations of the sounds imparted meaning in their own way. Many years later, I was at the Salle Olivier Messiaen in Paris, listening to the famous Groupe de Recherche Musicale's 'Acousmonium'. Composer Daniel Terruggi proudly explained it to me as an audio mixer that worked in reverse. Instead of taking many channels of sound and reducing it to one or two channels, the Acousmonium mixer takes two channels of sound and splits those signals out to 30 or 40 loudspeakers arranged on a large stage. The electronic music performances there were riveting and wild. The music that did play there was usually available to listen to on CD, but those CDs often felt claustrophobic and smaller when I played them at home on a stereo system.

In 2004, I assisted mastering engineer Paul Zinman on composer John Luther Adams' album *Strange and Sacred Noise*. John would soon write *Inuksuit* and *Ilimaq* as well. These were all concert-length percussion pieces, in which the scores specified that the players be situated all around the audience. *Noise* was meant for the concert hall, but *Inuksuit* was one of John's first immersive outdoor pieces. In the studio, John Luther Adams asked Paul and me how listeners at home could have the same experience that we had created in the performances. It was a good question. Unfortunately, at that time we didn't have a good answer. The excellent technology behind Sony's Super Audio CD (SACD) format had done nothing to keep it from failing in the marketplace, and the almost-as-good DVD-Audio format was rapidly falling into obscurity right behind SACD. The best we could do at the time was adapt the DVD-Video format for music, utilizing its compromised 5.1 audio functionality while ignoring all the unused bandwidth meant for playing video. John asked that each instrument be perceived with equal definition and weight anywhere in the horizontal plane of 5.1. But this was impossible to achieve. There were gaping holes on the periphery, and most home theater 5.1 systems didn't guarantee full frequency range in the surrounds. Later, while QC'ing the 5.1 DVD-Video release for the record label, I heard consistent problems in the sound reproduction which were actually inherent (and therefore unavoidable) artifacts of the Dolby AC-3 codec used at the time.

DOI: 10.4324/9780429297526-25

Work continued in this way for several more years. The 5.1 versions of John Luther Adams' award-winning *Become Ocean* and *Canticles of the Holy Wind* that I produced and mixed were unquestionably better than the stereo versions but lacked the even sound field one would experience at a live performance of these pieces. But then in 2012, everyone in the industry watched Stuart Bowling's Dolby Atmos introduction video. This marked the first mainstream application of object-based panning outside academic laboratories. It was very exciting, but it also looked to be prohibitively expensive and exclusively designed for movie theaters. But Dolby saw a much larger market, and by 2017, my colleague John Bowen and I were using an inexpensive version of Dolby Atmos technology not for theatrical releases, but for two-channel binaural podcast releases meant to be heard on headphones. This was a breakthrough. Dolby Laboratories' goal (and it wasn't completely clear to me in its early days) was to scale their immersive Atmos technology so the same mix made on a multi-million-dollar theatrical mixing stage with 64 loudspeakers could scale down to a $20 pair of headphones connected to a smartphone. The binaural headphone experience was quite compelling. Panning sounded more like the real world, and location cues regarding height and also depth along the *Z*-axis were reasonably convincing. John Bowen and I began to wonder if everything we mixed from that point on, be it a podcast, music album, television show, or film should utilize this technology; stereo started to sound flat by comparison.

## Recording

In September of 2018, John Luther Adams and I traveled to Seattle to record the Seattle Symphony's performance of his new piece *Become Desert*. Conductor Ludovic Morlot and recording engineer Dmitriy Lipay understood the piece completely and were eager to produce a follow-up to the Grammy-winning *Become Ocean* that we had done three years earlier. *Become Desert* features five separate instrumental ensembles. The staging is described in the notes to the score as follows:

> *Become Desert* completes a trilogy with *Become River* and *Become Ocean*. As in both of those earlier works the physical array of the instruments is a fundamental compositional element of this music. The orchestra for *Become Desert* includes five separate ensembles:
>
> - Woodwinds and crotales
> - Horns and chimes
> - Trumpets, trombones and chimes
> - Voices and handbells (or vibraphone)
> - Strings, harps and percussion
>
> These five ensembles should surround the audience.
>
> The strings, harps and percussion are on stage. The other four ensembles are elevated on high risers, or in lofts, boxes or balconies, around the house.
>
> - The strings are seated or standing, dispersed throughout the stage. The harps and percussion are interspersed among the strings, widely separated from one another, and slightly elevated.
> - The woodwinds and crotales are elevated at the front of the house, above the strings.

- The horns and chimes are elevated on one side of the house.
- The trumpets, trombones and chimes are elevated on the other side of the house.
- The voices and handbells (or vibraphone) are elevated at the rear of the house.

John writes above, '[T]he physical array of the instruments is a fundamental compositional element of this music'. John is not the first composer to regard the staging as fundamental to the composition, but in many interviews he has described his compositional process as *starting* with the instrumentation and its staging before moving into the actual notes and musical shapes.

On the day before the first day of recording, Dmitriy and the stage manager were arranging microphones in Benaroya Hall and asked for my input. This was years before any truly immersive music format would be available to consumers, but I knew that John would ask for the recreation of a truly three-dimensional sound space during the mix. The score called for the low brass and chimes to be heard directly to the left and right of the listener – locations that were a painful weak point in 5.1. And not only that, the score also defined different *elevations* for the ensembles, something that was nearly impossible in the flat plane of 5.1. I asked Dmitriy to hang as many additional mics as possible. Not because I knew exactly what I was doing, but because it would give us flexibility if we were to mix *Desert* in Dolby Atmos. We recorded for several days and nights and left Seattle with over 130 tracks to edit and begin the mix with.

In January 2019, I accepted an invitation from Ceri Thomas to bring the *Desert* mix to a full-spec, proper Dolby Atmos mixing studio. I brought my hard drives, scores, stage layouts, and recording notes and attempted to recreate the sonic experience of being in Benaroya Hall with Seattle Symphony in the studio. I was already an established music producer, and I knew how to make professional stereo and 5.1 mixes. But my early efforts in immersive mixing fell flat, nevertheless.

## Panning

In stereo, the all-important spaced pair located above and just behind the conductor is the anchor around which all the other orchestral microphones are placed. I've mixed albums in which those two microphones are 70% of the sound, despite having an additional (40) microphones to work with (see Figure 22.1).

I tried mapping the microphones in the hall to the equivalent locations in Atmos space, thinking that a precise recreation would translate the hall sound into the studio. Starting with the main spaced-pair, I set those mics as beds and panned them halfway back and toed-in (Figure 22.2). This correlated to their physical position in the hall. The sound was cluttered and a little vague.

Thinking that width was the problem, I pushed the main pair out to the side surrounds (Figure 22.3). In nearly every stereo production I've worked on, engineers strive for increased width. We have a litany of techniques and subtle hacks to open the stereo image and give listeners the impression that the sound field is wider than the two loudspeakers in front of them. In Atmos, with the main microphone pair pushed to the hard left and right surrounds, I had a sound that was way too wide. There was no field to speak of; it was as though I had pasted one sound on the left wall, and the other sound on the right wall. It was comically wrong.

The only real solution to placing the main pair was to fall back on conventional stereo techniques and push the microphones to the front left and right channels (Figure 22.4).

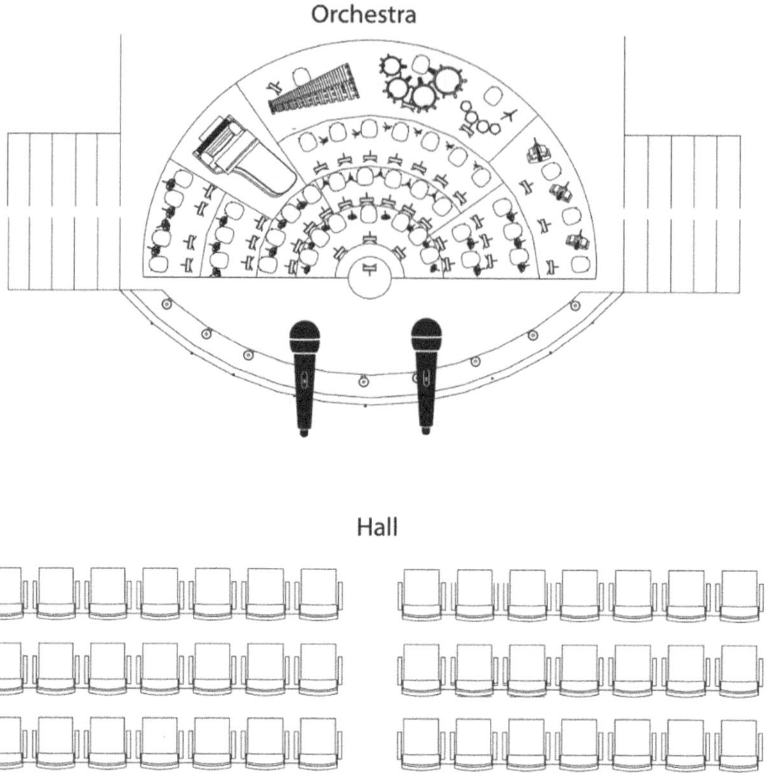


*Figure 22.1* Configuration of the spaced microphone pair

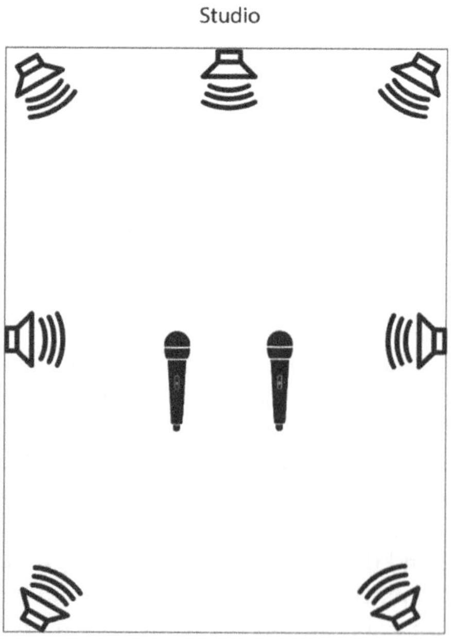


*Figure 22.2* Microphone mapping in a studio – setup A

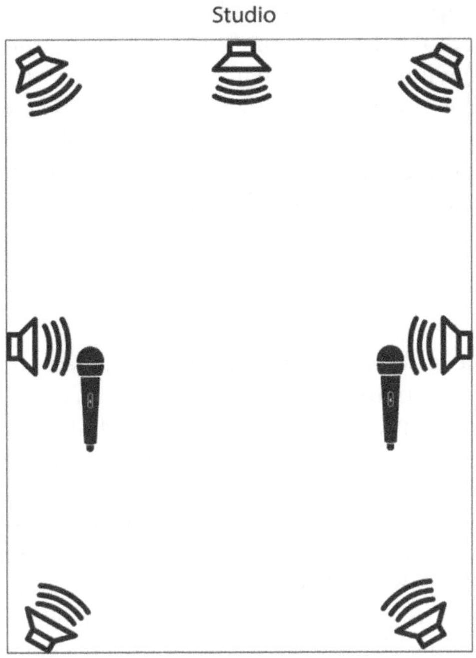


*Figure 22.3* Microphone mapping in a studio – setup B

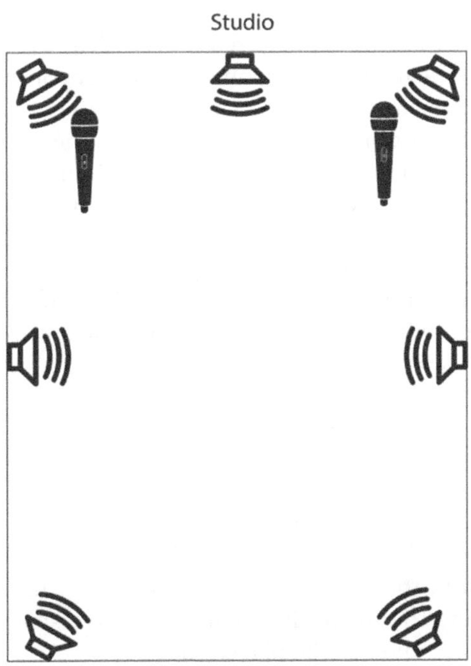


*Figure 22.4* Microphone mapping in a studio – setup C

Disappointingly, this compressed the space along the Z-axis that the instruments on the stage would occupy, and, to make the sound right, I had to time-delay the section microphones (i.e., percussion, winds, etc.) to match the distance of the main pair. Immersive recording techniques have improved considerably in the last few years, and one notable engineer, Morten Lindberg, has a cube design for microphone placement which is one solution to the Z-axis problem noted here. Fortunately, Dmitriy and I had additional microphones further back in the hall which helped us with the sense of depth along the Z-axis.

## Objects

One of the first experiments I made in the Atmos studio was to take dry, mono spoken word and pan it around the room as a bed and again as an object. In bed mode, the voice moved around the room in the familiar, broad way that a stereo panner moves. The energy was widely dispersed, and the localization was smooth but pretty vague overall. In object mode, the localization tightened up in a remarkable way. The sound seemed to hover in space, and it was much easier to point at exactly where in the room it was emanating from. The trade-off for this spatial accuracy seemed to be a subtle shift in the timbre of the voice that I attributed to the psychoacoustical algorithm running in the renderer. Dolby has continued to work on their renderer in an effort to minimize this timbral artifacting while maintaining the spatial accuracy of the rendered objects. More recent versions of the Dolby Atmos renderer exhibit less of this behavior.

Wondering if my main spaced-pair over the orchestra would benefit from increased spatial accuracy, I switched both microphones to object mode. This instantly wrecked the sense of physical space I had been hearing in this pair, and the timbral shift gave my previously silky-sounding strings an unattractive grainy quality. It was obvious that this pair had to remain in bed-mode for the album. After additional experimentation, I realized that the object option is meant to take discretely recorded, usually mono sound sources, and place them in three-dimensional space on loudspeakers or in synthetic binaural space on headphones. A spaced microphone pair in a concert hall is a natural binaural recording method. Making those tracks objects superimposed a synthetic (i.e., fake) binauralization over the naturally occurring binaural sound. In general, Atmos technology is exceptionally good at taking many disparate recordings and merging them into one perceived acoustical space. This is well-suited to pop music and the sound-effects tracks in a film. But when all the tracks have already been recorded in an existing real acoustic space, some of this technology is unnecessary, and, at worst, damaging. Fortunately, engineers can use the conventional bed channels when needing to preserve a natural acoustic.

## Decorrelation

By the time I had gotten to the Atmos studio, I already had a good rough stereo mix of *Desert*. Not all orchestral recordings need artificial reverbs, but *Desert* did. A small amount of artificial reverb helped blend the space between the ensembles, and it also gave me a margin for level changes throughout the piece. These level changes may have sounded unnatural without the added reverb. The reverbs I utilized were stereo-only, so the next step after placing the microphones in immersive space would be working out how these stereo reverbs would also be placed in space.

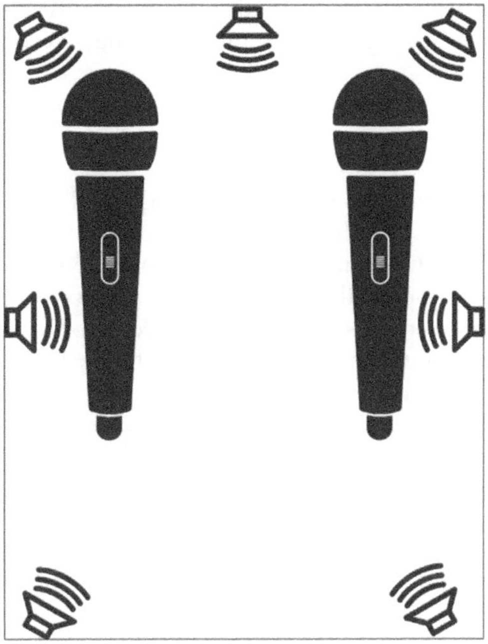


*Figure 22.5* Microphone mapping in a studio – setup D

Simply panning the two channels to the front or back or overhead loudspeakers left the rest of the room empty. But the Dolby renderer has a size parameter that increases the area where a single channel is coming from, bleeding that sound into the surrounding loudspeakers more and more as the size increases (Figure 22.5). I imagined this would be an easy way to translate my stereo reverbs into immersive.

I was wrong. Increasing the size parameter to any reasonable extent caused large areas of unappealing dead space in the immersive field. The effect is hard to describe, but even on the best-calibrated systems, the result felt unnatural and strangely flat. After additional experimentation, I realized that it was the increased signal correlation causing the 'dead space' sensation. In the real world, sound sources scatter and bounce off the objects around a listener in wild ways. This experience can be measured and thought of as highly decorrelated. The design of good recording studios and concert halls are purposefully asymmetrical, and avoid parallel walls or perfectly circular enclosures, designs which would increase correlation. So while I could make a stereo reverb play on most of the speakers in an immersive mixing studio, the result was awful. I quickly came to realize the benefits of this new generation of immersive – and highly decorrelated – reverbs. As of this writing, there are very few conventional immersive reverb plug-ins or applications. My hope is that programmers continue to fill this market space with new designs.

Interestingly, the correlation problem has been with us since the emergence of stereo. But when mixing a stereo program, increased signal correlation between the left and right loudspeakers has come to be heard not as a problem but as a more focused 'phantom center'. Phantom center is what we say to describe this very familiar sound. Blowing that sound up and onto three or more speakers, however, loses any appeal it may have had in stereo. Moreover, poorly calibrated loudspeaker systems can easily hide correlation problems. In these early years of immersive music mixing, excess correlation is a problem I hear frequently.

## Bass symmetry

John Luther Adams's aspiration to saturate the tonal field led us to make recordings in which instruments and voices are heard in their widest possible frequency range in any location in space around a listener. But before immersive sound technology, there was no guarantee that deep, room-shaking bass frequencies could be heard evenly around the room in 5.1. The inherent design of 5.1 favored the proscenium staging of classic theater and conventional film. Big sounds were up-front. Lighter, more fleeting sounds were occasionally heard behind or to the sides of the listener. 'Bass management' is the technical term to describe how economical loudspeaker systems share a subwoofer in order to reproduce a wider frequency range than the individual physical cabinets would otherwise be capable of. And while bass management is still technically in use today in most playback systems, the recent immersive playback specifications (such as the specification for Dolby Atmos), require completely even frequency response from any location in three-dimensional space. I like to think of this as bass symmetry.

## Listening

All of this technology allows performers, producers, and engineers to more fully realize a composer or a conductor's artistic intent. The final mix of *Become Desert* featured the ensembles completely surrounding the listener. I was very pleased to hear the winds and crotales slightly above me, the growling horns and chimes to my left and right, and the choir and handbells behind me, just as I had heard them in the concert hall. The result was richer, more powerful and more emotional in immersive playback compared to stereo playback. Stereo came to feel more and more like a small landscape painting hanging on the wall in front of us compared to the complete envelopment of immersive. And while *Desert*, with over 90 musicians, was designed for a three-dimensional listening experience, the more recent recording of Adams's *Waves and Particles*, for only four musicians in a string quartet, benefits hugely from immersive recording and mixing as well. In stereo, the string quartet sounds like it's coming from loudspeakers. In immersive, the string quartet sounds like it's in front of you. As my colleague Dmitriy said when we started the mix for *Desert*, 'Let's try to make the loudspeakers disappear!' That's been the goal all along.

Chapter 23

# Music composing for immersive experiences

*Piet Goddaer*

I stumbled into immersive sound quite by accident when I was invited to collaborate with the National Orchestra of Belgium on a project featuring a 90-piece ensemble. As the son of a classical composer, this was a remarkable chance for me to honor my father's teachings on composition, harmony, fugue, and counterpoint.

This opportunity allowed me to incorporate these principles into my score and showcase my understanding of the orchestra as a cohesive instrument. My father always believed that a well-crafted score reveals its own quality, so I put significant effort into ensuring my score was exceptional.

During rehearsals, the orchestra was arranged differently due to the unique layout of the room. Instead of the usual configuration, I found myself positioned among the strings, rather than in front of them. This setup was dictated by the room's small, narrow dimensions, which required a different arrangement from the norm.

## Sensation

The experience was profoundly impactful for me. The sensation went beyond just hearing the sound; it involved physically feeling the vibrations and the warmth of the room. The dynamic range was so intense that you could feel it in your body, and the richness of the sound was far beyond what you'd get from a standard recording. It was an immersive, uncompressed auditory experience.

I was determined to capture this in the highest resolution possible. I discussed this with Ronald Prent and Darcy Proper, with whom I had collaborated previously. We explored how to present this high-resolution experience to an audience, given that at the time, the only medium available was compact discs. The challenge was figuring out how to maintain the quality without compromising it for the limitations of existing technology.

They suggested recording in surround sound, which offered more spatial depth. However, I had never been particularly fond of the surround format; it often felt more like a gimmick than an emotionally immersive experience. This likely stemmed from how surround sound was often used to highlight technical possibilities rather than enhance the emotional depth of the music.

The next step was immersive sound. At that point, all I knew was that it involved more channels and higher resolution and that it promised significantly better sound quality with increased spatial depth. I understood that this would allow for more precise placement of sounds, but I had little grasp of how it would affect the listening experience. I hadn't yet considered aspects like directional cues, the impact of reflections, or how

DOI: 10.4324/9780429297526-26

the experience would differ from stereo sound. My primary expectation was simply a substantial improvement in sound quality.

## Emotional impact

Because I had the full cooperation of the orchestra, I promised to invite them for a playback in the rehearsal space whenever we finished a project. When we organized the playback in the rehearsal space, which had excellent sound, I noticed that most of the people who had worked with the orchestra for many years, some of whom were students of my dad, were quite emotional about it.

I was puzzled by the strong emotional reactions from others, who described experiences like, 'It becomes physical, and I recognize this space'. This sparked a curiosity in me: why was it so moving for them? What was making it different? They attributed it to recognizing the space, which led me to question how the brain processes these experiences.

Understanding this could provide me with an additional tool as an artist in creating music. This curiosity became the driving force behind my desire to explore immersive sound more deeply and see why it could be an ideal medium for my creative work.

I sincerely believe that for those who are experiencing immersive sound, it's a double-layered experience: there's the experience of what you're actively listening to, and there's the experience of what you're passively listening to. As a musician, most of the time, I'm outside the experience. I prepare everything, but when it happens, I sit outside the experience. This is necessary because you need to be in a place where your microphone doesn't pick up what's being played back. You're not in the same experience, but you see that people who are exposed to the experience have a double-layered experience.

## Space

When explaining immersive sound, I often reference the experience room in my studio. While my studio has a 7.1.4 setup, the experience room is a simple, open space with a roof to keep out the rain and equipped with a 4.1.4 speaker setup. The space is designed so that sounds from outside, like passing cars, are heard as if you were actually on the street, with no filtering from nearby sounds like those from the garden or street.

Sometimes, people are skeptical about immersive sound, especially if they haven't experienced it firsthand. Those who are most resistant often haven't encountered it in practice. In corporate settings, I've dealt with skeptics like the chief marketing officer (CMO) or chief financial officer (CFO) who view it as just another technology to dismiss.

To address this, I set up by recording sounds from my garden and the street, then play these recordings as they enter the room without revealing they're listening to a recording. As we sit down, I explain that my work with immersive sound aims to replicate the quality of how we experience the world. I use music and field recordings to create an engaging and dynamic auditory environment.

I then ask them to observe the space, noting its minimal windows and other features. I stop the playback and explain that the sounds they heard – representing the street and garden – are no longer present. This shift in perception helps them immediately grasp what immersive sound is all about.

To explain my work, I first help people experience how they are always passively listening to their surroundings. When you walk through a garden or forest, you're enveloped by natural sounds. If those sounds suddenly disappear, your senses become heightened. This is what I call passive hearing – you're constantly scanning your environment. Our brains are always on alert for unusual sounds or silences, which triggers other senses to investigate. This awareness adds an extra layer to the auditory experience.

In my immersive sound projects, I design experiences that engage both passive and active listening. By subtly shifting the environment to capture attention, I elevate the active listening experience. This often makes music and sound feel more physical, emotional, and vivid – an intriguing aspect of artistic expression. I also strive to synchronize the sound with the environment, blurring the lines between what's generated and what's naturally occurring. This interplay creates a unique experience that stereo sound simply can't achieve.

First and foremost, I focus on the space and the message that needs to be conveyed. For example, one of my top clients is a champagne brand. My task is to translate the story of their champagne into an immersive experience. This involves two aspects: the synesthetic translation of taste qualities into sound, and the narrative of how the champagne is crafted.

The key elements to incorporate are nature and time. For instance, understanding the Champagne region's environment allows me to transport people to that place. If it's a cold year, I consider how the world sounds when it's cold. These elements help shape the immersive narrative, reflecting the sounds of a specific place.

Even in familiar or defined locations, it's fascinating to work with the space's real acoustic reflections. People often recognize these sounds, and integrating them into the music can enhance the experience. Color also plays a role in this process.

Just as an artist considers the canvas before painting, I start by evaluating the space. The size of the canvas dictates the scale of the image. Similarly, the space determines how I approach the creation of the immersive experience. If the space is large and the canvas is small, the elements within must be appropriately scaled.

First, I assess the space. What are its dimensions? Is it expansive or confined? Is it a natural environment, a tense one, or perhaps an abstract setting? Once I understand the space, I choose the appropriate instruments. I consider which elements will complement or contrast with the space. If the goal is to evoke a natural feel, I select elements that blend seamlessly with that environment. Conversely, if the intention is to create contrast, I identify elements that starkly differ from the space's characteristics.

## Cultural connections

Working with clients from diverse cultural backgrounds, such as those in Japan, Korea, China, or Italy, has shown me the varied perceptions of sound across cultures. Despite these differences, I've found that if I accurately define and design the space for my work, people universally feel that I've captured the essence perfectly. When the space is right, the story I need to tell comes across effectively.

I think it has to do with the fact that I didn't just get stuck on taste. Just as people choose cough syrup based on its effectiveness rather than flavor. It's about the medicine, not the flavor. My primary focus is ensuring the core experience aligns with the environment and narrative.

When designing projects, I consider both the cultural context of the audience and the specifics of the location. For instance, in the desert of Marrakech, where sound reflections are minimal, I relied on the natural environment – primarily the local wildlife. I incorporated frequencies that the animals respond to, which are often inaudible to humans. This approach created a unique and immersive experience, as the animals reacted subtly, either becoming quiet or agitated, much like a dog whistle. The absence of reflections in the desert made these interactions feel organic and seamlessly integrated into the environment.

In Tokyo, I created an experience set in a park near an underground metro line. The metro's braking, which produced a distinctive harsh sound, was consistent and predictable. I timed this sound to synchronize with the music in my project, using the easily recorded sound to enhance the experience. We were in a park where the public usually isn't allowed, making it a special, almost sacred place for locals. The familiarity of the metro sound, despite its harshness, resonated with them. Visitors from France and Belgium initially found the sound intrusive, but by incorporating and acknowledging it, I provided a deeper connection to their environment. The most significant feedback was that recognizing and integrating local sounds made the experience feel personal and meaningful, making the space feel more authentic and connected.

Incorporating the environment into music is essential for creating a compelling experience. To me, music and environment are interdependent – one enhances the other. Just as scenery sets the stage and actors bring a story to life, instruments and music act as the narrative elements. They interact with the environment to tell a story, shaping the experience to fit a specific place and time.

When I began working on immersive experiences, I collaborated with a brand focused on creating memorable moments for people. Initially, I thought such moments happened by chance – like an ordinary day that turns into an unforgettable evening through chance encounters and great conversations. However, I soon realized that with the right canvas, one can intentionally craft a space with a specific quality. This allows for directing and creating a moment where a story comes to life, blurring the lines between reality and illusion, and making the experience intensely memorable.

The process of creating music and designing sound within a space can be compared to a dance. It involves balancing the need to engage and distract the audience while maintaining a dynamic flow. When successful, the result might not seem spectacular at first, but it resonates deeply, like a thunderstorm that captures your attention and evokes vivid memories. This kind of immersive experience captures a moment in time, leaving a lasting impact.

Sometimes, you select colors that evoke specific feelings, times, or cultural associations. These choices can include particular rhythms – whether consistent or unpredictable – that align with the journey or narrative you're creating.

Reflecting on my first travels, particularly to Mumbai, the most striking element was the unique soundscape. The noise in Mumbai, with its overwhelming density, creates a rhythm distinct from anywhere else. What's the rhythm of traffic? What rhythm are we sensitive to? Even close to the street, the sound is different from what you'd experience elsewhere, influencing how you perceive rhythms and auditory patterns. Capturing this rhythm, even in something as simple as a pop song, and placing it in a specially designed space can evoke memories and immerse you in the moment, creating a deeply introspective experience.

## Formats

Previously, working with sound meant fitting it into a constrained format, like a stereo CD. This often involved compressing and equalizing, which felt limiting and distorted the original experience. I missed the silence and the uncontextualized space it represents – something that is often lost when compressing sound into a small format.

With immersive sound, however, I found a new way to work. Instead of translating and compressing sound into a limited format, I could create a more expansive and accurate representation. It felt like stepping into a new space where the music remained true to its essence. This approach allowed me to preserve the integrity of my initial vision, avoiding the sensation of losing elements as I adapted to different formats.

For me, it is much more natural to work in immersive sound. You get in a room. You get on stage. There's a piano. You sit at the piano. You play one note . . . and the room is talking to you. In immersive sound, that interaction remains intact. Unlike traditional recording, where the essence of the room often gets lost once you start capturing and sharing the music, immersive sound preserves the room's presence. The way you play – whether with purity, softness, or dynamism – is influenced by the room's response. The space itself guides your performance, allowing you to express your composition more authentically, as the room is directly talking to you.

When you have a great piano in a great space, the sound naturally enhances the music, making it easier to perform. You can intuitively sense when everything is right or off. However, once you remove the space, the music might seem lacking. For instance, long, beautiful reflections that enhance the rhythm in the original setting may disappear, leaving the music feeling rhythmically empty. This illustrates how crucial context is: without the space, elements that once made perfect sense can appear out of place.

When I create new music, I do so within an immersive context, as I've described. Sharing music today still requires a stereo master, which I handle separately. This approach allows me to complete the work in its full immersive form before making any necessary translations. It's akin to creating a painting and then taking a photograph of it; the original work remains intact and unaltered. The immersive experience is preserved, ensuring that nothing is lost or diminished in the process.

Sometimes, you're satisfied with how a picture represents the original work, while other times it falls short. Yet the original work remains intact. Previously, I felt frustration because the shared version never fully captured the essence of the original. I would question whether the translation was done well and doubt whether I lost control over the process. This led to uncertainty about whether I could trust my own ears. With the current approach, those doubts have diminished. Now, when creating, I don't grapple with those issues as much; instead, I focus more on the context and the work itself.

## Directions of immersive sound in live performances

I believe it would be fantastic to translate the immersive experience created in the studio into live performances, benefiting both the performer and the audience. However, the financial and commercial realities where I live present challenges. In Europe and in many venues I encounter, there's limited interest in this approach. Agents often fill small arenas or venues of 2,000 to 7,000 seats with traditional setups, and there's little incentive to change because many aren't aware of what they're missing.

In 2024, during my tour, I made a point of engaging with the audience directly to share the experience and gauge their reactions. My goal is to bring the immersive quality of my recordings into live performances. I believe doing so would enhance the impact of the performance, allowing for a more memorable and emotionally resonant experience.

People attend concerts because they share a love for certain music, and being among others who appreciate the same sounds creates a unique connection. Experiencing that shared affinity in a tangible, physical way enhances the overall experience and deepens the connection. Recently, someone mentioned that, while the concept is intriguing, its inability to scale up might prevent it from becoming mainstream. Achieving the same authentic sensation on a larger scale does have limitations, but that's perfectly okay.

The quality of the work, the depth of the experience, and the intensity of the emotion constitute your cultural capital. When attending a concert or listening to music, you seek to enhance your cultural capital by connecting with others who share your taste and pursue similar experiences. This mutual recognition and enrichment of cultural capital is at the heart of these shared musical moments.

When working with a 360° environment, I experiment with how sounds move and interact. For example, many clubs where I began playing had poor acoustics, with sounds like a snare drum reflecting harshly off concrete walls. I enjoy using these familiar elements because they resonate with people's everyday experiences. Placing sounds behind the audience, for instance, can create a sense of safety.

I also strategically position vocals. If a vocal is central to the experience, I place it forward and in the center of the mix, then set sounds around it. Rather than simply placing instruments in specific locations, I focus on how each sound evokes certain feelings and how these sounds can move or connect within the arrangement.

The advantage of immersive sound is that it naturally creates a distinct space for each sound, reducing the need to push elements forward. This simplifies both the listening experience and the arrangement process, unlike stereo setups where you often have to guide the listener's attention to the main theme and emphasize certain elements more.

## Designing immersive sound for champagne

One interesting project I undertook was for a champagne brand. To start, it's important to understand the intricacies of their champagne. This particular champagne is crafted from grapes harvested from distinct plots on the brand's property. The champagne is made from three different wines. Each plot yields a different wine due to variations in soil, sunlight, and other factors, resulting in a unique taste for each component of the champagne.

The champagne brand I worked with harvests each plot separately and keeps the wines distinct. To ensure they can produce the highest quality champagne annually, what they refer to as the 'fullest expression of champagne', they maintain a comprehensive library of wines. This includes both the current year's wines and those from previous years. If a challenging harvest results in a wine not meeting their quality standards, they can draw from their library of older vintages to find a suitable match.

The process of making a particular Champagne often involves blending wines from various years. For instance, the Grande Cuvée consists of about 120 different Champagnes, with the oldest and youngest vintages spanning 21 years or so. This blend creates a complex profile, much like a musical score where different musicians – both seasoned and young – contribute unique qualities. Together, these elements produce what is considered the 'fullest expression of Champagne'.

To translate this into music, imagine creating an expansive soundscape with the complexity of an orchestra of 120 musicians, each showcasing unique qualities. Some might exhibit a rich, velvety sound, while others may convey a more mature or brisk tone. By arranging these diverse elements within a timeline, you craft a cohesive musical piece that mirrors the essence of the Champagne. Each sound element represents a different facet of the Champagne, coming together to form a harmonious and evocative representation.

Sometimes it takes a long time to achieve the desired result, and other times it may be just a year or even a single bottle of Champagne. Collaborating with the tasting committee, you analyze their feedback on the flavors they feel are missing. Then, you reflect on how these missing elements relate to the music you've created. You might ask, 'Which musician embodies this flavor?' This process deepens your understanding of both the Champagne and your music, allowing for a more nuanced and cohesive final product.

In this process, the composition evolves, taking on new forms and sometimes changing entirely. Elements that were once central themes may shift to supportive roles as the true theme emerges. The goal is to enhance the tasting experience, which is validated through scientific analysis. This involves not only assessing if the composition aligns with our expectations but also verifying if it actually improves the tasting sensation. When the translation is done correctly, it does enhance the overall experience.

For me, this body of work has been immensely enriching and enjoyable. It's taught me a great deal, especially as I learn from the complex craftsmanship of winemaking, which requires harmony with nature despite numerous uncontrollable factors. This experience profoundly influences how I approach music. When creating a score, envisioning it precisely can shift when someone introduces a new phrasing. Adapting fluidly and reinterpreting the work in an orchestral context is fascinating. It helps me recognize and appreciate qualities I might otherwise overlook.

## Lessons learned

As an artist, I've learned several valuable lessons, with the foremost being the importance of listening. Midway through my career, I discovered that, despite being a professional, my listening skills had room for improvement. Much like a chef who continuously tastes their creations, we must never stop listening, even if we feel confident in our abilities. Sensations evolve, and staying attuned is crucial to growth.

The second lesson I've learned is about connection. If your work doesn't establish a meaningful connection, it lacks substance for me. It's akin to preparing food that can't be eaten – without connection, it's as if it doesn't truly exist. A body of work must forge multiple connections; it can't be limited to just one-on-one interactions. If you play music and it fails to resonate or hold meaning, it feels as though it's not truly present. While language is a tool for communication, making a connection doesn't necessarily mean that your audience fully understands your intentions – understanding is a deeper, separate layer.

The third lesson is to never underestimate the listener or your audience. Don't make assumptions about what they will or won't appreciate, even though commercial interests often shape these perceptions. I've learned that the image of your audience defined by others is rarely accurate. People are generally more perceptive and open than anticipated. Every piece of work has its own value, regardless of personal preference. What resonates most with me isn't necessarily the best music – it's simply the best for me, which is a distinct perspective.

There's always more to learn, and feeling like you're constantly running out of time often means you're on the right track. To ensure you're making a connection, stay attentive to reactions and continuously listen to your work. Sometimes revisiting the same piece can reveal new insights and deepen your understanding.

My metaphors often revolve around food. Take ramen in Japan, for example. Mastery comes from repeatedly refining the same process, continually improving. This act of listening and refining teaches a lot. You don't have to endlessly tweak everything, but understanding that there are countless possibilities and variations can be overwhelming. It's about recognizing choices and the richness they bring.

One last, important lesson is learning to let go. Once you create something, you must allow it to exist on its own. It's natural to look back after some time and think, 'If I were to do this now, it might be better or different'. This is part of the creative process and reflects your growth and evolving perspective. However, this doesn't mean the work wasn't valuable or shouldn't have been shared when it was.

Letting go means accepting that your creation is a snapshot of your skills and ideas at a particular moment. It's essential to recognize that perfection is a moving target. Sometimes, the act of sharing and releasing your work is as important as the work itself. It allows you to engage with an audience, receive feedback, and continue evolving. The work exists in its own right, regardless of how you might perceive it in retrospect. By letting go, you free yourself to move on to new projects and ideas, knowing that each creation has its own place and purpose.

Chapter 24

# Immersion in 3D audio music productions

*Lasse Nipkow*

Creating an impressive immersion, i.e., feeling enveloped by sounds, is an overarching goal of 3D audio music productions. Lasse Nipkow shows this with two productions: Silent Work 'Pads Of Humility' (Electro Pop) and Shostakovich Symphony #10 (live recording from February 11, 2023, at Amare, The Hague).

On the one hand, the song 'Pads of Humility' consists of a complex soundscape comprising many synthetically generated sounds. On the other hand, the microphone system, Spatial Direct Sound Microphone Array, is used for the vocal recording to create envelopment with vocals. The recording of the Shostakovich Symphony #10 consists of 110 microphone signals. The unique feature is that 45 instruments are stereophonically captured: in the mix, the strings are placed around the audience, and their sound leads to intense envelopment – several instruments with the same sound play identical melodies simultaneously.

## Case study: 3D audio vocal recording – 'Pads of Humility'

### *Introduction*

In this case study, the author presents an innovative 3D audio vocal recording of the song 'Pads Of Humility' by the duo Silent Work. The production is part of the album '*Sonic Leaks*, 'released in December 2022 as a Pure Audio Blu-ray in Dolby Atmos and Auro-3D. The vocals were recorded using a five-channel Spatial Direct Sound Microphone Array, with the singer directly singing into one of the five microphones and passing by the other four nearby. The strategy behind this microphone technique is that the voice sounds similar in all microphones, and our hearing perceives it as a single sound source, even when the loudspeakers reproduce the signals from different directions in the playback room. This creates a more balanced perception of the vocals in the playback room, regardless of the listener's position. The Roomenizer plug-in by Pinguin is used to add a 3D audio convolution reverb as a second layer to the vocals, providing liveliness without the listener being able to pinpoint the source of the effect.

### *Background*

At the 2021 Tonmeistertagung in Düsseldorf, I met Patrick Schappert, the CEO of GROBI.TV. He gave a presentation on the topic 'Everyone Talks about 3D Sound Formats – What Has Happened in the Last 24 Months?' Patrick built high-end home cinemas

DOI: 10.4324/9780429297526-27

with 3D audio sound systems and noted a deficit in music content: there are few productions with convincingly sounding music in 3D audio. During subsequent discussions, I introduced him to songs by my music partner Olaf Strassen, and we decided to collaborate to produce a Pure Audio Blu-ray with songs in Dolby Atmos and Auro-3D. Since 2015, Olaf has been creating songs in 3D audio to integrate psychoacoustic phenomena using synthetic sounds in song fragments, which I present at international 3D audio workshops. One of these songs is 'Pads of Humility'. This song aims to create envelopment using pad sounds. The original version contained a sample voice from a library. Therefore, we engaged singer Anna Smith for the version on the album and recorded her vocals on June 15, 2022, at Olaf's studio.

### *Challenge of vocal recording in 3D audio*

Since the introduction of surround sound for music productions about 25 years ago (with the introduction of Super Audio CD in 1999), there has been a lively debate about whether lead vocals should be reproduced as phantom sources (L-R) or real sound sources (C). On the one hand, real sound sources sound more natural, but on the other hand, we are used to the somewhat duller sound of vocals as phantom sources with their wide spatial spread in stereo productions. However, phantom sources have a significant disadvantage: they are not stably localizable. The image becomes distorted when a listener moves away from the stereo symmetry axis. Therefore, it is advisable to include the center channel for lead vocals. The question is: How should an LCR signal be designed to combine the respective advantages of phantom sources and real sound sources?

Another challenge arises with songs for 3D audio that have only one lead voice: if only the center speaker reproduces the lead vocals, the balance between the lead voice and the rest of the music depends on the listener's position in the room. This is because the loudspeakers reproduce the music not only from one direction but from many directions. Near the center speaker, the lead voice is too loud, while in the rear third of the playback room, the lead voice is too soft compared to the music. How can signals for the rear speakers be created to achieve a balanced lead voice between the front and rear?

Finally, another question arises in the context of 3D audio: when a singer sings in a real room, such as an acoustically exceptional concert hall or a large church, the voice sounds much livelier due to the acoustics than the signal from a lead vocal microphone. Artificial reverberation usually does not create this impression in 3D audio playback. This is due to artifacts such as metallic-sounding signal components that make the room sound implausible and makes the sound seem absent of movement. How can such artifacts be avoided, and how can movements be created through room sound for the vocals?

## ***Vision and solution***

I made the following observation from numerous 3D audio recordings in churches and concert halls. A largely homogeneous sound field is established when multiple loudspeakers reproduce room sound from different directions with similar levels into the playback room. This means that a listener can freely move around in the playback room and have the impression of being in the original recording room. The higher the number of loudspeakers in the playback room reproducing room sound, the smaller the distance at which individual loudspeakers can be recognized.

Producers of vocal recordings for pop music usually do not record lead vocals in concert halls or churches but in studios with significantly smaller room volumes. However, room microphones in small rooms result in a high density of early reflections due to the proximity of the boundary surfaces. Our hearing interprets signals with a high proportion of early reflections as sound sources from a large distance, which differs from what producers and listeners expect from pop lead vocals. One solution is to capture direct sound instead of early reflections or reverberation, adhering to the following principles: the direct sound reproduced by multiple loudspeakers from different directions should not be identical to the direct sound from the center channel; that would lead to localization errors and comb-filter effects. In other words, the correlation between the signal for the center speaker and the other loudspeakers should not be equal to one, especially for mid- and high frequencies, but rather be around zero. Low frequencies are less critical as they contribute little to localization sharpness.

Furthermore, the signals should sound similar to allow fusion (Moore & Gockel, 2012); otherwise, the listener perceives not one but multiple lead voices simultaneously. Finally, it must be ensured that other signals do not compromise the direction of the lead singer. Since high frequencies play a significant role in direction perception, the direct sound from the center speaker should sound much brighter than all other loudspeaker signals.

### *Spatial direct sound microphone array for 3D audio vocal recording*

In 2016, I thought about how to create width and height for a lead voice using a multichannel vocal microphone system. The setup corresponds in proportions to the front of the Auro-3D speaker setup, with the side microphones positioned at approximately 17 cm from the center microphone. The center microphone was a Neumann M 149 Tube, alongside 2 × Neumann U 87 i and 2 × Neumann TLM 103. The idea was to introduce slight differences between the microphone signals using different large-diaphragm microphones. Together with my music partner Olaf (Silent Work), we created vocal recordings with singer Marena Whitcher for experimentation. Marena sang into the center microphone. The focus of the recording was to capture signals for the front loudspeakers. Upon analyzing the signals, I found that the four microphone signals, where Marena sang past, sounded similar but not identical; the higher the frequency, the more the correlation went to around zero. Despite the microphones being only about 10 cm from Marena's head, the signals sounded comparable to early reflections without creating a sense of distance.

In another experiment, I used the signals from the upper microphones for the surround rear speakers. Due to the similarity in sound, the side microphone signals lead to auditory connections between the loudspeakers reproducing the signals. The center microphone signal has two notable characteristics: first, it stands out distinctly because the central microphone signal's high-frequency content is significantly higher than in the other four microphone signals, resulting in fission at high frequencies (Moore & Gockel, 2012). This aspect leads to a stable sound source localization from the center speaker during playback. Secondly, the correlation is high for mid- and especially low frequencies, as all microphones are near each other, and the wavelength becomes longer as the frequency decreases. Therefore, those signal components sound very similar and contribute to fusion. Although the microphones were positioned within the critical distance in a dry studio room, our hearing interprets the signals from the four side microphones as room sound due to their similarity, even though they primarily consist of direct sound.

### *Using room impulse responses for vocal recordings*

As described previously, exceptionally sounding real rooms add signal components to a vocal track that our hearing perceives as lively and appealing. Due to the need for more suitable tools, it appears nontrivial to accurately replicate these components for artificial reverberation. Therefore, the following approach aims to achieve a realistic reproduction of room sound: Room impulse responses represent an acoustic fingerprint of a room's acoustics (Kessler, 2005). In other words, room impulse responses capture all auditory phenomena occurring in the room and transfer them to the playback room. It is not necessary to understand how the auditory sensation of liveliness is generated; instead, it is essential to hear in which presets of impulse response reverberation the liveliness is present.

At the AES Convention 2022 in The Hague, I had a long conversation with Ralph Kessler, the CEO of Pinguin, about his plug-in Roomenizer. We agreed to collaborate, where I would record impulse responses from various excellently sounding concert halls and churches. A unique feature of this plug-in is its ability to provide many output channels with impulse response reverberation, which was crucial for producing the song 'Pads of Humility', as the maximum format was intended to be Dolby Atmos 9.1.6.

### *Technical implementation of the vocal recording*

We recorded the vocals for the song 'Pads of Humility' on June 15, 2022, with singer Anna Smith in Olaf's studio. The recording room has a volume of approximately 36 $m^3$ (4.5 m × 3.8 m × 2.1 m) and is designed to have diffuse and absorptive elements. I installed four video cameras in the recording room to film the recording session for documentation purposes. Reto Fausch oversaw the recording, and Olaf handled communication with Anna. Additionally, Patrick and Jan Schappert created documentary videos for social media.

We set up the microphone array using the same microphones and arrangement as in the 2016 recording – all microphones were set to a cardioid polar pattern. For the positioning of the microphones, we used standard components (two standard stands, a table clamp, and several stereo rails) from König & Meyer, assembling the array in the center of the room. We placed a pop filter in front of the center microphone. We used Focusrite RedNet microphone preamps. Anna used Beyerdynamic DT 990 Pro headphones.

I asked Anna to avoid pronounced head movements while singing to ensure consistent acoustic conditions throughout the session. Due to Anna's disciplined way of working, no disruptive deviations affected the recording. For monitoring, we initially listened to each microphone individually with headphones, but during the recording, we only monitored the center microphone.

The recording and editing of the vocal takes were done in Avid's DAW Pro Tools. I performed the editing consistently across all simultaneously recorded vocal tracks to maintain synchronization and afterward rendered the edited version. I sent the result to Olaf for further processing with Melodyne and effects in Ableton Live.

### *Timing of vocal tracks for 3D audio playback*

After Olaf processed the vocals, he exported all vocal tracks with effects on separate tracks and sent them to me along with all the song's individual tracks for the 3D audio mix. As described, I assigned the recorded vocal tracks to the loudspeakers: C, L-R, Lrs-Rrs. Up to this point, the vocal tracks were in sync. The goal of the 3D audio mix is to create a large sweet spot where the vocals sound plausible. The timing of the individual vocal tracks changes for the listener depending on their position in the playback room. The following applies to the center and L-R signals: The center signal should provide directional cues. Due to the spectral composition of the three front signals, this is already the case. It can be optimized for lateral listening positions by slightly advancing the center signal compared to the other signals during playback. Therefore, I applied a 5-ms delay to the L-R signals.

From my observations, most listeners prefer positions in the middle or toward the back of a playback room. When loudspeakers reproduce vocal tracks synchronously, the first sound wave from the rear reaches listeners in the rearmost third. This can lead to a confusing auditory impression. Therefore, I applied a 15-ms delay to the Lrs-Rrs speakers. It needs to be small enough not to be perceived as an echo yet large enough for the first sound wave from the center speaker to reach each listener in the playback room first.

### *Technical implementation of the mix with Roomenizer*

First, let us be aware of the signals already present for 3D audio playback and their function analogous to a situation in an acoustic room. The center signal corresponds to the direct sound. The signals from the other vocal microphones correspond to early reflections, resulting in the vocal signal's densification in the room. The last element we need is the reverberation of a large room to utilize its advantages. The source for the Roomenizer is the direct sound. The other microphones are not used for this purpose. I chose St. Michaelis Church in Hamburg preset with a room volume of approximately 100,000 $m^3$ (71 m x 51 m x 27 m). One instance of Roomenizer does not provide the total number of signals required for a 9.1.6 speaker setup. Therefore, I used two instances and a stereo output from the first instance as the input for the second instance to compensate for the missing tracks.

I adjusted the balance of the resulting reverberation tracks based on the following criterion: the church room should not be distinctly recognizable in the overall mix. Nevertheless, the liveliness of the room sound is clearly audible during the 3D audio playback of the mix. This auditory phenomenon becomes particularly apparent when the room sound is tested by turning it on and off during playback. It demonstrates the strong effect of room sound, even if we are not consciously focusing on it or cannot consciously recognize the auditory phenomenon from the overall mix.

## ***The result***

Olaf produced the song 'Pads of Humility' to create an enveloping effect in a pop song similar to that of acoustically sensational sounding rooms. He achieved this using pad sounds and 3D audio effect tools. The spatial quality of the lead vocals impressively supports this impression.

After completing the mix, we traveled to the MSM Studios in Munich, Germany, to carry out the 3D audio mastering with Tom Ammermann, an experienced pop music producer in 3D audio and an advisor. The sound engineers of the house were Stefan Bock and David Merkl. The album was released in December 2022.

The chosen concept for recording the lead vocals has proven to be effective. However, there are a few aspects to consider: The presented Spatial Direct Sound Microphone Array requires discipline from the lead singer to maintain the direction toward the center microphone while singing; otherwise, undesirable effects may occur that cannot be easily corrected during post-production. Tools that alter the timing and pitch of the lead vocals should be used with caution: the changes must be applied evenly on all tracks. Otherwise, this may lead to fission between pairs of tracks. Lastly, there are limitations when down-mixing 3D audio to stereo: the vocal tracks are partially correlated and slightly time-shifted. This can cause disruptive comb-filtering effects during down-mixing. We created two separate mixes: 3D audio 9.1.6 with a down-mix to 7.1.4 and surround sound, as well as a stereo mix containing only the center signal of the five microphone signals.

## Case study: MCH recording Shostakovich Symphony #10

In the present case study, the author presents an innovative 3D audio orchestral recording of the work Shostakovich Symphony #10, which was made in the Amare concert hall, The Hague, with the Residentie Orchestra. A special feature is the concept and the miking technique used: the concept is based on a separation between the recording of the direct sound of the orchestra, including applause from the audience, and the room sound. The room sound is recorded with the help of the 3D audio room impulse response reverberation processor Roomenizer from Pinguin and the impulse responses of the concert hall where the recording took place. The recording includes each register of the work individually and quasi-binaurally for 45 registers. The string instruments are captured several times per register. The case study shows that an impressive envelopment can be created using direct sound.

### *Starting point*

At the AES Convention in The Hague in 2022, Paul Jeukendrup, Head of the Art of Sound Department at the Royal Conservatoire, asked me to conduct a 3D audio master seminar. After a first exchange to shape the content, Paul had a conversation with Sven Arne Tepl, the director of the Residentie Orchestra, The Hague. He got permission to record a concert for this purpose on February 11, 2023. As a result, I developed a miking technique concept with the following approach: listeners in the playback room should be surrounded by string sounds when reproduced. In contrast to conventional recordings, the envelopment is affected not only by room sound but by direct sound – the playback should give the listener the impression of being in the middle of the orchestra.

As a next step, we sat down with the sound engineer Daan van Aalst, who was responsible for the recording for other uses (radio, TV, and streaming) and who cooperated with us. The challenge was to define the recording strategy and at the same time to clarify the restrictions that had to be considered by the framework conditions. Since this was their first collaboration with me, there were two points at the beginning:

- How can I convey the sound aesthetic and technical vision to my partners?
- How can I build up the confidence to successfully realize the project under significant time pressure while achieving a convincing result?

#### *The challenge of capturing for 3D audio orchestral recording*

One of the biggest challenges of this project was the planned miking technique, which deviates significantly from the conventional setup. In the traditional setup, in addition to a main microphone system, some supporting microphones for groups of instruments are used, which capture the direct sound, especially instruments that are far away from the main microphone; this allows the sound engineer to adjust the depth of the mix and optimize the balance of the instrument groups. The concept of my miking technique has a different approach, which will be explained in detail later. The concept is about capturing each instrument individually, if possible, in stereo (quasi-binaural). As a result, the distances between instruments and microphones are much smaller, and many more microphones are used. In addition, it is to be expected that the captured instruments will sound different depending on the miking technique concept: there is, therefore, a risk for Daan's recording that the new recording does not fit the range of productions that have already been made.

### *Vision and solution*

As described, the vision is to create envelopment for listeners with string sounds. This raises the question of exactly how to create envelopment with string sounds. The chosen approach is based on the auditory phenomenon of fusion in the case of sounds that sound very similar and arrive at the listener from different directions (Moore & Gockel, 2012). A signal that corresponds to this auditory phenomenon is room sound. A recording room has the property of smearing the direct sound of a sound source in terms of time and direction. The result is a multitude of reflections (similar sounds) arriving at the listener from all directions in the room.

Let's look at the string parts of the orchestral composition Shostakovich Symphony #10. All first violins play the same part, and the same applies to the second violins, violas, cellos, and double basses. The instruments in a section sound very similar, so the ear performs a fusion when the individual instruments play together. The term 'sound body' describes this hearing phenomenon very aptly. Fusion occurs not only with different instruments of the same kind and the same tones but also with individual instruments that are stereophonically captured. Stereo spot microphone signals with small AB (17 cm) for strings are known to sound broad and spatial, especially when placed close to an instrument. The reason for this behavior lies in the radiation characteristics of instruments, especially those with resonance bodies; they emit sound differently in all directions depending on the frequency – the sound of the microphone signals is not identical but very similar.

#### *Quasi-binaural spot miking technique*

When we make recordings, we can use microphone techniques that differ more or less from binaural recording techniques, such as the dummy head. If stereo signals deviate significantly from binaural signals, there is a risk that the result will sound unnatural.

With binaural hearing, the two ear signals are crucial: They have to meet specific criteria so that we perceive them as natural sounding. If we omit one of the two channels for headphone playback, this leads to a very unpleasant listening experience. If the two signals are identical, everything appears from the same direction; in nature, however, all reflections never come from the same direction.

We need signals that are different from each other in order to get binaural signals. The signals must fulfill specific rules that we know from natural hearing: time and level differences and spectral differences. So we come back to the dummy head. As the main microphone, this is relatively inflexible: the instruments must be placed such that the balance is right. A subsequent correction of the balance is not easily possible.

According to Günther Theile's association model (1980), binaural signals can be superimposed. This means we can capture sound sources one after the other from different angles with a dummy head and then mix them together. However, we can also mix with several dummy heads in different recording room positions. The condition for this is that there is a sufficiently large signal separation; otherwise, the signals from the sound sources will overlap and lead to a change in the perception of the sound source. We also encounter this phenomenon with multiple-microphone techniques of large sound bodies. However, recording with several dummy heads is not practical for various reasons. That's why we use a microphone system that preserves some of the relevant signal properties of the dummy head.

If we place a dummy head in front of a sound source at about 1 m, neither the auricles nor the head shadowing leads to any appreciable spectral signal differences. In this situation, a reasonable approach to a dummy head is a small AB system with a stereo base width of 17 cm. With this method, we do not binauralize the loudspeaker signals but take them from the sound sources directly. Thus the recording room of the sound sources and the playback room for the binauralization are identical.

This procedure differs in some respects from a dummy head recording. In contrast to other miking methods, however, it does not show any significant errors that our hearing classifies as unnatural. One of the most essential points in this context is the quasi-binaural transmission of the room sound. According to Günther Theile's association model (1980), this miking method does not contain any pronounced spatial cues but primarily transmits the 'Gestalt' of the sound source. Thus any movements of the captured instruments do not strongly affect the image.

A key advantage of this process is flexibility in mixing. Mono signals from spot microphones very quickly become disturbing in a stereo mix. Conversely, we perceive binaural signals as natural, so they don't bother listeners even at a relatively high level in the mix.

### *Fusion and fission on captured signals*

Stereo spot microphones (small AB 17 cm) are intended for one instrument each in the planned concept and are placed close to the instruments. The captured stereo (quasi-binaural) signals are assigned to two adjacent loudspeakers in the playback room; thus fusion is guaranteed between these speakers as long as the affected instruments are playing. Captured signals from two different violins also sound very similar if they simultaneously play the same notes. However, this method is conditional. Fusion only lasts as long as the violins play the same notes simultaneously with the same intensity or timbre.

Otherwise, fission occurs, i.e., the auditory connection between the two loudspeakers disappears (Moore & Gockel, 2012).

For orchestral recordings, this means the following: the various string sections consistently play the same notes with the same intensity. The better the interaction within a register, the more the register appears as a homogeneous body of sound. The same applies to the auditory connection: the more precise the interplay, the stronger the auditory connection.

The question remains as to where the threshold lies to achieve the desired effect for the recording and whether the orchestra concerned will be able to muster the necessary precision. This question will be answered in the 3D audio control room during the recording.

### *Close miking of individual instruments and influence of leakage*

The miking technique concept aims to have maximum freedom for sound design during post-production. The stronger the separation between the individual instruments during the recording, the lower the leakage and the more precise the individual sound elements can be represented separately during playback.

This strategy can also be used on the musical layer. If several instruments in a group are combined on spot microphones and a musical error occurs, this cannot easily be corrected during post-production. The reason for this is the generally higher leakage due to the larger distances between the instruments and the microphones. If the distances are significantly smaller, i.e., less than 1 m away, the assigned instruments can be heard almost exclusively in the microphone signals. This allows for much stronger interventions, such as muting instruments that play imprecisely or time corrections to each other in rhythmically tricky passages.

Another aspect concerns the spatial assignment of the instruments for playback. Suppose the leakage between the spot microphones is high. In that case, a specific instrument can be heard at the location of the assigned spot microphone and from all those directions where the leakage with the instrument is exceptionally high. This reduces the localization sharpness of the affected instruments. Conversely, if the leakage is low, all instruments can only be heard from the direction of the assigned loudspeakers; this leads to a realistic auditory impression since leakage does not occur in nature but only strongly pronounced individual early reflections in acoustically inadequate rooms.

### *Technical realization*

There were two concerts with Symphony #10: February 10 and 11. The recording of Daan, including the video recording, took place on February 10. We agreed to do that sound recording with a conventional miking technique and the second concert with the new miking technique concept. Thus we were able to avoid the risk just mentioned regarding the sound of the recording.

Due to the high number of 102 spot microphones and a tight time frame on the part of the Amare administration, it was important to create a clear logistics concept. A central point is the wiring and placement of the microphones on the stage: how can so many microphones be set up and wired up within 90 minutes so that all the equipment is hardly visible? The logistics concept envisages setting up the microphones in an adjoining room the day before and placing them on the stage within the specified time window. We used

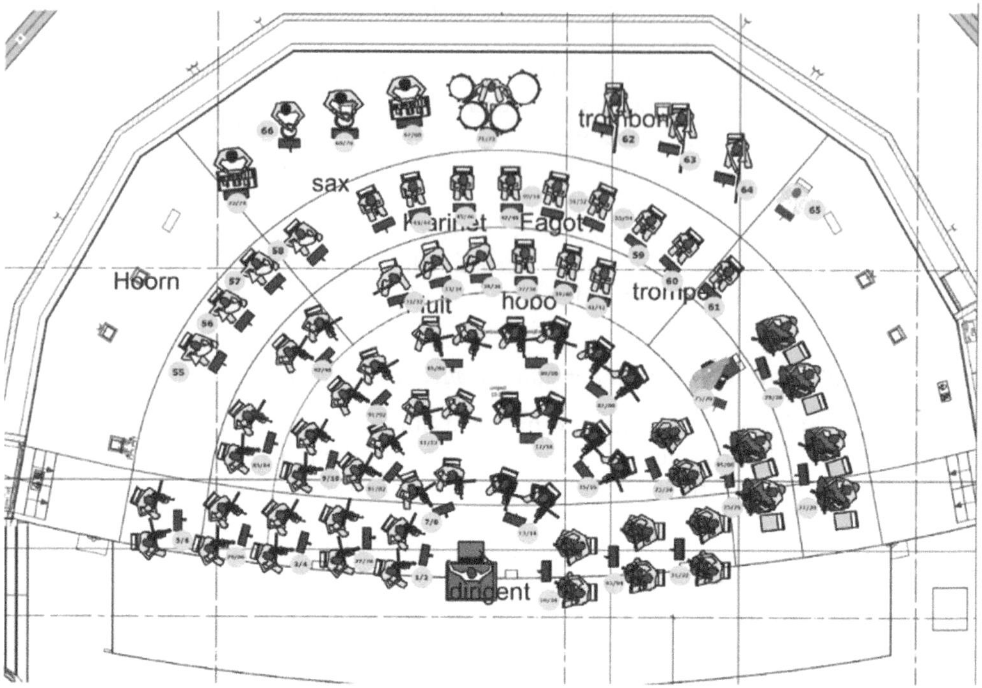


*Figure 24.1* Diagram of the microphone setup for the Shostakovich Symphony #10 recording

adapter cables for all 45 stereo spot microphones (2 × 3–pin female to 1 × 5–pin XLR male) with a length of 2 m, which are color-coded on the female XLR plugs; thus errors in channel assignment L-R can be avoided. In this way, the technical team – consisting of the master students from the Royal Conservatoire and my two assistants from Switzerland, Nicolle Hummel and Reto Fausch, who are familiar with the logistics concept from earlier recordings – place all microphones on stage within the specified 90 minutes and connect them to stage boxes with 5-pin XLR extension cables (see Figure 24.1).

### *Assignment of microphone signals to speakers*

The concept is to create envelopment for the audience with string sounds. The question is how should the microphone signals be assigned to loudspeakers to maximize the effect? It is obvious to use room sound as a model for envelopment. However, there is one aspect to consider: room sound has a strong masking effect. This is easy to understand: only direct sound can be heard in the immediate vicinity of a sound source. At the critical distance, the energy ratio of direct sound to room sound is 1:1; in other words, the –3 dB point is at that position in the room, and the critical distance is usually only a few meters in concert halls and churches. The masking effect is due to the room sound's strong auditory connection.

In real rooms, the auditory connection occurs equally in all three dimensions since the room sound is distributed evenly in all directions. A 3D audio loudspeaker setup is usually located in a rather dry room, so it is not too noticeable. Therefore, creating auditory connections where they are masking is not mandatory. In my experience over 14 years, working with 3D audio is this: masking occurs when a sound projection passes through the audience. This is the case when connections hit the audience at eye level, comparable to a light projection from above toward the floor. Masking can thus be avoided by creating connections predominantly on the sides between the front and back and above the audience. This leads to the following concept for the present orchestra recording: the string registers extend between the front and back sides. In this way, they form auditory connections on the sides without creating a projection through the audience. Also note that when there is no auditory connection between the two sides, the music falls in half, compared to the original Beatles stereo mixes of the 1960s. This can be avoided by creating a spectral balance between the two sides. Specifically, this means: first violins on the left, second violins on the right, violas on the left, cellos on the right, and the double basses in the front. The remaining instruments are placed in the 3D audio mix according to their placement on stage.

#### *Recording impulse responses of the concert hall*

As mentioned at the beginning, room sound is important for a convincing result. But how should the spatial sound be captured? A crucial point is the correlation of the individual microphone signals to each other: the correlation should be around zero between all microphone signals at low frequencies. This requirement means that the microphones must not be placed close to each other – otherwise, the correlation cannot approach zero at low frequencies. Due to the lateral arrangement of the outer ears, our hearing is particularly sensitive to the correlation with signals arriving from the side, i.e., the signals according to the Dolby Atmos terminology L-R, Lfh-Rfh, Lss-Rss, Lrs-Rrs, Ltm-Rtm, and Lrh-Rrh should have a correlation around zero. Conversely, we are much less sensitive to the correlation between front L and rear L or front R and rear R (Griesinger, 2018). Therefore, providing the largest possible distances between the first-mentioned group of channels makes sense. In contrast, the correlation of the second-mentioned group can have higher values. In order to meet this requirement, the following arrangement of the room microphones makes sense: the microphones are arranged on the two side edges of the parquet floor; the distance in the Amare concert hall is around 15–20 m. The distance between the rows' microphones is only 3–4 m.

The impulse responses were measured using a Neumann KH-150 DSP loudspeaker and eight Sennheiser MKH 8020 omnidirectional microphones. Another eight omnidirectional microphones were hung in a line along the edge of the stage for the concert recording. In addition, we used a B&K 4006 as a center microphone. The eight MKH 8020s hung from the ceiling. Three rounds of impulse recordings followed. The Sennheiser microphones hung at a different height above the ground for each round: 2 m, 7 m, and 12 m. In this way we measured 33 positions in the concert hall using sine sweeps as the excitation signal.

After the measurements were made, Ralph Kessler, Managing Director of Pinguin, processed the impulse responses and integrated them as a preset in the Roomenizer. The concert hall was therefore available to the master's students for the 3D audio mix.

*Consequences for mixing when using concert hall impulse responses*

The idea behind using room impulse responses is to use a real room to generate room sound. A few points to consider that are not obvious at first glance. First, the room impulse response is recorded in an empty room, i.e., no sound-absorbing audience. Furthermore, a room impulse response plug-in does not generate applause; this requires microphones in the room. The same goes for the background noise an audience creates; this is an advantage in most cases because it means that annoying noises, such as coughing are not added to the mix. Finally, there is a striking difference to signals from room microphones in that there is no direct sound in the generated room sound. Thus the room sound obtained in this way does not lead to possible misrepresentations.

## The result

One of the master's students in the Art of Sound program at the Royal Conservatory of The Hague assigned the signals to the loudspeakers according to the concept and optimized the balance. The result was already apparent in the 3D audio control room during the concert recording. The following aspects of the recording are particularly noteworthy:

- The auditory connection using string sounds occurs according to the concept, leading to an impressive envelopment. It has a powerful effect, mainly when the orchestra played with high precision.
- The transparency of the recording is exceptionally high. Sven Arne Tepl said, 'I can hear the fingerings the violins play with! 'The high level of transparency affects the level of detail but also the timbre of the instruments; they sound very natural.

The points described in the concept were all confirmed. The way the strings are played influences the auditory connection: the more precisely the strings play, the more homogeneous and unobtrusive the spatiality created in this way appears. If the precision of the way of playing decreases, the string sound becomes livelier: the sheer-off instruments stand out from the overall sound. Overall, 3D audio playback gives the impression that the audience is in the middle of the orchestra, which is both spectacular and pleasant.

The presented concept stands out from other concepts in tonal naturalness and is setting a precedent. Patrick Schappert, managing director of GROBI.TV played excerpts of this recording to his customers in his 3D audio home cinema, and the reactions were astounding: the music also appealed to people who have no direct connection to classical music.

More recordings of this kind are planned. On the one hand, this concerns collaboration with Paul Jeukendrup and Sven Arne Tepl; on the other hand, further orchestral recordings are planned for the summer and autumn of 2023.

## References

Griesinger, D. (2018). *Reproducing Low Frequency Spaciousness and Envelopment in Listening Rooms*. AES Convention: 145 (October 2018) – Paper number 10063.

Kessler, R. (2005). *An Optimized Method for Capturing Multidimensional 'Acoustic Fingerprints'*. AES Convention: 118 (May 2005) – Paper Number: 6342.

Moore, B. C. J. and Gockel, H. E. (2012). Properties of auditory stream formation. *Philosophical Transactions of the Royal Society B*, 367, 919–931. http://doi.org/10.1098/rstb.2011.0355

Theile, G. (1980). *On the Localization in the Superimposed Soundfield*. Doctoral dissertation, Technische Universität Berlin. https://hauptmikrofon.de/theile/1980-2_Diss._Theile_englisch.pdf

Chapter 25

# Creating and producing music in immersive audio

*Justin Gray*

This chapter explores the transformative role of immersive audio, illustrating how spatial technology can influence music creation at each stage of music production. Over the past decade, immersive formats like Dolby Atmos and Sony 360 Reality Audio have expanded the boundaries of music production, allowing for new creative dimensions and a deeper narrative experience for listeners. This study examines the evolution of immersive audio, tracing the impact of technological advancements in surround sound and object-based audio on the artistic process of music creation and production.

This chapter outlines a comprehensive approach to creating and producing music for immersive audio environments through a detailed exploration of recording, mixing, mastering, and spatial orchestration. Key elements include developing spatial intent in the composition phase, the role of immersive monitoring, and the unique demands of mixing music in a three-dimensional space. The chapter highlights how immersive techniques, from capturing natural ambiences in recordings to using advanced digital tools, enhance musical realism and create captivating soundscapes.

The core philosophy emphasizes that technology should serve the music's narrative, focusing on artistic intent as the compass for every technical decision. Through a case study on an immersive audio album created by the author, Justin Gray, this work demonstrates how spatial orchestration can reshape traditional music structures, pushing the boundaries of sound in pursuit of an enveloping, three-dimensional experience for listeners. This chapter ultimately contributes to the evolving discourse on how immersive audio can redefine how creators and audiences engage with and experience music.

## Background

As we explore the history of music, we observe a continual dialogue between musical creation and technological innovation. Over the past century, rapid advancements in audio recording technology have consistently acted as catalysts for new forms of musical expression, playing a vital role in the art form's evolution.

My ongoing engagement with immersive audio stems from a personal search for ways to express music that authentically captures my creative vision. Fifteen years ago, I installed a 5.1 system in my studio to experience music in surround sound. At the time, although I primarily worked in stereo as a bassist, composer, producer, and engineer, I was deeply fascinated by spatial music. My extensive collection of recordings (DVD-A, SACD, Blu-ray, and binaural) facilitated active listening, which ultimately led me to realize that spatial audio represented a creative palette I wanted to explore. My

DOI: 10.4324/9780429297526-28

initial experiments involved remixing my own music and productions in 5.1 surround. Although many of these projects were never released, they were essential creative exercises. These projects also helped me connect with a community of composers, for whom I began mixing music and film scores in surround (5.1 and 7.1). This professional experience was inspiring and instrumental in deepening my exploration and understanding of spatial audio.

Over time, my fascination with immersive audio has only deepened. I have spent considerable time studying the history of spatial music, exploring aesthetic concepts developed by early composers, acousmatic artists, sound recordists, film mixers, and contemporary music artists and mixers. As I synthesized these influences into my creative expression, I encountered the challenge of practically sharing my music with an audience – a common issue for those working in this field. In 2012, I first learned about Dolby Atmos through colleagues in film post-production. Drawn to its adaptive, object-based technology, I became increasingly interested in its potential. In 2015, I discovered Dolby was expanding this technology into music production. Consequently, in 2016, I transformed my studio into a 7.1.4 mixing environment, confident that this new technology would enable immersive audio to reach listeners in unprecedented ways.

By 2017, I began using Dolby Atmos technology to mix music in immersive audio. At that time, Blu-ray was the only commercial outlet for Dolby Atmos music, and few artists worldwide were aware of the creative potential of this technology. Thus my initial projects in the format involved producing my own music and remixing past productions, revisiting the original multitracks to craft new immersive experiences. During this early phase, few engineers were dedicated to music in the Dolby Atmos format, so I refined my approach to creating and producing music within this ecosystem through trial and error.

While my current work primarily resides in contemporary music, my early inspirations came from immersive classical productions, early quadraphonic and surround productions, binaural recordings, and immersive acousmatic and electroacoustic works.

In 2019, I became one of the early engineers to mix music in Sony's 360 Reality Audio ecosystem. I have long admired NHK's 22.2 Multichannel Sound Reproduction System, particularly its use of bottom-layer loudspeakers, which offer a unique sound reproduction experience. Naturally, I was drawn to this format as it emerged, seeing it as another immersive tool for creative expression.

By the time major music streaming platforms began offering immersive formats like Dolby Atmos and Sony 360 Reality Audio, I had already developed a deep artistic relationship with these technologies and a thorough understanding of their mechanics. I am now fortunate to work regularly across multiple genres in immersive audio. My studio has dedicated 11.1.6 and 13.0 speaker arrays for work in Dolby Atmos and Sony 360 Reality Audio. While I have had the opportunity to be at the forefront of these object-based formats, it is crucial to emphasize that they represent only branches of a much larger immersive audio ecosystem. These formats serve as digital containers of intent; ultimately, the music they hold matters most.

## Immersive design philosophy

Establishing intent is fundamental in music production, serving as a creative compass that guides decisions at every stage of the process. My work in immersive audio spans from remixing legacy recordings to creating original music designed for immersive environments. When spatially remixing popular music originally conceived for stereo, the intent may differ significantly from when I am mixing in immersive for the first time or composing new music with spatial audio in mind from the outset. Regardless of the approach to utilizing immersive technologies in music production, the key factor is the clear definition of musical intent. This clarity ensures confident decision-making, leading to the successful realization of the desired artistic outcomes.

I am frequently employed by artists, producers, mixers, and record labels worldwide to spatially remix music that was initially produced in stereo. In these cases, the intent is often to create an immersive experience that both honors the original stereo master and delivers a uniquely engaging immersive experience. Achieving this requires a deep understanding of the essence of the original production, combined with trusting my instincts and making decisions that align with this objective. Some may argue that this approach risks limiting the potential of immersive production. Still, I take a different view: the artistic intent is embedded within the stereo reference (master), and my role is to re-orchestrate the music to preserve this intent while fully utilizing the expanded possibilities of three-dimensional space. This is a highly specialized task with its own specific set of success criteria. In the current industry landscape, there is a significant focus on this type of work; however, it is crucial to recognize that this is just one branch of an ever-expanding spatial audio ecosystem.

In my own music productions, I am passionate about composing and producing music with specific spatial intent that thrives in an immersive playback environment. Given my work in immersive audio for a considerable period, spatial considerations are now an integral part of my compositional process. Incorporating spatial orchestration into compositional frameworks has been explored for centuries, and I am committed to advancing this evolving practice in my work.

Producing music for an immersive presentation involves numerous considerations. Next is a list of key elements that form the foundation of my immersive design philosophy, applicable to both native immersive productions and spatial remixes.

### *Music first*

While technology plays an essential role in immersive audio, my primary focus always remains on the music itself. Ultimately, what matters most is whether the music conveys a compelling story. The more I work with immersive audio, the more I am reminded that the purpose of this technology is to serve the music. Every technical and artistic decision should support the goal of enveloping the listener in a captivating sound world that enhances the narrative the music seeks to communicate.

For me as a composer, rhythm, melody, harmony, form, and improvisation have consistently been the foundational elements of my work. Over the past decade, spatiality has also become an integral part of my compositional process. However, I have found that spatiality must be carefully balanced with these other elements, as spatial orchestration can fall flat if there is not an inspiring story at the core of the composition. Of course, this is a personal observation based on my experience, and I do not claim that this relationship applies to all compositions.

As an immersive audio producer, I have worked on numerous music projects conceived explicitly for immersive presentations. Spatiality in music production can be a potent storytelling tool, and I am deeply inspired to collaborate with artists and composers to help them integrate these concepts into their creative practice. I aim to support their artistic vision by leveraging technology to enhance their narrative and musical expression.

As a mixing and mastering engineer, I am continually reminded that the foundation of all great recordings, beyond the performances, lies in effective sound capture. Using spatiality in music production greatly benefits from capturing sound in ways that align with the immersive intent. Therefore, I am passionate about approaching recording with techniques that allow performances to naturally come to life in an immersive environment.

### *Perspective*

Defining the spatial perspective we want the listener to experience is essential. If I aim to create something that feels realistic, it will inspire different decisions than if I am striving to create a surreal environment. Once we understand the intended perspective, it can significantly inform decisions at both the compositional and production levels.

In studying the history of spatial music, we observe that early composers who incorporated spatiality into their works clearly understood the perspective they intended to present to their audience. Similarly, producers, recordists, and engineers should have a well-defined sense of the listener's intended perspective in effective immersive music production. This understanding is crucial, as it guides decisions on how to record and spatially orchestrate the music.

Immersive audio expands the potential for spatial expression beyond the traditional proscenium-based presentation. I am personally inspired by the opportunity to create surreal listener experiences. However, many of my favorite immersive productions maintain a proscenium-based perspective while adding a deep sense of realism. Ultimately, what matters most is defining the artistic intent and using the technology to serve that vision.

### *Spatial storytelling*

One of the most powerful opportunities in immersive music production is the ability to use spatiality to support and enhance a musical narrative. Understanding the arc of a song before mixing is invaluable, as it allows for thoughtful decisions about how to use space to reinforce the story. There are moments where the use of spatial elements – or the deliberate absence of them – can further support the song's progression, while at other times, creating a consistent immersive environment helps the listener fully engage with the music.

The term 'exit sign mixing', borrowed from the film world, describes a situation where an immersive mix decision distracts the audience from the story. In immersive music production, there are moments where we must remind ourselves to avoid drawing attention away from the narrative. However, there are also instances where I believe that pointing to the 'exit sign' may be the right creative choice. Ultimately, the key is to define the story we are trying to tell and then use the immersive soundscape to support and guide that journey.

### *Commercial goals*

The promise of current object-based formats, such as Dolby Atmos and Sony 360, lies in their ability to create a single file that translates across various playback scenarios. While these delivery formats adapt to different playback environments, compromises often need to be made to ensure the production inspires listeners across multi-channel playback systems, soundbars, automobiles, and headphones. Understanding the intended playback scenarios and delivery formats is therefore valuable, as they can influence decisions at various stages of production.

For example, if the goal is to present the production in a bespoke multi-channel installation, I might approach the production differently than if the aim is to ensure congruence with its stereo counterpart in binaural on a consumer-facing streaming platform. This is not to suggest that the intended playback environment should impose creative limitations but rather that understanding the full scope of the artistic vision can inform decisions throughout the production process.

## Main lessons learned

I will share some key lessons I have learned at different stages of an immersive production, including composition, recording, editing, mixing, mastering, and translation.

In my experience, when an element in a production succeeds or fails, it is typically not due to its inherent quality but instead because it was used in the appropriate or inappropriate context. As a result, I am cautious about suggesting definitive techniques that 'work' or 'don't work', as what proves effective in one production may be ineffective in the next.

### *Immersive composition*

There are many diverse approaches to composing music, so I can only speak to my compositional experience and aesthetics. As a composer, I have realized that certain fundamental aspects of my music are format-agnostic. When creating music for an immersive presentation, I have found it essential to understand the music's rhythmic, melodic, and harmonic relationships before orchestrating it spatially. Once the core elements of my composition are established, I can then conceive how spatiality might enhance its development. This includes using space in relation to form, orchestration, and melodic, harmonic, or rhythmic counterpoint.

In my experience, spatiality can be engaging when used appropriately, but it is only as strong as the core musical content being expressed. For example, if a piece is meant to have a strong groove, it must first groove in mono – an elaborate multi-channel array will not resolve underlying issues. Similarly, if my goal is to build interest through counterpoint, spatial techniques can augment the emotional impact, but they cannot replace the need for strong voice leading.

### *Immersive recording*

I continue to learn at the recording stage with every production I undertake. While there are countless ways to record a sound source, the key is selecting techniques that align with the musical intent. In immersive audio, various recording techniques have been and continue to be developed to create enveloping immersive experiences. Common examples include multi-channel microphone arrays and Ambisonic capture.

Contemporary recording engineers should familiarize themselves with these methods to have them readily available when appropriate. As such, I remain committed to learning from the masters of immersive recording. In my own productions, I focus intently on source capture and on finding ways to integrate spatial recording techniques into contemporary music. However, I am frequently reminded that advanced spatial microphone techniques cannot replace solid mono and stereo capture when the production aims to blend direct and ambient sounds. Ultimately, making the right decisions at the recording stage requires the engineer and production team to have a clear understanding of their artistic goals, enabling them to apply the most suitable capture techniques.

### *Music editing*

At the editing stage, I have repeatedly learned that a good edit is format-agnostic. An immersive presentation will not correct these issues if a performance is out of tune or poorly timed. I encourage editors to use headphones – or even mono fold-downs – to ensure the precision and accuracy of their edits in an immersive environment.

The same principle applies to take selection, which often overlaps with the production role. When selecting captivating performances, I am reminded that an immersive presentation can enhance a great performance, but no number of speakers can compensate for the lack of an inspiring take. The foundation of a compelling immersive experience always lies in the quality of the original performance and editing.

### *Immersive mixing*

Before addressing specific mixing elements in a three-dimensional environment, it is important to recognize that understanding the application of spectral, dynamic, and temporal processing is largely format-agnostic. For those beginning their journey in immersive music mixing, it is often beneficial to remember that foundational mixing skills are sometimes best developed and explored in mono or stereo scenarios.

It is equally crucial for mix engineers entering the immersive audio space to evolve their aesthetic preferences beyond the norms of stereo mixing. Listening to a broad range of immersive audio recordings and critically discussing them with peers is essential for developing a personal aesthetic in this new medium. While immersive audio shares many aspects with mono and stereo production, it is ultimately its own art form. Therefore, mix engineers should cultivate a deep relationship with the immersive audio ecosystem before working critically in this format, particularly regarding spatial orchestration. In native immersive productions, spatial orchestration is typically the responsibility of the composer or producer. However, in spatial remixes (where stereo is the original format), the mix engineer often makes these decisions. Consequently, developing a strong spatial awareness is necessary to arrange various production elements within the immersive environment intuitively.

For me, immersive mixing is primarily about the distribution of sound energy, with volume being the most critical element. I encourage engineers to prioritize volume adjustments before using plug-ins or hardware, as volume changes often hold the key to achieving the desired outcome. This includes proper gain staging, inner balance adjustments, and volume automation. As a mix engineer, I have spent most of my time automating volume levels on a mixing desk or within a DAW.

Spectral processing in music production shapes sound sources, refining their individual presentation and ensuring they fit cohesively within the mix. In immersive audio, the wider soundstage can provide greater spectral flexibility when presenting sound sources. As a result, I often find that less spectral shaping is necessary for individual elements to fit within the mix's sonic landscape. The auditory masking commonly encountered in stereo mixing may be less pronounced in immersive audio, depending on the spatial arrangement and the number of loudspeakers used during playback. Furthermore, immersive mixers should be aware that our spectral perception shifts depending on the location of a sound source. Consequently, spectral decisions must be informed by the spatial positioning of the music. For instance, a sound source behind the listener requires a different spectral treatment than one beside or above.

Dynamic processing in immersive audio follows the same fundamental principles as other playback formats. Compression, a key element of contemporary music production, can range from subtle to aggressive depending on the desired effect. However, our perception of dynamic range in an immersive environment differs significantly from stereo or mono. Compression that enhances a stereo mix may have the opposite effect in an immersive mix due to the altered perception of loudness when surrounded by sound sources. If those sources are too loud, the result can quickly become overwhelming. For this reason, compression plays a less prominent role in my immersive mixing practice. While different musical styles have distinct aesthetic preferences, even songs that benefit from heavy compression in stereo often come to life in immersive formats with reduced compression.

In immersive audio, temporal processing, such as reverb and delay, is closely aligned with aesthetic preferences and artistic intent. When creating a realistic listening environment, I may use a combination of mono, stereo, or immersive digital reverbs to simulate a natural acoustic space, paying careful attention to reflections and timing. Conversely, if the objective is to craft a hyper-real, engaging experience, the possibilities for temporal processing are virtually limitless, creating unique sonic environments that fully envelop the listener.

Several key principles contribute to effectively using temporal processing tools in immersive audio. One such principle is the importance of decorrelated sound sources. In nature, identical sound sources do not emerge from multiple locations, as this is physically impossible. In immersive audio, while it is possible to send the same sound source to multiple output destinations, this can lead to issues such as a lack of envelopment, volume buildup, and phase discrepancies. The solution is to distribute decorrelated audio in immersive mixing consistently. This can be achieved for temporal processing by using different settings on multiple mono or stereo reverb sources or by employing immersive-capable digital tools, such as immersive reverb or delay. Over time, I have increasingly recognized the value of capturing decorrelated sound at the source. While this is not always feasible, as it requires careful attention during the tracking phase, one solution is to play the sound source back into a suitable acoustic environment and record the decorrelated immersive ambiance.

### *Immersive audio mastering*

Mastering immersive audio is an art form that requires an engineer with a mature aesthetic sense and advanced technical knowledge, well-versed in immersive music, to make informed decisions at this critical stage.

Mastering remains an essential stage in immersive music production. With the recent emergence of object-based encoding and decoding technologies, such as Dolby Atmos and Sony 360 Reality Audio, it has become common for some music productions to bypass a dedicated mastering stage. The reasons for this are complex, but as the ecosystem and best practices evolve, I am hopeful that the mastering stage will once again become a consistent component of the process. Just as with stereo music, immersive music greatly benefits from the perspective and refinement of a dedicated mastering stage.

My intent as a mastering engineer remains consistent across all playback formats. Whether mastering stereo, surround, Dolby Atmos, or Sony 360, I aim to make decisions that ensure the successful translation of the artistic intent to the listener. This often involves a combination of spectral, dynamic, and temporal processing to refine the overall balance of the mix. Similar to stereo mastering, the role of the immersive audio mastering engineer also includes quality control, sequencing, authoring, delivery, and often archiving.

Since the mastering engineer's goal is to ensure effective translation, an immersive mastering engineer must work with a mastering-grade playback system and environment that meets the format's precise requirements. This requires using the proper number and arrangement of speakers in a well-tuned critical listening environment.

## Case study: Justin Gray's Immersed

### Introduction

Throughout my 20-year career as a professional bassist and composer, I've had the privilege of performing and recording with musicians worldwide, working across various musical styles. My compositions, primarily instrumental, are deeply influenced by the diverse musical traditions I've studied and performed. These experiences have continually shaped my creative approach, inspiring me to imagine music beyond conventional sonic structures and explore new ways of expressing sound.

In 2023 and 2024, I produced an album of original music titled *Immersed*, conceived from the outset as an immersive audio experience. The album features eight original compositions performed by an ensemble of 36 musicians, incorporating diverse instrumentation and musical styles from across the globe. This project represents my deep passion for spatial audio and my ongoing exploration of how music can be transformed into a three-dimensional sonic experience, blending traditional composition techniques with immersive technology.

### Spatial composition

#### Creative intent

My goal with this project was to compose new music designed for immersive audio. As a composer, this project pushed me to explore the intersection between my creative and technical practices more deeply. I set out to compose each layer of music with spatial intent in mind.

During the initial stages, I discovered that my compositions' core melodic, rhythmic, and harmonic elements remained largely format-agnostic. For example, if a melody didn't inspire or a groove didn't move, there was little value in spatially orchestrating it. As a result, many of the initial compositional ideas were developed using my instruments, voice, a looping pedal, and manuscript paper, focusing on the fundamentals before moving into the immersive audio realm.

Additionally, collaborating with two ensemble members, Ed Hanley and Shawn Rompre, on two compositions was an inspiring opportunity to expand the musical material further.

### Spatial orchestration and documentation

Once the initial melodic, harmonic, and rhythmic ideas were solidified, I could better envision how these elements would unfold in a three-dimensional space. Spatial considerations thus became integral to the compositions during the arranging and orchestration phase. As each piece developed, I created visual diagrams representing the instrumentation and spatial orchestrations. These mock-ups were invaluable throughout the recording phase, serving as a guide and helping to shape the compositions so that they would come to life in an immersive audio environment.

### Immersive monitoring

Once the compositions were established, I recorded and sequenced them in Pro Tools, utilizing a combination of piano, guitar, bass, voice, and digital sample libraries. This method enabled me to refine the music using a 9.1.6 speaker array. Incorporating immersive monitoring into my creative process was pivotal – it enhanced my ability to conceive the spatial arrangement precisely and sparked new ideas that further evolved the compositions.

I chose a 9.1.6 setup because I had already decided that Dolby Atmos would be the primary production format for the project. I intended to present the immersive experience with a video component in theatrical settings and on Blu-ray. My experience producing, mixing, and mastering music for Blu-ray and theatrical Atmos playback strongly influenced this choice.

### Notation and spatial instructions

I documented each part using Sibelius notation software to prepare for the recording process. I also included spatial instructions in the scores based on the speaker names within a discrete 9.1.6 channel orientation:

- **Front Speakers:** Left (L), Center (C), Right (R), Left Wide (LW), Right Wide (RW)
- **Surround Speakers:** Left Side-Surround (LSS), Right Side-Surround (RSS), Left Rear-Surround (LRS), Right Rear-Surround (RRS)
- **Overhead/Top Speakers:** Left Top-Front (LTF), Right Top-Front (RTF), Left Top-Middle (LTM), Right Top-Middle (RTM), Left Top-Rear (LTR), Right Top-Rear (RTR)
- **LFE:** Low-Frequency Effects (LFE)

These speaker layouts provided a clear and practical legend for spatial intent throughout the recording process.

I also considered two alternative methods for incorporating spatial information into the scores, which could have been beneficial had I not used a predefined speaker layout. These alternatives included:

- **Number system:** A method to indicate horizontal and vertical positions
- **Spherical diagram:** A visual representation with spatial points plotted on a sphere

Ultimately, I chose to use speaker locations, as this approach was the most universally understood by the entire production team.

## Pre-production

### Recording design

The next step was to design a recording process that would fully realize the music as I envisioned it. I initially considered two approaches: capturing the entire ensemble positioned specifically around an immersive microphone array or recording parts separately and spatially orchestrating them during the mixing stage.

These compositions placed the listener at the center of a 20- to 30-piece ensemble, with musicians arranged in front, beside, behind, and above. The instrumentation's dynamic range varied significantly, including acoustic and electric instruments.

For this project, I merged acoustic immersive recording techniques with a pop/film-style production workflow, incorporating overdubs and isolating individual layers. Over the years, I've studied and implemented multi-channel array techniques to capture acoustic ensembles in optimal sonic environments. I've also developed immersive studio recording techniques tailored for individual instruments that allow the sources to integrate more naturally into an immersive production. Combining these methods allowed me to integrate various immersive techniques into a contemporary studio recording process, ensuring that the final mix fully supported the compositional intent.

### Location and setup

Choosing and testing the recording space was critical to the success of this project. The room at Humber College in Toronto, where I teach, proved ideal for its high ceilings, clean transient response, short reverb decay times, and exceptionally low noise floor. Approximately six months before the recording began, I conducted immersive tests with my colleagues Alex Gamble and Ryan McNabb, using a drum kit and piano to evaluate different capture approaches. Through these tests, we identified an eleven-channel microphone array consisting of seven main-layer cardioid microphones and four upper-layer cardioid microphones (Schoeps CMCU-6 + MK4 capsules). This configuration proved optimal for the room's dimensions and the playback angle of a 7.1.4 Dolby Atmos speaker array.

### Refining the recording techniques

The multi-channel microphone array was found to be suitable only for instruments capable of generating enough amplitude to engage the space fully. For quieter instruments, immersive capture techniques were developed based on the sources' characteristics, such as their dimensions and sound output. We also experimented with layering overdubs and combining various immersive recording methods. Maintaining the precise placement of the array across layers became essential to avoid timing and phase issues, and we took extensive measures to prevent noise floor buildup throughout the process.

**Studio preparation**

I equipped the studio with a 9.1.4 speaker array, allowing real-time immersive monitoring throughout production. The speaker array was calibrated, and we designed a *Pro Tools* session at 96 kHz/24 bit that could seamlessly switch between discrete 9.1.4 channels, stereo, and mono. This setup allowed us to quickly focus on elements like tuning and timing in lesser channel counts when desired.

**Rehearsal**

Once the control room monitoring was set up, I rehearsed some of the core band members in the studio. This prepared the musicians for the recording sessions and allowed my co-engineers, Alex Gamble and Ryan McNabb, to test and fully optimize the immersive recording system.

**Filming**

One additional layer of complexity came from the decision to capture the entire recording with multiple cinema-grade cameras. My ultimate goal was to present the project as an audiovisual immersive experience in a movie theater. Since the music incorporated improvisational layers, filming every studio performance live was necessary, which required a robust video production process to run parallel with the audio recording. This involved setting up the recording environment and staging the room for visual capture. An entire studio space was dedicated to the video team for asset management. This process was carried out in collaboration with Canadian cinematographer Michael Fisher.

**Recording**

**Immersive recording**

Once the recording process commenced, each day was dedicated to different artists and instrumentation. A tailored recording approach was developed for each sound source, designed to address the instrument's unique characteristics and role in the spatial orchestration. Alex Gamble was my co-engineer for this project, and studio coordinator Ryan McNabb provided additional assistance.

For instance, while recording the brass quartet, I arranged the four horns beside and behind the listener. The trombones were orchestrated to the side surround speakers, while the trumpets were in the rear speakers. We positioned the musicians accordingly within the room and oriented the microphone array so that the players directed their sound toward the rear of the array. Additionally, we incorporated supplementary microphone placements based on our studio listening tests, which helped provide additional reflection points to enhance the instrument presentation. This setup included four omnidirectional microphones positioned in the room's upper corners and bidirectional microphones aimed at the side walls.

This multilayered approach allowed us to capture close mic recordings of each instrument while adding immersive layers, which were later mixed to create a sense of depth and realism. We applied the same level of detail to each instrument, resetting the studio every night over the span of 30 recording days to ensure that every recording session approached the sound with a tailored process suited to both the performance and the spatial arrangement.

**Immersive production**

As a producer, my responsibilities for this album were seemingly endless. Spending a month straight in the studio was both energizing and exhausting. As with all productions, success depended on meticulous organization, clear communication, maintaining a positive atmosphere, and surrounding myself with a fantastic team. It was truly a dream to work with such talented individuals, as everyone came together seamlessly to ensure the success of this extensive recording process.

I was also fortunate to have the invaluable support of my co-producer, Drew Jurecka. His feedback at critical moments was instrumental in refining the arrangements as they developed.

The immersive component of this production was a seamless blend of creativity and technical precision. On the technical side, I focused on establishing recording techniques that would honor the spatial intent of each layer. Artistically, my main role as a producer was collaborating closely with each musician to ensure we captured both inspiring and accurate performances. As an immersive producer, I was constantly engaged, listening as the music evolved in the studio, and ready to explore new creative possibilities in real time with the artists.

One of the most powerful aspects of this production was the opportunity to bring the artists into the control room to collaborate on the immersive production in real time. Even though many of the musicians were unfamiliar with immersive audio technology, they were extraordinary listeners and world-class improvisers. Their creativity thrived as they heard their takes in the immersive environment, quickly grasping the spatial context of their parts. This led to deeply inspired performances as they adjusted their playing and interpretations based on real-time feedback, resulting in genuinely dynamic and context-aware recordings.

**Live floor monitoring**

On the studio floor, we opted for stereo headphone monitoring rather than introducing binaural setups, which could complicate the process. Our main priority was ensuring the artists felt comfortable and confident, and we didn't want to risk unsettling them with unfamiliar monitoring systems or additional latency.

**Post-production**

**Editing**

Editing the music required a highly organized approach, with some songs containing over 400 audio channels. While editing in immersive formats was valuable, some moments required editing in mono or stereo for pinpoint accuracy. Studio assistant Andrew Chung and I edited the music in Pro Tools over a number of months.

**Mixing**

Once the music was fully edited, the mixing process began. I mixed this album in Dolby Atmos, utilizing Pro Tools and an external Dolby Atmos Renderer. My studio is equipped with an 11.1.6 Dolby Atmos speaker array. My objective with this mix was to blend contemporary mixing techniques within the framework of a discrete acoustic immersive presentation. The intent behind every decision was to bring every detail of the compositions to life as intended, creating an enveloping world for each piece. The spatial orchestrations were quite dense, so the most integral part of the mixing process was the careful volume automation of each layer. This precise control ensured that every element maintained clarity and balance within the complex immersive soundscape.

A mix's architecture should be dictated by the music's needs rather than personal mixing preferences. I designed a specific layout of objects and beds, maintaining the discrete-channel reproduction angles of the various microphone arrays while allowing for more flexible object-based panning for mono sources. This layout respected the spatial intent captured during the recording phase while allowing for the flexibility to augment certain layers of the music with more precise placement in the soundscape when needed.

I also planned the session layout to ensure that the full album could eventually be mastered as a single file, which is ideal for cinematic remixing and Blu-ray production. This required a highly organized structure, employing consistent routing and naming conventions across all songs.

My approach to processing sound in a mix involves a hybrid analog/digital workflow. I enjoy using carefully selected analog hardware on appropriate sources, typically committing these analog decisions early to establish a sonic foundation. This approach allows me to benefit from the distinctive qualities of the analog domain while also maintaining the precise recallability and flexibility that the digital realm offers.

Once the analog stage of the mix is completed, I continue working digitally ('in the box'). The mixing process varied from song to song, adapting to the unique requirements of each piece. Given the intentional nature of the capture, many of the key sonic decisions had already been made during the recording process. Therefore, my primary focus during mixing was on precise balance adjustments and refining the overall sound based on my artistic preferences.

I also manipulated several sources using advanced digital immersive tools, both within and outside Pro Tools. One of the most prominent tools I used outside of Pro Tools was Sound Particles, which allowed me to further transform certain sounds into three-dimensional immersive gestures. These manipulated sounds were then seamlessly integrated into the mix, enhancing the spatial depth and detail of the composition.

I utilize a wide range of tools for dynamic and spectral processing, each specifically chosen to refine a particular sound or group of sounds. The tools I use range from conventional stereo processors to immersive audio tools, depending entirely on the context and what I aim to achieve within the mix.

When it came to temporal processing, I was highly intentional about using digital reverbs and delays, selecting them either to enhance the realism of the sound or to expand it into surreal, enveloping soundscapes. This careful approach ensured that each layer contributed meaningfully to the immersive experience.

### Immersive re-recording

During the final stages of mixing, I encountered a challenge in finding a digital reverb or combination of tools that would allow the lead melody instruments and soloists to envelop the listener while seamlessly blending with the ensemble. Despite having access to virtually every tool available to immersive mixers, none could achieve the precise sound I was after.

The solution was re-recording these audio sources in a concert hall to capture the natural ambience. I've always admired the acoustic depth and rich quality of the Aga Khan Museum's concert hall in Toronto, where I've performed numerous times. I collaborated with Patric McGroarty, Alex Gamble, and the museum's recording team to determine the optimal speaker and microphone placement.

We positioned a single speaker on the stage, facing the backstage wall, and set up three immersive microphone arrays throughout the hall. The primary array, a 15-channel equidistant setup, was designed to mirror the 9.1.6 mix architecture, featuring nine microphones on the main layer and six on the upper layer (Schoeps CMCU-6 with MK4 capsules). The second array, consisting of five DPA 4006 microphones, was placed on the upper balcony. The final array, a four-channel Omni setup of Line Audio OM1 microphones, was suspended from the ceiling corners, over 100 feet above the stage.

The ambiences captured during this session were stunning and seamlessly integrated into the overall mix.

### Mix perspective

I was also fortunate to receive valuable mixing advice from a few colleagues throughout the process, including my co-producer, Drew Jurecka, and one of my personal idols in the immersive audio world, Morten Lindberg. In addition, I took the opportunity to check these mixes in additional immersive studios, which allowed me to study the translation and gain a new perspective as a listener. Ultimately, this album was mixed over the span of a year.

### Final steps

Once the initial immersive mixes were complete, the project entered five additional stages: immersive mastering, stereo mixing, stereo mastering, Blu-ray authoring, and cinematic remixing.

### Immersive mastering

For the immersive mastering, I sequenced the album into a single ADM BWF Dolby Atmos master file, ensuring alignment with the video sequencing for both Blu-ray and cinematic presentations. This step was crucial for maintaining continuity and enabling efficient workflows during remixing and authoring. Michael Romanowski at Coast Mastering handled the mastering in Dolby Atmos and stereo. As a dedicated mastering engineer, Michael can refine mixes while preserving their original intent, ensuring balanced tonality and engaging dynamics.

**Stereo mixing**

For the stereo mix, I created a dedicated version of the album to capture the full intent of the compositions and productions within a two-channel format. While Dolby Atmos technology offers highly capable fold-down tools, I crafted a unique stereo mix for each song. This approach allowed for precise control over every layer and fully preserved the original artistic vision.

To achieve this, I worked within the original Dolby Atmos session in Pro Tools, monitoring the audio through a real-time stereo fold-down. I then made specific mix decisions, including adjustments to inner balances, automation, and spectral, dynamic, and temporal processing. Additionally, I simplified some immersive microphone sources to suit the stereo format better, ensuring that the music translated effectively in a two-channel playback environment.

**Theatrical remix**

The final stage was the theatrical remix of the album. Movie theaters provide powerful spaces for collective listening to immersive content, and I have worked on numerous immersive music productions specifically remixed for cinematic release. I have also spent significant time studying the differences between Dolby Atmos productions created for home entertainment and those intended for cinematic environments, as these formats have key technical distinctions that influence decisions on spatial orchestration and the implementation of objects and beds.

For this project, I collaborated with senior re-recording mixer Steve Foster and mix assistant Mike at Company 3 in Toronto to create dedicated Dolby Atmos and 5.1 mixes. We began by reviewing the music in my studio and discussing the best approaches to preparing it for the theatrical remix. From the outset, I utilized objects and beds in the Dolby Atmos format to ensure that the production would translate effectively to a cinematic setting while providing enough control to adjust the mix for larger playback environments.

Once the cinematic mixes were finalized, they were combined with the video assets and authored into a Digital Cinema Package (DCP), allowing for seamless playback in theaters worldwide.

**Blu-ray authoring and digital distribution**

For the commercial release, the Dolby Atmos and stereo master files were sent to Stefan Bock at MSM Studios for Blu-ray authoring and distribution. Blu-ray was chosen as the ideal commercial format for this production because it supports high-resolution audio and includes the video component, an integral part of this immersive experience. The Dolby Atmos master files were also distributed to participating streaming platforms.

## Conclusion

Through the creation of *Immersed*, I have deepened my understanding of how technological innovation and artistic intent intersect. This project allowed me to merge traditional composition techniques with immersive audio production, blending spatial orchestration and immersive mixing with musical storytelling. The journey from composition to cinematic release was as much a learning process as a creative one, inspiring me to push the boundaries of how music can be experienced in three dimensions. For this composer, performer, producer, and engineer, *Immersed* represents a culmination of my years of obsession with spatial and immersive audio and an evolving exploration of how sound and technology can come together to bring new music to life.

Chapter 26

# Immersive audio case studies for immersive audio in rock and pop music

*Miles Fulwider and Vincent Gabriel Antonini*

Most often when working in pop/rock genres mixing, production, and recording techniques center either around close (direct) sound recording and/or room-environment (diffused) sound recording. To enable complementary immersive sound production, different microphone techniques should be used in the recording process. Such applications would be made up from common microphones found within the studio. Additionally, common mix techniques and approaches must be altered or reconfigured when working in object-based immersive formats, to provide the listener with the desired listening experience. Such considerations include changes to mix-bus processing, 'up-mixing', relative microphone techniques, and key mix techniques to ensure proper aesthetic punch and impact are delivered in immersive playback.

We will be referencing different case studies across two main sections. The first, dealing with how common and well-preferred microphone techniques can be augmented within the studio environment to better present source material in immersive formats. The second section will reference and provide additional examples from case studies where key mix techniques will be outlined. Such techniques include how to use and how to avoid mix-bus processing inside of an immersive mix; how to 'up-mix' mono, stereo (and surround) material that was not originally recorded for immersive formats; and how to provide substantial low and low-mid punch that is often found in stereo pop/rock mixes in an object-based mix. Additional information will be provided with tips and tricks regarding dynamics processing and the use of equalization and how that impacts an immersive mix.

## Introduction

Transitioning into immersive technologies has felt like a natural progression. Advances in consumer playback technology have facilitated much easy enjoyment of spatial audio to both the professional and the consumer.

Despite the challenges posed by early Ambisonic, multi-channel (channel-based), and immersive formats with cumbersome tool sets, our desire to utilize these formats in allowing our productions to cause a listener to be more engaged with the listening experience and the artist's intention.

Today, we benefit from greater uniformity in formats, enabling us to create and share immersive content more readily. While there's ongoing discussion and work needed to further develop these technologies, the ability to create music where spatial sound cues

DOI: 10.4324/9780429297526-29

are preserved, whether in a near-field studio environment or on headphones during a subway commute, represents remarkable progress in our view.

The delivery methods have become easy to stream to conventional stereo headsets and to smart devices. This has caused immersive capture a profitable endeavor for a lot of companies that were confused on how to capture this audience. The demand is now evident in areas such as gaming, film, music, live sports, and even theater. The wave has begun, and you, sitting at home in a home studio environment, can create immersive content too. Having great accurate microphones and knowing how to use rejection to your advantage when placing them, will give you – dare we say – your immersive unique place.

## Immersive design philosophy

Our Immersive Design Philosophy originates from our backgrounds as musicians. As performers, our goal is to replicate the intense experience of ensemble performance in immersive audio productions.

When dealing with formats like 5.1 surround sound or 7.1 surround sound, we encountered specific limitations inherent in traditional channel-based systems. The primary challenge arises when relying solely on direct speaker channels to convey source information. While sounds may emanate from all directions, they often become focused as point sources, maintaining intensity but lacking envelopment. This can result in perceived gaps in the soundstage and difficulties in integrating material positioned between speaker locations, leading to a disjointed listening experience.

Ultimately, the aim is *immersion*. Whether striving for a lifelike representation or a purposefully crafted audio environment, the goal is to evoke an emotional connection with the source material. Achieving the proper balance of immersion is evident when the program material resonates deeply with us on an emotional level.

## Production rules and guidelines

What are the rules? Oh wait . . . there are no set rules. Now that we have that out of the way! As humans, we often like rules, guidelines, or expectations when creating something. A builder likes blueprints. A visual artist may have a model. When it comes to audio production and recording we often hear . . . 'There are no rules!'

In the realm of immersive audio formats, we firmly believe that there are no rigid rules dictating how one *should* mix and produce audio. Allow me to elaborate on this conviction.

First and foremost, immersive audio is an ever-evolving field, characterized by constant innovation and experimentation. Unlike traditional stereo mixing, where established practices and conventions often guide the process, immersive audio offers a wide-open canvas for creative exploration. As such, attempting to impose strict rules would stifle innovation and limit the potential for groundbreaking sonic experiences.

The very nature of immersive audio encourages a departure from conventional thinking. With the ability to place sound sources in three-dimensional space, artists and engineers are empowered to craft immersive experiences that transcend the limitations of traditional stereo imaging. This newfound freedom allows for unprecedented levels of creativity and expression, unbound by the constraints of established norms.

Moreover, the diversity of immersive audio formats – from Ambisonics to object-based systems – further underscores the absence of one-size-fits-all rules. Each format presents its own unique set of possibilities and challenges, requiring a flexible and adaptive approach to mixing and production. Rather than adhering to rigid guidelines, practitioners must leverage their creativity and technical expertise to harness the full potential of each format.

The subjective nature of perception plays a significant role in immersive audio. Unlike traditional stereo listening, where the listener's position is fixed relative to the speakers, immersive audio allows for dynamic spatial interactions that vary based on the listener's position and environment. Consequently, what may sound optimal to one listener may differ from another's experience. Thus attempting to impose universal rules would overlook the inherent subjectivity of perception in immersive audio.

The absence of strict rules in mixing and producing audio in immersive formats is not a limitation but rather a testament to the boundless potential of this exciting frontier. By embracing experimentation, innovation, and individual expression, we can unlock new dimensions of sonic artistry and create immersive experiences that transcend the boundaries of traditional audio production.

Yes, it is wise to approach each performance and space as if it is the first time you are designing. Doing so creates an innocent and *open-minded* approach that gives us creativity and a license to break rules created by others. We are not suggesting that the hard work and guidelines that have been created are not great ones but rather that, when we rely solely on a 'paint by numbers' approach we might have a fantastic result, or perhaps one that is documentative – but not exciting or inviting or accurately immersive with respect to the project we are focusing on. We work to stay away from *MMM* (mic muscle memory) as much as possible. This is when we rely too much on habits because we must get the job completed! Take a step back and experiment with every position the microphones will be in when they capture the performance. Put on a pair of headphones and move the microphone as the artist performs – find that sweet spot by using proven methods and innocent curiosity.

Now, here is where we become self-contradictory.

In any given space are fundamentals based on room dimensions and sonic perspectives we strive to capture. So, while devising a strategy for creatively miking all the dimensions of a space such as the floor, the height, the mid-level, the below, the back, the front, etc., one might begin by mathematically measuring the space and placing microphones in an equidistant way so we can capture the bones of the space with respect to the position of the source. Thus we ensure good coherence while controlling comb filtering/phasing issues. This will become the skeleton to creative miking. Together this gives the 'Feet on the ground – head in the clouds' setup. We can rely on *static points* and also experiment within the spaces between them.

During the early days of multi-channel 5.1 recording and mixing (and even early stereo, for that matter) we saw the excitement and liberation one has aesthetically and creatively in that format. Being able to present the listener in an environment, pan sound sources around the listener, and ultimately strive to 'surround' listeners and bring them into the experience they were listening to.

One of the most common approaches was to utilize each speaker for a specific sound source. For example, drums in front (L-C-R), guitars in the background, etc. What was discovered with this process is that, by doing this, though there is a feeling of some *immersion*, the challenge is that relying solely on specific speaker channels to provide the source information caused the speaker/sound source to feel like a persistent point-source and that what was often lacking was the feeling of cohesion across the soundscape with the sound sources feeling quite separated one from another.

Another discovery was how, early on, the approach was often, in order to record the natural sound in a large hall or environment, multiple microphones were placed in the environment's extremities. As a possible result, microphones were sometimes placed in the rear of a hall at opposite ends of each other. The problem here is that the microphone pickup becomes so decorrelated that the sense of envelopment is lost without a focal point of the microphone in relation to the other microphones and the environment itself. Said another way, our head is simply not as big as the concert hall (egos, possibly), where a more coherent microphone implementation will provide a stronger sense of environment and spatial cues.

When mixing in immersive formats, without the ability to rely on mix-bus processing as one might normally do, an effort to maintain the preferred 'punch' within a mix by often bringing a sound object into the center of the immersive field is counterproductive due to the divergence across the speakers. Some reasons why positioning a sound course in the center of the listening environment will render less desired dynamic 'punch and glue' include:

- **Phase interference:** Placing a sound object in the center can lead to phase interference, especially if the object has strong low-frequency content. This interference can cause cancellation or reinforcement of certain frequencies across the speakers, leading to an unpredictable and potentially less impactful sound.
- **Spatial imbalance:** Immersive formats rely on spatial placement of sound objects to create a sense of depth and immersion. Placing everything in the center can lead to a lack of spatial balance and can diminish the overall impact of the mix.
- **Listener fatigue:** Constantly having sound objects in the center can lead to listener fatigue, as the brain may struggle to process and separate the individual elements of the mix. This can result in a less engaging and enjoyable listening experience.
- **Creative opportunities:** Immersive formats provide a wide range of creative opportunities for spatial placement and movement among sound objects. Limiting everything (*or key sound objects intended to deliver impact*) to the center can restrict the creative possibilities and may not fully utilize the capabilities of the format.
- **Dynamic range:** Utilizing the entire immersive field allows for better management of dynamic range within the mix. Again, placing key sound elements in the center can lead to a compressed and less dynamic sound, impacting the *perceived punch* and impact of the mix.

A simple demonstration can be performed to help understand how loudspeakers and the human ability to localize centrally located sound sources in a 7.1.4 playback environment.

To set up such an experiment, take a single mono object and place it at the center (listening position) in the three-dimensional playback environment (renderer, etc.). Notice the lack of reliance of the front speaker channels Left and Right. Instead, most amplitude will come from the side speakers and possibly the center channel.

Our auditory response systems are more acute in spatializing sound that is in front of us. This is a great way to create movement and sound sources may *envelope* the listener, but for sounds to bring impact and punch, bringing them into the center of the listening environment will result in greater divergence and usage of side and rear channels, which will decrease the presence-punch often desired for things in the rhythm section, such as bass, drums, and even vocals.

## Effective approaches

Let's examine a few key solutions to these challenges. One of the strongest *recommendations* we provide when working in any multi-channel and immersive format is the realization that, with the benefit of multiple loudspeakers for playback, these should be thought of as a number of 'stereo pairs'.

We use more than one transducer (our ears) to hear and provide adequate spatial resolution. While thinking about stereo pairs in immersive environments, it's easy to simply assume stereo groupings such as Left-Right, Left Side Surround–Right Side Surround, Left Rear Surround–Right Rear Surround, Height Front Left–Right, Height Rear Left–Right. What becomes really powerful and much more capable in creating an immersive soundscape, regardless of a mono sound source or stereo sound object, is the realization that each loudspeaker positions becomes part of a stereo pair in any one direction.

When applying this approach in object-based immersive formats, such as 7.1.4 or 9.1.6 setups, the metadata containing object spatial information is preserved. This ensures that playback maintains the intended orientation and travel path of objects, regardless of the specific loudspeaker configuration or whether the listener is using a loudspeaker playback system or a binaural listening setup. The renderer works to deliver an enveloped solution that honors the object's spatial characteristics.

A practical example of how this can dramatically improve one's immersive mix and production process and be implemented in an immersive audio mix workflow is outlined by the technique made popular by the Grammy-nominated recording engineer Steve Genewick (2020). In fact, several DAWs have now implemented this methodology into their routing, which will allow the user to modify and expand on this for their own creative and artistic purposes.

The *Object Bus* can be applied to most any DAW, as shown below in Table 26.1. (In the table, notice that all channel derivatives of a 7.1.4 Object Bed are available; however, when working with Objects, you will be limited to stereo and mono objects.) You will see the routing specification and 'subpaths' within a traditional 7.1.4 output. Building an Object Bed in this manner allows for any variety of path reductions or expansions of said paths that are routed to a specific object path that feeds the renderer.

Now that the internal routing is set up, our mix workflow can begin to take shape. In the simplest form, Figure 26.1 shows how the stereo and mono objects that are feeding the renderer will be accessed.

Additional emphasis needs to be placed on the fact that this is the Object Bed in its simplest form. What you see here is that this practice can (and should) be expounded upon by providing additional paths as needed. These paths may include diagonal pair-relationships or creating 'side-walls' with stereo pairs just on the left side or right side. Applying these practices will offer increased control and flexibility in manipulating the desired spatial experience.

*Table 26.1* All channel derivatives of a 7.1.4 Object Bed

| *OBED* | *Channel width* | *Channel assignments* | | | | | | | | | | | |
|---|---|---|---|---|---|---|---|---|---|---|---|---|---|
| OBED.7.1.4 | 7.1.4 | L | C | R | Lss | Rss | Lsr | Rsr | LFE | Lft | Rft | Ltr | Rtr |
| OBED.7.0.4 | 7.0.4 | L | C | R | Lss | Rss | Lsr | Rsr | | Lft | Rft | Ltr | Rtr |
| OBED.7.1 | 7.1 | L | C | R | Lss | Rss | Lsr | Rsr | LFE | | | | |
| OBED 7.0 | 7.0 | L | C | R | Lss | Rss | Lsr | Rsr | | | | | |
| OBED.5.1.4 | 5.1.4 | L | C | R | | | Lsr | Rsr | LFE | Lft | Rft | Ltr | Rtr |
| OBED.5.0.4 | 5.0.4 | L | C | R | | | Lsr | Rsr | | Lft | Rft | Ltr | Rtr |
| OBED.5.1 | 5.1 | L | C | R | | | Lsr | Rsr | LFE | | | | |
| OBED.5.0 | 5.0 | L | C | R | | | Lsr | Rsr | | | | | |
| OBED.Quad | Quad | L | | R | | | Lsr | Rsr | | | | | |
| OBED.LCR | LCR | L | C | R | | | | | | | | | |
| OBED.Stereo | Stereo | L | | R | | | | | | | | | |
| OBED.Sides | Stereo | | | | Lss | Rss | | | | | | | |
| OBED.Rear | Stereo | | | | | | Lsr | Rsr | | | | | |
| OBED.Ht.Front | Stereo | | | | | | | | | Lft | Rft | | |
| OBED.Ht.Rear | Stereo | | | | | | | | | | | Ltr | Rtr |
| OBED.L | Mono | M | | | | | | | | | | | |
| OBED.C | Mono | | M | | | | | | | | | | |
| OBED.R | Mono | | | M | | | | | | | | | |
| OBED.Lss | Mono | | | | M | | | | | | | | |
| OBED.Rss | Mono | | | | | M | | | | | | | |
| OBED.Lsr | Mono | | | | | | M | | | | | | |
| OBED.Rsr | Mono | | | | | | | M | | | | | |
| OBED.LFE | Mono | | | | | | | | M | | | | |
| OBED.Ltf | Mono | | | | | | | | | M | | | |
| OBED.Rtf | Mono | | | | | | | | | | M | | |
| OBED.Ltr | Mono | | | | | | | | | | | M | |
| OBED.Rtr | Mono | | | | | | | | | | | | M |

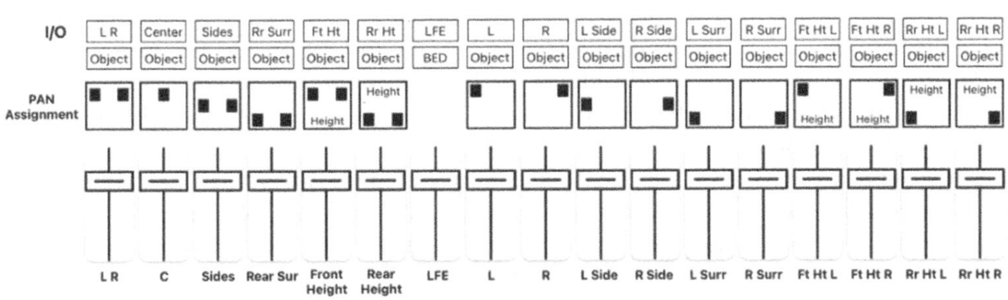


*Figure 26.1* Object bed mix matrix: Here all (except the LFE) channels are mapped to specific objects. Individual tracks or stereo busses now can all be routed to the various Object Bed mix matrix channels that will provide incredible flexibility within the mix process.

## Where am I? Pop/rock philosophy on mixing and immersive playback

As mentioned, the common technique of using mix-bus processing to bring 'punch' and 'impact' to a (stereo) mix does not function the same way when working with object-based formats. The lack of a central mix-bus and the variability within the renderer to accommodate a variety of playback environments (including headphones), the question is how to present the listener with the same punch/impact they are expecting.

Two solutions that one might consider will depend primarily on the status of your source material: are you mixing from stems, or are you working with a full multitrack?

### *Working from stems*

Often the challenge with working from Stems is if the track has already been released in stereo. If this is the case, or if you are working from stems to 'remix' or 'up-mix' the track for Spatial Audio delivery, to accompany a stereo delivery, one must take great care to ensure the tonal properties of the immersive or 'spatial' mix is appropriate and inline with the stereo mix tonally. Often this approach can be referred to as '*spatial remixing*'. This is often necessary when a label, artist, or producer is desiring an immersive mix that follows the soundstage of the stereo mix as closely as possible – but is requiring an immersive mix that follows suite to the stereo. Thus 'spatial remixing'.

An approach shared by multi-Grammy-winning Mastering Engineer Brian Lucey, is what he refers to as 'Cardinal Points Quad'. Here Lucey explains how when working with stems and focusing the rhythm section and essentially getting the low-end into the cardinal points – the main four speakers. The reason for this is that with the low-end being focused into the four key speakers, this will provide punch and impact in the middle position, as this is where also the DSP is able to provide you with punch over headphones (as well as loudspeakers) (2)(Lucey,2023).

The recommended 'weight' within the mix should be 52% front of the middle position and 48% behind the chair position. This will work to ensure that the low end and intended punch are similar to, if not even more effective than, the stereo mix.

While the exactness to this 'low-end' approach (or rather where the rhythm section could be placed) may be argued, it's worth suggesting that biasing the balance simply where there is stronger representation in front of the listener, than behind the listener, even if by just a little, allows one to naturally utilize the biological system humans use to localize sound and will provide a stronger 'punch and weight' to the mix as opposed to the inverse, where more amplitude from the rhythm section is coming from behind the listener. By doing the latter, the perceived 'punch' will be decreased due to how we naturally localize sound.

When working with this approach, it is very common, in fact, to just duplicate the stems necessary to reach the four corners with low end elements. Assuming you are duplicating stems for this process, low-end elements located in the rear corners clearly require a high-pass filter, as it's undesirable to rely solely on each playback device's high pass. A range of 40–60 Hz is recommended.

Also if duplicating bass and drum stems 'add a high shelf cut approximately –1.5 dB at 1 kHz in the rear channels, to eliminate phase issues in the middle of the room and the headphone center of the DSP' (2).

After this foundation has been laid, additional sound objects can be added and moved forward and backwards and up (height). Working in this way, positioning sound objects at 100%, 50%, and 0% to the four cardinal points in any one direction will result in an immersive resolution in the center position.

It goes without saying that it is still important that one is 'mixing' these balances. As one begins to duplicate stems, the level will increase by approximately +6dB. That being said, this sound energy will be coming from different locations (and not out of the same speakers), but consideration must be given in how to handle the increased signal level that will be created as one duplicates stems.

A good example of the *Cardinal Points Quad* technique in use is Mac Miller's song 'What's the Use' on the *Swimming* album (Miller, 2018) (Atmos mixed by Greg Keller, mastered by Brian Lucey).

The current state of immersive audio production in pop and rock music is certainly in a bit of a transition. Often catalog work (music we all know and have loved) that was released in stereo is now being remixed for spatial audio. This process is generally happening one of two ways. The first, just as explained, where one is working with printed mix-groups or tracks, or stems, to create an immersive mix.

This process is no-less challenging. Working in this manner often requires working to have the immersive mix sound as similar as possible to the stereo mix. In many cases, there are few liberties to be taken where sound sources may move or how the envelopment may happen. Instead, the immersive mix needs to be a very strong representation of the stereo mix stems, in an immersive environment.

This is all very common due to the effort of allowing these catalog recordings that we know and love to live on in the new formats, without sounding different.

However, when we do have a multitrack to work with, here we are enabled to provide a more precise and explorative listening experience by these sound sources being available to us.

### *Working from full multitracks*

When working from a full multitrack, one might prefer to 'print' stems and create the immersive mix as just outlined or choose to work with a full multitrack; the preferential routing to access the environment can be accomplished via a routing matrix that can be created by channel outputs and sends.

Consider the internal input/output routing as explained earlier. Here you can see in Figure 26.2 that the Kick Drum is being sent out the main output to the OBED Stereo path, while also being sent to the Side, rear, and even a bit to the LFE.

This approach works for individual tracks, busses, subgroups, and most other common mix practices. However, there is one word of caution. Depending on your DAW's plug-in delay compensation, the plug-ins you use, and the various latencies they may cause in your session, pay special attention to any strange timbral shifts of phase issues that may occur. These can be tricky to chase down when they happen within a busy mix project. Hardware inserts, plug-ins not reporting delay compensation properly, and any other complex mix routing may all be culprits in creating phase problems that will smear the sound source. There is no one perfect way to work, so keep a listening ear open to any of these changes.

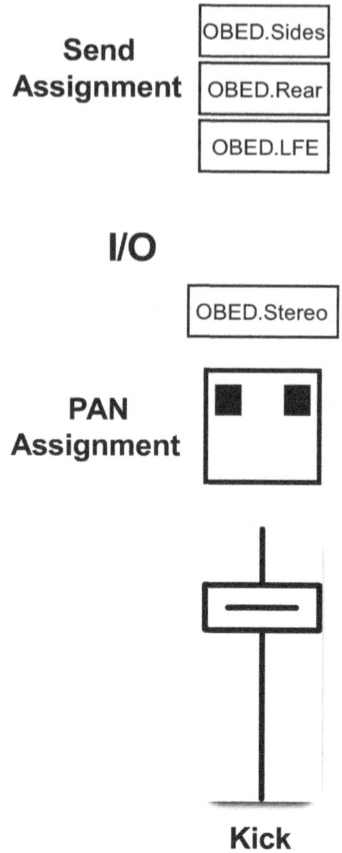


*Figure 26.2* Kick Drum track is being routed to the Stereo OBED path, along with the OBED sides, rear, and LFE paths to ensure that the desired impact from the kick drum is felt at the center listening position.

### *Four on the floor*

Pop/rock recording uses familiar microphone methodologies and procedures for adapting them to immersive audio formats.

Just as we have spoken about the importance of mono and stereo sound sources, the same is true for recording and capturing key instruments that are intended to have spatial priority in the immersive mix.

The use of coincident and near-coincident microphone techniques can be extremely useful in pop/rock productions that are intended to be mixed in an object-based format. Because of the relationship between correlated material and decorrelated material, the source can be expanded in the object-based environment to allow the engineer to 'naturally' manipulate the size, space, and impact of an individual sound source.

Though large multi-microphone arrays are wonderful in capturing a sound within a source or serve a necessary purpose in recording classical or jazz ensembles in a desirable space. However, when we work with pop/rock or other contemporary genres where we are able to dictate the space and soundstage to a certain degree, here common coincident and near-coincident microphone techniques can be used, such as:

- X-Y technique
- Blumlein technique
- ORTF (Office de Radiodiffusion Télévision Française) technique
- NOS (Nederlands Omroepproduktie Bedrijf) technique
- A-B technique
- Mid-Side (M-S) technique
    - Double Mid-Side (M-S) technique
    - Double Mid-Side + Z (Mid Side + Z (Height)
- Decca Tree

Consider Mid-Side, or double MS + Z in front of a kick drum, or ORTF across guitar or re-amping a synth pad and capturing that with NOS.

The power here is that the standard microphone tools and techniques we already are familiar with will transition into the Object-Beds that we created and will allow us to manipulate size and presence naturally by how those are placed and processed.

*Note:* Most DAWs that offer object-based immersive workflows have a 'size' parameter in dealing with the object in how 'big' it is in the environment. As size is increased, the sound source experiences more divergence and its sound is emitted from neighboring speaker locations. It is advised that exceeding a percentage or rate of greater than 10% (such as on the panner within Pro Tools) can become detrimental as there is enhanced spearing of spatial and timbral information. To begin, simply keep this between 0 and 10 depending.

### *Immersive mixing*

#### *Using a stereo reference*

An essential aspect of mixing multi-channel and immersive audio productions is the incorporation of a stereo reference. We've found that maintaining a stereo reference throughout the mixing process is very useful for achieving optimal results.

When working with formats like 5.1 or 7.1 surround sound, it's easy to get caught up in the intricacies of channel-based audio. However, relying solely on the direct speaker channels can lead to challenges in spatial coherence and overall immersion. That's where the stereo reference comes in.

Constantly referencing the stereo image while mixing in multi-channel formats will ensure that the integrity of the spatial soundstage is preserved. The stereo reference serves as a point of comparison, allowing one to gauge how elements translate across different audio setups and ensuring that the immersive experience remains consistent across various playback environments.

Moreover, the stereo reference provides valuable insights into the balance and cohesion of the mix. It helps me identify any discrepancies in imaging or localization that may arise when expanding the audio to multiple channels. This ensures that the immersive experience is not only technically precise but also emotionally engaging for the listener.

Particularly if we are working from a source that has already been released and we are working to 'make the stereo version *immersive*', the stereo mix is vital for direction for both balance and the intent and artistic vision of the artist and producer.

Also worthy of mention is the old way – mono. Mono-summed channels also can go a bit further by allowing us to check the sonic image of all channels on a single monitor. This is an ear-opener. When we have all that information together, it is a quick deduction that we need to minus certain channels in order to retain the intimacy and intelligibility while maintaining the expansive landscape we have worked so hard on. The stereo and mono also gives us a look into the overall phasing of our mix. This can be the *discrepancies* mentioned earlier. These phasing issues can be caused by simple elements like microphones in front and behind the drums for only one example. After summing, we may be dealing with phasing concerns. So this gives us a chance to make it right regarding phase and also balance the levels to maintain musical coherence in a given space. Another aspect to be aware of is the on/off axis linearity of the microphones used. If we use microphones that are linear in both on/off axes at all angles and frequencies – then, with a good transient quality – can we battle phase and non-equidistant room dimensions with better sound armor.

When we unfold the mix, it will be more coherent and in phase with correct polarity correlation. Our bodies can feel when all is correct, we are wired for accuracy.

Should there be more flexibility, where the immersive mix is not required to be complementary to the stereo version, more drastic mixing steps can be taken to ensure a mix that carries artistic intent but may serve to create a much more immersive listening experience that may differ from the stereo experience. Both are valid approaches but are largely dictated based on the artist, producer, and label direction.

In essence, the stereo reference acts as a guiding light throughout the mixing process, helping strike the perfect balance between intensity and envelopment in multi-channel and immersive audio productions.

In the case where there is no stereo reference and the *immersive mix* is leading the direction, or when there stereo mix will be derived from a re-render or stereo render or fold-down from the immersive mix, this stereo mix must be referenced during the process to ensure that technical issues do not arise such as comb filtering, frequency masking, or problematic balance issues or egregious cancellations or accentuations that will be distracting and extremely problematic.

### *Forget about mix-bus processing*

Mix-bus processing, a staple technique in traditional stereo mixing, involves applying effects and processing to the entire mix before it reaches the stereo output. However, in the realm of immersive audio production, the application of mix-bus processing presents unique challenges and limitations that make it less effective or even counterproductive. Let's delve into why mix-bus processing may not work optimally in immersive audio production.

### *Spatial considerations*

Immersive audio formats, such as Ambisonics or object-based audio, introduce a three-dimensional spatial dimension to the mix. Unlike stereo mixing, where the focus is primarily on left-right panning, immersive formats allow for precise positioning of sound sources in a spherical space around the listener. Mix-bus processing, which treats the entire mix as a single entity, can disrupt the spatial integrity of the mix by affecting the relative positioning and movement of sound sources. This can lead to spatial smearing, phasing issues, and a loss of immersion.

### *Individual object processing*

Immersive audio production often involves working with individual sound objects or channels that can be independently positioned and processed within the spatial environment. Unlike stereo mixes, where all sounds are summed to a single stereo output, immersive audio allows for the independent treatment of each sound source. Applying mix-bus processing to the entire mix can compromise the ability to treat individual objects separately, limiting the flexibility and precision of spatial processing.

### *Dynamic spatial interactions*

In immersive audio, the spatial relationships between sound sources dynamically change based on the listener's position and orientation within the virtual environment. Mix-bus processing, which applies effects uniformly to the entire mix, may not account for these dynamic spatial interactions. As a result, the spatial coherence and realism of the mix may be compromised, leading to inconsistencies in localization and immersion.

### *Loss of control and flexibility*

Immersive audio production often involves complex spatial processing techniques, such as binaural rendering, spatial reverberation, and object-based panning. Mix-bus processing can limit the control and flexibility over these spatial processing parameters, as it applies effects globally to the entire mix. This loss of granularity can hinder the ability to achieve precise spatial effects and may result in a less immersive listening experience.

### *Compatibility with rendering platforms*

Immersive audio formats are typically rendered using specialized playback systems or plug-ins that are designed to decode and render the spatial audio content in real-time. Mix-bus processing applied to the entire mix may not translate effectively across different rendering platforms, leading to inconsistencies in spatial rendering and compromised audio quality.

While mix-bus processing has proven effective in traditional stereo mixing, largely because of the desire to have all the sound sources represented (and 'fit sonically') into two loudspeakers or headphones, an additive process, such as compression, often presented the desired 'glue' to bring these elements together in two speakers across a mix. In immersive audio production, this type of mix-bus processing mindset often poses significant challenges and limitations. To achieve optimal results in immersive audio production, it is essential to adopt a more nuanced and spatially aware approach to processing, focusing on individual object processing and dynamic spatial interactions within the immersive environment.

### *How to reintroduce mix-bus processing*

There may be instances where one does desire to use the type of compression or limiting that one might often use on a stereo mix-bus. Though few 7.1.4- or 9.1.6-based dynamics processors exist currently, this does not mean one cannot use the tools they already have to a similar effect.

This is achieved by creating an Object Bed like Image 5 consisting of mono objects. Across all the mono objects, place the desired dynamics processor. In this case you will need to make sure that it is capable of an 'external' sidechain input because this will be needed to get them working together.

Here each of the object masters in the Object Bed are sending to an unused mono bus (in this case called SC [sidechain]).

Create another mono master trim control fader that has the input from the Object Masters Sends (SC). The output is sent to another unused Mono bus, here named To-Comp.

Now that the desired dynamics processor is across all the Object Bed Masters, the external sidechain for the dynamics plug-in should be sent to 'To Comp'.

Now on playback we have dynamics processing that is linked across all the channels in the Object Bed (Figure 26.3).

One may quickly realize that, though this is very functional, it is a bit cumbersome.

There are continuing developments in tools to simplify this process that offer more dynamic control. Two tools from PSP Audioware offer one such option.

- **PSP auralControl:** Allows for individual adjustment of level, polarity, bandwidth, and timing for each individual channel in the multi-channel set.
- **PSP auralComp:** This multi-channel compressor and limiter presents up to 16 channels of compression per multi-channel input.

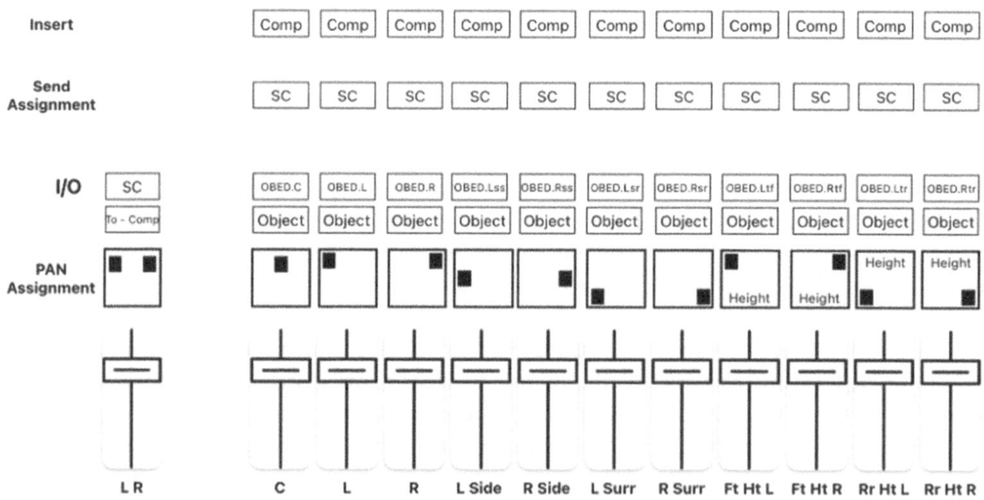


*Figure 26.3* Example of the dynamics processing linked across all the channels in the object bed

Both of these tools are brilliant and offer exceptional control and refine the workflow within your DAW.

Regardless of the workflow you choose, the desired results and control can be achieved by the use of purpose-designed tools, which offerings are continuing to grow, or by using one's own DAW's routing flexibility and tools already at their disposal.

## Future of immersive audio in pop/rock production

While there is a great focus on remixing back catalog music for immersive formats, the ability for artists, composers, arrangers, producers, music directors, etc. to begin writing and exploring an immersive soundscape and how this can support and accentuate their music and composition is just in its infancy.

New creative compositional tools are being developed and released which allow an artist to compose in entirely new ways. Spatial processing available in headphones and binaural renders that can be experienced over standard headphones all make the technology much more accessible.

Immersive audio is changing the way we create, experience, and share our music with others. What better than with pop, rock, and other contemporary genres to push these boundaries?

## Afterword/appendix

### *Strategies working between stereo and immersive audio formats as a whole*

Working between stereo and any immersive audio format requires a nuanced approach that considers the unique characteristics and capabilities of each format. Let's dive into the strategies and considerations involved in seamlessly transitioning between stereo and immersive audio.

### *Spatial awareness*

Immersive audio formats introduce a new dimension of spatiality, allowing for precise placement of sound objects in three-dimensional space. When transitioning between stereo and immersive formats, it's crucial to maintain spatial awareness and ensure that the spatial characteristics of the mix translate effectively. This involves careful positioning of sound sources within the immersive environment to preserve the spatial integrity and immersive experience.

### *Object-based mixing*

One of the key differences between stereo and immersive audio is the concept of object-based mixing. In stereo mixing, sounds are typically panned within the stereo field using pan pots or automation. In immersive formats, sounds are treated as individual objects that can be independently positioned and manipulated within the three-dimensional space. When transitioning between stereo and immersive formats, it's important to leverage the capabilities of object-based mixing to enhance spatialization and create a more immersive listening experience.

### *Adaptive mixing techniques*

To ensure a seamless transition between stereo and immersive audio, it's essential to employ adaptive mixing techniques that cater to the specific requirements of each format. This may involve dynamically adjusting levels, spatial positioning, and processing settings to optimize the mix for different playback environments and formats. By adapting the mix to the characteristics of stereo and immersive audio, you can maintain consistency and coherence across different listening setups.

### *Spatial effects and processing*

Immersive audio formats offer a wide range of spatial effects and processing techniques that can enhance the immersive experience. When transitioning between stereo and immersive formats, it's important to explore these spatial effects and leverage them creatively to enhance the spatialization and depth of the mix. This may include using spatial reverbs, ambience effects, and dynamic panning to create a more enveloping and immersive soundstage.

### *Monitoring and playback systems*

Effective monitoring and playback systems are crucial for accurately assessing and optimizing the mix in both stereo and immersive formats. When working between stereo and spatial audio, it's important to use monitoring systems that allow for accurate spatial representation and playback of immersive audio content. This may involve using a standard loudspeaker monitoring environment, headphones, or virtualization plug-ins that simulate the spatial characteristics of the target playback environment.

Transitioning between stereo and immersive audio formats requires a comprehensive understanding of the spatial characteristics, mixing techniques, and playback considerations involved. By adopting adaptive mixing strategies, leveraging object-based mixing techniques, and optimizing the mix for different playback systems, you can create a seamless and immersive listening experience that translates effectively across stereo and immersive formats.

Let's imagine that you recorded a project a few years ago with a stereo master. Now, you wish to have a product that can be considered 'immersive'. Well, to record an immersive recording, you will need to plan ahead and ensure that you record multiple microphones in key locations along with direct and indirect locations. This is required to produce a discreet product that accurately puts the listener in the space you were in when you recorded it. It can be difficult to get that older stereo recording and enhance it in order to magically become an immersive project that can be delivered to streaming platforms and multi-channel discrete playback systems. Despite this challenge, all is not lost.

Perhaps this can be 'immersive-ish' If we have the original stems from the older session, we could make sure they are discreet regarding all sources such as drums, guitars, vocals, brass, etc.

From here we can set up a stereo or mono loudspeaker in a space that we want to place the listener. We then re-record the stems back into a new session with fantastic ambiance from a local church, gymnasium, library or a room in your house that sounds cool. Why not?

Immersive is when something surrounds you. In sound, it could be more like being enveloped by something. Like being underwater without getting wet perhaps.

When we play back, for example, the drums through our playback speakers, we can set up an array of omni and perhaps a few directional microphones that capture the sonic signature of the space we are in.

One example is stereo playback, and the other is mono playback. The stereo playback can be captured in a multi-channel way so that the room becomes part of the enhanced immersive landscape. Place 4–6 omni microphones equidistantly on tall stands and record the playback, then bring it back into the session. We can also place a few microphones lower in the space to capture that energy that happens at this lower position (reflections off the floor and other possible boundaries). This source can be fed from a stereo aux feed. Take notice of timing or delays that are not lining up. Move the microphones or speakers until the sound feels natural, combine different levels of the closer microphones and the ones that are further from the source speakers. Also be mindful not to double the low end. If there is a lot of low-end energy in the source, the re-recording could cause smeared muddy sound in this frequency range at around 40Hz > 800Hz. If the re-recorded low end is desirable – consider backing off the source stem low end in the final mix so that they are not competing and causing a blur in these frequencies.

We can also place binaural microphones closer to the source to give us a great angle. Placing two first-order cardioids in an ORTF will also add another great aspect to the sound arsenal. The grid of omni microphones often brings back the low end accuracy and will deliver the sound sculpture that we desire to bring our basic mix to another realm.

This can also be done with a mono playback speaker, and we should be reminded that the playback speakers should be of good quality with full flat frequency response so that the sonics in your original tracks will not be altered in the frequency domain. We want to keep things neutral and flat during this process. The mono capture affords us the option to place the microphones in a similar way as just mentioned, or place them in a way that mimics the immersive playback system they will be heard in. Like 7.1.4, for example. Place the microphones in those (relative speaker) positions and record the mono source. The mono source could also be thought of as the listener position. If the playback speaker were a drummer, we would consider placing a more coincident multi-channel microphone near the musician with hopes of capturing his/her perspective regarding immersive capture. You can see that we are reversing this because we are using a playback speaker to mimic the musician. This source can be fed from a mono aux. In this example, there are 12 channels, and we will place the microphones in the 7.1.4 positions. This approach can be beneficial as it results in a *real feel* since we are really enhancing stems in a more *immersive* environment.

So now we have re-recorded some stereo sources and a few mono vocal performances. We now have new tracks in our session that have all the original source material, with another track that has benefited from all the sonics from the stem re-record.

We can now remix our older session in an authentic immersive way. If care is taken and attention to phase and timing – we can make a beautiful version of your work.

Experiment and do not be afraid to break the rules while also staying true to proven methods.

One word of advice while applying an 'immersive re-recording technique'. Keep the gain of your microphones all the same setting in such a microphone array. This will add to the accuracy and intimacy of your recording. If the microphones are further apart, there will be a feeling of more space but also sometimes at the cost of less directionality. Experiment with each project to find that balance of direct versus indirect sound. One solution never works for all, but practice will always yield a good starting point!

No doubt we are in an exciting time, and the tools and workflows continue to improve for immersive audio. There is no shortcut, and experimentation is called for, as we really are just at the threshold in all of this. As we employ our skills and knowledge in what we already do to create and capture great sound, we may find some new techniques and best practices that may shape our creative work moving forward.

## References

Genewick, S. (2020). 'Steve Genewick on Creating Music in Dolby Atmos at Capitol Studios | Hear from the Masters.' YouTube, 20 April 2020, https://www.youtube.com/watch?v=AsLTczWCHkM.

Lucey, B. (2023). 'Cardinal Points Quad – The Fastest and Most Musical Approach for Beating the Stereo Master in Headphones.' *Magic Garden Mastering*. http://magicgardenmastering.com

Miller, M. (2018). *What's the Use. Swimming*. REMember Music.

Chapter 27

# Producing and recording acoustic music in immersive

*Morten Lindberg*

## Background

I began my journey as a recording engineer in the early 1990s, specializing in classical, traditional folk, and jazz music. My passion revolved around capturing the essence of natural acoustic environments. Unlike other regions, Norway never embraced industrialized music production with rigid, compartmentalized categories. Consequently, it felt only natural for me to adopt a more holistic approach to the art of phonography. As a music producer, I now engage in the entire process, from conception and production to effectively communicating the final product to consumers. Establishing 2L as a music label was a seamless progression in my career.

During the 1990s, I took a leap and founded my own production company. At that time, the classical music production scene in the Scandinavian countries was thriving. My company offered production services to established labels like Universal Music, Naxos, EMI and others. However, in the late 1990s and early 2000s, the labels began to retreat, showing little interest in investing in new productions within the Scandinavian region. Faced with this dilemma, we had two choices: either close down or forge our own label. We opted for the latter, and thus 2L was born. Initially, our focus was on stereo recordings. However, being an independent label allowed us the freedom to explore new formats, such as surround sound and later immersive audio. We didn't require permission to experiment with different formats; we simply pursued them. This creative freedom allowed us to make decisions that served the projects best. This approach extended to our collaborations with composers and artists, as we fostered a 50–50 cooperation rather than enforcing formal artist agreements. Together, we jointly own the master recordings for their lifetime, only deviating from this arrangement if one of the partners departs. We have witnessed the captive nature of the record industry in many instances, and it's a true disappointment. Additionally, we do not allow for inheritance since our cooperation is built on the foundation of two or more parties. Historical experience has shown that children and grandchildren are not always the best caretakers of a legacy. Once the contract is fulfilled, we move forward and focus solely on what benefits the project at hand.

In my professional capacity, I wear two hats: that of an audio engineer and a producer. Prior to commencing a recording session, I assume the role of an audio engineer. However, as the session progresses, I seamlessly transition into the producer role. While I continually monitor from an audio engineer's perspective, any adjustments made stem from my position as a producer. This demanding yet fulfilling process requires a tremendous

DOI: 10.4324/9780429297526-30

amount of energy and focus. After a week of recording, I often find myself shedding around six pounds, not due to physical exercise but due to the energy expended.

Presently, we produce 10–15 new releases each year, encompassing Hybrid SACD, Pure Audio Blu-ray, downloads, and streaming services. Each release entails one to two weeks of recording sessions, allowing me to alternate between recording and post-production work. Sustaining a recording session every week for 48 weeks a year would be overwhelmingly exhausting. I have found that this alternating rhythm is the only viable way for me to maintain productivity. During these intense periods, I enter a mode where everything aligns seamlessly, and my creative faculties are firing on all cylinders.

My transition into immersive sound originated from a frustration with stereo recordings. Despite my efforts within the stereo format, I couldn't fully realize my artistic vision. However, at the time, there was no viable alternative. It was only when I encountered surround sound in 2003 and further met Wilfried Van Baelen from Auro-3D in 2011 that I glimpsed a potential solution. We had an ongoing project planned in two parts, with the first part already completed. Two months later, we were scheduled to record the remaining portion. Wilfried's demonstration left me deeply inspired, prompting me to construct a 3D microphone array during the summer break, which we employed for the second part of the album. During that period, I lacked a comprehensive monitoring system, relying solely on a theoretical framework to transform my microphone technique from stereo to surround by expanding it into the third dimension. We approached the recording, editing, and mastering in layers. With only a surround sound monitoring system at my disposal, I possessed an understanding of the contents of the upper and lower layers. A couple of months later, I met Wilfried at Galaxy Studios, as he was among the few individuals equipped with such a system. When we listened to the recording, the revelation was astounding. Adding the third dimension transcended the realm of intellectual exercise; it evoked a profound emotional experience that resonated within my being. The additional height saturated the music with a newfound emotional depth.

## Design philosophy

The beauty of the recording arts is that there is no fixed formula and no blueprint. It all comes out of the music. Every project starts out by digging into the score and talking with the composer, if contemporary, and the musicians. It is not our task as producers and engineers to try to recreate a concert situation with all its commercial limitations. On the contrary, we should make the ideal out of the recording medium and create the strongest illusion, the sonic experience that emotionally moves the listener to a better place.

With the 2L label we often record in spacious acoustic venues: large concert halls, churches, and cathedrals. This is actually where we can make the most intimate recordings. The qualities we seek in large rooms are not necessarily a big reverb but openness due to the absence of close reflecting walls. Making an ambient and beautiful recording is the way of least resistance. Searching the fine edge between direct contact and openness – that's the real challenge! A really good recording should be able to bodily move the listener. This core quality of audio production is made by choosing the right venue for the repertoire and by balancing the image in the placement of microphones and musicians relative to each other in that venue.

No method is available today to reproduce the exact perception of attending a live performance. That leaves us with the art of illusion when it comes to recording music.

As recording engineers and producers we need to do exactly the same as any good musician: interpret the music and the composer's intentions and adapt to the media where we perform. Immersive audio is a sculpture that you can literally move around and relate to spatially; surrounded by music you can move about in the aural space and choose angles, vantage points, and positions. Rather than reproducing a concert situation, we consider the Recording Art a discipline on its own. It gives us the possibility to place the listener in an ideal position and become an actual party to the event. Through a dedicated production of the music, we can maximize energy, reveal all the small nuances and avoid distractions. The emotional impact can be made more massive than ever. The conductor's position is the seat no audience can afford – until now with these dedicated recordings.

Our sonic image is created in the recording with dedicated microphone techniques. The composers and musicians need to perform to the extended multidimensional sonic sculpture, allowing more details and broader strokes. Then immersive audio and surround sound is just a matter of opening up the faders. When the music is created, performed, and recorded in immersive audio, then stereo is our most time-consuming challenge; figuring out how to preserve the total impact and level of details from this sculpture into a flat canvas.

All my work is dedicated recording sessions and my roles are both as an engineer and as a producer. This means I get to directly interact with the musicians with the sole purpose of creating the recording. All classical musicians are trained to project their sound all 150 feet down the aisle of the concert hall. The first thing we do is to move ourselves toward a more intimate communication with the listener. This type of sound-making affects both texture, volume, timbre and articulation. Then volume is the next step. It's the same with both singers and instrumentalists, but a string player is best for demonstration. The first 80% of added energy results in volume and enhanced dynamics in a musical sense. But adding further vertical power from the bow to the string only results in distortion and a shrill harshness. With this in mind, the musicians manage to produce a more beautiful sound for me to record. In those situations where I succeed with this approach also with a full symphony orchestra, magic occurs, and the string players start pulling out their earplugs. Combating the acoustic loudness war makes all the difference in the world on how they control their own sound-making.

### *The height dimension*

When I delved into the world of immersive sound, the incorporation of the height dimension was an experimental leap into the unknown. Uncertain of what to anticipate, the conscious design of the album I was working on had already outpaced my previous experiences in surround sound, predominantly focused on the horizontal plane. I never intended to specifically design for the vertical plane; rather, I focused on the horizontal plane, recognizing that the addition of the vertical dimension naturally infused the music with an emotional essence. It simply wouldn't make sense within the musical style I worked with to place a trumpet in the top layer. Instead, my approach revolved around the intentional utilization of specific spaces, such as churches, for inspiration.

To me, designing the audio is inseparable from incorporating the composition into that design. While I readily assume that role when dealing with old classical repertoire, working with contemporary music always involves inviting the composer and conductor to our studio well in advance of commencing any projects. This allows them to experience the dimensional space and its possibilities, irrespective of its immediate relevance to their specific undertaking. At times, there is already a composed piece, while in other instances, it's merely an idea awaiting development. Often, the composer will return and rewrite the music to embrace the potential of the vertical dimension. It is within this collaboration that we unearth truly intriguing results.

A growing number of composers demonstrate a genuine interest in exploring the vertical dimension. Nowadays, composers are raised and educated within a stage–audience orientation, entrenched in its traditions and the accompanying restrictions. They are steeped in the stereo world, accustomed to sounds being layered atop one another. While they have dabbled in the depth of field, the sounds still remain stacked. Our initial task is to liberate them from these constraints. Hence, it becomes imperative to disengage from the notion of a frontal image, as it significantly influences their compositional approach in terms of economy. They have grown accustomed to prioritizing sounds for the listener within their compositions. For instance, if they place a line in one instrumental section, they would then be unable to reuse that section due to potential overshadowing. I demonstrate to them how they can distribute the sonic experience, arranging the lines opposite each other rather than on top of each other.

Consider, for instance, passages involving instrumental sections where one line ascends while another descends. In their original configuration, the lines intersect at a specific crossing point. By separating them and assigning distinct spaces, they no longer intersect at that juncture. The lines gain newfound independence, forging their own melodic and harmonic trajectories. In the stereo plane, we were compelled to make choices for the listener, constrained by the limited window of auditory perception. We had to select what should take precedence in the foreground at any given moment. In the immersive realm, such choices are rendered unnecessary. Listeners enjoy greater navigational possibilities. This becomes apparent when welcoming first-time listeners into my studio. Seated in swivel chairs, they initially assume a typical, classical position. However, within moments, they begin to move organically. Their movement is not haphazard; rather, they are genuinely impacted by the musical focal points. They instinctively gravitate towards moments that captivate their awareness, fluidly and naturally responding to the unfolding sonic landscape.

## Production process

I approach each project without preconceived notions, allowing the scores to unfold before me in a distinct manner. While a conductor comprehends the structure and guides the musicians through the music, my perspective is different. I seek structures, balances, and colors within the score. My objective is to meld elements together or, alternatively, to break them apart and create transparency. Contrasts and convergences are what captivate me. The essence lies in separation and openness. It is from this point that I begin to sketch, ensuring the listener remains at the core, as the centerpiece, while these intricate structures materialize around them.

Every instrument possesses a specific volume and place within the overall balance. However, I find the texture to be of even greater importance. For example, with string players, the sound produced by the bow carries significance. By observing the score, one can discern whether a light, fast, slow, or heavily accentuated vertical force bowing technique is intended. Such details can often be corroborated with the conductor. In the case of a violin concerto, where the violin assumes a highly soloistic role, I would separate the soloist from the rest, even placing them in a distinct location within the sonic space. However, in a classical violin concerto like Mozart's, where tradition dictated the soloist's integration within the group, the solo part should simply rise above. In such instances, I align the groups together in space to maintain this delicate balance – a prime example of separation and blending.

I strive to achieve a sense of freedom while remaining well-informed. Though my recordings, which can be quite daring, have garnered recognition within our community, I also engage in sessions where I explore a more concert-realistic sonic image. Since I ceased performing live concerts two decades ago, I have embraced the dedication of recording sessions for various reasons. One important aspect is the freedom to decide the creative path we wish to pursue. Working closely with musicians during recording sessions is highly rewarding. Through multiple takes and focused attention on specific parts of the music, we aim to channel energy and deepen our understanding of each musical segment. This approach is not meant to manipulate their performance but rather to explore the music's profound depths. Unlike during live performances, where orchestras must maintain an overall perspective to deliver a complete concert experience, dedicated recordings grant us the time to delve into every aspect of the music.

## Working on a project

I have been truly fortunate to collaborate with individuals who possess a strong determination to create the finest recordings possible. The impetus for such excellence usually originates from either the composer or the musicians themselves. Everyone involved remains fully committed to the project until its completion, ensuring that everyone involved is truly satisfied. Time constraints are typically not a factor, even when working with a full symphony orchestra. This level of dedication is truly exceptional, and the resulting projects are remarkable from both a performance and technical standpoint.

In any project, you can achieve a good result with proper planning, preparation, and resource allocation. However, there is another level beyond good – a level that can only be described as magic. Magic is elusive and cannot be meticulously planned. It happens spontaneously, emerging from the depths of the creative process. Although I have not yet deciphered the secret behind this magic, I wholeheartedly believe in its existence. Magic occurs when everyone involved in the project is fully focused, energized, and in complete agreement on the direction of the creative journey. That's when true magic happens.

I have had the privilege of experiencing this magic on numerous occasions. However, it cannot be summoned or predicted. It occurs when the group's collective energy and interpersonal connections align perfectly. Yet it remains delicate and fragile. For instance, any distraction earlier in the day that could disturb mental preparation toward the session could shatter the connection and break the magic.

An illustrative example of this phenomenon is the album *Reflections*.[1] During one morning session, we meticulously worked on Ralph Vaughan Williams's *Fantasia on a Theme by Thomas Tallis*. We dissected the piece part by part, diligently fine-tuning each segment. It sounded good, but it was fragmented. After three hours, we took a lunch break. Upon returning, we decided to perform the entire work in one take (perhaps there was something magical in that lunch . . .). The energy, tension, and sound were simply perfect. It resonated beautifully, reaching a state of absolute perfection. We had progressed to approximately five and a half minutes into the piece when, unexpectedly, a single wrong note disrupted the flow. In that moment, the entire performance unraveled. The rest of the take became unusable. Those first five and a half minutes were used in the release, untouched by editing.

There exists a notion in our industry that the 'first take is always the best'. However, I perceive it as a process of construction and deconstruction. In my experience, the first take captures the initial spirit and reveals challenges within the ensemble. This initiates the process of breaking down and identifying areas for improvement. It is this perspective that dampens the energy. Consequently, subsequent takes become unusable as they serve the purpose of resolving those challenges and problems. Once the issues have been addressed, we must embark on the journey of rebuilding the energy. In my opinion, those who claim that the first take is always superior are settling for the initial good and spirited take. Instead, we should persist, work through the process, and rekindle the energy that fuels greatness.

## Technical process

Our 2L-cube is only remotely inspired by Decca- or Mercury-tree. The microphone array is really a direct consequence of the speaker configuration in the immersive playback systems. Time of arrival, SPL, and on-axis HF texture are directly preserved in this 7.1.4 microphone configuration. Proportions are cubical, and the dimensions could vary from 150 cm for a large orchestral array down to 40 cm in an intimate chamber musical setting. I always use true omnidirectional microphones in the main array. Depending on the room, the music, and the instruments, I alternate between the DPA 4003 and the 4041, the latter with the larger membrane providing a more focused on-axis texture. My very first microphones were a matched pair of B&K 4003, recording straight to DAT. Rather than exploring a wide range of makes and models, I stayed with DPA and spent my time getting to know them intimately. To maximize the potential of the DPA High Voltage concept, we have a custom-built power supply feeding 200 V and 130 V on 4- and 7-pin XLR totally transparent to the microphones.

Refining the microphone array is a constant evolution. We started with a 2L-cube for 5.1.4 then expanded it with side-fills within the cube for 7.1.4. Lately, we tend to stretch the cube into a 2L-prism where the sides are now the surrounds, and the surrounds slide further back to become rears. The typical dimensions are 1 m wide, 2 m deep, and 1 m high.

When preparing for distribution, we target the playback situation with separate mixes. For Auro-3D or Dolby Atmos, then, all 7.1.4 microphones go directly to each according to their loudspeaker. With diminishing numbers of loudspeakers, we do not sum or fold down. We take away sources. So for 5.1, only the lower bed, without the side-fills, of microphones are active. Then it is usually only front left and right microphone playing in stereo. Possibly with a slight texture added from the rear microphones. Pure, clean, and minimalistic. Our philosophy is simple; one microphone straight to one speaker. The important aspect is to configure the array so that time of arrival is captured and released in natural order.

### *Microphone technique*

My approach to microphone technique is very simple. I prefer to keep the setup minimal, typically utilizing only the main array and rarely using support microphones. The main array, being consistent and fixed, is the cornerstone of my approach. Its essence lies in predicting the playback environment. In today's context, the most prevalent playback configuration is 7.1.4. Therefore, my microphone setup aligns with this configuration, albeit on a smaller scale. Essentially, the microphone arrangement mirrors the speaker placement but in a reduced form. For each project, I determine the dimensions of the array. For instance, in a full orchestral setting, the dimensions might be 1.2 m wide, 1.2 m high, and 2 m deep. Conversely, in a chamber setup, this configuration would be scaled down by half. However, the order and configuration of the microphones remain essentially the same. The primary distinction lies in how the musicians are positioned in the room relative to the microphone array.

During the recording process, I occasionally make minor adjustments. As the day progresses, I seize opportunities to listen attentively, and if I sense that the microphones should be positioned 10 cm lower, I implement the change. I am confident in managing these slight variations, even when it involves editing across different sections. However, substantial changes are avoided. I limit myself to minor modifications of just 5 cm in elevation, lateral movement, or forward/backward positioning. In some instances, when we feel certain instruments need to be more prominently featured, instead of altering the microphone placement, we adjust the position of the musicians themselves. For example, we may request the violins to move forward by 30 cm. Moreover, within a piece, the mood can transition from one movement to the next, prompting a desire for more spaciousness. In such cases, we might raise the microphones by 15 cm to achieve the desired effect.

### *Method*

My key to a good recording is to keep the signal path clean and short, with as few components as possible. I don't spread as many microphones as possible for someone else to pull the faders in post-production. I make my choice there and then, so I have a result that will serve the fundamentals of decisions to follow. A simple and pure signal path is the means to true high-resolution audio – not only at recording but also preserving the purity all through editing, mixing, and mastering.

On venue recording sessions, I find that headphones give me exactly the kind of information I need, which is what the microphone is picking up at its current position. You can do pretty much anything and have full control on headphones. I emphasize this because tradition says, 'You can't record or mix on headphones'. Yes, you can, but you need to relate it to a speaker environment by experience. Build a sense of transfer functionality, which makes you secure on headphones. You also must have a very good mastering studio so you can make the final production sound as it did while recording on headphones. I think one aspect of quality production is to step back and just relax and actually allow yourself time to experience what is really happening. Only then can you make the right choices on how to move forward. Stop, listen to what's happening around you, reflect, and then act.

We record at 24-bit 352.8 kHz and preserve this resolution all through into our distributed master files. There has been much focus on word length, but to me the sample rate is more important as impulse response connects directly to our primal sonic perception on a subconscious level. The most important aspect of post-production is to not destroy the fine qualities captured at recording. Instead of EQ in post, I much prefer to rather shift the angle or the distance of a microphone already in recording. It just takes planning. With a good recording, I never use any EQ or dynamic processing at all. Editing is an important tool. It makes it possible to combine the highest level of energy and details into an intense performance. To do that, you need to use a sonically transparent workstation. For the past decade we've worked with Merging Technologies on their Pyramix system. Their latest development allows for an immersive workflow totally independent of destination format. I don't mix for a codec. I record for a playback environment. Any codec should then ideally provide for a transparent transport to the consumer.

## Session preparation

When preparing for a recording session, my first priority is to ensure the system is well-prepared. This involves visiting the recording venue a day in advance to set up the physical rig and conduct a live system check. The initial task I undertake is what I call 'noise hunting'. I focus on eliminating all extraneous noise, such as radiator noise or faint buzzing sounds. It is crucial to create an environment that resembles a black canvas, where all the light generated in the recording comes solely from the instruments themselves.

One notable project, *Tuvayhun – Beatitudes for a Wounded World*, was recorded at Nidaros Cathedral in Trondheim, Norway. This magnificent cathedral, constructed in the 11th century, has undergone various installations over time, including amplifiers, electrical systems, and fire alarms. Unfortunately, there is no coherence among these systems due to their installation at different times. Therefore, in collaboration with the church manager, we have developed a comprehensive checklist that encompasses over 120 devices throughout the church, including the balconies, roof, and lounge. It takes approximately 45 minutes each day to disable these systems before we start recording. Consequently, this process results in the loss of lighting, prompting us to bring our own lighting equipment to the venue.

Completing the physical setup and preparation the day prior to the session is essential to ensure a smooth start in the morning. The setup requires a significant amount of energy and serves as physical exercise for the team, which is why we endeavor to have everything ready in advance. As for finding suitable spots for musician placement, I don't follow any specific routine. Over time, I have spent extensive periods at various venues, visiting and listening in those spaces. Additionally, I have had musicians join me to perform and evaluate the acoustics. This accumulation of experience has provided me with insights into a wide range of venues. Larger rooms, like cathedrals, are not merely one unified space; rather, they consist of chambers and pockets with distinct acoustic characteristics. Depending on the desired sound, we can choose from at least five different acoustic signatures within such spaces. For instance, when working with a grand organ, we naturally position ourselves in that specific area of the church. Similarly, for an a cappella choir, there may be a preferred spot at the opposite end due to the ornate surroundings and sculptures.

The placement of musicians within the venue is a process rooted in the score. Everything I do is driven by the score's musical logic. It is essential for each musician to instantly grasp why they are positioned where they are. This approach is guided by the score itself, and ideally in collaboration with the composer.

For a project with the London Philharmonic Orchestra at the Watford Museum, focusing on a contemporary piece, we carefully planned the session. The group leaders and stage manager were well-prepared, but the musicians themselves were unaware of the plan. Interestingly, the tuba part coincided with that of the double bass. Consequently, we decided to position the tuba player in the middle of the double bass section. When I initially observed the tuba player's reaction upon taking his seat, it was apparent that he was not entirely pleased. However, as the downbeat approached, our eyes met, and he instantaneously understood the musical reasoning behind his placement.

In my process, I begin by thoroughly studying the score and creating a drawing that depicts the musicians' placement. This drawing is then shared with the composer, who may provide insights and opinions, typically focusing on the structural, architectural, and aesthetic aspects that the configuration might emphasize. Based on the composer's input, revisions may be made to the drawings. These revised plans are subsequently shared with the group leaders of the orchestra, who typically do not express strong opinions regarding aesthetics but rather concern themselves with practical aspects of communication, such as sight lines and proximity between different sections. Taking their input into account, further adjustments may be made. This planning process occurs months in advance to avoid any surprises in the days leading up to the recording.

### *Preparation with composers*

During my preparation sessions with composers, I emphasize the importance of letting go when it comes to translating their musical ideas into a recording. I explain that relying solely on performing musicians to recreate their thoughts will only achieve partial results. At this stage, composers must let go of their control and recognize that they have created a framework and hopefully provided inspiration for the musicians. When it comes to recording, the musicians need to assume responsibility and not simply act as executors on behalf of the composer. This is less of an issue when working with older compositions because the composer is not present, and musicians are accustomed to interpreting the music on their own. However, in the presence of a composer, it's common to see musicians constantly seeking guidance from them about their preferences.

Therefore, composers should step back but still remain engaged during the recording process. They should be present as a source of inspiration and guidance. There are instances where the performance may evolve in different directions, and having the composer present becomes crucial in determining which direction they prefer the performance to take. It's a delicate balance between allowing the musicians to infuse their own interpretation while having the composer's input to guide them.

Ultimately, my aim is to convey to composers that their role shifts during the recording session. They have laid the groundwork, and now it's the musicians' turn to bring the music to life. The composer's presence is valuable, but it should be as an inspiration and guiding force rather than micromanaging the performance.

## Final thoughts

If your goal is to create a hit album, cater to the consumer market, or please a specific audience, chances are you'll end up failing. The key to navigating the creative process successfully is to aim for what you personally would like to experience. Developing this instinct doesn't take long. I have built that confidence. I trust it. I act on it. I firmly believe that a sense of good taste in experiences develops early on, as evidenced by young students who know what they enjoy but may feel insecure or question whether their preferences will please others.

In recent times, we've witnessed a resurgence of composers driving the artistic direction and pushing for technical solutions, some of which may not have been previously considered from a pure audio engineering standpoint. This dynamic has led to the creation of new tools that expand the possibilities in music production.

Regarding immersive audio, we have made significant progress in creating captivating experiences. However, even with the emergence of streaming services, the ease of use and accessibility for consumers in the immersive audio realm still has room for improvement. While dedicated fans are willing to invest in multiple speakers and equipment, my desire is for everyone to have access to a more natural and live audio experience. I envision a future where the delivery of high-quality, transparent audio becomes effortless and seamlessly integrated into daily life. Users should be able to enjoy immersive sound without needing to worry about preparation or technical details. This natural model, which prioritizes both quality and ease of access, is the direction in which the industry should evolve.

## Note

1 https://www.2l.no/pages/album/125.html

Chapter 28

# Perspectives on immersive recordings

*David v.R. Bowles*

## Starting with stereo

When I started recording in stereo, I liked the spaced pair technique best because of the time of arrival differences, and the very good sense of envelopment I got when using omni microphones. I had always associated directional microphones with radio recordings that I heard growing up: a coincident pair or semi-coincident pair placed far enough away so that it would pick up the biggest configuration of an orchestra or opera production. But that resulted in a loss of immediacy. Therefore I thought about broadcast recording as more distant, and commercial recording as more up close. I always asked myself the question, 'What makes a recording sound better, or is it really a combination of engineering and production?' What I didn't realize at the time was that I was gravitating toward recordings which had this sort of 'omni envelopment'. And not surprisingly, it turns out these were ones that were based on three microphones: either a Decca tree or RCA's 'Living Stereo' techniques, both of which had a left, center, and right microphone. Back in those days, RCA used the center mic for the mono release, and the left and right mics for the stereo release. Unfortunately, in the RCA releases, those mics were spread out quite a bit, so you tended to get the hole in the middle, but you got this incredible sense of detail from these classic Neumann M-50 microphones.

Going back to my performing days, I was recording for a label called Harmonia Mundi. Legendary recording engineer Tony Faulkner explained about the M-50 as we were doing a project with a soprano whose voice was a little bit edgy. He changed one M-50 pair with another one: and I asked very innocently, 'You have more than two?' and he said, 'Yes, I have a whole suitcase full'. I then asked, 'Why do you need to match these up', and he said 'Because the tubes wear differently, and I need my pairs to sound the same'. That stuck in the back of my head – this means that one's choice of microphone isn't just polar pattern or manufacturer, but there's also this sort of undefinable thing: the quality of each particular mic. Later, when I started recording, I would study a lot of mic specs and, when possible, would put up a couple of pairs together and do a recording concurrently. I started just recording mixdowns to stereo and very soon moved to an eight-track recorder with a very nice conversion system. That way I could mix in a better environment than on location. What struck me immediately was how these different mics would respond to different acoustics.

DOI: 10.4324/9780429297526-31

### *The room*

That gets me to the second point, which is about the rooms I was recording in: they are often not ideal. I had experienced some of these acoustic spaces as a performer, and we always knew that, when we moved from one venue to another, we'd have to make some adjustments as performers. Later on, it was really quite the same with sound engineering. I was once working with a client recording all four of their concerts in different venues, and so I thought, 'Okay, I'm going to play this game of not having the mic stand always at a constant height: for the more resonant acoustic, I'm going to lower it and put it a little closer to the group (and then do the reverse for the less resonant spaces). It turned out that I actually had to mix some of these performances together for a commercial release, so I was glad I did that! But this leads to another issue: when we are in these different acoustics, we have to be able to change our techniques accordingly to get acceptable results by adapting our setup (not to say always use 2.75 m up etc.) Going back to my training at the Aspen Music Festival, I worked with John Eargle. He always advocated not just putting an ORTF pair straight up and down, but tilting down a little bit, so the back of the ensemble was on axis. And I thought, 'Okay, that actually makes sense, because then you get a little bit more of the front of the ensemble, and more of the ambience coming in from the audience'.

## Working with surround

In 1998, I began experimenting with an eight-track machine, focusing on surround recording to understand the spatial relationships among microphones. I explored not only the front-to-back distances between the main array and surround mics but also the positioning of the rear microphones. According to the ITU specification, the left and right channels are positioned at 30° off-axis, while the rear left and rear right mics have more flexibility (100–120° off-axis). I suspected this flexibility was due to head or body shadowing, which indicated that precise positioning wasn't always necessary. Early surround sound and movie soundtracks used rear mics more for ambiance than for directing the audio.

While judging a student recording competition, I listened to a surround entry with very few instruments, with each placed in different channels. At home, it sounded fine, but in the AES listening session with large speakers spread widely apart, the whole mix fell apart. What was intended to be an immersive experience instead resulted in excessive direct sound and an unclear spatial environment.

In both stereo and surround recordings, we can create illusions of depth. However, immersive audio can also provide a method of capturing height. When I began height recordings, I lacked playback options at home, so I simply played them back in surround. During a session with a client, I included the height layer in a 5.1 surround mix. After playing it, I asked them to listen again after removing the height channels. They immediately noticed something was missing, and once the heights were reintroduced, they perceived sounds seemingly emanating from the ceiling, highlighting that our perception can exceed our playback capabilities! This experience underscored psychoacoustics as well as the benefits of object-oriented mixing (where specific placements create a more accurate spatial audio experience).

### *A flexible approach to ensemble recording*

In 2010, while at Skywalker Sound, I recorded a chamber orchestra performing Vivaldi's *The Four Seasons*. I positioned the solo violinist at the center, with the rest of the orchestra arranged in a horseshoe around her and the music director on harpsichord facing toward her. I placed the first and second violins in the left and right rear channels, and utilized spot mics for moments in the spring concerto that featured bird-like effects and solos. This arrangement created a triangular musical dialogue within the surround image.

Although I faced a channel shortage, I set up two height microphones for both the front and back of the group. Initially, I intended to use them for ambient sound but later remixed the recordings with height information, realizing that immersion works well for some groups while others may not lend themselves to it.

Typically, I record in an arc, but during a lockdown video I needed to record a 12-member chorus spaced in a circle. I needed an adaptable recording setup, as the video reversed position at the end of the program! I devised an array using four mid-side pairs arranged in quadrants, allowing me to mix in stereo or 7.0 format effectively. I also reversed the left-right-front-back elements for that last session and retained integrity of the ensemble.

In general, large ensembles sit on a stage which bulges in the middle toward the audience. I've also recorded some of those performances, adjusting the group from a crescent facing the wrong way to the right orientation for better immersion. This flexible approach improved the overall experience, suggesting that creative use of space and mixing can transform recordings from a literal documentation. My goal is to create a listener experience that is neither from the conductor's nor the audience's viewpoint but rather something accessible and inviting.

I always consider how performers adapt their playing beyond traditional expectations. Recently, at Coast Mastering (with Michael Romanowski), I worked with the same orchestra and microphone arrays, but in the immersive mix I encountered vastly different sounds from individual players. Even after remixing to achieve some similarity, the inherent differences remained apparent. In such cases, I focus on the unique characteristics of the ensemble, considering individual sounds, the type of music, and whether the desired outcome is a lush atmospheric quality or a detailed and crisp presentation. In this case, we opted for the lushness in the instruments and chorus, yet detailed presentation from the vocal soloists. The interplay of these factors ultimately shapes the final recording. The same applies to binaural re-renders in Dolby Atmos: these present a different listening experience than a stereo mix!

Before a recording project, if I can't listen to a group in person, I often ask them to send me recordings, or I research their work online. For one particular group, we had collaborated on a composer's CD, which included other works performed by different ensembles. We got along well and began discussing a new project, but then the pandemic caused delays. When we finally reconvened, that project had evolved into something quite different: they had commissioned new music now paired with pieces written about 150 years earlier.

During the session, I noticed the commonality in the music – it often had material that felt slightly off the beat, creating unique rhythmic moments. This influenced my approach; I aimed for atmospheric backdrop but had to ensure it didn't blur those rhythmic distinctions. We decided to record in a location I had previously worked in, which helped me anticipate potential challenges. When we tested the piano, I found its low end sounded fantastic but was too strong for the acoustic space, indicating I would need to ask the musicians to balance differently (*and* I'd have to adjust this in post-production).

I recalled a previous session at Skywalker Sound with another piano that lacked a low end. In that case, I was constantly encouraging the pianist to play out more, only for the player to express concerns about volume. This highlighted that each piano has its unique sound profile, and there isn't a one-size-fits-all recording technique. I've recorded multiple pianos at different venues, resulting in varied sonic experiences with each setup.

Years ago, I worked with a soprano in a newly built concert hall that looked promising on paper. However, upon entering, I discovered one wall was entirely glass, leading to a highly reflective acoustic environment. This considerably changed the sound quality for the soprano, producing results that none of us expected.

What resonates with people is the balance I achieve between the musical voice, the instruments' presence, and the surrounding ambiance. My recordings prioritize the score's connection and engagement rather than strict concert realism. I find beauty in how the score speaks directly to listeners, creating a magical experience.

### *Production process*

As for my production process, I focus increasingly on pre-production as time goes on. While I always recognized its importance, I've appreciated it more since the COVID-19 pandemic, especially with limited setup time or the need for greater spacing among musicians. This flexibility in placement has proven beneficial – musicians often adapt better post-pandemic, when asked to spread out or change angles for better ensemble cohesion.

In any acoustic setting, my approach is to quickly assess the space, understanding its characteristics through my own voice, stomping my feet, and by clapping my hands to assess resonance and possible slap-back echoing. Setting up microphones can sometimes cause performers to worry about hearing each other, requiring adjustments to maintain sound quality. Occasionally, I've worked with chamber orchestras positioned in a way that departs from traditional concert setups, allowing for better sound recording.

I've always aimed for a balance between direct sound and ambiance, recognizing that standard stereo techniques often result in a compromise. However, with immersive recordings, I've discovered that I can position the main array closer while still using ambient mics effectively. Early surround recordings sometimes struggled with poorly correlated rear channels, but I learned to create a coherent sound image for both front and rear channels, bringing depth to the listening experience.

Now, as I work more with immersive recordings, I've found more ideal configurations for capturing envelopment. I recorded a 12-person chorus arranged in a half-circle, which worked exceptionally well. The center microphone setup for large chorus recordings has evolved: I typically use three front mics arranged in a straight line to ensure that the sound reaches each channel simultaneously, creating a natural and balanced sound field. For the rear microphones, I position them wider apart. My initial height setup involves placing four supercardioid microphones above left/right/left surround and right-surround main microphones, ensuring that left and right distances matched those of the main left and the front left microphones. If height microphones were pointed completely vertical, the null would be angled down at 30°. Pointing them 30° down effectively aims the null towards the soundstage (rejecting those signals). This arrangement works well, especially with height speakers installed in the ceiling.

Before the pandemic, Paul Geluso and I discussed experimenting with an array using a specific imaging technique for left-to-right height. This concept involved using a simple stereo configuration for the rear mics, which could be conveniently mounted on a central stand or stereo bars. For both choral works, the side microphones were placed in different configurations, allowing for various recording approaches. Bringing side channels into an immersive recording brings additional challenges: where are those speakers in your studio – and will they be in the same place for various listening environments? If they are designated as objects, then they will reproduce more or less correctly. As with rear channels, there is considerable flexibility with placement . . .

One significant improvement was the use of more than four height microphones. I successfully used height outriggers during a recording at Skywalker Sound. The Scoring Stage has a high ceiling that doesn't return much sound. However, that was never released immersively, and I haven't had the opportunity to replicate that setup.

### *Cappella SF: a laboratory of immersive recordings 2014–2018*

This chorus of 24 members was composed of the top musicians of the San Francisco Symphony Chorus and ran as a separate organization with a separate board of directors. Music director Ragnar Bohlin was amenable to my recording using an immersive setup; I connected them with Delos Records, and we made a total of four recordings during this period. All recordings were released in stereo (CD and downloads up to 24-bit 192 kHz). There was a successful initial release of Christmas music, followed by a funded project of the music of Conrad Susa and David Conte. The third project, titled *Timeless*, ranged from Hildegard von Bingen to contemporary music, and really captured the essence of the space we recorded in. There are moments where you can hear the sound carry beautifully, particularly during loud notes followed by silence, allowing the nuances of percussion and high soprano parts to shine. This recording earned top ratings in *Stereophile* magazine, favoring this over three other recordings made by my distinguished colleagues!

### *Recording Mason Bates's* Mass Transmission

This project, which is currently released as an immersive download on Tidal, Amazon and Apple Music, features composer Mason Bates. He has a rich background in electronica, is active as a DJ, and works extensively with symphony orchestras. I emphasized to the team that everything needed to work in stereo, which influenced my decision to use traditional microphone arrays. For a test recording, we opted to position the organ at a distance, which improved the overall sound balance. However, the organ was meant to be an obligatory instrument, integral to the whole. The virtuosic writing of the piece often made it necessary for the organ to stand out in the mix; that particular instrument lacked the 'punch' we desired. Also, we encountered challenges because Mason needed contact with the conductor (who was conducting a 24-voice chorus in a resonant church). Unfortunately, the setup didn't function as planned, prompting us to rethink our approach. Mason's electronica setup was noisy, and the main microphones picked this up all too accurately! I discussed the need for a pop-style recording style with the ensemble's executive director and the music director of the chorus. Since the score includes tempo and meter changes, a standard click track wouldn't suffice; I had to create a customized click-track, one using music scoring software to ensure proper beat emphasis. The first track we recorded was of the organ. Hence, we recorded the elements separately: organ, chorus, and electronica. The conductor was able to listen to the click track while working with the organ registrations, allowing for more focused attention on the chorus as the session progressed. To address issues during the session, we recorded the pieces in larger sections and planned for pickups, enabling us to navigate around timing challenges. The organ track proved virtuosic, which resulted in a clean recording. For the chorus session, I provided the conductor with the edited organ track, plus the click, allowing us to synchronize their performance more effectively. In post-production, I utilized time-stretching software to ensure the chorus and organ lined up perfectly, achieving a more cohesive sound. Mason than recorded his electronica with the organ, chorus and click-track. However, my only regret was not being able to record electronica elements separately, which would have allowed for a better immersive listening experience during mixing.

Instead of placing all elements in front, I repositioned the chorus front and center while placing the organ toward the back. The rear channels focused on the organ, simulating the experience of moving toward the organ pipes. This configuration enhanced the stereo effects, allowing for better perception of depth and sound placement. Both the organ and choir were tracked in a 3D format, and the electronica was captured in stereo. To compensate somewhat for some limitations, I created a reverb track of the electronica to add depth to the sound. The electronic elements started in the front, gradually transitioning to the back and sides, allowing listeners to become familiar with each component as it was introduced. By carefully crafting crossfades and spatial movements within the mix, I maximized the auditory experience. However, I still wish to incorporate more elements that move in both vertical and horizontal directions during future projects.

For this album, we opened with his 12-part piece *Sirens* that introduces listeners to his innovative style. This piece explores themes such as the siren song, drawing parallels between that and *Mass Transmission*. This piece explores the historical context of the Dutch telegraph office, which facilitated radio calls between the Netherlands and Java – a remarkable achievement in the early 1920s, given the time delay involved in such communication.

I was cautious about not revealing too much about my plans for immersion. The initial stereo mix performed well, earning an Editor's Choice feature in *Opera News*. As I transitioned to immersive mixes, I began to understand both the opportunities and challenges they presented. If given the chance to redo certain aspects, I would consider adjusting the chorus balance and communicating clearly with Mason about the electronic elements, ensuring they complement rather than overpower the organ and chorus.

### *Realizing artistic vision and immersive sound*

It can be difficult to get buy-in from others on creative ideas if they lack familiarity with the concepts. Whenever I present work at Dolby, those attending are often impressed by the immersive quality of the recordings. Feedback from different projects has led me to evaluate my microphone placements, with variations based on input from music directors and my own evolving skills.

It's essential for sound engineers to revisit their past projects, listening without harsh self-critique but rather reflecting on growth and changes they would make today. This includes not only technical choices like microphone placement but also artistic decisions regarding how to collaborate with the musicians. Moving forward, I plan to engage more explicitly with artists about the possibilities of immersive audio.

With the advent of Atmos releases, I can now present these concepts practically, allowing potential collaborators to experience what immersive audio can offer through specific headsets or home theater setups. This practical approach feels like a significant step forward compared to where we were just a few years ago, around the time of the first releases on Amazon.

For an immersive project, I aim to create a clear sense of space, allowing listeners to identify the performers around them and feel their presence. I dislike overly scattered sound; some orchestral recordings fail to convey where the focus should be or the location of each instrument. Specificity in microphone placement and musician positioning is crucial, and I believe these elements must be honored. Unfortunately, some pop remixes of immersive audio do not respect this specificity. If a producer has enough multitrack parts to reimagine a piece, that's one thing; however, working with only stereo material or a few spotlight tracks without a coherent vision often leads to unsatisfactory results.

I respect professionals like Jean-Michel Jarre, who revisited and remixed *Equinoxe* and *Oxygene* for surround sound. He approached his work anew with a clearer vision, fully reimagining the pieces rather than merely engaging in an electronic reprocess of stereo (which often feels like a marketing gimmick without true artistic purpose). I found another pop recording that was effective in its immersive remixing: it used front and side speakers to create a rough arc of the performance. This approach honored the original intent and consequently worked well for me. The same goes for immersive remixing: if the original team retained all the materials from sessions, they should be the ones to decide how – and if – an immersive mix is possible.

I believe recording should be viewed as an integrated process, combining the acoustic environment, performers, and equipment. Decisions about the acoustic space must be made early on. I consider whether the acoustic feedback is too dominant in the low or high end and explore ways to manage it – like using platforms, rugs, adjusting curtains or moving panels.

Once the acoustic environment is addressed, the focus shifts to performer and microphone placement. Sometimes, I enlist a second person to assist with placement, as in a recent project where I wanted to record in a specific location. After discussing it with the director, I suggested repositioning the chorus based on my past experience with the violins in that space. We collaboratively devised a basic positioning strategy while considering both our inputs. This reflects the importance of working with performers to ensure they feel comfortable and effective in their setup.

I've encountered sessions where producers dictate placement without considering the musicians' perspectives, leading to discomfort and confusion. It's vital for performers to have visibility of the conductor and to be able to communicate with one another during a performance. Unfortunately, in some cases, it's challenging to balance effective microphone placement with health and safety concerns, especially when the mic stands can't be positioned appropriately without obstructing the audience. These are the compromises every sound engineer must navigate.

Chapter 29

# The vertical AB system, content correlation, and the David Bowie exhibit

*Gregor Zielinsky*

## Background

My first contact with 3D recordings was at a presentation in Berlin. It was interesting but not totally convincing. But then one or two years later, I thought I could give it a try. During an orchestral recording of *Pictures at an Exhibition*, I had the idea of setting up an immersive system, and we made an event out of it, and many people visited us there. I remember when I switched on the top two speakers, we were completely struck. Tears nearly ran down my face because I had never heard anything like it. From that moment on, it was clear to me that immersive recording will be a very important part in the future of audio engineering. At the time, we did not know whether we were going to be able to really make this a relevant platform. Today, it certainly is, but one has to remember that 5.1 surround sound was never all that successful in the music area. And, of course, we all knew people who would say that 5.1 is just another format that will go away.

The format that we used at that time was Auro-3D and it was mainly an issue of how the top speakers were placed. Then later Dolby Atmos, which was cinema based, had some height speakers above you instead of directly above the main layer, as in Auro-3D. We were also allowed with Dolby Atmos to put the top speakers where the Auro immersive top speakers were as well. The main question then was how do we find an immersive playback platform for customers as we didn't have a good connection to the final consumer? Home cinema enthusiasts were the ones most willing to set up an immersive speaker system at home. A very important step was to figure out how the consumer can get immersive sound.

## Designing an immersive experience

When I think about designing an immersive experience, the main goal is to get an emotional result. That is absolutely the most important point, although how we achieve this can be through completely different ways. In the beginning, we had the idea to produce the feeling that you're sitting in a concert hall for an orchestral recording. With all the translation going on, you may need to compensate for the difference between real life and a playback through loudspeakers. You may also need to emphasize certain things to bring back the emotion that you have in real life. This is also true in stereo recordings, but there is now the opinion that, with immersive, we should obtain a true neutral photo of what is happening. In reality, it is not always possible.

DOI: 10.4324/9780429297526-32

In the beginning we experimented with the orchestra coming from the front, and the rest was only reflections. This sounded already very convincing, very emotional. Eventually it became clear to me that this is not enough. We can have lots of spot microphones and put them all around, and then we have lots of mono sound sources, so now we're experimenting with using multiple signals from each source to get away from this multi-mono effect. If your speakers are in close proximity, such as in a standard studio control room, your mix may perform well. However, the challenge arises when transitioning to larger rooms. In such cases, the mix may not translate effectively. Complicating matters further is the unpredictability of home setups, where the specifics of people's audio equipment remain uncertain. I want presence, but it must all be embedded in immersive.

There are many people producing immersive music that don't have the traditional classical recording background that I have. For these people, it's much easier in a way to approach the topic of immersive production because they don't have all these rules in their mind based on the more traditional stereo classical recordings. Many of these rules are not relevant anymore. You can still save the first strings from the left, but you also have music where the strings can come from back or all around. The Beatles worked with strings in a very non-classical way at the time, but it just sounds great.

Our customers' expectations and likes have changed over the years, as we do not have as many classical listeners anymore. Many people experience immersive sound by watching movies. Many of those are interested in 3D music. Some, however, still prefer to buy Blu-rays rather than streaming. Incidentally, a similar effect happened when the CD was launched. The first half of what was put out on CD was strictly classical music. Therefore, many people bought classical CDs, although they were not necessarily into classical music – they just wanted something for the CD player. They probably also bought early pop albums on CD like Phill Collins '*Hello, I Must Be Going*' at face value as well. But the main content on CD was classical.

I aim for a sound that envelops the listener, seeking a balance where strong mono signals or standout objects don't dominate the experience. For instance, the sudden shout of a singer from a single speaker can be disruptive and undesirable. The goal is to achieve a close, present sound without resorting to a mono-style effect. Additionally, maintaining a sense of the room adds complexity to the challenge. The objective is twofold: to convey a feeling of presence while preserving the spatial characteristics of the recording environment, as if capturing an acoustical performance in a quality room.

It's important to consider that, in the original setup of a hall or room, repositioning the ensemble around the main immersive microphone may not always be feasible, especially in live concert situations, such as with an orchestra. Attempting such changes could compromise the cohesion of the performance. It's important to recognize that good immersive sound doesn't inherently enhance the quality of musical performance – it serves to intensify the playback experience. If musicians fail to play together effectively, immersive audio may even accentuate existing issues rather than improving the listening experience.

## Complementary spot microphone technique

We use various techniques to enhance our recordings, such as doubling microphones for instruments and strategically placing them to create an AB-style sound. This approach serves a dual purpose: maintaining the instrument's proximity and preventing a monophonic effect. These production methods work well in both small and large rooms. I've observed that, regardless of the setting, getting the direct sound is crucial, but the challenge lies in ensuring that the direct sound doesn't become overly mono. To address this, we utilize two microphones per instrument, panning from left to right or across the spectrum to create what I call the 'complementary signal".

In recent experiments, we explored using stereo microphones with panning, encircling the listener. This idea proved quite successful. One microphone captures the instrument's original position, while the other serves as a complementary signal, positioned elsewhere in the immersive mix. The core concept revolves around maintaining a close sound without having a mono effect. This proved very good results.

Sometimes proposing to move musicians within the space during a recording session can prove to be impractical or get a reluctant response, especially from classical musicians. Jazz musicians appear to be more open and tend to embrace the idea. For example, during a recording project, we initially placed the percussion player behind the listener, but it proved musically and sonically unsatisfactory. Adjusting the group's configuration to a half circle still didn't yield a cohesive performance. Eventually, we brought them closer together in front of the microphone. Utilizing spot microphones and the 'complementary microphone' technique, we successfully maneuvered the sound within the space, achieving a pleasing outcome.

## The David Bowie exhibit

The David Bowie exhibit toured internationally for five years with its final installation at the Brooklyn Museum in July of 2018. The exhibit included costumes, original album art, photos, and videos chronicling David Bowie's prolific career. The exhibit included an immersive multimedia experience with continuous video and audio. The sound experiences were created by Sennheiser.

For this exhibition, 9.1 made the most sense initially. That was a standard system at that time. It was an Auro immersive setup in what we call nowadays 5.1.4. Later, we actually used a 7.1.4 setup. But the side speakers had a mix of the front and back signals, so no new side channels were created. It was just to fill the space because some of the exhibition rooms were very large. We also noticed that we had a hole in the side of 5.1.4. We used tracks from famous Bowie songs and mixed them together in a new way. I was given the stereo version of these songs and was able to produce a very good 3D sound out of it based on my up-mix method. Tony Visconti, Bowie's long-time producer, arranger, and engineer, was very happy with the results. My up-mix algorithm, which was patented by Sennheiser, can work with these kinds of signals very well and can move them around anytime.

We used the standard Auro-3D setup from that time, which made for an easy setup. The elevation of the top speakers was supposed to be half of the stereo base width. However, it didn't always work, because sometimes we had lower ceilings. But that was the kind of thing I would work with, of course, and we had enough energy to produce some really good bass in the sound.

The goal of our immersive room extended beyond the 3D effect; we aimed for a loud, clear, and bassy sound which would be pleasing and exciting to the listeners. As a result, we expanded from a basic 5.1 to 7.1, addressing a perceived gap in the audio experience. Attending various exhibitions, each with unique setups, required a lot of rebalancing of the immersive up-mix. Notably, during the London exhibition, we encountered issues with harsh sound from the speakers used, prompting a recalibration of the up-mix to account for the heightened high frequencies present at that time.

So, basically, the immersive mix was based on my up-mix algorithm, which for the most part worked really well. And then the whole thing had to be adapted to each exhibition room, which sometimes was very difficult to do. We would listen to three of the most challenging sound examples that we had. The first one was originally mono. It was a half-live recording from Bowie's 1972 Top of the Pops appearance. Thankfully, the basic sound was very good: it had good bass, it was present, and it had enough high-frequency content. So the only thing to do was to make it lay well between left and right and then use my up-mixer. The second two examples were very poor-quality live recording, which were probably taken from the live mixer and also recorded poorly. We could really hear these effects like that of an old cassette recorder. These examples were not chosen based on how they sounded and what the sound quality was; rather the curators chose these sound examples for the exhibition.

## *The vertical AB system*

The use of the vertical AB system as well as aspects like correlation and content-oriented correlation is mainly relevant for recording of acoustic performances, like any kind of classical performance, live performance, etc.

In 2010 and the following years, there were a heavy discussion and strong competition between the so-called channel-based system and the object-based system. The channel-based system was featured by Auro industries, and the object-based system was featured by Dolby.

In reality, these discussions were not very helpful, as Auro was talking more about music recording, while Dolby was evidently talking about cinema sound. So, finally, this competition was mainly a business competition. We all know how it ended up. From an artistic and sound-orientated point of view, meanwhile, both systems are mixed up as two complementary systems. It is possible to do all kinds of configurations in DAWs and work in any way needed or preferred.

The other discussion that happened pretty much among producers of classical music. Some people were of the opinion that the ensemble itself should only be reproduced on the lower speakers, while reverb and reflections should only come from the top speakers. Others, like myself, were of the opinion that there should be a certain correlation between the lower and the top speakers. I called this kind of correlation 'content correlation'. This will be explained more precisely later in the chapter.

In order to achieve this content correlation effect, I suggest using what I call the 'vertical AB'. The advantage of the vertical AB is the improvement of timbre. The sound or timbre of an instrument or specially of voices is extremely improved by using the vertical AB. Instruments and voices sound much more natural and much warmer than using a two-dimensional system with a 5.1, 7.1, or a system like this plus reverb on the top speakers.

Furthermore, instruments and especially orchestras have a vertical extension as well. Also, they have a three-dimensional size. All of this can only be reproduced by a vertical AB or an equivalent microphone setup. This is not to forget compositions like Mahler's *Symphonie of the 1000* or Saint Saens's *Organ Symphonie*, which are using the whole room horizontally and vertically around the audience.

In general, the reproduction of an acoustical event with the help of the vertical AB is improving the complete performance, and is giving the listener a much better impression of sitting in the hall and really experiencing the performance.

Before discussing the vertical A-B system, it is important to understand the background of an AB system, correlation, and what I call the 'content correlation'. Let´s do a short excursion to AB, XY, and correlation. In a stereo AB setup, usually two omni pattern microphones are used that are set at different distances, depending on the ensemble and depending on what you want to achieve.

This can be from 1 foot (or less) up to 10 feet or more. The advantage of the AB system is that it produces a very nice and 'open' sound. It is mainly based on the running times between the two mikes plus the level differences. Problems might be negative phase behavior, which is especially difficult for analog broadcast or vinyl records.

The XY setup is using two cardioid microphones that are placed in an XY position. The sound is based on the differences in level between both capsules. The result is not as 'open' as an AB. However, the localization is very precise, and there are no phase problems at all.

So let's have a look at the resulting correlation of these signals.

### *Mono*

The correlation of a mono signal results in a '1', meaning, there is a 100% phase correlation between the left and right channels.

The XY-based system still delivers a very strong correlation between left and right.

An AB system can deliver a phase correlation approaching a '0'. It can vary between less than zero but can also move into the negative region, depending on the space between the two mikes and the individual situation.

## *Technical correlation versus contentcorrelation*

The term 'zero correlation' evidently refers to a pure technical measurement of the phase relationship between the two left and right signals. As an example, take a grand piano in a larger room or hall with a larger AB in front. If there are 9 feet of distance to the piano and 3 feet of space between the two microphones, the technical correlation may approach zero. However, if you listen to the two single microphones just by switching between left and right, you will very likely find that, for our ears, there is no big detectable difference; thus the content correlation could be described as 'one' between the two microphone signals. Nevertheless, the stereo signal will produce a good stereo result.

In another opposite example, imagine you have a totally isolated signal on one channel and another unique isolated signal on the other channel. The technical correlation will be zero as well. But the main difference between our first example and our second example is that, in our second example, both the technical correlation and the *content correlation* are zero (see Figure 29.1).

In order to understand the vertical AB, we need to understand this difference, as it helps us to understand how we use the content correlation in the lower and higher level of the microphones and finally in the reproduction of the 3D result. The content correlation in 3D is important for correlation in elevation, the elevation of instruments themselves, the sound/timbre of instruments, 3D positioning, and the 3D dimensions of

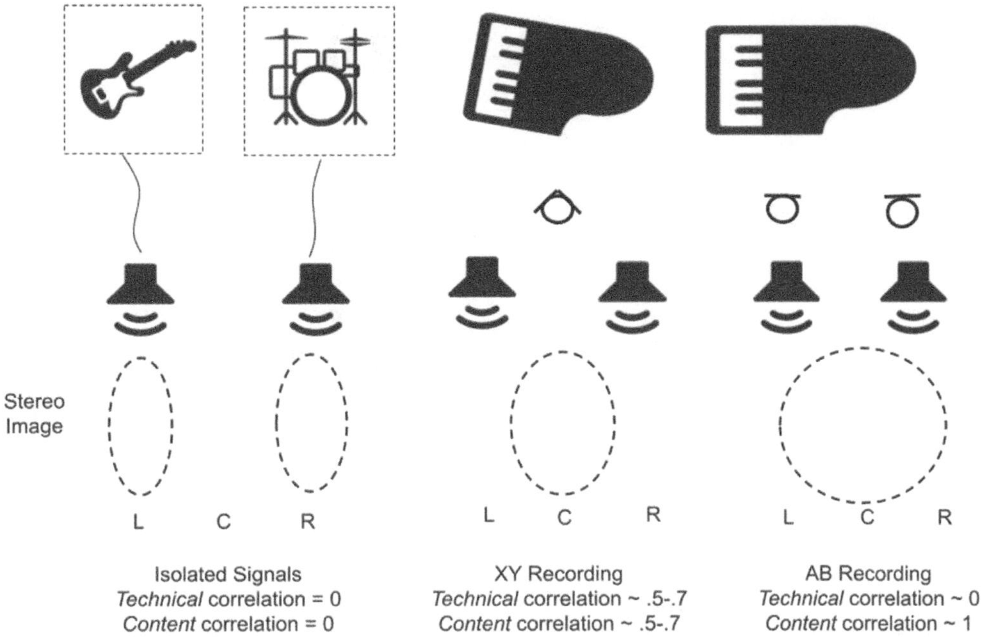


*Figure 29.1* Isolation, XY, and AB recording techniques technical (phase) Correlation versus content correlation: The dashed areas represent the perceived images.

instruments themselves. The vertical AB will deliver a technical correlation of around zero, while the content correlation will be around 0.5 or higher.

As described in the beginning of this article, this is what we want to achieve. One way to achieve this result is using what is called 'the Cube' or 'the Square'. The cube is based on the idea that every microphone of the 3D array has its dedicated speaker, as in a stereo AB system. (Sport microphones are panned in a different way, of course.)

The microphones can be used in different ways. The easiest way would be to use all cardioid for all channels. You could also use wide cardioid at the front. Wide cardioids deliver a better-sounding result concerning the width and depth of sound. However, they can be more dangerous in larger halls or locations with strong and long reverb like churches. More sophisticated and more expansive is the use of double capsule microphones like the Sennheiser MKH 800 Twins. Both capsules of these microphones are led outside, so you can decide and fine-tune the pattern in any way during post-production. You can use a mix of Twins and single capsule mikes. Minimum would be two Twins preferably for the stereo lower (main) signal. The square needs the use of four Twins, alternatively the use of eight cardioids.

In this case, the capsule of the mikes are *hardwired* to the equivalent speakers. So the left front lower capsule is sent to the left front lower speaker. The left rear lower capsule is sent to the left surround lower speaker, and so on. This system has proved to work very easily and reliably. The amount of content correlation and also the amount of reverb coming from the top microphone can be adjusted by angling those mikes in different ways (see Figure 29.2).

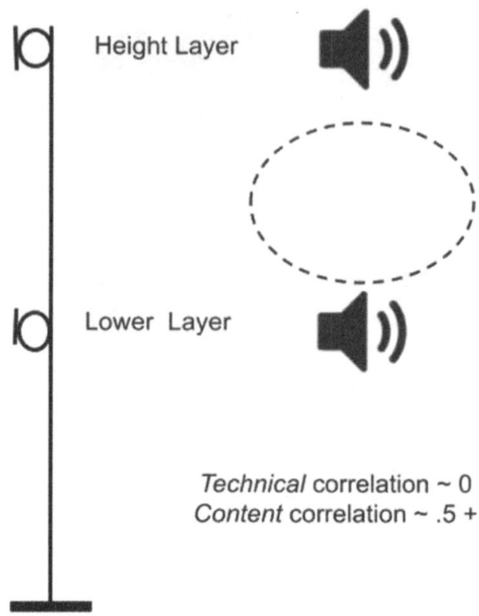


*Figure 29.2* Vertical AB recording technique and resulting technical (phase) correlation versus content correlation: The dashed area represents the perceived image.

### *Monitoring the 3D array*

In many recording situations you may not have a 3D speaker system available. (Whether you use a headphone/binaural 3D speaker simulation is up to your taste and experience.) In order to judge the correlation between the lower and the higher signal, switch between them, and try to judge the differences by critically listening. When you do this, compare them as stereo pairs.

The best result would be that the higher signal sounds about 50% less direct than the lower. Of course, this is an individual judgment, and it needs some experience, as always. In case the higher signal sounds better than the lower, move the array upwards, so that the lower mikes go where the higher ones had been originally. The elevation should be kept, of course. In case the higher signal sounds pretty much like the lower, increase the elevation. In case the higher signals sound too reverberant, decrease the elevation and check the angles of the higher mikes. Finally you need a properly aligned 3D speaker setup to work on the final result. There are many roads to Rome, as the saying goes. That means the described vertical AB system is not the only way to achieve a good 3D signal. We found that the 'so-called High Z' system by Paul Geluso works very well also. And the combination of both systems works even better. There are hard rules about how to work in 3D. In stereo there are a few, that, if you follow them, give you certain very predictable results. Working in 3D is different, though; it leaves a lot of room for creativity.

## Conclusion

Working in immersive, one needs to be flexible and adapt to more unusual ideas. In immersive, one should not stick to the old basic rules that we know from classical acoustic recordings. Try not to adapt to the taste of the main audience that we have for immersion. That is a big lesson that I needed to learn. As you can imagine, 13 or 14 years of doing classical rule-based stereo recording alone can burn you out. I was always happy to do experiments, but there was little to no chance to experiment, especially not at Deutsche Grammophon. It was always a more traditional approach there, such as strings always to the left and double bass always to the right. It's good to stay open to experiments all the way to the end. Ultimately, it needs to sound good, and we always want to have that emotional impact of the sound.

It's not enough to be happy with the sound quality of your 7.1.4 system in your studio, thinking it sounds fantastic. The real test is entering another room with your mix and listening to the experience. You don't want it to feel like a mere ball of sound around you. This is precisely what needs to be avoided.

While there are still valuable basic rules to follow, it's essential to set some aside and explore new approaches. This is why I've incorporated a 3D audio system and mix system into my home studio recently. Experimenting with different setups takes time, but the insights gained are well worth the effort.

Chapter 30

# Recording and mixing *Picturing the Invisible: focus 1*

*Jane Ira Bloom, Ulrike Schwarz, and Jim Anderson*

This contribution is based on an interview with Jane Ira Bloom, Ulrike Schwarz, and Jim Anderson, to share their experience working on the Immersive Grammy–nominated recording project titled *Picturing the Invisible: focus 1.*

*Picturing The Invisible: focus 1* is a selection of compositions by saxophonist Jane Ira Bloom inspired by the science photography of legendary New York City photographer Berenice Abbott, arranged for duet during the pandemic and recorded remotely in high-definition immersive sound by audio engineer Ulrike Schwarz. Award-winning jazz soprano saxophonist Jane Ira Bloom pairs with renowned improvisors drummer Allison Miller, koto artist Miya Misaoka, and bassist Mark Helias to play a series of eight compositions performed in real time connected remotely from their homes in New York City. Using the most up-to-date technology, Grammy-nominated tonmeister Ulrike Schwarz has recorded and mastered the acoustic beauty of these extraordinary musicians' creativity as they duet across the internet. Audio science and improvisational art collaborate to make this recording a one-of-a-kind musical event.

## Jane Ira Bloom (composer, performer, co-producer)

*Picturing the Invisible: focus 1* is a significant project. For years, I've been recording exclusively with Jim Anderson as an improvising jazz artist. It was during *Sixteen Sunsets* at Avatar that I noticed more microphones appearing. Unbeknownst to me, Jim was recording in surround sound. He was ahead of the curve, and that marked the start of our journey.

Fast forward to 2022 and we're working on *Picturing*. Was it a part of the conscious process? I recall sitting and having coffee, discussing this project with Jim and Ulrike. From the composer's point of view, I pondered what compositional choices might be best suited to this. As an artist, I'm very interested in how sound moves, and I move when I play. The motion and spatial placement of sound have always intrigued me.

However, the choice of music in a duet format and the choice of compositions that have a certain spare quality were deliberate. I refer to it as 'internal orchestration'. Even though we don't explicitly discuss it, Jim, Ulrike, and I are orchestrating on an aesthetic level. I'm orchestrating in my mind how to leave mental space for a spatial dimension. They are orchestrating it in a technical way, but it's the same thing. It's an artistic process to determine why some things might work in immersive and why some might not. It's something we think about, and it's part of the process.

DOI: 10.4324/9780429297526-33

### *Experience with surround sound*

I'll share a significant experience. It was when we did *Early Americans*, and Jim played the surround sound mix for the very first time. I'd never heard surround sound speakers before. My experience was like hearing your insides outside. It's not just about being in the room and hearing my band play. It's more than that. What Jim and Ulrike have done is delve inside my aesthetic. I'm not only hearing other musicians in a spatial format, but I'm also hearing myself in a way that I have never experienced before. Is that intoxicating? Yes. So, I'll leave it at that for now. That's just a big overview. But internal orchestration and the idea that you're hearing yourself from the inside out are key. And to do that with artist technicians who are not just engineers, but artists themselves who know me intimately, know my music intimately, and we're on the same page conceptually about what we're trying to achieve, is truly special.

### *Musical impact*

The ultimate goal is to create a musical impact that affects and moves people. And that's what we feel we've achieved with our work. It's a testament to the power of sound, technology, and artistic expression.

During the COVID-19 pandemic, when musicians couldn't play with one another because it was too scary, we had to find a way to connect. No matter the technology, we got it as close as we could in terms of how accurately we were hearing one another. The most important thing is that we were connecting in ways that improvisers do, and it's magical. I don't know how we do it.

Our ears in these situations are actually out like radar scopes, listening and connecting even more intensely than sometimes when we're in person. Why? It's because we needed to. We were desperate to play together.

One thing that Ulrike was able to do was to get the latency down to the point where it was almost as if you were in the room. That would be the same latency that you would just hear naturally acoustically from one person to another. I think that's another reason why this all works. Technically, Ulrike was able to achieve what you would actually experience if you were in the same room. With people having the right connections, we got the latency down to a level that is comparable to the sound propagation delay of being separated by 5 or 6 m.

### *Musical connection*

The musical connection is very familiar. The phase lock, which is a way of playing time where we are completely locked with one another. There is a kind of groove that we do – for example, how I can lock in with a bass playing together. It shouldn't be possible remotely, but somehow we do it. I believe that if you can clearly hear each other, that's the groove. That's the groove you just fall into. And either you are locked, or you're not locked. If you're not locked, it doesn't groove. It really doesn't matter what kind of music it is: it could be classical, it could be jazz. You can tell when a band is swinging, and that is really when everybody locks in, that's when you're grooving. Off you go!

We describe it as being in the zone. Even the repartee that was going on with me and Miya on the koto, it's almost like a ping-pong. And something mentally is going on. It's mysterious to me. But because we have a lot of history together, and we have vocabulary, and we can anticipate and know other people's musical vocabulary, we can make different kinds of anticipatory choices in an up-to-the-moment decision-making that are what improvisers do. We're just making it up out of the air. How do we do that? It's a process of some kind, and it's facilitated by technology that's gotten better and better. It's really interesting to consider our initial assumption that you always had to be in the same room with the other person to synchronize. However, this has been proven to be possible even over the internet and through headphones. Initially, you think you have to be in one room. Then you transition to a studio where you're in separate rooms but at least have sight lines. Now, you don't even need a sight line. The idea is to keep whatever the delay is, whatever the latency is, stable so that you can lock into it once you have understood where it comes in.

In terms of research, it's really interesting that you don't even need to be in the same city. It's more about the history that you have as persons, probably over decades, that has nothing to do with being in the same physical space. It's really about being in the same *thinking* space.

When we set it up, though, we had a visual image. You still needed to know that the other person was there visually. That triggers something in the brain that the person is there.

Maybe because we're improvisers, but we need to see one another. We didn't think that it was unimportant at all. It was very important. It could also be the number of people. You had duos, so it was just a picture of you and a picture of Mark Helias (double bass) or a picture of you and Allison Miller (drums, percussion). If it had been 15, and there will be 15 little icons, maybe that is not as helpful. But I think if you're just a duo and have one stable other image, I'm pretty sure that helps a lot. I actually really do know that the person is there and not just breathing.

### Ulrike Schwarz (co-producer, recording engineer, and editor)

The tracking process for *Picturing the Invisible* requires breaking down into several stages that Jane hasn't mentioned yet. Initially, the idea was to record in a studio setting like Power Station in New York City or even in a large space like a church. The concept was to picture the invisible, essentially the space between the notes. We had initially received a substantial grant for a larger ensemble and a larger recording room. However, the COVID-19 pandemic hit, and no one was leaving their houses. We were asked to deliver the project we had promised, so it had to be rearranged. Fortunately, Jane distilled her entire idea down to duos, and even one trio, which was an overdub with the bass recorded first and then the koto and saxophone recorded later. Due to the pandemic, we realized it wouldn't be a large studio or anything of that sort, we had to devise a concept for recording apart but being together. This happened via the internet. We also wanted to maintain the highest quality possible, which at the moment was 384 kHz sampling rate. The challenge was to make the personal internet so fast and stable that they could play duos, but we would have recording systems on both sides that I was controlling for one side and recording.

#### *Recording in small rooms*

The idea then was to record people in small rooms in their homes but still make it feel expansive. I'm familiar with how Jim records Jane, which usually involves nine microphones. However, nine microphones wouldn't even fit into Jane's office. So my idea was to figure out how to record her with enough microphones to provide a spatial image, but with the fewest microphones possible to avoid capturing her room acoustics in the recording.

On the other hand, Allison's room, which is close to us, was really nice. She had been recording and giving lessons in her basement studio for about a year at that point. Her drum kit was tuned to her room, and she had a couple of microphones that worked very well in connection with what Jim usually does with the drummers for Jane. My thought was to capture not too much of the acoustics of that room but still be able to mix it immersively. In essence, the process was to make the least but then take out the most. This was done in Jane's office, in Allison's basement studio, and, before the koto, in Miya Masaoka's living room.

### *Room acoustics treatment*

Indeed, the challenge was significant, but it seems we navigated it well. The room acoustics treatments were makeshift, involving big jackets on sofas and towels on the desk to prevent early reflections. Jane was recorded with two of her usual nine microphones. The drum kit had a full set of about 10–11 microphones, which is a standard setup. The koto was recorded with approximately eight microphones, enabling the sound to be spatially represented.

Merging Technologies systems were employed on both sides. I was running these systems at 384 kHz. We had one set of computers solely to be used for the visual connection so that the musicians could see each other. This was entirely out of sync, but it gave the artists the impression that someone else was in the room with them.

We also used communication software on a third set of computers, which we tried to keep stable. The trick here was a bit different from what others have done. I attempted to use the highest bandwidth and the highest sampling frequency because it would use the same samples, and they would just come faster, instead of working with codecs that would slow down the process.

### *Communication*

From my years in broadcasting, I knew that, while having recording systems at both ends was essential, the heart of everything was the communication between musicians. The recording itself was almost secondary, even though it would become our final product. If I didn't focus my energy on getting that connection right, nothing else would matter – the whole project would fall apart.

So that's why I was really intent on having chains of recording, like having multiple systems in multiple locations, so that that was achieved. Unfortunately or fortunately, there were no three outside broadcasting (OB) trucks that could be parked everywhere. It was just one person. But I think that came out very well, and especially the communication part was the one to concentrate on.

I talked a lot with Karlheinz Brandenburg when he measured our room at the Clive Davis Institute at New York University. He said that, with immersive audio, the research is much more focused on how the room you're actually sitting in visually affects what you hear. I think it takes some very masterful musicians to be able to forget that they're in an office or in their home basement or somewhere else. They need to feel as if they are actually playing music with another master and to completely forget the circumstances and then listen to each other or even anticipate what is coming.

Even when there was latency, particularly as Allison and Jane played together or changed musical direction, I didn't need to adjust their timing in post-production. They were remarkably synchronized, requiring minimal editing. It takes true masters to mentally process – or rather, transcend – technical delays while performing at such a high level. That's part of what makes their musicianship so extraordinary.

Finally, we took it to Skywalker Sound, where we initially planned to do just a 5.1 mix. This is where the mixer comes in. I had a very experienced mixer who did the mix at Skywalker. This was a unique and rewarding experience.

## Jim Anderson (mixing engineer)

One thing I think would be good to refer back to is Jane's original inspiration. She had approached us with the idea of Berenice Abbott's scientific photographs. Looking at those photographs led to the title, *Picturing the Invisible*. At the time, I was examining all of her black-and-white science photography, which was filled with these gorgeous, starkly contrasted images of light waves, particles, and motion. This was the inspiration for the composition. As Ulrike said, this is a good turnaround. She talked about what's in the silence? What's in between the notes? These were visualizations of exactly that, the inspiration for the entire piece. That is what drove all of us: Jane in the composition, Ulrike in the recording, and me in the mixing. We always had these images in the background.

Since we didn't have access to the copyright of the Berenice Abbott photography, a friend of Ulrike's was able to recreate one of the pictures. The Berlin designer Assen Semov took an image of two waves in water interfering with each other. He didn't only do the cover with it, he actually did a 3D rendering, so it is a moving image. We have a video of the image moving very slowly with the Atmos mix for about 30–40 minutes. It has a bit of a mesmerizing and addictive quality because we just look into the abyss there for the whole time. It just slowly morphs from one motion to the other very slowly.

### *Mixing at Skywalker*

So we take the tracks up to Skywalker Sound and the nice thing is, the tracks are pristine and they're recorded so well. I don't need to do anything to the track. I just put it up and start working with it. I've got the elements that I need. Essentially, what I need is a great direct sound on Jane since she's the star of the recording. We need to put everything around her.

Then we have, essentially, a trick that I've been doing for a long time, which is on the drum overhead to have a widely spaced pair of cardioids you'd normally see. But at the same time, an MS pair. With that MS pair, I can put that behind you, and I can put the wide cardioids at the front. If Allison would move around naturally with a bell or other percussion elements, she would move naturally between these two microphones. So you get this natural movement that just sits above you.

### *Recording height*

I usually would take four microphones and arrange them in a Hamasaki-like square. I would then place Jane in the middle of it, and position the microphones roughly about 1 m above her head or where she's playing. Then, I could pin down the four corners in the room where she's playing. Now, we couldn't do that in Jane's office, as she mentioned. Part of the reason we wouldn't want to do that is that it would highlight the acoustics of her office, which we don't want. We just want a good, clear signal from her.

At Skywalker, over the course of 12 years, we have been developing this technique. The first time I went out there to do a surround mix, I was not very happy with most of the digital surround reverbs. Looking from the control room at the wonderful soundstage, I thought, why don't I use that as my live chamber? So, I originally just set up a quad chamber; there I would have four channels, four microphones, and two speakers blasting into diffusion panels. This setup allowed me to utilize the wonderful room that has about 2.5 s of reverb.

Indeed, our setup is now much closer to a real Hamasaki-Square because we're actually recording what comes back from these baffles, not someone standing in the middle of it. This setup has now morphed into a cube, and then into a 12-channel cube. We've been expanding it, moving from surround to immersive over the past 12 years.

We took the original four microphones and added another four on top, creating a cube of eight microphones. The additional four are exclusively for our upper level channels. In the mastering of the 5.1 mix that we did, we sent Morten Lindberg (Mastering Engineer) the tracks of the upper level, and that's what he added to the top.

It's interesting how, with the double microphones of the overheads, you get the sensation of a lid on top of the mix. But that really comes from the microphones at the front and back; you feel like you're on top of that. Then, above that, we have the room of Skywalker, the soundstage, being layered on top. This whole combination works out very nicely. The recording process can feel like four-dimensional chess. Imagining what the end result can be, when we went out to mix there, is crucial. It's as if you have to know what you'll end up with to understand what's possible during the recording and mixing. That's the perspective I had.

Another thing to consider, is that we mixed the entire piece in a day, both the stereo and surround mixes. This was possible because the concept was already there. We started with the stereo, got that done, and I had a good idea of what the pieces were like and how they would flow. Then we stopped, set up the surround system, and started exploring what was possible there.

I may have even used most of the automation on the console. This mix wasn't done in the box; it was mixed on a regular Neve VR console, the analog, and then re-recorded back to Ulrike in 384 kHz. I believe that having an analog chamber, a live chamber, and also mixing in analog is what helps the mix breathe and feel like a real, cohesive piece.

### *Creating a timeless piece*

It's interesting to consider the longevity of a piece. If done right, a piece could be listened to 50 years from now and still feel natural. It wouldn't be tied to a specific era or trend. I find it fascinating how you can almost pin the year of a mix based on the effects being used at the time. But a truly timeless piece, like the one you're describing, transcends these temporal markers. It feels like something that has always existed and could have been created 20 years ago or 20 years from now.

### *What is natural?*

In the future, all of this may just disappear, and we may have instantaneous communication, visual communication, and all sorts of things. So I think it really comes down to the artistry involved and the acoustics. I'm always on the lookout for good tracks, tracks that don't have a footprint of where they were recorded. This is because I can then take them and manipulate them any way I like, either in space, being three-dimensional space, or just like Jane's sound and reverb that I put on top of it.

The idea is to have enough of it that it feels realistic but not so much that it feels gimmicky. Jane would always look at me when we would do a recording. She would say, 'I don't feel like I'm flying'. So I would have to interpret that and wonder, 'How can I make her fly?' Usually, what that meant was it needed more air, needed more lift, or more reverb added to the sound. On first listen, it just didn't take off. So I am always interpreting her metaphors when we sit next to each other during mixing.

The whole idea about what is 'natural' is interesting. None of this recording, especially this one, was at all natural, but it still somehow feels natural. Once you hear it, you almost think, 'Oh, there's a real room around it'. It could have never been that way. For instance, there's a big saxophone that is now 3 m, 4 m tall, and the drum kit that's all around you. You're sitting somehow in the drum kit, but you're looking at the drum kit. So none of this is real, but somehow it is believable that it could have been.

I think a part of it is that it is actually a real room. The difference between an artificial reverb and something that was in a real room, like the Skywalker room, makes all of this feel very real somehow. Although, nothing about this is real. The second layer was recorded in Brooklyn, the Upper West Side in New York City, and the mix was done in California, and the final touch was done in Oslo, Norway. None of this has anything to do with reality, and once you even get the editing process in there, which makes it even more unreal because now you're even messing with the timeline, none of it is real, but it's being made to feel real. It's quite amusing when you think about it.

It's funny that, especially in immersive audio, everyone talks about the natural room or the real room. You start to wonder, what are we even talking about when we are recording? This is just getting really complex. It's like you are creating an illusion, and now it's an immersive illusion.

When you shape the sonic environment around the original performance, you're creating a space that enhances the music itself. While this might be technically artificial, it's no different from any other artistic choice – everything we do in production is about crafting an immersive experience for the listener. We're crafting an immersive experience, even if it's based on an illusion.

## References

Jane Ira Bloom (2016). 'Early Americans'. [Outline Records].
Jane Ira Bloom (2022). 'Picturing the Invisible: Focus 1'. [Anderson Audio New York].

Chapter 31

# Recording *Soundtrack of the American Soldier* and remixing *Star Wars*

*Leslie Ann Jones*

My first exposure to surround sound was through film, and a lot of my music mixing for film would be for 5.1. But my recording work for records has always been in stereo. And some of my TV work has been stereo too. But my film work was always 5.1 surround.

The primary release format for the record work I do is still stereo. A lot of my clients are independent clients, so they don't necessarily have the budget to do an immersive mix. But I do tend to set up my recording to include height channels, so that they have an opportunity afterwards to mix for an immersive format. Other clients know they're going to release something for an immersive format. In those cases, from the beginning, we plan the microphone placement and setup accordingly. In those cases, I have a lot more control over my vision of what it's going to sound like.

## Vision

I think that immersive, at least post-5.1, is totally market driven. It's really the only time in my career where it seems that we, as engineers, and producers are not deciding what the format's going to be. Somebody else is deciding what the format is going to be, and then we just have to conform. This is particularly true with Dolby Atmos. My philosophy would be to provide a more intimate listening experience, where you feel it's multidimensional, as opposed to something speaking to us from two speakers because this is where your experience becomes more profound.

Not every project I do can be immersive because it depends on many factors, including the budget. I don't work with a lot of major label artists so I don't do a lot of catalog work. Most of what I do in immersive is intended to be immersive from the very beginning. For example, the US Army Field Band record. From the very beginning, the intention was that it would be an immersive album. We designed how the musicians were going to be laid out in the room based on the piece we were recording, whom we wanted to feature, and the speaker layout. There was a very thoughtful and deliberate approach to it. Similar to Morten Lindberg, who typically tends to do a pretty static microphone setup, and then arranges his artists around that setup to achieve the listening experience that he wants – that's what we did The result was that, when we went to mix, it sounded exactly the same as it did when we recorded it. There were no surprises in terms of who was featured, who was in the back and who was in the front. Mixing was more a matter of raising a solo level or lowering the woodwinds, for example. It was about rebalancing rather than reinventing the wheel.

DOI: 10.4324/9780429297526-34

## Intimacy

When I'm listening to immersive music, I feel like I'm more in touch with the performance aspects than when I listen in stereo. This is particularly true with popular music. When I listen in stereo, I tend to react to so much of it: the beat, the rhythm, the lyrics, what the chorus sounds like, and whether or not I want to turn it up really loud. In immersive, I'm much more drawn into the experience. I feel like I can get more inside the music and appreciate the performances more without necessarily thinking about the rhythm, the beat, or the lyrics. It's a more intimate and three-dimensional experience to me.

I did a live project for the Cabrillo Music Festival in 2012 that involved a lot of prerecorded girls' voices. It was done with composer Laura Karpman and the Kitchen Sisters, who had a radio show called the *Hidden World of Girls*. They had a way for girls to call in and talk about the secrets that no one else knew about them. A performance was created out of that with live orchestra music and these girls' voices. This was my first experience doing a live immersive experience. It just struck me . . . all these girls were talking in the composition, and some were singing. I didn't want it coming *at* people as it normally would with speakers from the stage and everybody on the stage; instead you should be looking at that and experiencing it from a spatial plane. So I called my friends at Meyer Sound, and we got some speakers brought into the Civic Auditorium in Santa Cruz. I placed speakers all around, and we were able to place the girls' voices in different locations. Every time one of the girls spoke, it came from a different place. To me, that was a much more intimate experience than just having everything come at you while you're trying to pay attention to everything. I felt that it was much more engaging for the audience that way.

My goal was to make the audience feel like they were listening in: they would hear a girl at one o'clock, another at eight o'clock, and another at five o'clock, and every girl comes in and talks about what they're doing. You're almost listening in to the conversation, as if they are in a park somewhere, sitting around chatting by themselves, and you're overhearing their conversation. That was my goal.

## Storytelling

I don't really have a film background except for having worked on music for films, and I'm actually not a big movie buff, but I do think storytelling is one of the most important things that we do, even in music. One of the big challenges doing catalog remixes is how to convey what made that piece of music a hit that had people put their convertible top down and turn up the volume. In the *Barbie* movie that came out in 2023, there's a scene where an Indigo Girls record is being played and people are singing along with it. How do you capture that with the same feel in immersive so that you're still experiencing your favorite guitar line or vocal harmony without completely tearing it apart? That's very challenging.

## US Army Field Band project

Dan Mercurio is a wonderful producer on the East Coast who called me about recording a project with the US Army Field Band called *Soundtrack of the American Soldier*. The project consisted of film music that all had something to do with the military. They wanted to do it at Skywalker Sound. I was somewhere on the East Coast, and I flew into Raleigh Durham. Dan and I met at a coffee shop and did napkin drawings and talked about the different pieces. The field band is basically winds, brass, percussion, harp and upright basses. We talked about the different pieces that were being recorded and what they featured. He had not been to rehearsals, yet so we just sketching out rough drawings of what it could look like, and then he went to rehearsals. The band was extremely willing and happy to participate in this. The musical director, Colonel Jim Keane, is a real force in doing new things with this band.

Colonel Keene was involved from the beginning. After Dan went to rehearsals, he refined all the drawings. Then I had to figure out how to mic it. Typically, when I record a large ensemble, we use a Decca Tree and wide surrounds and heights for immersive. But this was a little different. I decided to use room mics and height mics, but I used stereo pairs for each section. When we planned the placement, sometimes in the middle would be trumpet, sometimes it would be saxes, sometimes there would be a live drum kit, and sometimes the Soldiers' Chorus was included, which we decided to have in the back (except for 'God Bless America', for which you can never have the chorus in the back). In general, the microphone placement stayed pretty consistent, but the musicians moved around.

The biggest difference was that we had intended to do it in 7.1.4. When we recorded it, we had side speakers. But at that time, receivers were still defaulting to 5.1.4, even if it was 7.1.4 material. Between recording and mixing, we decided that we were going to do native 5.1.4. The biggest challenge was figuring out what to do with the material that was on the side speakers, and how to make a phantom side. Everyone is arranged around the conductor. If the clarinets are in the back, that's what you hear. This allowed the intention of the compositions and arrangements to really shine because you could hear the meaning of everything around you, and it contributed to the listener's enjoyment of the music.

One of the elements that worked very well in this project was the use of stereo miking. Many of my colleagues have commented about there being no phase problems. I used a stereo mic for each section. Not necessarily a single mic that is stereo, but two mics that I can space apart and cover a section. First of all, this allowed me to cover the section. Second, if we needed to go from 7.1.4 to 5.1.4, it allowed me to pan that stereo image and create the phantom side that I needed, which was very beneficial in that respect.

There was one tune that had two drummers, which created a challenge. We have three isolation booths at Skywalker, so we were able to put one of the drum kits in one of the rooms, and the other one out on the stage.

Isolating one of the drum sets allowed us to put those drums in the back speakers. A lot of times what we'll do is we'll set up the Decca Tree and get the room sound through the speakers. It allows us to maintain the same kind of sound. In this case, however, we needed a different kind of reverb just because of the nature of the space.

### Height

The elevation of the heights significantly varies based on the content I'm recording. When considering what I want to hear from the top speakers, it's intriguing. Particularly on this project, during the mixing process – transitioning from a full 5.1 setup to stereo – the impact of the heights on the sound was notable. I wonder if others have similar

# TUSAFB - Soundtrack of the American Soldier
## Recording Layout - DRAFT 20181011

## ARMY STRONG

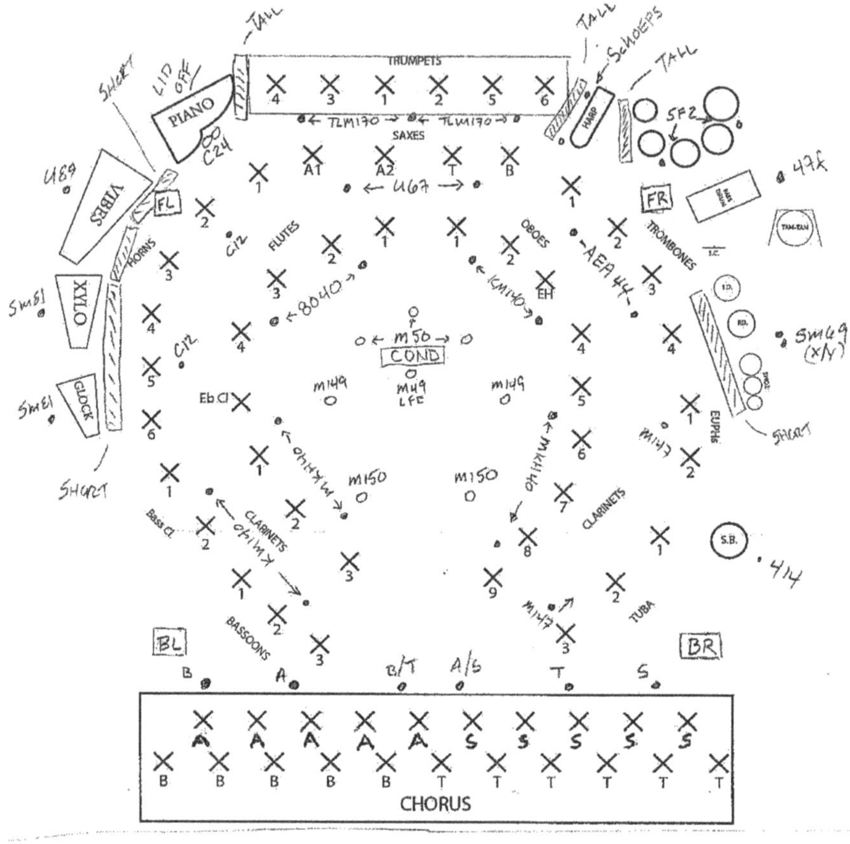


| MAINS | | HEIGHTS | | CHORUS SPOTS | |
|---|---|---|---|---|---|
| L | M50 | FL | MKH800 OMNI | ALTO | 4011 |
| C | M50 | FR | MKH800 OMNI | AL/SOP | 4006 |
| R | M50 | BL | MKH800 OMNI | SOPRANO | 4011 |
| LFE | M49 CARDIOID | BR | MKH800 OMNI | BASS | 4006 |
| L SIDE | M149 OMNI | | | BASS/TEN | 4006 |
| R SIDE | M149 OMNI | | | TENOR | 4006 |
| L BACK | M150 | | | | |
| R BACK | M150 | | | | |

*Figure 31.1* Soundtrack of the *American Soldier* recording layout

experiences. In the more expansive 5.1 version (originally 5.1.4), determining how much of the heights should be integrated into the mix was crucial. The same applied to stereo; sometimes, the heights proved distracting, while in other instances, they contributed a pleasant, expansive quality.

This variability was consistent across multiple projects I've worked on. There's no one-size-fits-all approach; it's not a plug-and-play scenario. Each song demands a unique understanding of how it should sound and how the remixing process should adapt for different formats.

There is a tune called 'Army Strong', composed by Mark Isham. The location of Colonel Keane, the Decca Tree, and wides are presented in Figure 31.1. The choir was in the back with section mics. The height mics were back left and right, front left and right. We had stereo pairs on the piano, stereo on the flutes. Everything really was a stereo pair. When you're listening to it, you are listening to it from the perspective of where Jim Keane was. You're hearing the clarinets behind you, and the chorus is behind you. There's nothing really dead center in the back, except for the chorus, and all percussion and piano. I like to record the piano with the lid off because then the piano fills the whole room. The only thing that really had mono mics were the mallets.

We monitored this project in Atmos. However, that is not always the case. When I'm working on something, and I'm unsure whether there's going to be an Atmos version, I hang the mics and listen to make sure that they're working and that they're capturing what they need to capture. I generally listen in stereo. But for this project, we had a 7.1.4 setup. It was an extensive setup.

## *Star Wars*

Another project is one I worked on with Shawn Murphy where we archived all of the *Star Wars* music. Dann Thompson, one of our engineers, did the transfers from analog to digital. We were able to remix those originally for 5.1 and then do stereo mixes. A lot of the newer recordings were in 5.1, but the older ones weren't. This allowed Lucasfilm to repurpose the music and mixes. Shawn went back and did the Atmos versions of these as well.

The original *Star Wars* was not in 5.1 surround sound. I believe it was the original Dolby four channel: L-C-R and mono surround. It was not an easy thing to do. We constantly referred back and had to make the same edits because there weren't any notes. We worked with Mark Willsher. Mark was able to A/B the material and match all the edits.

The early *Star Wars* recordings were all analog. So the sound was different. There are many fans out there who are used to the way that music sounds, and they don't want to hear any difference. These discriminating listeners will pick out something that's not quite right also. This type of work teaches you the value of taking copious notes about everything. This project was around 2018. The first thing we did was to edit the takes together to create the take that was in the soundtrack. We took that and remixed it, so that it could be used for 5.1.

One of the things that I learned about working in this field is the creative process. Not everything was meant to have a spatial component to it. It really should be all about the song and storytelling and about what the composition requires, not about what the marketplace wants.

Chapter 32

# Immersive music production for Latin American artists

*Andres Mayo*

## Design philosophy

My fundamental philosophy is that technology shouldn't merely support music. Rather, there should be a reciprocal relationship between technology and music, where music is placed at the forefront. This realization dawned on me as I delved into the immersive audio field, drawing from my extensive experience in music production. Initially, I believed it to be an entirely new concept, but I soon discovered that many principles from my past, working with stereo and even mono, remained relevant.

Surprisingly, the transition wasn't a complete overhaul of my knowledge. It involved adapting my mindset and philosophy, yet the fundamentals were rooted in my previous experiences. The key revelation was that, irrespective of the format (stereo, 22.2, or other), the focus should be on enhancing music. This was reassuring as it meant that much of what I knew was still applicable, especially when it came to mixing in immersive environments.

At the outset, it felt like navigating a new toolset and concept, but the core remained constant – is it musical? Does it contribute to the music, or is it merely a showcase of technology? Distinguishing between the two took time, and my collaborator Martin Muscato and I began experimenting by rearranging speakers around a microphone. This raised eyebrows, but our goal was to identify spots where the experience felt more authentic, believable, and aligned with the music's needs.

Reflecting on those early days, we now apply our learnings with advanced tools, recognizing the significant progress made in design since we started in 2016. Despite the evolution in tools, the guiding principles we learned remain crucial. It's not just about what you can do but discerning what benefits the music. There's a fine line between choices that enhance and those that don't, and I've learned to navigate this line by understanding why I would make certain decisions.

These insights range from subtle adjustments like applying EQ to a moving object, to considering the height of a moving element to enhance plausibility. Learning how distant objects can be perceived in relation to closer ones has been crucial. Objects working together help establish distance effectively. Some lessons are nuanced and were only uncovered through experimentation – there was no alternative method.

In essence, my learning journey has been about defining the boundaries of what I wouldn't do for the music and understanding the myriad possibilities of what I would do, always driven by the goal of serving the music in the best possible way.

DOI: 10.4324/9780429297526-35

## Experimentation

In Latin America, experimenting with immersive audio isn't easy due to the limited number of practitioners in this field. The luxury of time and basic elements are prerequisites for meaningful experimentation. Despite these challenges, we were determined to try as many tools as we could. Our journey began by moving speakers and progressed to experimenting with Ambisonics, binaural heads, and beta testing multiple solutions and applications.

The invaluable outcome was a broad expertise, not necessarily making us better mixers but providing clarity on the boundaries of our exploration. We now understand when we're deviating from the mainstream, and that's a significant lesson. While we predominantly work with mainstream artists, venturing off-road occasionally is acceptable, as long as we stay within the defined boundaries of the highway.

The highway, representing the mainstream, is vast, allowing for experimentation without straying too far. However, I refrain from producing content that might be perceived as a mistake by the majority. Courageous experimentation is welcome, but I avoid creating something that people would find incomprehensible or dismiss as a mere showcase of technology.

Personally, I prioritize a high-quality stereo mix over an extravagant immersive experience. Working with classical genres like Tango or folk music reinforces my commitment to maintaining a connection with the audience. I don't want to risk alienating them with overly experimental content.

An instance where I encountered a limitation was when I decided to experiment with moving elements in a classical production conducted in Mexico with an orchestra. I believed it would enhance the music based on my interpretation, disregarding the audience perspective. Unfortunately, it didn't resonate with the listeners, prompting me to reevaluate and make adjustments. On the flip side, I took a daring approach in binaural audio, a realm I had been immersed in for years before acquiring my Atmos room.

Working in binaural taught me that I needed to exaggerate certain elements since many listeners wouldn't have the opportunity to experience Atmos. This meant creating noticeable differences for headphone users compared to a standard stereo mix. However, I sometimes missed the mark by implementing extreme changes that deviated too far from the essence of the music. For example, in a piece with dynamic elements like a door banging as an effect, the surprise factor was there, but it didn't contribute meaningfully to the music.

I've learned that effects or additions should seamlessly integrate into the production concept, avoiding the temptation to showcase new techniques like displaying the features of a newly bought car. Such elements must align with the core of the production rather than being stand-alone features. It's about ensuring that every addition serves the music and isn't just a flashy display of technology or creative experimentation for its own sake.

## Working with artists

### *Luzmila Carpio*

I appreciate artists who actively engage in the immersive audio production process. I like working with artists who not only occupy a space in the mix but contribute with a mindset aligned to the immersive experience. I've been fortunate to collaborate with an exceptional artist named Luzmila Carpio and her producer Leo Martinelli. I'd like to highlight her outstanding work. Luzmila is from Bolivia, and her music defies categorization, being distinctly different. She sings in a Bolivian dialect, adding to the uniqueness of the project.

In this immersive audio project titled *Inti Watana – El retorno del Sol*, the producer, coming from an electronic background, blended seamlessly with the unconventional nature of the music. Our collaboration was transformative, creating an astonishing sound. From the outset, I conveyed the vision of producing something utilizing a minimum of 12 speakers, emphasizing the need for numerous high-quality tracks that would enhance the immersive experience rather than being haphazardly placed. The producer delivered content perfectly suited for our immersive concept.

Throughout the production, we strategically placed objects, allowing them to move dynamically or remain static, creating a vast and immersive space. Luz Mila's vocals, enriched with echoes, effects, doubling, and tripling, were masterfully woven into the audio landscape. The beauty lies in the subtle intricacies – layers of vocals responding to one another, creating a truly mesmerizing effect. The result is an immersive experience where her voice and the accompanying elements seamlessly blend and captivate without overtly dominating the sonic landscape.

### *Duki*

We've had the opportunity to collaborate with Duki, undeniably one of the major figures in Latin America, known for his work in trap and its variants. Despite his commercial success, what sets him apart is his unique mindset. While his primary focus is on commercial success, he comprehends the intricacies of immersive audio. In a groundbreaking project, he brought his production team to work on a documentary for theatrical release, entirely mixed in Atmos. Understanding the importance of pre-production, he joined us early in the process, bringing about significant changes. My role has evolved into an educational one, where I engage with producers well in advance, often a year or six months before the project starts.

Whether it's for streaming or live sound, the driving force behind the choice of an immersive format is whether it serves the music or merely showcases technology. So, rather than discussing formats, I prefer to delve into the overarching concepts and the unique spatial experiences that immersive audio has to offer.

## Space

I was working in Mexico on the production of the album titled *Todo este campo es mio* – a tribute to the Venezuelan composer and singer Simón Diaz. While working on this project, I realized that it could have been a stereo mix, but there was an element unattainable in stereo – the sense of space. The concept of working in a sphere, rather than a traditional circle, dawned on me when I started using Atmos. The idea of the sphere was relatively new, and there are multiple ways to create space in audio, but what stood out to me was the size that couldn't be achieved in a stereo mix. It's where a single axis falls short.

The crucial difference lies in how you use height. Most times, I don't place discrete objects in the back or top left; instead, I manipulate the air around these objects, creating a sense of elevation without moving the object itself. This method works exceptionally well when the recording captures all the essential elements in a large venue with good microphones.

For example, in a Tango recording titled *Astor Piazzolla – Tangos All Around*, while some elements are strategically placed in the back, I would never place a piano on the ceiling. Instead, the positioning, coupled with reverbs and the natural venue acoustics, allows me to convey a much larger space than what is physically present. The key is maintaining balance across three axes – left to right, rear to front, and top to bottom.

In the rear axis, the balance is different, with brightness in the front and the low end and supporting elements in the back. The top axis is reserved for believable elements, avoiding anything that might seem out of place.

When I think of space and how to produce it, I focus on achieving balance in these three axes. Most times, the air of the objects or music becomes essential, often processed in a way that, when heard from the top speakers, conveys a filtered and processed version, creating a sense of air and resonance. This approach creates a space and depth that wouldn't be possible in stereo.

My attempt is to preserve this balance, using the ground for anchor elements, the air for atmospheric elements, the left and right for balance, the front for brightness, and the back for the low or middle end. While it may not be a fixed formula, this balance is my guiding principle in creating an immersive mix. It's a result of years of experimentation, and I stick to it not for the sake of trying something eccentric but because it aligns with the direction the music is asking me to go. Adapting to the music's needs is crucial, but the underlying principle is to maintain a balance that amplifies space, making it distinct from traditional stereo mixes.

Regrettably, the compromise lies in catering to the expectations of less experienced listeners who anticipate an extraordinary experience due to their investment. This compromises my artistic preferences, pushing me to make unconventional moves merely to meet these expectations. It's a response to the inflexibility of the stereo mix, which producers are reluctant to deviate from.

I find myself admiring productions like Morten Lindberg's, where he captures beautiful sounds by strategically placing the right microphones in the right locations with the ideal orchestra or instruments. He creates beautiful recordings without the need for extensive post-production adjustments. Personally, I lean toward minimizing movements whenever possible. My preference is to keep adjustments to a minimum while striving to craft an expansive space.

However, the reality is that crafting a vast and believable space isn't a straightforward task. Contrary to common belief, working in immersive audio doesn't automatically guarantee a rich spatial experience. It's a complex and challenging process to authentically create a space that the listener can believe in – one that goes beyond a simple click of a reverb button.

For me, the primary and most challenging aspect is achieving a genuinely great space. This difficulty contrasts with the misconception that space is readily available when working in immersive audio. I emphasize the intricacy of the task; it's not something that effortlessly materializes. While flashy elements can be incorporated, particularly in genres like urban music, it's considerably challenging in other genres.

## Urban tracks

A significant portion of my current work revolves around urban music, which operates with a distinct mindset. Urban genres demand a flashy and utterly unbelievable approach. However, they often provide a limited number of stems – typically around eight – restricting the creative process. While I express my concerns and seek additional elements from recording companies, the outcome isn't always favorable.

In the urban realm, we often adhere to a formula: a powerful anchor at the bottom and a stereo mix with a predetermined balance of vocals and the beat. Producers are focused on these proportions. This strict structure imposes limitations on my creative freedom. Consequently, I find myself rearranging elements more frequently compared to classical productions.

Urban tracks are relatively short, around ten minutes at most, leaving little room for extensive experimentation. To overcome these constraints, I focus on identifying brief moments of quiet within the song, often devoid of beats or with subtler drums. During these intervals, I leverage all available elements, transforming them into objects for dynamic movements and unconventional arrangements.

## Listener's perspective

Immersive audio, in my view, offers the best of both worlds – an exceptional audio listening experience and the perspective of the best seat in the house. Ideally, everything should sound great, delivering the optimal audio experience. Personally, I believe the ultimate seat would be in the center of the stage, especially when an orchestra is performing. Sitting in the middle, surrounded by the orchestra, allows you to be part of the story, almost like being the director amid the action. While this may not be feasible in a traditional theater, providing the audience with this perspective, together with outstanding sound quality, creates a unique and personal connection, similar to having your private choir or orchestra. This perspective, however, may not be universally applicable, as it may not suit urban music or other genres. It particularly shines in classical music, providing an excellent experience.

Additionally, there's a very important element: dynamics. Dynamics play a pivotal role in elevating the immersive audio experience. In pieces like 'Libertango', specific moments leverage dynamics to enhance the sense of space and the immersive audio encounter. Identifying these moments might require attentive listening, as there are instances where instruments unite or play in unison, showcasing how dynamics contribute to hearing the entire orchestra or ensemble. This ability to perceive the full story, surrounded by the power of instruments, enhances the immersive audio experience. The dynamics serve as a crucial element, adding depth to the space. In my perspective, dynamics and frequency range are intertwined with the concept of balance. Achieving balance across these elements is what defines a great mix for me.

When everything sounds spacious but lacks dynamics, the experience falls flat. Hence, I emphasize paying close attention to dynamics and the frequency range. It's about striking the right balance between these elements to create a mix that truly stands out. Visualizing this as a graphic with amplitude on one side and frequency on the other, the interplay of balance, dynamics, and frequency range is what shapes a truly exceptional mix in the realm of immersive audio.

## Dynamics

In the realm of immersive audio, the first crucial consideration is dynamics due to the vast dynamic range it offers. This extensive range presents opportunities across all three axes, creating a profound impact. To illustrate, take a moment to revisit 'Libertango', specifically toward the end or at certain pivotal moments. Two remarkable instances stand out where a crescendo builds up, defying intuition as the notes descend while the volume ascends. It's a counterintuitive yet effective approach that adds a unique layer to the immersive experience, lasting for around 10 seconds but significantly enhancing the overall experience.

These crafted moments are intentionally designed to elevate the immersive audio experience, creating peaks of enjoyment within the broader musical journey. An interesting case to explore further is Luz Mila Carpio's work, which offers a distinct fusion of genres. It's a departure from conventional sounds and genres, incorporating a mix that introduces a wealth of dynamics, expansive space, and subtle effects. The complexity of this production involves over 150 layers meticulously grouped to optimize available resources.

The result is a captivating experience that defies easy explanation, requiring a dedicated listen to fully appreciate its intricacies. The fusion of dynamics, spatial elements, and subtle effects contributes to an immersive journey, making it a noteworthy addition to the diverse world of immersive audio.

## Proximity

I'd like to give an example of a project we worked on to demonstrate how the effect of proximity (near and far) can be accomplished. This project involved audio with no video. When we began working from the drummer's perspective, the aim was to create a distinct auditory experience. Each element, like the snare and high-hat, was positioned to convey proximity. It wasn't only about adjusting volume; it involved layering sounds strategically. Initially, the layers surrounded you, then shifted gradually to create a sense of movement, alternating between near and far.

For instance, transitioning from the entire drum kit as a cohesive source to individual microphones demanded precision. As we moved away, the drum kit became a single source. Separating these sound elements aimed to articulate proximity – showing that something could be either near or far, often juxtaposed for comparison.

I sometimes like to use a technique involving changing the height of an object when moving it. In fixed scenarios with moving instruments, the challenge is less pronounced. However, when the listener is at the center and surrounded by dynamic elements, spatial localization can be tricky. People's ability to discern the exact location of sounds, particularly in the front or rear, is different from person to person. To address this, we adopted a subtle approach – sketching the entire movement first. By slightly adjusting the height of an instrument – raising or lowering them – it added a natural touch and significantly enhanced believability. This small change, coupled with minimal EQ adjustments, contributed to a more convincing experience. The process was somewhat empirical, driven by exploration rather than adhering to a rigid recipe, but it was quite effective.

## Workflow

Preparing for a recording session involves visiting the venue whenever possible. In the case of an interview with the urban artist Cazzu for a documentary, the use of Ambisonics and direct mics provided dual recording options. This practice extends to various scenarios, ensuring flexibility in post-production. Drawing from experiences in the movie industry, where I recorded sound for films, I learned the importance of always having a second option, even if it might not be used.

The commitment to having two recording options applies whether working on a single interview or with an orchestra. This approach stems from the unpredictability of future needs; what may initially seem unnecessary might become crucial. This dual-option philosophy extends to recording Ambisonics, even if it may not be used due to external noise issues. It's about being prepared for any circumstance, recording not only from the main source but also capturing ambient sounds, acknowledging that unexpected elements may need attention.

Creating the desired space begins with envisioning it during a venue visit. Understanding the venue's acoustics and the musical needs for that specific space is crucial. A cathedral setting, for example, may pose challenges in avoiding excessive reverb. Sometimes, despite being in a less-than-ideal venue, certain elements are used creatively, while in other cases, compromises may lead to contemplating post-production adjustments.

Dealing with unexpected challenges, such as excessive natural reverb in a museum recording, may require negotiating with the director. Being present during recordings is important, as decisions made on-site significantly impact the outcome. In a solo project, proximity to the recording engineer and the main event producer provided the necessary control and collaboration for a successful recording.

Ultimately, the key is being present, not only physically but also mentally, during the recording process. It involves making decisions on-site, envisioning the desired sound, and navigating challenges collaboratively. While software tools offer infinite options, clarity about the desired outcome is essential, drawing from years of experience. If uncertainty persists, it's crucial to reevaluate and, if necessary, explore alternative paths.

## Lessons learned

Here are some lessons we've learned along the way regarding the do's and don'ts when mixing immersive audio:

- It works to stay true to the original concept of compositions, especially in classical music (considering *Tango* as highly specialized classical music for this purpose). If the mix doesn't require flying objects and excessive paraphernalia, don't incorporate them. If the music can stand on its own, creating a space like a big sphere should suffice for all instruments, vocals, and effects to coexist harmoniously.
- It does not work, however, to add numerous moving objects that might distract the audience. If the song doesn't engage by itself, the immersive mix is unlikely to do so. Producers should prioritize understanding the musical and lyrical aspects of the song before making decisions on object placement, sphere size, movement, color, dynamics, and overall balance.

- It works to hire an assistant to organize everything before starting the mix, given the complexity of immersive mixes involving numerous assets that need proper accommodation in tracks, busses, and outputs.
- It works to use objects (instead of beds) for instruments placed in the ceiling, especially when full separation on the front–rear axis is needed. Limitations arise when Dolby provides a 7.1.2 bed, while a 7.1.4 speaker setup is in use, particularly when addressing specific locations in the top of the sphere.
- It sometimes works (although not extensively used in this project) to employ a doubler for a specific instrument (e.g., a guitar) and to pitch-shift the resulting double down, placing it on the opposite side of the room. This technique is useful for balancing a high-pitched instrument left alone in a corner, contributing to balance in the *x-/y*-axes.
- It does not work to make fast movements across the sphere as they are challenging to perceive and generally not well-received by the audience. Real-time rendering of rapid position changes can also create frequency notches, degrading the audio quality of the object.
- It works to subtly change the height of an object while moving it slowly around the listener. Since our perception of distance and position is not linear, especially in the front–rear axis, these slight height changes enhance accuracy.

## Case studies: recording projects

### Astor Piazzolla in Dolby Atmos: *Tangos All around*

Martin Muscatello and I started together exploring the world of immersive audio back in 2016, right after the first AVAR Conference, which I co-chaired. I remember coming back to Argentina (where I live) and telling Martin (my colleague and business partner in my studio) that a huge new path was opening for us. From that moment on, we put our hands in just about every immersive audio toolset we could. We have been Beta Testers for the (later discontinued) Dolby VR Suite, the very early days of the Facebook 360 toolkit (right after they bought Two Big Ears), and Dear VR (which in turn became Dear Reality) before they got acquired by Sennheiser. We tested dozens of workflows, incorporating binaural heads, Ambisonics microphones, software solutions, etc. After five years of exploring, the fog settled a bit, and we decided to choose Dolby Atmos as our main platform for production and delivery. Since then, we never stopped producing, mixing, and mastering all sorts of artists from very different parts of the world, and music genres ranging from urban all the way through classical.

In 2021, we celebrated the 100th anniversary of Astor Piazzolla's birthday. Astor was undoubtedly the greatest Tango composer and bandoneon performer in history. His renown extends far beyond borders, as his legacy has become deeply ingrained in the heart of Argentina's musical culture and resonates worldwide.

In this context, we set off to produce *Tangos All Around*, a first-of-a-kind immersive album that would gather a set of his most iconic pieces.

The magnificent Centro Cultural Kirchner in Buenos Aires hosted a series of concerts as part of Astor Piazzolla's tribute. Inaugurated in 2015, this venue features one of the finest concert halls in the world, with a capacity of around 2,000 seats, and boasting a massive Klais Orgelbau organ.

Martin and I arranged with the Artistic Director of the venue that this would be a joint production, and so at the end of the tribute they sent us all the multitrack recordings of the concerts that took place there. A massive amount of data to listen to! The ensembles varied from duets to full orchestras, and everything in between. We listened carefully to a large number of performances and ultimately decided to work with pieces from two concerts: the Opening Concert and the Quinteto Astor Piazzolla Concert.

**The Opening Concert**

This concert featured a large number of artists and ensembles, so the stage layout changed multiple times throughout the set list. There were a total of 40 channels to capture four bandoneons, three violins, two pianos, drums, two marimbas, one vibraphone, two guitars, a harmonica and an upright bass. In addition to that, three more channels accommodated a Decca tree, to capture the ambience and the audience. The detailed list of the microphones used for this concert was not available, but we chose to work with some of these pieces based on their unique performances and the quality of the recordings.

The first piece we took from this performance, 'Libertango', was an improvisation at the very end of the show, gathering all the musicians that performed on stage at the same time. For this reason, and given the overall complexity of the show, this was quite a challenging mix. We invested a good amount of time editing and cleaning up the session to make sure we had a clear sound. Also, we cut over 6 minutes of improv by different musicians to achieve the final length of 3 minutes and 43 seconds. The concept of the immersive audio design for this album was clear from the beginning: we wanted to show the audience how you would hear a Tango orchestra from the center of the stage, without any of the problems (booming, floor noises, etc.) that you would encounter if you were actually in that spot.

The other piece taken from this concert was 'Oblivion'.

> In contrast to 'Libertango', this was a piano and harmonica duet, by worldwide known Maestros José Colangelo and Franco Luciani. We used only the close microphones for the piano and the harmonica. At first anyone would think that having just a few channels to work with would mean that the mix should be an easy one. However, in search of a deep, immersive listening experience, we took some time to explore a few different panoramas. Ultimately we decided to create the illusion that the harmonica is walking around us, at a very slow pace, in sync with the music mellow flow.

**The Quinteto Astor Piazzolla Concert**

This is a quintet composed of piano, bandoneon, violin, electric guitar, and upright bass. The recordings took place in the main stage of the venue, named The National Auditorium. Standard close miking techniques were used, in addition to stereo room microphones and a Decca Tree.

- The piano had two stereo pairs. 2 × Neumann U87 and 2 × DPA 4099.
- The bass was captured with a DPA 4099.
- The guitar had a small amp, miked with an SM57.
- The bandoneon had two AKG c414, one on each side.
- The violin was picked up by a Neumann KM184.

Tracks mixed in Dolby Atmos:

- 'Adios Nonino'
- 'Fuga y Misterio'
- 'Muerte del Angel'

All three pieces had been impeccably performed by the Quintet, in a concert with a highly respectful audience, so the signal-to-noise ratio was as good as in a recording hall. But on top of that, the performances had the warmth of the live session, so we couldn't ask for more.

The mixing process took place in the main mixing and mastering suite at my facility in Buenos Aires, which is the first Atmos–Neumann room in Latin America. As we started working in the process, we had not yet settled on a specific delivery format. We took some time to experiment with Ambisonics, but, after all considerations, we decided to go with the Dolby solution, taking into account the better distribution and market outreach possibilities. Also, some key reasons for this decision were playback flexibility, streaming services compatibility, and overall workflow streamlining.

When we went into production, we were also at the verge of refitting our mixing room with a new speaker setup, provided by Neumann. The incoming new system was a Dolby Atmos 7.1.4 array, consisting of 3 × KH310 for LCR, 4 × KH120 for sides and rear speakers, 4 × KH80 for the tops, and a KH870 subwoofer. The main hub for the system is an Avid MTRX Studio, which serves as the speaker calibration and bass management unit.

Some of the more relevant decisions in the mixing process had to do with editing out some long passages of music, which ended up working perfectly well for the live performance. We wanted to make sure that the listener would not lose the excitement, and we always made the next bit more ear-catching.

After that, we experimented with different surround scenarios and movement, aiming to build a sphere as big as possible but also (most importantly) 100% believable. We always tried to make our decisions based on the idea that the music wasn't obscured by the 'hype' of the immersive technology. A moving piano or a flying upright bass definitely didn't feel 'natural'. But a walking harmonicist could easily happen. Therefore, we didn't move objects around in most of the cases, but rather we concentrated on optimizing the focus of each instrument to make them all sound as if coming from very different cardinal points.

We also used the height quite intensively, not to place discrete sound sources as we would normally do in other music genres (pop, rock, urban, etc.) but rather to create the illusion of an even bigger venue. Even though the Centro Cultural Kirchner is really big, the use of the top speakers really added a nice air to the reverbs and allowed us to make use of the natural room that resembled very closely what you could feel on stage.

We worked on the project for several weeks before sending out the masters to a couple of friends for a second opinion. Ceri Thomas (still working at Dolby at that time), George Massenburg, and Michael Romanowski were very kind to have a first listen, and their opinions have been invaluable to us, given the circumstance that we were also new to the sound of the Atmos room. All in all, we were able to send the masters for distribution after a few tweaks, and they were released on Apple Music, TIDAL, and Amazon Prime shortly after.

Chapter 33

# Mastering for immersive

*Michael Romanowski*

## Background

I got into engineering, as most of us do, fueled by my passion for music. As a musician, I played various instruments during my time in elementary and high school. I experimented with different instruments and eventually spent my high school years playing trombone in the marching band. Interestingly, this experience led me to the world of playing bass.

Although I was born in Michigan and moved to Nashville at a young age, it wasn't until college, at Bowling Green, Kentucky, that I truly immersed myself in the Music City scene. I transitioned from trombone to bass and joined a band in town, eventually playing at the local club. This venue, situated about an hour north of Nashville, attracted musicians traveling to and from the city. It provided me with the opportunity to witness a diverse array of music and contributed significantly to my musical journey.

And then, when my turn finally came to play at that club, it was exhilarating. Looking back, I noticed someone in the back tweaking knobs, affecting the sound from the stage. I was intrigued. I was curious about what he was doing, and after persistent inquiries, he was kind enough to say, 'Okay, let me show you some stuff'. I was captivated, and after some time, I got the opportunity. 'I'm taking a break; you run it for a bit'. Then, 'I'm taking the night off; you run it'. Eventually, it turned into, 'I quit, you're the new house sound person'.

So I stumbled into engineering through live sound, and I still believe it's one of the best ways to learn ear training. Live sound taught me valuable lessons. You have to pay attention, stay focused, and really listen to what's happening. Making adjustments on the fly based on what you hear becomes crucial. Identifying and solving problems in real time hones your problem-solving skills. You're already equipped with the listening skills because you're attentively engaged. For instance, seeing the singer accidentally kick the kick drum mic, you know it's going to cause issues, and you adjust or do whatever it takes.

Reflecting on those moments, I realize how pivotal they were, not just for refining my listening skills but also for developing temperament, attention span, and various other attributes.

As I continued with my involvement in bands, I ventured into the realm of recording. It dawned on me that there existed a similarity between live sound and recording – both have consoles, and I thought that was very cool. Intrigued by this realization, I delved into the studio side of music production, initially working with the bands I was part of.

DOI: 10.4324/9780429297526-36

As time progressed, others took notice of my work, approaching me with opportunities to collaborate on their projects. This evolution marked the birth of a genuine passion for studio work.

Meanwhile, during my college years, I pursued a double major in math and computer science, immersing myself in both the technical and artistic sides simultaneously. I cherished the duality of these aspects and developed a love for both. It was the late 1980s, and the advent of computers fascinated me. After college, I returned to Nashville, and did bartending and various odd jobs to sustain myself while still engaging in music and recording projects. Then, on a whim or a desire for a change, I embarked on a month-long solo motorcycle journey across the country. Packed with a backpack and a tent, I traveled and looked for new experiences and connections. My travels led me to San Francisco, California. There, a serendipitous call from the owner of Rocket Lab changed my course. They were looking for another engineer, and although I was familiar with mastering, I had yet to develop my skills.

I took the initiative to visit them in person, spending extra days in the Bay Area. This marked the beginning of my friendship and mentorship with Paul Stubblebine. Our collaboration eventually evolved into a venture called the Tape Project, focusing on analog releases using quarter-inch tape. This period brought clarity to the alignment of my interests in computer science, math, art, and music. I soon realized that mastering was the culmination of my diverse background.

Offered a job, I took a couple of weeks to return and pondered the prospect. Ultimately, I decided to embrace the opportunity. In July, I made my way to California, initially planning a two-year stint. However, my love for the area and its deep and broad culture grew rapidly, turning two years into five, then ten. After ten years, I finally changed my license plates. Shortly after, my car was stolen, but instead of deterring me, it reinforced that California had become my new home, and I embraced the journey ahead. I just loved it.

Paul Stubblebine is not only a fantastic friend and mentor but also a sharp individual with a great listening ear. He introduced me to a lot of people, introducing me to the audiophile world while continuing my journey in mastering. During that period, I got into various music genres from hip-hop, which thrived in the Bay Area, to classical. My passion for music led me to explore different facets, listening attentively, and collaborating with fellow enthusiasts.

As my interest evolved, I became fascinated with surround sound. Paul and I, having previously worked at Rocket Lab, pursued this avenue. Following Paul's venture to start his own studio at Hyde Street, I remained at Rocket Lab until changes in the business landscape prompted me to establish Plant Mastering at Sausalito's The Plant. After a couple years as chief mastering engineer there, it eventually became evident that it was time for a change.

Subsequently, I rejoined forces with Paul Stubblebine. After a year of collaboration and brainstorming about ideas and studio spaces, we decided to embark on the journey of building our own facility. Serendipitously, we were able to acquire the old Coast Recorders name and facility. This prompted the creation of two rooms, one of which retained the legacy of the Coast's tracking room, named the Bill Putnam room.

We kept the original Bill Putnam tracking room at Coast, pretty much untouched, while the other two rooms underwent a comprehensive reconstruction. Our goal was to mold the rooms according to our vision, and to meet our requirements. Part of this transformation was to set up a surround sound system. Surround sound has always fascinated me. We walk through life in an immersive world. Even when you go see a band play on stage, there's the sound around the club. You go see an orchestra, you've got the concert hall and all these other elements as part of the performance.

This fascination with surround sound started in 2001 when we initially moved into the facility. We had a Sonoma system and started working with SACDs. As time progressed, Paul semiretired, and I took over the studio's leadership, partnering with Matt Boudreau, who ran his studio in the Coast Recorders space. After many years, I decided a change was needed, and an opportunity presented itself at Fantasy when I discovered the long-unused Creedence room, Studio C. Once a Foley studio for Saul Zaentz Media Center, where notable projects like *Amadeus* were brought to life, the room had remained unused for nearly a decade. I seized the opportunity to relocate Coast Mastering into this space. Within a year, I expanded the studio's capabilities by incorporating height channels, realizing the immense potential it offered.

I began consulting with Dolby, engaging in product evaluations and listening tests for their codecs and processes. I got into height channels and was fascinated with immersive. But my journey started in analog – mastering and mixing records. During my tenure at the Fantasy building, I was deeply immersed in various aspects of audio.

A series of issues at the Fantasy studio led to the time for a change again. I sublet the space, reflecting on the lessons learned from different rooms. The Coast rooms on Mission Street were originally post-production and isolation rooms, each with its unique physical challenges – an angularity issue in one and a feeling of constriction in the other. Fantasy's Studio C had its own set of problems, being excessively large with challenges in speaker localization due to the height channels' distance.

I decided to do what ended up being a two-year project to build my dream room. The result is a 9.1.6 room with a flexible track system for speaker positioning, allowing me to adapt to various formats. The goal was to stay current while future-proofing the space as technological advancements happen.

The underlying philosophy is not solely about the tools used but centers on a forward-thinking mentality. I appreciate the aspect of being surrounded by music, of hearing vibrations, of feeling vibrations come at you from different directions. I love cues that come to you and give you different sounds, and the interaction of elements of sound going on around you and within you. I love the aspect of immersive music that we can do all this without feeling as though we are hammering it into people's eardrums. Hearing the pieces a little bit better and avoiding the pitfalls of forcefully channeling it through one or two speakers offers insight into the composition and the artistic process.

## Design philosophy

My philosophy centers on being a service provider. Whether tracking, mastering, or mixing, my goal is to serve as a conduit, aiding artists in becoming, by the time they leave, more authentic expressers of themselves and their art than when they arrived. With over 30 years of experience in mastering, my journey began in live sound, moved through recording, mixing, and producing, eventually leading to mastering.

I firmly believe that mastering requires a different engineer than the mix engineer, as it demands a distinct perspective – the 10,000-foot-high view versus the microscope. The focus shifts from 'what is it?' to 'how is it?' Rather than approaching my work like a

Swiss army knife with various tools, I specialize, deeply immersing myself in a particular skill.

My diverse background in different roles as an engineer has contributed to the unique viewpoints I bring to mastering. I don't just perform tasks; I guide and interpret. When clients express their desired sound or vibe, it's my responsibility to understand and deliver. I solve problems, ensuring they achieve the sound they seek, allowing them to concentrate on creativity, expressing thoughts, emotions, feelings, or stories.

At its core, music is human expression. My role is to distill it to its purest form, minimizing technological barriers. I see myself as a facilitator, dedicated to bringing out the best in the musicians I work with. It's about honoring their identity and sound.

There are essentially two different perspectives when it comes to presenting one's expertise. Some might say, 'Come to me because I excel at this particular sound, and I do it better than anyone else'. While that approach has its merits, I personally lean toward uncovering the unique aspects of an artist's work, those little nuggets that make it special.

To delve a bit deeper into this, the comfort and appropriateness of the setting are crucial. For instance, when tracking, creating a space that aligns with the genre is essential. A big rock record might not thrive in a vast, empty church due to mismatched reverbs, delay times, and tonal reflections. Conversely, what suits a string quartet might not be ideal for a drum ISO booth. It's about context, understanding, and experimentation – building a mental library of sounds and discovering what combinations work harmoniously.

It's a complex and multifaceted process. My approach varies based on the uniqueness of each situation, always in collaboration with the clients. Communication becomes a blend of sociology and psychology, making the process dynamic and tailored to the specific needs of the artist or project.

As we navigate the evolving landscape of audio technology and immersive techniques, artists are also on a journey of discovery. They grapple with questions like, 'Is this right for me? Should I focus on the past, the present, or the future? How can I bridge the gap between where I started and where I'm headed?' Adapting music to align with an artist's evolving vision involves education, iteration, and sometimes a complete creative overhaul.

## Fantastic Negrito, Bonnie Raitt, and Peter Framtom

Two examples from my experience highlight this approach. First, working with the artist Fantastic Negrito, who was unfamiliar with immersive audio. I invited him to the studio, played a selection of immersive tracks, and engaged in discussions about the nuances of immersive sound. This hands-on experience allowed him to conceptualize how his music could benefit from this format. The result was a unique, immersive version of his record that diverged from the stereo version.

A similar process unfolded with another artist, Bonnie Raitt, who wanted to explore Atmos but wasn't sure how it would fit her music. By inviting her to the studio, playing various immersive tracks, and facilitating a dialogue about her artistic vision, we customized the immersive experience to align with her preferences. In her case, it meant recreating the live concert atmosphere, placing the band in front of the listener, just like a live show.

Both experiences underscore the importance of understanding an artist's intent and tailoring the immersive format to complement their creative vision. It's about providing a conduit for artists to express their unique identity and ensuring that the technology serves the art, not the other way around.

Handling back catalog material poses a significant challenge because many records are being revamped without the involvement of the original engineer, artist, or producer. This often leads to subpar results, causing problems that require corrections and ultimately doubling the label's expenses. Taking the time to do it right from the start would save resources and ensure everyone's happy with the result.

Each project is unique, shaped by distinct circumstances, production paths, and end goals. Some artists enthusiastically embrace immersive audio, viewing it as an opportunity to expand their creative horizons. Others reluctantly adopt it to meet distribution requirements, feeling compelled by platforms like Apple Music that prioritize Atmos releases.

In certain cases, resistance to immersive audio stems from suboptimal listening experiences, often involving unauthorized alterations to the artist's work. Seasoned professionals with decades of established techniques find themselves needing to unlearn their accustomed methods and embrace a fresh approach, which can be a jarring experience.

A prime example is my current work on *Frampton Comes Alive* with Chuck, facing the challenge of recreating iconic crowd moments. Chuck must meticulously restore those well-known elements that define the record's character. Whether it's a live or studio album, the process involves a delicate balance of honoring what people expect while reimagining the material in an immersive context.

Engaging in a dialogue with artists and producers throughout the process becomes crucial. Questions about goals, intentions, and desired outcomes help clarify the working path and objectives for the album. By fostering clear communication and understanding the artist's vision, we can navigate the unique challenges posed by each project.

## Space

Space plays a unique role, offering non-verbal cues that guide us through environments. If you close your eyes, the localization of sounds, the reflections, the contrast between brightness and dullness, nearness and farness – these elements are as crucial to me as scenes in a movie are to storytelling. In a film, scenes provide information beyond dialogue; they encompass colors, time settings, atmospheres, and more. Similarly, space serves as an active element, indicating where an event is unfolding and what surrounds it.

Even in the stereo world, the use of space is critical. For instance, an awkward record might feature a small room jazz trio with a drippy reverb on the singer, creating a disjointed feel. The singer seems out of place with the band, impacting the sense of space around the music. Space can either disrupt expected experiences or create cohesion, guiding listeners into a specific atmosphere.

In movies, when you recognize green screen or computer-generated elements, it pulls you out of the experience. Similarly, if you're distracted by sounds that don't align or enhance the music, you're missing the essence of the song itself. The goal is to create a sense of space that complements the music, enriching the overall experience. It's like a well-plated dish – you should savor the music itself, not just the elements around it.

The sense of space should be seamless, becoming a character in the story. When done right, it becomes an integral part of the narrative. The challenge lies in making it noticeable without being too much or too little – achieving that delicate balance where it effortlessly blends into the overall experience.

## Process of preparation

As I delve into this, I'll focus on the creative aspect rather than mastering, considering my role as a pioneer in this field. I constructed a room that earned a TEC nomination for studio design, likely the first immersive mastering facility. Collaborating with Dolby, Sony, Fraunhofer, Auro-3D, and others early on, I provided feedback and developed methods to define what mastering means in this immersive context.

Now, mastering holds even more significance in the immersive world than in stereo. While we understand its importance in stereo, the immersive landscape presents challenges. This includes mixing various songs within an album and ensuring a cohesive release that feels like a complete body of work. The real challenge in the immersive world lies in deciding how much to adhere to the stereo mentality of presentation. For instance, Fantastic Negrito encouraged experimentation, embracing a different approach. However, some artists prefer maintaining a closer connection to stereo conventions. Consider *Frampton Comes Alive* and the recent *Frampton at the Royal Albert Hall*. Although they share many songs, they differ significantly, with changes in miking techniques over the decades.

Let me share an interesting anecdote. In *Frampton Comes Alive*, the lack of a prominent kick drum is not intentional. The kick drum went sideways at the start of the show, resulting in poor mic placement. Consequently, the album's kick drum sound is synthesized from overheads and other sources, showcasing the unpredictability of live performances.

When contemplating projects, it's essential to consider the various stages from conception, mixing, and recording to the mastering side of things. This encompasses understanding whether the project is analog, hybrid, or fully digital. It involves figuring out the intricacies of immersive audio, the listening preferences of people, and the importance of mastering in this context. However, it's crucial to recognize that, by the time a project reaches the mastering stage, most decisions have already been made.

For instance, I worked on a record for a Bay Area rock band named Devil in California, known for their dynamic live performances. When we discussed creating both a stereo and an Atmos version of the record, I approached it with an experimental mindset. The challenge arose: should we focus on the stereo or the Atmos version first? This dilemma prompts a broader question for albums moving forward. Should the stereo and immersive mixes be done simultaneously, especially if it's the same person handling both? It's a decision that requires consultation with the artist. Some may prefer hearing the current state first and then exploring future possibilities, while others may have a different approach.

With the band Devil in California, we embarked on a unique experiment, aiming to push the boundaries of recording techniques. Reflecting on a previous project, a power pop trio, I recalled placing all the guitar amplifiers in a semicircle, employing a single stereo mic in the center. Similar experimentation was extended to the drums, while the bass was handled using a DI, with later adjustments to eliminate bleed. This setup involved multiple amplifiers, each with different switches activated at various points. Unlike conventional close-miking, we opted for a holistic approach, creating sonic landscapes that aligned with the envisioned musical narrative.

The project, which we did at Coast Recorders, allowed us to utilize the room's natural acoustics. My goal was to guide the listener by physically placing sounds in distinct positions, thus enhancing the sense of space. A notable variation involved using SF-24 ribbon mics turned sideways in an X-Y configuration for drum recording. This approach aimed to capture both the upper and lower spatial elements, experimenting with unconventional techniques to achieve diverse auditory experiences.

Venturing to a studio in Stinson Beach for a five-day residential recording session, we delved into nonlinear arrangements and unconventional miking of guitar cabinets from various distances. For vocals, we implemented multiple microphones at different locations, emphasizing distance to enrich the playback experience. The process involved meticulous measurement and experimentation to achieve a balanced and spatially dynamic mix.

The immersive project with Devil in California was a captivating exploration, presenting opportunities to experiment with drums, guitars, and distances. Placing an omnidirectional microphone on a guitar amp from a distance, for instance, gave us intriguing results, adding depth and lower octaves to the sound. The goal was to create a full waveform development without relying on extensive post-production distance-building.

In comparing the work on stereo versus immersive audio, I emphasized the efficiency gained in immersive audio production. Unlike stereo, where extensive EQing is often required to fit multiple elements into a single speaker, immersive audio allows each component to exist in its full resolution. The approach of starting with a natural sound and then adapting for stereo has proven to be a more effective production flow, maintaining the authenticity of the sound.

Furthermore, I discussed the challenges of referencing immersive audio for artists accustomed to stereo playback. Bridging the gap between agreed speaker playback standards and the novel experience of immersive playback involves careful consideration. Playing back immersive mixes for artists becomes crucial in aligning their expectations with evolving industry standards.

While working on Kenny Wayne Shepherd's record, we made a deliberate choice to avoid listening to the stereo mix initially. This approach allowed for a more open perspective, and, when comparing it with the stereo mix later, the differences were striking. The ongoing effort is to strike a balance between openness and immediacy, respecting the artist's preferences and pushing boundaries in the immersive audio realm.

## Experimenting

My 'bag of tricks' essentially encompasses the wealth of experience I've accumulated, drawing from techniques that have proven effective in various situations. It's like reaching into a toolbox and pulling out a specific procedure or approach based on past successes. However, flexibility is key – it's crucial not to be rigidly committed to any particular method. You evaluate if, for instance, over-compressing the floor mic beneath the drums produces the desired sound. Is it suitable for this project? Yes, for one, but not necessarily for another. The 'bag of tricks' also involves assessing the appropriateness of a technique for the specific context of the project.

I enjoy experimenting, constantly seeking novel approaches. Recently, I tried an unconventional method: placing the Neumann head (KU-100) in the hallway while leaving the studio room open. The echoic nature of the hallway, with its scattered reflections due to angles and surfaces, added a fascinating dimension to the drum sound. When blended into stereo, it resembled a subtle reverb, but on headphones, a unique dimensionality emerged. Though not fully translatable to headphones due to mic and binaural limitations, the interest it added for headphone listeners differed from that experienced by those using speakers. This interplay between spatial dimensions is something I find intriguing and enjoyable to explore.

To illustrate further, I had a discussion with Josh from Tiny Desk a couple of months ago about creating Atmos versions of Tiny Desk concerts. During the Music Expo in Mexico City, we were exchanging stories, and I shared an incident involving a jazz artist I recorded on a back deck during the pandemic. While hastily setting up for a live broadcast stream, I forgot an outdoor microphone for ambient sounds. A week later, I replicated the environment, capturing the same birds and atmosphere, and blended it back into the recording. An unexpected and amusing addition was a fly that had entered the studio during the session. Intrigued by the idea, I placed the Neumann head at my location, closed the door, and let the fly buzz around. Eventually, I incorporated the sound of the fly back into the live recording and sent it as a reference. The surprise element of the fly became a humorous touch, proving that experimentation can lead to unexpected and enjoyable results.

## Do's and don'ts

Artistically, I believe there are no strict rules. The mantra is simple: if it sounds good, go for it; if it doesn't, skip it. This might sound like a bumper sticker, but it holds true. Challenge conventional norms. For instance, someone might say you can't place a kick drum in the ceiling – it doesn't make sense. But I decided to experiment and see what happens. I envisioned a scenario where the kick drum, if used creatively, could mimic the sound of an upstairs neighbor annoyed by loud music, stomping on the floor to turn it down. It's about thinking outside the box and finding creative ways to use unconventional methods.

Consider bands like the Flaming Lips in their album *Yoshimi Battles the Pink Robots*. They manage to shift the sound around, keeping drums dynamic and mobile while other elements stay put. It defies the typical notion that drums need to be anchored. It's all about creative composition and thinking differently to achieve something interesting.

Now, a piece of advice – if you're working with LFE (low-frequency effects), ensure you're actively filtering the LFE channel. Don't solely rely on bass management or room calibration. LFE contains a full range, and if you're creating a binaural reference, that full range is captured. Filtering it in the monitoring environment is crucial because what works in speakers might not translate accurately to binaural, and vice versa.

Speaking of LFE, in the realm of music, I'm not a big fan of relying on LFE for bass. It's an effect, not a channel for continuous bass. Depending on LFE for bass can lead to unpredictable results. If someone's playback system lacks a subwoofer, they lose all the low end. Moreover, differences in calibration, especially in systems like Atmos, can create inconsistencies in LFE playback. My recommendation is to use full-range channels for left, center, and right speakers in music. Push the limits into the 20–30 Hz range, especially since most headphones can handle those frequencies.

Considerations also extend to elements like vocals. Should they be in the center channel, spread out, or pulled into the room? It's a case-by-case decision, and often the 'it depends' answer is quite appropriate. Always think about how the audience will experience your mix – whether on headphones, speakers, or immersive setups.

Finally, the importance of client involvement in the listening process cannot be overstated. Recently, when Peter Frampton went to listen to mixes at Blackbird audio in Nashville, he discovered that Apple Music had been playing the wrong files for years. Listening on different systems can reveal discrepancies that might be missed on headphones alone.

In essence, dialogue with the client is crucial. Understand their vision, whether it's recreating the ambiance of a Cuban percussion circle or exploring unconventional placement of sounds. This dialogue is an opportunity for expansion and an essential aspect of navigating the evolving landscape of music production.

## Looking forward

There are abundant opportunities in the realm of immersive audio, particularly when considering creative expression. The primary focus is on transporting the listener, encouraging active engagement rather than passive consumption. This involves intentional listening, where the audience seeks out an experience, as opposed to the passive scenario of having background noise while doing other tasks. By immersing oneself in the sonic environment, the listener becomes an active participant.

In stereo setups, a single speaker in a room might cater to one side, missing out on the full spatial experience. Much like a movie scene's lighting and composition, sound in a song can serve as a character, contributing to the narrative. Much like interpreting a book into a film or translating a lyric sheet into a record, the artist's intent is subject to various interpretations. The key is to evaluate whether the envisioned vision aligns with the listener's experience.

Reflecting on the origins of recorded music, the goal has always been to convey the emotional essence of human expression. With the advent of recording technology, the challenge was to recreate the feeling of being present during a live performance, sans external cues. Immersive audio now allows us to capture and reproduce the fundamental elements of experiential life with sound, marking a return to the roots of recording.

Developing best practices is pivotal in this evolving industry. Immersive audio serves as a stepping stone toward a more interactive human experience, expanding beyond just auditory engagement. As technologies like Apple goggles merge sound with sight, and medical applications help those with sight or hearing impairments, the potential for innovation is immense.

In conclusion, immersive audio presents an array of possibilities for sharing and experiencing sound. The industry's progress relies on honing best practices in capturing and reproducing audio, fostering a cleaner and more pristine auditory experience. Moreover, the creative use of tools can lead to new sonic landscapes, reminiscent of past innovations like intentional microphone distortion or creative guitar techniques using DIs. The potential for pushing the boundaries of technology to create unique sounds is a driving force in this dynamic field.

# Index

For Product Safety Concerns and Information please contact our EU representative GPSR@taylorandfrancis.com Taylor & Francis Verlag GmbH, Kaufingerstraße 24, 80331 München, Germany